Introduction to

Environmental Remote Sensing

Introduction to Environmental Remote Sensing

(Third Edition)

E.C. Barrett
Director
Remote Sensing Unit
Department of Geography
University of Bristol

and

L.F. Curtis
Formerly Exmoor National Park Officer and
Honorary Senior Research Fellow
Department of Geography
University of Bristol

CHAPMAN & HALL
London · Glasgow · New York · Tokyo · Melbourne · Madras

Published by Chapman & Hall, 2–6 Boundary
Row, London SE1 8HN

Chapman & Hall, 2–6 Boundary Row, London SE1 8HN, UK

Blackie Academic & Professional, Wester Cleddens Road, Bishopbriggs,
Glasgow G64 2NZ, UK

Chapman & Hall, 29 West 35th Street, New York NY10001, USA

Chapman & Hall Japan, Thomson Publishing Japan, Hirakawacho Nemoto
Building, 7F, 1-7-11 Hirakawa-cho, Chiyoda-ku, Tokyo 102, Japan

Chapman & Hall Australia, Thomas Nelson Australia, 102 Dodds Street,
South Melbourne, Victoria 3205, Australia

Chapman & Hall India, R. Seshadri, 32 Second Main Road, CIT East,
Madras 600 035, India

First edition 1976
Second edition 1982
Third edition 1992

Typeset in 10/12 Plantin by Best-set Typesetter Ltd, Hong Kong
Printed in Singapore

ISBN 0 412 37170 7 (PB)

A catalogue record for this book is available from the British Library

Library of Congress Cataloging-in-Publication data

Barrett, E.C. (Eric Charles)
Introduction to environmental remote sensing / Eric C. Barrett and Leonard F. Curtis.—3rd ed.
p. cm.
Includes bibliographical references and index.
1. Geography — Remote sensing. 2. Remote sensing.
I. Curtis, L. F. (Leonard Frank) II. Title.
G70.4.B37 1992
621.36'78—dc20 91–36829
CIP

Dedication

To the memory of
Leonard F. Curtis, 1934–1991,
remote sensing pioneer,
distinguished teacher,
effective manager,
skilful diplomat,
colleague and good friend.

Contents

Colour plate section appears between pages 220–221
Preface
The authors
Abbreviations and acronyms

PART ONE: REMOTE SENSING PRINCIPLES

Chapter 1	**Monitoring the environment**	**3**
1.1	Growth in awareness of environmental problems	3
1.2	*In situ* sensing of the environment	5
1.3	Remote sensing of the environment	6
1.4	Economic and practical benefits of remote sensing, past, present and future	7
1.5	The growth of remote sensing in geography and the environmental sciences	12
Chapter 2	**Physical bases of remote sensing**	**15**
2.1	Natural remote sensing	15
2.2	Technologically assisted remote sensing	16
2.3	Electromagnetic energy	18
2.4	Radiation at source	20
2.5	Radiation in propagation	23
2.6	Radiation at its target	25
2.7	Radiation observed	27
2.8	General conclusions	28
Chapter 3	**Radiation characteristics of natural phenomena**	**29**
3.1	Radiation from the Sun	29
3.2	Radiation from the Earth	36
3.3	Spectral signatures of the Earth's surface	39
Chapter 4	**Sensors for environmental monitoring**	**47**
4.1	Introduction	47
4.2	Passive sensors in the visible wavelengths	47
4.3	Passive sensors outside the visible wavelengths	55
4.4	Active sensors outside the visible wavelengths	66
Chapter 5	**Sensor platforms, sensor packages and satellite data distribution**	**73**
5.1	Introduction	73
5.2	Ground observation platforms	74
5.3	Airborne observation platforms	75
5.4	High-altitude sounding rockets	79
5.5	Satellite platforms	80
5.6	The Evolution of Earth Resources satellite observations	82
5.7	The Space Shuttle	85
5.8	The SPOT satellite system	90

5.9	The meteorological satellites	93
5.10	Geostationary meteorological satellites	94
5.11	Special purpose sensors and satellites	95
5.12	The ground segment and satellite data distribution	97
Chapter 6	**Calibration, evaluation and validation in remote sensing**	**101**
6.1	Introduction	101
6.2	Selection of ground data sites	102
6.3	Collateral data for crop studies	105
6.4	Field measurements of spectral reflectance	107
6.5	Fully integrated local area studies	110
6.6	Data sampling and tests for accuracy	112
6.7	Data collection systems for transmissions to satellites	113
Chapter 7	**Remote sensing data interchangeability, preprocessing and processing**	**119**
7.1	Introduction	119
7.2	The interchangeability of basic remote sensing data types	119
7.3	Stages in the handling and use of remote sensing data	124
7.4	Preprocessing and processing examples	126
Chapter 8	**Analysis and interpretation of aerial photography**	**133**
8.1	Introduction	133
8.2	Methods of photo-interpretation	133
8.3	Measurement and plotting techniques: selected aspects of photogrammetry	138
Chapter 9	**Digital data handling**	**147**
9.1	The need for numerical data manipulation	147
9.2	The quality and quantity of remote sensing data	150
9.3	The analysis and interpretation of selected features	158
9.4	Automatic decision/classification techniques	161
9.5	End products of numerical processing	165
9.6	Systems for digital data analysis and interpretation	165
9.7	Remote sensing and GIS	168
	PART TWO: REMOTE SENSING APPLICATIONS	
Chapter 10	**Weather analysis and forecasting**	**175**
10.1	Remote sensing of the atmosphere	175
10.2	Representative weather satellites	182
10.3	Satellite data applications in applied meteorology	188
Chapter 11	**Global climatology**	**203**
11.1	Climatology emerges from the shadows	203
11.2	Global climate monitoring: the nature of the problem	203
11.3	The Earth/atmosphere energy and radiation budgets	204
11.4	Global atmospheric moisture distributions	214
11.5	Wind flows and air circulations	220
11.6	The climatology of synoptic weather systems	222
11.7	The climatology of the middle and upper atmosphere	224
Chapter 12	**Water in the environment**	**227**
12.1	The importance of water	227
12.2	Hydrometeorology	227

12.3	Surface hydrology	238
12.4	Hydrogeology	243
12.5	Oceanography	244
Chapter 13	**Soils and landforms**	**257**
13.1	The nature of soils	257
13.2	Conventional mapping of soils	258
13.3	Soil parameters detected by remote sensing	259
13.4	Remote sensing of soils and landforms by photographic systems	259
13.5	Remote sensing of soils and landforms by non-photographic systems	268
Chapter 14	**Rocks and mineral resources**	**281**
14.1	The use of air photo-interpretation	281
14.2	The use of non-photographic airborne sensors in geology	287
14.3	Space observations for the study of rocks and mineral resources	293
Chapter 15	**Ecology, conservation and resource management**	**303**
15.1	Introduction	303
15.2	Reflectance from vegetated surfaces	305
15.3	Phenological studies	309
15.4	Conservation in National Parks	311
15.5	Resource management: the example of forestry	313
15.6	Disease and fire hazards	320
15.7	Wildlife studies	320
15.8	Marine conservation	323
Chapter 16	**Land use and crop production**	**325**
16.1	Introduction	325
16.2	Aerial photographic surveys	327
16.3	Multispectral sensing of crops	329
16.4	Radar sensing of crops	340
16.5	Crop yield modelling using infrared sensing	341
16.6	Land use in the past	344
Chapter 17	**The built environment**	**349**
17.1	General considerations	349
17.2	Rural structures	350
17.3	Urban areas	353
17.4	Industrial complexes	356
17.5	Demography and social change	358
17.6	Remote sensing for civil engineering	364
Chapter 18	**Hazards and disasters**	**367**
18.1	Definitions	367
18.2	Disaster assessment	367
18.3	Hazard monitoring	370
18.4	Severe convective storms	372
18.5	Hurricanes	375
18.6	Insect infestations	379
18.7	Future developments	384
Chapter 19	**Problems and prospects**	**387**
19.1	Satellite remote sensing as a global pursuit	387

19.2 Remote sensing as a public service or a commercial activity 389
19.3 Handling large data sets 390
19.4 Archiving and distribution problems 390
19.5 Relations between technologist and user 392
19.6 Future developments 393

Bibliography **403**

Index **413**

Preface

In the Preface to the second edition of this volume, published in 1982, Dr Curtis and I referred to an 'explosive growth' of remote sensing of the environment, for such a description was fully merited. Ten years later the same description is still appropriate, indeed all the more so, for the continuing explosive growth of environmental remote sensing involves an ever greater mass of relevant materials: it is probably true to say that the accomplishments of the past decade approach, or perhaps even exceed, in scope, magnitude and significance all those that had been achieved previously.

Whilst such impressive vigour proffers and promises much to those who, for any one of many possible reasons, are now interested in remote sensing, it also poses some difficult problems, not least for book authors and publishers. One such problem is how to produce an introductory text that is still broad enough to fulfil its tutorial purposes yet compact enough to remain within the purchasing power of students: whole libraries of remote sensing texts and journals are now springing up! A second problem is how best to update a text in this field within all the necessary but seemingly stringent sets of accompanying constraints, rather than to wipe the slate clean and start again. In the present case so much new material has clamoured for inclusion that we were often on the point of giving up the revision task.

Fortunately such problems are offset by special opportunities. One of these is that, as any field of study becomes broader and deeper, so the need for self-contained books to describe it as a whole – from first principles, through to present uses and on to future possibilities – becomes not less, but greater than before. Another is that entirely new books run the risk of painting pictures of scientific and/or technological trees without their roots – often failing to do justice to the all-important ways in which key philosophies, principles and practices have grown to what they are today. All of us who have been involved with the first and second editions of the present title have been pleased and encouraged by the wide circulation of this book, and its acceptance as a standard text in its field. The preparation of this third edition has proved a fascinating task, challenging, and for several reasons much more difficult than originally expected.

Since the field of environmental remote sensing is now so extensive, even more than before this new edition must be seen as a *beginning* for students who can afford to purchase one book but not several, a *basis* for staff who have broad courses to design and deliver, and a *lifeline* for career professionals whose formal training did not include any or enough information on this youthful discipline to fit them for present-day decision making. In preparing it we have been able to draw from particularly broad personal experiences: both Dr Curtis and myself became actively involved in remote sensing at very early stages in its development as a recognized field of study, and have devoted large parts of our working lives to remote sensing research and applications. We have been able to monitor closely the training needs of both students and professional users of remote sensing through our everyday work, and have good reason to believe that this new volume will not only do justice to modern remote sensing as a whole, but, above all, communicate its essence to the growing numbers of those who wish or need to understand what it is all about.

It came as a great shock to me in April 1991 to learn of Len Curtis's sudden, unexpected and untimely death – less than one week before we had planned to meet to select new illustrations for this book. Fortunately, we had already agreed to many of the revisions to the text, for otherwise this third edition would never have been completed. Thus the contents of this third edition owe as much to my co-author as they do to me, but responsibility for any mistakes or errors is mine alone. Mrs Diana Curtis and Dr Sarah Curtis (Len and Diana's elder daughter) were keen that the third edition should advance to publication, and gave practical help without which its publication would not have been possible.

A great debt of gratitude is also owed to the many friends and colleagues around the world from whom we have learnt much about remote sensing, and with whom we have discussed matters covered by this book. Where direct use has been made of the work of others acknowledgement is given through captions and references; the latter have been augmented by other selected items also, which will greatly widen the horizons of the reader.

Particular thanks are due to our past and present friends and colleagues in the University of Bristol, the Exmoor National Park Office, the UK Meteorological Office, the National Rivers Authority, the European Association of Remote Sensing Laboratories, the European Space Agency, the National Aeronautics and Space Administration, and the US National Oceanic and Atmospheric Administration among others for their support and encouragement over the years, and access to material used in this book. We trust that they, and all others whose work we have reviewed will feel their findings have been represented fairly – if much more briefly than either they or we would have liked! Mrs Shirley Sparks and Mr Tony Philpott of the Bristol University Department of Geography deserve special mention for their practical help with word processing and photography respectively. My wife Gillian also has lent a hand (no, *two* hands!) with the word processing of the text, my son Andrew with several aspects of it, and my daughter Stella with sympathy and understanding, and great evening meals when both parents were often late home: I am indebted to them all.

For the sake of Diana, Sarah and Ruth, and for the memory of a great colleague and very good friend, I hope this book will be a successful, and therefore a fitting, tribute to Len.

E.C. Barrett
Backwell
Avon

The authors

Dr Eric C. Barrett

Dr Eric Barrett is Director of the Remote Sensing Unit in the Department of Geography in the University of Bristol. He is author or editor of over 20 books on geography, climatology, hydrology or remote sensing, in addition to over 200 papers, articles and reports mainly in these fields. His interest in remote sensing was aroused first by a presentation in London in 1963 by the chief of the TIROS Project, which was then in its infancy. This prompted Dr Barrett's first book, entitled 'Viewing Weather from Space' to be published in 1967, and led to his PhD on 'The Contribution of Meteorological Satellites to Dynamic Climatology' in 1969. Since then, Dr Barrett's interests and involvements in remote sensing have been wide and varied. He has undertaken many consultancies for commercial companies, national governments, and international agencies, including FAO, UNDRO and WMO, and for 6 years was Rapporteur to UNESCO on remote sensing applications in hydrology and water management. He has served on numerous expert committees on remote sensing in the UK, Europe, and the USA, and has played an active role in many pioneering projects, including the Landsat 2 Program (as one of only two UK Principal Investigators), the US AgRISTARS Project, NASA's WetNet Project (currently serving as Precipitation Working Group Chairman), and ESA's ERS-1 Programme. In 1982 he was awarded the Hugh Robert Mill Medal and Prize of the Royal Meteorological Society for the development of methods for rainfall monitoring by satellites, and the DSc 'advanced doctorate' by the University of Bristol for a 'sustained and distinguished contribution to geographic science'. In 1983 he was appointed Director of the newly established Remote Sensing Unit in that university, and has built it up to become one of the foremost centres of its kind, internationally recognized for its work particularly in the area closest to his own heart, applied satellite hydrometeorology.

Dr Leonard F. Curtis

Dr Leonard Curtis was Reader in Geography and Head of the Joint School of Botany and Geography in the University of Bristol before leaving in 1978 to become National Park Officer for Exmoor. He was invested by HRH Queen Elizabeth II with the Order of the British Empire on his retirement from the Exmoor National Park office in 1988. His early experience of remote sensing began with wartime air photography in the Far East during service with the RAF, first as a Navigator, then a Squadron Leader by the time of demobilization. Commercial experience of applied photo-interpretation followed during a period of secondment from the Soil Survey of England and Wales to Huntings Surveys, involving work on irrigation development surveys. After joining the Department of Geography in Bristol University as a Lecturer in 1956 his interests in remote sensing broadened rapidly, and he served on a number of early applications committees, e.g. of the European Space Research Organization and its successor, the European Space Agency. He later served as consultant to many UK and European programmes and projects in remote sensing, and recently acted as Secretary-General and Vice Chairman of EARSeL (the European Association of Remote Sensing Laboratories), a body which he had helped to found in the late 1970s. Dr Curtis authored or edited more than ten books concerned with remote sensing applications, soil studies and land use inventories, and contributed greatly to the development of uses of remote sensing in environmental monitoring and management in protected landscapes. Very sadly, Dr Curtis died suddenly in April 1991 at the age of 67. The *Earth Observation Quarterly*, published by ESA, said this of him in its subsequent Obituary: 'Len Curtis, without fuss or bother, carved a niche for himself amongst the pioneers of European remote sensing.' This book is one of the many monuments to his life and work.

Abbreviations and Acronyms

ACT	Automatic Collection and Telemetry
ADEOS	Advanced Earth Observation Satellite
AEM-A	Applications Explorer Mission-A
AgRISTARS	Agriculture & Resource Inventory Surveys Through Aerospace Remote Sensing
AID	Agency for International Development (USA)
AIS	Airborne Imaging Spectrometer
AMI	Active Microwave Instrument
AMSR	Advanced Mechanically Scanning Radiometer
AMSU	Advanced Microwave Sounding Unit
ARTEMIS	Africa Real-Time Environmental Monitoring and Information System
API	Air photo-interpretation
ATS	Applications Technology Satellite
ATSR	Along-Track Scanning Radiometer
AVHRR	Advanced Very High Resolution Radiometer
AVIRIS	Airborne Visible/Infrared Imaging Spectrometer
BDR	Bidirectional Reflectance
BF	Banding Feature (of hurricane)
BRDF	Bidirectional Reflectional Distribution Function
BRF	Bidirectional Reflectance Factor
CAD	Computer Aided Design
CAT	Clear Air Turbulence
CCT	Computer Compatible Tape
CCEW	Countryside Commission for England & Wales
CDO	Central Dense Overcast (of hurricane)
CEC	Commission of the European Communities
CNES	Centre Nationale d'Etudes Spatiales (France)
CO_2	Carbon dioxide
CORINE	Co-ordination Information, Environment
COSPAR	Committee for Space Research (UN)
CPU	Central Processing Unit
CRT	Cathode Ray Tube
CZCS	Coastal Zone Color Scanner
DBMS	DataBase Management System
DCP	Data Collection Platform
DCS	Data Collection System
DEM	Digital Elevation Model

DLCC	Desert Locust Control Commission
DMSP	Defense Meteorological Satellite Program
DoE	Department of Environment (UK)
DPA	Drainage Pattern Analysis
DTM	Digital Terrain Model
DVP	Digital Video Plotter
EARSeL	European Association for Remote Sensing Laboratories
EBR	Electronic Beam Recorder
ECU	European Currency Unit
EDC	EROS Data Center (USA)
EDIS	Environmental Data and Informative Service (USA)
EDU	Earth Data Center
EEC	European Economic Community
EMI	Electro-Magnetic Industries
EOS	Earth Observation System
EOSAT	Earth Observation Satellite
EPOP	European Polar Operational Platform
ERBE	Earth Radiation Budget Experiment
ERBS	Earth Radiation Budget Satellite
ERDAS	Earth Resources Data Analysis System
EROS	Earth Resources Observation System
ERS	Environmental Remote Sensing
ERS-1	Earth Resources Satellite
ESA	European Space Agency
ESMR	Electrically-Scanning Microwave Radiometer
ESOC	European Space Operations Centre (Darmstadt, Germany)
ESSA	Environmental Sciences Services Administration (ESSA)
ESTAR	Electrically-Scanning Thinned Array Radiometer
EUMETSAT	European Meteorological Satellite
EURASEP	European Association of Scientists in Environmental Pollution
FAO	Food & Agriculture Organisation of the United Nations
FAS	Foreign Agricultural Service
FFT	Fast Fourier Transform
FGGE	First GARP Global Experiment
FIFE	First ISLSCP Field Experiment
FLIR	Forward Looking InfraRed Radiometer
FOV	Field-of-View
FRONTIERS	Forecasting Rain, Optimized using New Techniques of Interactively Enhanced Radar & Satellite data
GAP	Geographic Applications Program (USGS)
GARP	Global Atmospheric Research Program
GATE	GARP Atlantic Tropical Experiment
GDD	Growing Degree Day
GEMS	Global Environment Monitoring System
GGEM	Gravity Gradiometer Explorer Mission
GHz	Giga-Hertz
GIS	Geographic Information System

GLAI	Green Leaf Area Indices
GLU	Grazing Livestock Unit
GMS	Geostationary Meteorological Satellite
GMT	Greenwich Mean Time
GOES	Geostationiary Operational Environmental Satellite
GOES-IO	Global Operational Environmental Satellite, Indian Ocean
GOSSTCOMP	Global Operational Sea Surface Temperature Computation
GRE	Ground Resolution Element
GSFC	Goddard Space Flight Center (Greenbelt, MD)
HCMM	Heat Capacity Mapping Mission satellite
HIPLEX	High Plains Experiment
HRIS	High Resolution Infrared Sounder
HRPT	High Resolution Picture Transmission
HRV	High Resolution Visible
IBP	International Biological Programme
ICSU	International Council of Scientific Unions
IFFA	Interactive Flash Flood Analyser
IFOV	Instantaneous Field of View
IGBP	International Geophere-Biosphere Programe
IMC	Image-Motion Compensation
IPIPS	Infrastructure Planetary Image Processing System
IRLS	InfraRed Line Scan
IRIS	InfraRed Interferometer Spectrometer
IRSA	Indian Remote Sensing satellite
IR/VIS	Infra-Red/Visible
I^2S	International Imaging Systems
ISLSCP	International Satellite Land Surface Climatology Project
ITCB	Inter-Tropical Cloud Band
ITCZ	Inter-Tropical Convergence Zone
ITOS	Improved Tiros Operational Satellite
JERS	Japanese Earth Resources Satellite
JPL	Jet Propulsion Laboratory (USA)
JPOP	Japanese Polar Operational Platform
JRC	Joint Research Centre (EEC, Ispra)
LAC	Large Area Coverage (NOAA data)
LACIE	Large Area Crop Inventory Experiment
LAGEOS	Laser Geodetic Earth-Orbiting Satellite
LAI	Leaf Area Index
LARS	Laboratory for Agricultural Remote Sensing (Purdue University, USA)
LAWS	Laser Atmospheric Wind Sounder
LFC	Large Format Camera
LFMR	Low Frequency Microwave Radiometer
LIMS	Limb Infrared Monitoring of the Stratosphere
LISS	Linear Self-Scanning
LRR	Laser Retro-Reflector

MCC	Mesoscale Convective Cluster
MESSR	Multispectral Electronic Self-Scanning Radiometer
MFE	Magnetic Field Explorer
MIEC	Meteorological Information Extraction Centre (ESOC)
MLC	Monitoring Landscape Change
MODIS	Moderate Resolution Infrared Sounder
MOMS	Modular Opto-electronic Multispectral Scanner
MOS	Marine Observation Satellite
MSFC	Marshall Space Flight Center (Huntsville, AL)
MSS	MultiSpectral Scanner
MSR	Microwave Scanning Radiometer
MSU	Microwave Sounding Unit
MTEM	Magnetosphere/Thermosphere Explorer Mission
MTF	Modulation Transfer Function
MWS	Maximum (sustained) Wind Speed
NASA	National Aeronautics & Space Administration
NDVI	Normalised Difference Vegetation Index
NERC	Natural Environmental Research Council (UK)
NESDIS	NOAA Environmental Satellite Data Information Service
NEXRAD	Next Radar (operational system, USA)
NIR	Near Infra-Red
NMC	National Meteorological Center (USA)
NMS	Nimbus-5 Microwave Spectrometer
NOAA	National Oceanic & Atmospheric Administration (USA)
NPOC	National Point Of Contact
NPOP	North American Polar Operational Platform
NROSS	Naval Research Oceanographic Satellite System
NRSC	National Remote Sensing Centre (UK)
NSCAT	NASA Scatterometer
NSF	National Science Foundation
OCS	Ocean Colour Scanner
ODNRI	Overseas Development National Resources Institute (UK)
ONR	Office of Naval Research (USA)
OSC	Outer Space Committee (of UN)
PMI	Passive Microwave Imager
PMR	Passive Microwave Radiometer
POEM	Polar-Orbiting Environmental Monitor (ESA)
PPI	Plan Position Indicator
PRARE	Precise Range & Range-rate Equipment
QPB	Quantitative Precipitation Branch (of USWB)
RADAR	Radio Direction And Ranging
RAE	Royal Aircraft Establishment (UK)
RAF	Royal Air Force
RAM	Random Access Memory
RAMP	Radar Mapping of Panama

RAR	Real Aperture Radar
RBV	Return Beam Vidicon
RHI	Range Height Indicator
RPV	Remotely Piloted Vehicle
SAM	Stratospheric Aerosol Measurement
SAR	Synthetic Aperture Radar
SBUV	Solar Back-scattered Ultraviolet
SCAMS	Scanning Microwave Spectrometer
SCMR	Surface Composition Mapping Radiometer
SCR	Selective Chopper Radiometer
SDD	Stress Degree Day
SDSD	Satellite Data Supply Division
SeaWifs	Sea-viewing, Wide Field-of-view Sensor
SIR	Shuttle Imaging Radar
SIRS	Satellite Infrared Spectrometer
SLAR	Side Looking Airborne Radar
SMMR	Scanning Multichannel Microwave Radiometer
SMS	Synchronous Meteorological Satellite
SPAM	Spectral Analysis Manager
SPOT	Systeme Probatoire d'Observation de la Terre
SOM	Space Oblique Mercator
SSM/I	Special Sensor Microwave Imager
SSM/T	Special Sensor Microwave Temperature
SST	Sea Surface Temperature
SSU	Stratospheric Sounding Unit
THIR	Temperature Humidity Infrared Radiometer
TIREC	TIROS Ice Reconnaissance
TIROS	Television & InfraRed Observation Satellite
TM	Thematic Mapper
TOMS	Total Ozone Mapper System
TOPEX	Topography Experiment (Ocean)
TOVS	Tiros Operational Vertical Sounder
TREM	Tropical Rainfall Explorer Mission
TRMM	Tropical Rainfall Monitoring Mission
UARS	Upper Atmosphere Research Satellite
UKAEA	United Kingdom Atomic Energy Authority
UNDRO	United Nations Disaster Relief Organisation
UNEP	United Nations Environmental Programme
UNESCO	United Nations Educational, Scientific & Cultural Organisation
USA	United States of America
USAF	United States Air Force
USB	Unified S-Band
USGS	US Geological Service
USWB	United States Weather Bureau
UTM	Universal Transverse Mircator

VHRR	Very High Resolution Radiometer
VIRR	Visible Infra-Red Radiometer
VIRSR	Visible and Infrared Scanning Radiometer
VISSR	Visible & Infrared Spin Scan Radiometer
VTIR	Visible and Thermal Infrared Radiometer
VTPR	Vertical Temperature Profiling Radiometer
WEFAX	Weather Facsimile
WMO	World Meteorological Organisation
WWW	World Weather Watch

Part One Remote Sensing Principles

1 *Monitoring the environment*

1.1 Growth in awareness of environmental problems

Most people today are aware of the existence of environmental problems. Many people are concerned about the quality of life that they can enjoy in their own locality. These thoughts may be triggered by various events affecting the air they breathe, the water they drink, or the open green space they are accustomed to using for leisure. In the main the problems arise from human use of land, air and water, for pressures on land and resources are growing fast.

Some people even live in constant fear of hazardous environmental conditions, which can lead with frightening speed to life- and property-threatening disasters. These environmental threats may involve earthquakes, volcanic activity, flooding, storms, droughts or other factors. Each of these may be of sufficient severity to cause considerable damage to property and/or loss of human life.

As people look beyond their immediate localities they are becoming increasingly aware of larger scale problems facing particular regions of the global environment. The Earth is the only planet in the universe known to sustain human life. Yet it is human activity that is now conspiring to reduce the life supporting capacity of planet Earth. In part this is due to the disproportionate consumption of world resources by an affluent minority, and to the damage caused by high-tech systems and ways of life. But, at the same time, the majority of the world's population is poor and struggling to achieve better standards of living and adequate food supplies. And there is growing evidence that many nations are degrading or even destroying natural resources in their struggle for betterment.

Clearly we as a species have developed the capability of subjugating all other biological species for our own ends. One result of this power is that the human population of Earth has grown from about 200 million some 2000 years ago to about five billion today. It is estimated that by the end of the twentieth century, six billion people will inhabit the Earth (Fig. 1.1). Population pressures will continue to build in the next century and the question arises as to when the global population may stabilize. Some authorities have suggested this may not happen until figures of between ten and 14 billion have been reached.

Through activities resulting from this population growth we have reduced the biological and genetic diversity of Earth. An estimated further 25 000 plant species and more than 1000 vertebrate species surviving today are now threatened with extinction. Because of these trends, concern is now being expressed that Man's relationship with the *biosphere* (the thin covering of the planet that contains and sustains life) will continue to deteriorate unless a new environmental ethic is adopted.

Although the human race and other living creatures inhabit the biosphere, the global environmental system contains other important components that interact with it (Fig. 1.2). These components include the atmosphere, geosphere and hydrosphere, each of which makes important contributions to the conditions of the environment affecting human life. The principal processes influencing the development of the global environment are the energy, hydrological and biogeochemical cycles.

Although it is often hard to dissect out the relationships linking us with our environment that are distinctly natural, technical or cultural in type, our species is today a key, possibly *the* key, participant in planetary processes. Many chains of

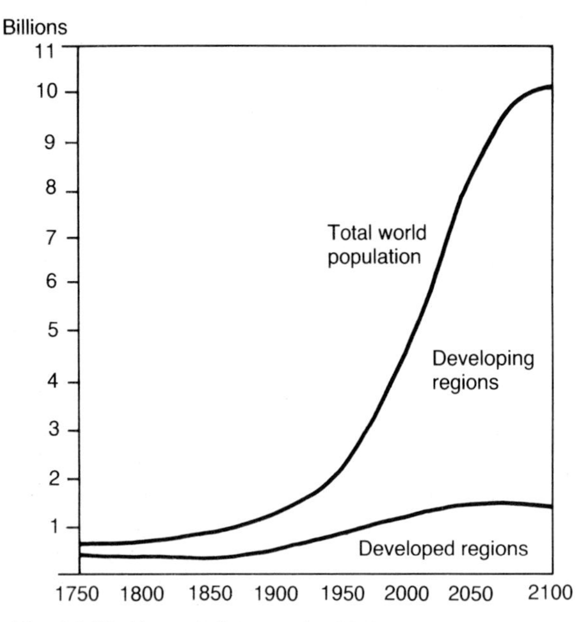

Fig. 1.1 World population growth with future projections.

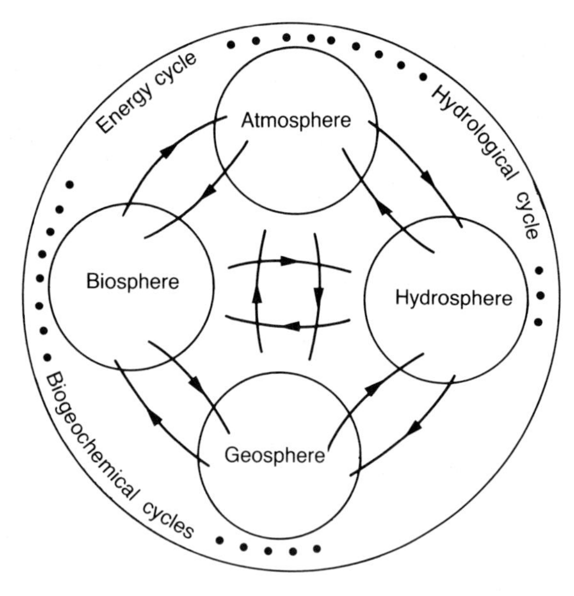

Fig. 1.2 Components of the global environmental system. (Courtesy, NASA.)

cause and effect are either intricately ramified or inadequately known or understood. The great majority involve so-called 'feedback' effects. Thus one person's 'improvements' to the environment may become another person's environmental problems – or, even ultimately our own!

Fortunately alongside a recent growth in environmental concern there has been a growth in our awareness of the need for *conservation*. Conservation, like 'development', is beneficial to people. However, whereas development aims to achieve benefits through use of the biosphere, conservation aims to achieve benefits by ensuring that such use can continue. There is a growing realization that the world economy must move towards *sustainable development* rather than exploitation and destruction of natural resources. In this respect, living resource conservation has three specific objectives:

1. To maintain essential ecological processes and life-support systems.
2. To preserve genetic diversity.
3. To ensure the sustainable utilization of species and ecosystems.

Many reports and organizations relating to our environment and its everyday use have been brought into play since the first *World Environment Day* in 1974. The International Council of Scientific Unions (ICSU) has been a major co-ordinator of international environmental research programmes. Likewise, initiatives of the United Nations Educational Scientific and Cultural Organization (UNESCO) and the United Nations Environment Programme (UNEP) have played important parts in developing knowledge of environmental problems and how to solve them. Programmes such as the International Geophysical Year (1957–1958) and the International Biological Programme (IBP) running from 1964 to 1974 were early studies of significance. The UNEP has been active in promoting the Global Environment Monitoring System (GEMS), which provides databases for the international community on environmental matters. In recent years the lead in developing an integrated concept of research on the global environment has come from the USA. The National Aeronautics and Space Administration (NASA) and the National Science Foundation (NSF) have made significant contributions to this. One result is that the International Geosphere–Biosphere Programme (IGBP) has now been established by the ICSU.

The IGBP is one major project that is attempting to integrate a wide variety of disciplines and areas of study within a global environmental research programme. Particular emphasis is placed on the need for development of an adequate global data and information system. Without this it will be impossible to make sound policy decisions for the maintenance of a stable and productive Earth environment. In assessing environmental matters, it has become commonplace to follow a well-established sequence of steps. These include:

1. The *recognition* of forms, structures, and/or processes of significance.
2. The *identification* of such phenomena in their real world situation(s).
3. The *recording of their distributions*, often through both space and time.
4. The *assessment of these distributions*, sometimes singly, but more often in some combinations.
5. Attempts to *understand the nature and cause* of any specially significant relationships between and amongst the key phenomena and processes.

It is only when such background studies have been completed successfully that their results can be used to benefit man and his environment. Commonly their use is in one or more of the following ways:

1. The *preparation and execution of schemes* of environmental or resource management.
2. The *planning and development of future projects* to modify and improve the environment.
3. The *prediction of forthcoming events*, especially those over which we have less direct control.

At the heart of the all-important basic or background studies are problems concerned with *observation*. Data must be obtained by appropriate means, they must be put into permanent forms, and there are great advantages if they lend themselves readily to computer processing for rapid analysis, interpretation and intercomparison. *It is here that environmental remote sensing has its most important part to play* and it is with such activities that this book is mainly concerned. By common consent, remote sensing is beginning to fulfil a long recognized need for both intensive and extensive monitoring of Earth, simultaneously in much more detail yet also with much greater uniformity than has ever been possible before. But first, a summary of the development and roles of environmental observation by longer-established means.

1.2 *In situ* sensing of the environment

Humans have made measurements of key aspects of our environment since an early stage in the development of our civilization and culture. Examples of measuring devices include the famous Nilometer by which water levels in the River Nile were noted, and rudimentary rain gauges used by natural philosophers in the city states of Ancient Greece. The European Renaissance of art, science and literature marked a new surge of interest in the need for, and design of, monitoring instruments. Attention began to be paid not only to environmental variables or effects which were readily visible, but also to others less directly evident in the world of nature. The invention of the thermometer by Galileo at the end of the seventeenth century is particularly noteworthy. A steady deepening of interest in environmental factors and conditions ensued in the eighteenth and nineteenth centuries.

The scientific rebirth of the Renaissance period and its aftermath was followed by a rapid acceleration in our concern for the environment as the twentieth century unfolded. Not only could the environment be monitored by a wide range of sophisticated instruments, but advances in the related technologies of communications and recording also ensured that data could be collected, collated and processed, even from quite remote locations, in 'near real-time' (i.e. very close to the time of observation). At last environmental monitoring could be organized on a wide scale so rapidly that short-term event prediction, and dependent practices of environmental management and control, were possible.

Most recently, the advance of other support technologies, especially in the related fields of electronic computing and the microchip, has begun to exercise what will doubtless grow to be a profound influence upon the design of what are often called '*in situ*' sensor networks, and through them, upon the types of questions we may hope to solve through the analysis and interpretation of such 'conventional' data. Unfortunately, despite great advances in *in situ* technology by the mid-twentieth century, significant problems of environmental monitoring by such systems remained. Not least significant is the fact that most conventional data are related to point locations or transects: for areal assessments, spatial interpolation procedures must be applied. These are more satisfactory when point observation data are numerous. They remain unsatisfactory for many purposes today because many *in situ* instrument networks are thin, or even virtually non-existent in some regions of the world. Furthermore, many interesting, even important, environmental parameters do not readily lend themselves to *in situ* measurement at all. Remote sensing, especially from satellites, is rightly seen as a growing answer to such problems.

1.3 Remote sensing of the environment

Remote sensing can be defined as *the science of observation from a distance*. Thus it is contrasted with *in situ* sensing, in which measuring devices are either immersed in, or at least touch, the object(s) of observation and measurement. Some authors have spoken of remote sensing systems in connection with instrument packages which are remote only in the sense that they are placed in relatively inaccessible locations, or are connected to central data processing facilities by automatic data acquisition and transmission links. These arrangements do not accord with the definition above, nor with the generally accepted view of remote sensing by those who practice it. Others have spoken of remote sensing in terms of a lack of physical contact between the sensor and its target. Again, strictly speaking this is incorrect, since where no physical contact exists, no observation is possible: while no 'touch-type' contact exists between a remote sensor and its target, some physical emanation from, or effect of, the target must be found if aspects of its property and/or behaviour are to be investigated. The most important of the physical links between objects of measurement and remote sensing measuring devices involve *electromagnetic energy*, *acoustic waves* and *force fields*. These will be discussed in greater detail in Chapter 2.

For most surface and atmospheric remote sensing, electromagnetic energy is the supreme medium, and the home base of our species is located at the interface between Earth and its atmosphere. Therefore this book is much more concerned with the exploitation of electromagnetic energy than all other energy forms together, and is more concerned with the bottom of the atmosphere and the top of Earth than with other places or regions in the Universe.

In Chapter 2 we will discuss ways in which we all practice naturally some methods of remote sensing: by these we observe the world in which we live. But because the science of remote sensing is concerned as much with *recording* as with observation, the remote sensing era may be said to have dawned with the invention of photography in 1826. Devices, such as telescopes, had been invented much earlier to extend our personal observing capabilities. But it has been only in the last century and a half that means have been devised whereby the environment can be both observed and recorded objectively by artificial devices. In this respect, the potential of photography was quickly appreciated, especially for recording scenes of special significance to the observer. New cameras and types of films were invented, even to extend our view beyond the relatively narrow waveband of visible light. Then began a more systematic search through the radiation spectrum, and into the realms of acoustical, chemical, gravitational and radioactive energy to discover fresh means whereby many other aspects of our natural and cultural environments might be investigated. This search was stimulated greatly by the military demands of two World Wars, especially the second. The search has gathered even more momentum recently, and is by no means yet complete. (See also section 1.5).

It was in 1960 that reference was first made by name to 'remote sensing' either as a distinctive field of study, or a set of approaches to the human environment. Since then it has passed take-off point, deriving great impetus from the opening of the satellite era and the space race between the remote sensing superpowers, the USA and USSR. In particular, NASA has played a gigantic part as a result of American Government policy to make remote sensing data (much of it from satellites) readily available to the scientific community throughout the world. The success of NASA has led to the establishment of national space agencies in other countries, as far afield as Brazil, India and Japan, and international counterparts, most notably in the case of the European Space Agency (ESA), whose Member States are Belgium, Denmark, France, Federal Republic of Germany, Ireland, Italy, The Netherlands, Spain, Sweden, Switzerland and the United Kingdom. Austria is now an Associate Member of ESA, and Canada and Norway have Observer status.

Remote sensing is now accredited with a policy-making Sub-Committee of the United Nations Committee for the Peaceful Uses of Outer Space, a Sub-Committee more concerned with Earth

Observation than first appearances suggest: in UN terminology the boundary between 'Inner' and 'Outer' Space is merely 50 km above the surface of Earth. The research-orientated UN Committee for Space Research (COSPAR) also contributes significantly to environmental remote sensing through its international conferences and working groups.

Opportunities for discussion and exchanges of research results in remote sensing have multiplied dramatically since the early 1970s. National remote sensing societies have been established by now in most major countries of the world. International links are also growing, for example through the European Association of Remote Sensing Laboratories (EARSeL), which is also serving as a catalyst in internationally collaborative research. Spurred on by an associated upsurge of interest in the present and potential applications of remote sensing in everyday operational use, national and international remote sensing centres are multiplying fast.

Since 1983 a working group of NASA has been examining the mission requirements and major science goals for what is now conceptualized as 'Earth System Science' – the study of Earth as an integrated, dynamic whole. In this, remote sensing has a key role to play, through a new 'Earth Observing System' (EOS), whose basic structure is set out in Table. 1.1.

In the light of such activity, it is not surprising that remote sensing instruction has already spread widely through the tertiary education sector. Here it is often correctly portrayed as the kind of integrative study that is needed in these days to bring together specialists and specialized information from many fields of environmental science, broadly defined. The result is that graduates with interests and expertise in remote sensing of the environment can find employment in such seemingly diverse areas as meteorology, pedology, hydrology, geology and geophysics, agriculture conservation and protection, pest control, fishery development, land use planning, civil engineering and computing, to mention but a few.

The needs of the growing community of remote sensing students and scientists are being met by an ever widening range of germane books and journals. The first area-specific journal, *Remote Sensing of Environment* was published in 1969; the *International Journal of Remote Sensing* first appeared in 1980. Newsletters have proliferated in this field, appropriate for one which has been characterized by such rapid growth and change. Some long-established societies have come to terms with the precocious newcomer by taking it under their wings: thus, for example, the *Photogrammetric Engineering* journal has been renamed *Photogrammetric Engineering and Remote Sensing*, and *Transactions on Geoscience Electronics* has become *Geoscience and Remote Sensing*.

1.4 Economic and practical benefits of remote sensing, past, present and future

It was in meteorology that satellite remote sensing first achieved fully operational status in 1966. In that year the significance of satellite data inputs to meteorological data-pools was acknowledged practically through the inauguration of a system of American satellites designed to yield information to any suitably equipped and relatively modestly priced receiver anywhere in the world – a system within which stand-by satellites were to be available for launching at short notice so that the supply of data to routine users could be reasonably assured. This system has worked well, and almost continuously ever since. Being a relatively self-contained programme, this provided the first convincing figures for cost/benefit assessments of satellite operations. Figures published in 1971 put the annual cost of American meteorological satellite research at around $200 million, and the annual cost of adverse weather to the American community at some $10 000 million. Of this huge sum, it was estimated that some 20% could be eliminated given better weather information, equivalent to ten times the annual expenditure on weather satellites. Already by that time the dramatically increased flow of weather data that satellites had made possible had led to benefits of about $75 million per annum through improved hurricane forecasting alone – a figure four times the annual outlay on weather satellite operations.

Subsequently evaluations of satellite meteorology were made through a study contract of the European Space Agency, relating both to general issues and the specific issue of the cost/benefit viability of the ESA geostationary satellite system, Meteosat (Chapter 5). At the outset of that study

Table 1.1 The recommended programme for a new 'Earth Observing System' as initially conceived by NASA in 1987. (Source: NASA, 1988)

Immediate	
Development of new management policies and mechanisms to foster co-ordination among NASA, NOAA, NSF, others Strengthening of international agreements and co-operation necessary for worldwide study of the Earth	
Near term: 1987–1995	**Observing Programme for Earth system science: 1995 and beyond**
Continuing and operational Earth observations (NOAA, NASA, others) Specialized space research missions (ERBE, LAGEOS, UARS, TOPEX/Poseidon, NSCAT/N-ROSS) Establishment of NASA Earth System Explorer mission series (initiate with GREM) Flight of other demonstrated instruments at modest cost Co-ordinated, interdisciplinary programme of basic Earth system research and *in situ* measurements (NASA, NOAA, NSF, USGS, DoE, ONR, others) Development of information system for Earth system science Instrument and technique development	New era of integrated, global Earth observations: 1. Earth Observing System (EOS) on polar-orbiting space platforms (USA, other nations) 2. Complementary measurements from the ground and from aircraft, balloons, and ships 3. Continuation of NASA Earth System Explorer missions (TREM, MFE/Magnolia, MTEM, GGEM) 4. Advanced geostationary platforms Expansion and utilization of information system for Earth system science Sustained support for expanded, co-ordinated, interdisciplinary programme of basic research and *in situ* measurements (NASA, NOAA, NSF, USGS, DoE, ONR, others)

several difficulties were identified, of which some are common to most or all satellite remote sensing studies of Earth phenomena. The most significant of these is that, whilst the costs of satellite remote sensing can be evaluated rather readily, it is more difficult to estimate its returns and benefits. One reason for this is that many 'customers', e.g. members of the agricultural or construction industries who are advantaged by improved weather forecasts, do not see themselves as users of satellite data and do not pay directly for satellite services and products as such. Figure. 1.3 illustrates the 'study logic' required for such studies.

Results revealed significant differences between Europe and Africa. For the developed countries of Western Europe, conservative estimates of expected benefits from the operational use of Meteosat-type satellites exceeded the corresponding costs by a global factor close to 3, rising in individual countries to as high as 7.8 in the case of Spain. Compared with the USA these figures are on the low side, reflecting especially the absence of tropical cyclones from, and the lower incidence of severe convectional storms in, Western Europe. The small sizes of some of the nations involved also influenced the comparison with the USA. When

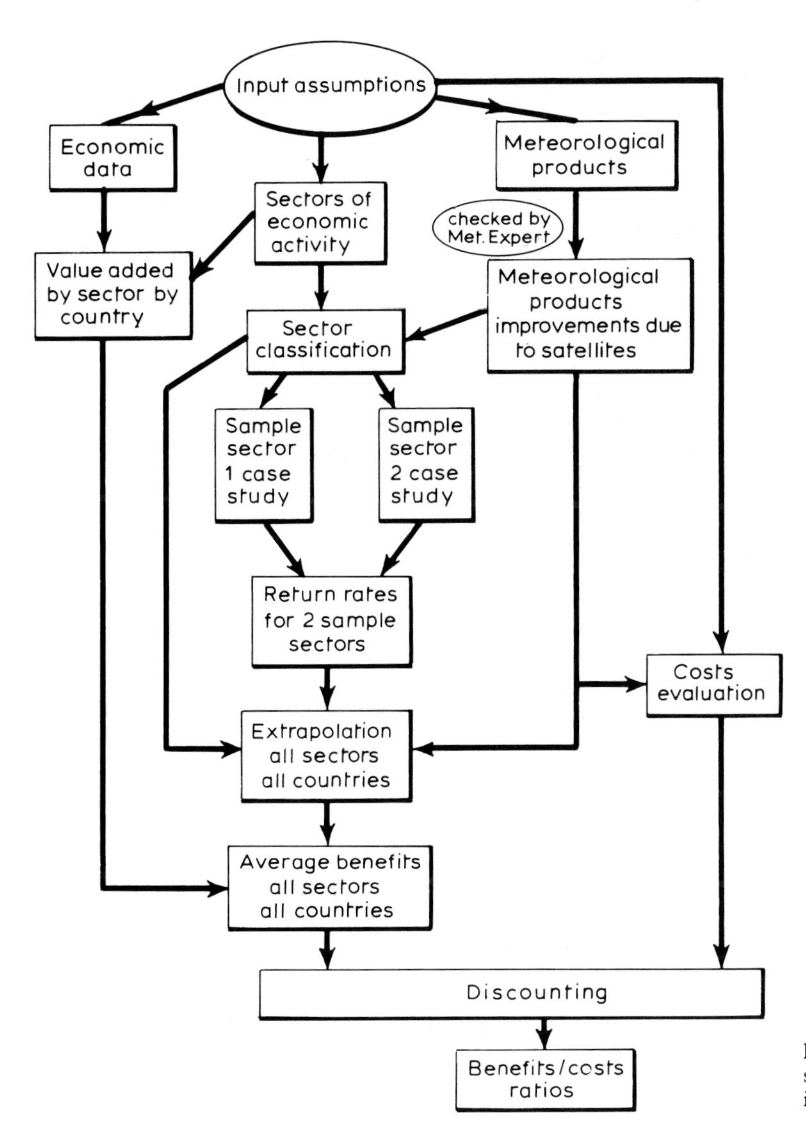

Fig. 1.3 Study logic for an evaluation of satellite remote sensing cost/benefit ratios in meteorology. (Source: Lagarde, 1980.)

Africa and the Middle East were taken into account also, the global benefits/cost ratio for Meteosat-type operations rose to the impressively high level of 16 : 1. This reflects especially the comparative sparseness of the conventional weather-observing network outside Europe and the less oblique views obtained from geostationary satellite altitudes over the very large land mass of Africa. Assessments of several different sectors of economic activity revealed that, almost universally within Europe, Africa and the Middle East, those sectors which were the most sensitive to the weather were Agriculture, Transport, Construction, and Utilities (Electricity, Gas, Water) in that rank order. It was concluded that clear potential benefits could accrue from all of these throughout the Meteosat image area.

It is hardly surprising, therefore, that Meteosat, which now represents one of the main sources of weather data for Europe, has proved a heavily used satellite system. Its cloud images are familiar to tens of millions of viewers on the TV screens across Western Europe. Whilst originally an exclusively ESA-derived mission, the administration of Meteosat moved from the European Space Agency in the mid-1980s to a quasi-independent managing agency named EUMETSAT based in Darmstadt, Germany. With this move the metamorphosis of

Table 1.2 Customer and cost profiles for Landsat data, 1974 compared with 1984 (Source: US Congress, 1985)

Customer category	Financial year 1974			Financial year 1984		
	Number of items	Total costs ($)	Unit costs ($)	Number of items	Total costs ($)	Unit costs ($)
Federal Government (less NASA Investigators)	28 493	87 156	3.06	16 017	1 696 710	105.93
Stage/Local Government	2 534	10 920	4.31	1 222	122 163	99.97
Academic	18 611	63 964	3.44	2 578	181 433	70.38
Industrial	35 890	114 140	3.18	8 213	985 362	119.98
Individuals	17 266	67 127	3.88	1 848	84 498	45.72
Non-USA customers	37 038	120 499	3.25	8 128	741 962	91.28
Non-identified	17 346	64 708	3.73			
Totals (columns 1, 2, 4 and 5) and averages (columns 3 and 6)	157 178	528 514	3.36	38 006	3 812 128	100.30

Meteosat from a research and development (R and D) project to a fully operational system funded by the national meteorological services of Western Europe had become complete.

In the realm of Earth surface monitoring it is much more difficult to assess the benefits of remote sensing. This is partly because, unlike weather data, surface data are not usually required in real time, i.e. within a short time period. Also, it is often difficult to compare the costs of data collection by conventional ground survey methods with those of remote sensing: the costs of conventional data gathering are often difficult to obtain and are frequently absorbed within other administration costs. Likewise the accuracy of conventional data is often uncertain. Last but not least, the prices charged for satellite products are different from source to source, and have changed sometimes quite erratically through time, as evidenced by Table 1.2, which contains much interesting food for thought.

However, for large-scale enterprises covering large land areas the costs of conventional methods are high, and the sizes of the overall budgets demand that survey accuracies should be as high as possible. For example, it has been recognized that subsidies for European agriculture, which have absorbed up to 70% of the European Community budget, could be reduced by millions of Ecus if improvements in the accuracy of European agricultural surveys could be achieved. Thus substantial efforts are being made to improve their accuracies by using remote sensing data to estimate crop yield and crop acreages (Chapter 13), parameters that conventional surveys may not evidence well. Such combined or *synergistic* uses of conventional and remote sensing monitoring methods are becoming more and more widespread today.

It is clear that the advents of high resolution data from the Landsat Multispectral Scanner and Thematic Mapper together with SPOT data (Chapter 5) have brought operational mapping and monitoring of land use and land cover features within both the practical and financial reach of many potential users. This is particularly true since a parallel development of spatial data computer systems (section 1.5) now makes it possible to combine data from satellites, aircraft and conventional maps both effectively and efficiently.

The potential benefits of environmental monitoring of oceans and coastal waters were first assessed more than ten years ago (Table 1.3) prior to the development of satellite systems for ocean monitoring, and were found to be already very large. Subsequently, the Coastal Zone Colour Scanner (CZCS), the Japanese Marine Observation Satellite (MOS) family and European Resources Satellite-1 (ERS-1) have been designed and developed (Chapter 5), with the Sea-Viewing, Wide Field-of-View Sensor (SeaWIFS) following by the mid-1990s.

By the end of the 1980s the attention of the world

Table 1.3 Potential aggregate gross benefits of environmental surveillance (including satellites, aircraft and *in situ* sensor systems) for marine transportation (1975 $ million) (From MacQuillan and Clough, 1978)

	1986–1990 (5-year total)	1991–2000 (10-year total)
Ocean routing:		
Pacific ports	120	370
Atlantic ports	100	320
Great Lakes ports	9	26
Ice surveillance:		
St Lawrence	75	230
Atlantic	45	150
Arctic oil and gas	135	745
Arctic supply, minerals	95	230
West coast log towing	10	20
Totals for comprehensive surveillance systems	589	2091
Contribution of satellite systems	402	1543

community had come to be focused on a new range of global environmental problems more complex, and potentially even more important than any mentioned earlier in this section, and whose costs are particularly difficult to quantify. These include:

1. Climate change. Possible changes in climate owing to the increasing concentration of the so-called 'greenhouse gases' in the atmosphere, such as CO_2, whose increases are caused by such activities as the burning of fossil fuels and deforestation.
2. Ozone depletion. The release of man-made chlorofluorocarbons into the atmosphere is thought to be the major cause of the catalytic destruction of stratospheric ozone, leading to the so-called 'ozone holes' recently recognized over both polar regions at certain times of the year. These are dangerous because of the increased fluxes of ultraviolet radiation which then reach the surface of Earth.
3. Acid rain. Increased acid deposition (acid rain) is causing damage to lakes, trees, and soils. This increased deposition mainly is a result of the release into the atmosphere of sulphurous compounds contained in fossil fuels burned in many power stations and domestic fires. The compounds so released are subsequently washed out in precipitation some distance from the point of origin.
4. Photochemical oxidant formation. These oxidants, which cause damage to plants and human health problems, result from photochemical reactions involving hydrocarbons and nitrogen oxides released by the exhausts of motor vehicles and other mechanical systems.
5. Desertification and deforestation. Overgrazing and deforestation are now widespread, with the result that the vegetative cover of Earth is being permanently affected. This leads to disruption of the ecological balance at the surface and to changes in the reflectivity of the Earth surface. As a consequence there is a danger that both the hydrological cycle and carbon dioxide balance will be seriously, perhaps even irreversibly, affected.
6. Earthquakes and volcanoes. Earthquakes and volcanic activity have great destructive potential. The possibility that an increased understanding of the Earth's crustal condition and its interior could permit an early warning system of such potentially cataclysmic events cannot be ignored.

Scientists are convinced that Earth observation from space provides the only viable and cost effective means of acquiring many of the necessary data to study, monitor, and hopefully contain or even mitigate these problems. These data are now required on global, regional and local scales so that theoretical numerical models can be tested against the background of actual changes in a large number of parameters and across a wide range of scales.

The families of satellites required for a fully effective 'Earth Observing System' (Table 1.1 and Chapter 19) will produce vast amounts of data from existing and future (planned) satellites. In this book we shall concentrate on the use of existing satellites for studies of environmental problems, but the principles by which data from future systems will be used will be at least basically the same. We shall find that, even if we are dealing with existing satellites only, many problems of handling data from different sensors at different scales arise. Some of the solutions to this and related problems are discussed later in this volume: the challenges are large, and the opportunities for remote sensing careers are growing rapidly.

In every respect there is no doubt that Environmental Remote Sensing is here to stay.

1.5 The growth of remote sensing in geography and the environmental sciences

Today the geographical and other environmental sciences make particularly broad and co-ordinated uses of remote sensing data from many different sources. Therefore it is appropriate that we conclude this chapter with special reference to the growing role of remote sensing in these germane sciences. In some ways the present high level of interest in remote sensing in geography and other environmental science disciplines can be seen as just the latest stage in a long history of their utilization of data obtained remotely from their physical sources. Historically the use of remotely sensed data in these fields of science can be summarized as follows:

1. Pre-1925: a period of slowly increasing study of the use of air-photos in topographic mapping. The use of air survey reconnaissance in the first World War accelerated the process to such an extent that systematic photography from the air became practicable for survey purposes.
2. 1925–1945: a period of widespread use of aerial photography in photogrammetry and map making. During this span of 20 years, air-photo interpretation became highly developed for military intelligence purposes. Many remote and harsh environments (e.g. Antarctica) were photographed, thus providing information for inaccessible areas.
3. 1945–1955: a period of development of interpretation and photogrammetric techniques. This was a period when geographers, soil scientists and geologists elaborated on photo-interpretation and photo-analysis techniques.
4. 1955–1960: a period of extensive application of aerial photography. In addition to topographic mapping, aerial surveys were used for regional planning, geological exploration, forest inventory, agricultural studies, terrain analysis, civil engineering, soil mapping and other tasks associated with temporal and spatial changes in the landscape.
5. 1960–1980: a period of active development of sensing devices and rapid establishment of satellite platforms leading to the first fully operational satellite remote sensing systems. The first meteorological satellite in 1960 heralded the opening of a period of intense activity, investigating the potentialities of balloons, rockets, and especially satellites for remote sensing, not only by conventional photography but also by a wide variety of new sensors.
6. 1980–present: a period through which countries other than the USA and USSR (notably France, India, China and Japan) launched Earth observation satellites, and laid plans for greatly increased future programmes related to remote sensing as a whole. This also has been a period marked by a systematic examination of specific sensor characteristics for particular applications, and intense development of automated image analysis hardware and software packages for computer analysis of remotely sensed data on a 'distributed' as distinct from a more 'centralized' basis.

Geographers in particular have long sought to synthesize information relevant to a particular area or region; often such syntheses used overlays of maps of different sets of data. Such types of syntheses are now made easier by the development of *Geographic Information Systems* (GIS), which use computer hardware and software. A Geographic Information System is designed to store large amounts of data and make it available on demand. Although initially computer-based systems for handling geographic information were little more than digital mapping systems, they are becoming increasingly sophisticated today. A classification of GIS functions is shown in Table 1.4.

At the same time, many users with computer experience expect nearly instantaneous responses to even relatively complex requests. The search is on, therefore, for new methods of storing vast quantities of data from both remotely sensed sources and conventional surveys simultaneously. Simultaneously new methods of data integration are sought using rapid access techniques. *Database Management Systems* (DBMS) are seeking fast access to spatial data out of large data collections.

Table 1.4 Classification of GIS functions (NERC, 1989)

Feature	Function
Data input and encoding	Data capture
	Data validation
	Interactive editing
	Quality control
	Data storage and structuring
	Remote access
Data manipulation	Conversion (e.g. vector to raster)
	Geometric correction (e.g. map registration, overlays)
	Generalization and classification
	Enhancement
	Abstraction
Data retrieval	Data selection criteria
	Browse facility
Data analysis	Spatial analysis
	Statistical analysis
	Measurement (e.g. line length, area)
	Error analysis
	Extension into three spatial dimensions and time
Data display	Graphic display
	Map generation
	Text
	Report writing
Database management	Database organization
	Multi-user access
	Maintenance and security

Already systems are being marketed for such disciplines as land evaluation, hazard monitoring, watershed management, environmental management, crop yield prediction and urban analysis.

Hopefully, the next decade will see remote sensing 'come of age' as a mature discipline with a balanced development of theory, practices and truly operational applications. The extent to which this will take place in the context of the environmental sciences will depend on a number of factors. The principal one may be the ability of organizations to accept the changing data formats as one sensor generation succeeds the other, with different spectral and spatial resolutions, making consequent demands upon database management systems. But other difficulties will face the pursuit of world-wide use of remote sensing data too, e.g. difficulties of technology transfer. The principal barriers for remote sensing technology transfer to developing countries, which are amongst the parts of the world which need, and stand to benefit most, from remote sensing have been said to include the following:

1. High costs of data in the form of computer compatible tapes or optical disks.
2. High costs of data processing equipment for computer interpretation.
3. Lack of personnel experienced in digital remote sensing.
4. Difficulties and high costs of maintaining hardware/software systems.
5. A shortage of foreign exchange to pay for remote sensing data, data processing equipment and technical support.
6. Lack of knowledge and/or interest amongst decision makers in developing countries so far as remote sensing is concerned.
7. Uncertainties concerning the future of Earth observing satellites provided by the major remote sensing nations.

The first four of these barriers may be said to apply equally to many so-called developed nations also. In particular there is a lack of experienced personnel at middle and high management levels in the 'user' agencies. As a result the introduction of remotely sensed data into operational administrative systems is taking place more slowly than the facts deserve. Nevertheless training and processing opportunities are steadily expanding. In the UK for example a National Remote Sensing Centre was set up in 1985. This provides a wide range of facilities, including some general-purpose high technology data processing and analysis systems. In addition it provides a browse file of available satellite images and a library of books and periodicals dealing with various aspects of remote sensing. Some such national remote sensing centres have been adopted as National Points of Contact (NPOC) for the dissemination of satellite data from European and USA satellites. Furthermore, some countries such as the UK have seen the need for regional data processing centres as well as a National centre. These centres are usually located in departments in Universities or Polytechnics and act as service points in various regions, offering facilities for training and use of data analysis equipment.

Much remains to be done before remote sensing becomes widely used as an operational tool for environmental mapping, monitoring and management throughout the world. It is hoped that, through reading this and other books on the subject, you yourself will want to play an active part in helping to bring about such a use of remote sensing.

2 *Physical bases of remote sensing*

2.1 Natural remote sensing

We all use our natural senses to observe and explore the environment in which we live. Certain senses – smell, taste, and, more often than not, touch – permit us to assess environmental qualities directly, through our neurophysical responses to the gases, liquids and solids with which we have immediate contact. The others – sight and hearing – can make us aware of more distant features through the patterns of energy propagations associated with them. Feeling, manifested through the sensitivity of skin to heat, also enables us to assess some characteristics of distant phenomena. These last three appreciations of the behaviour of energy sources some distance from ourselves are natural forms of remote sensing.

Since our visual powers are perhaps the most valuable we possess for gathering information about an object or phenomenon with which we are not in direct contact, we may usefully examine them in more detail as an introduction to the principles involved in remote sensing by artificial means. The key components of the remote sensing system which makes sight possible are the eye and the brain. Visible energy, in the form of light emitted by, or reflected from, an illuminated object is detected by sensitive cells in the eye. The eye is linked by the optic nerve to a high speed, real-time (i.e. near instantaneous) data processor – the brain.

The human eye is sensitive both to the *intensity* of the light energy received, and the *frequency* of the wave-like perturbations which may be taken to characterize such flows of energy. As a consequence we can differentiate both ranges of brightness and colour tone respectively. In the brain the quickly-processed data are compressed and presented as visual images. The brain also serves as a data bank in which earlier images can be stored, albeit within frustratingly narrow time limits, and with considerable loss of accuracy and definition. However, some qualitative pattern matching can be carried out by drawing mental comparisons of, say, the present image and selected images recalled from the past.

Our mental pictures can be modified by simple means, such as optical lenses, to correct vision defects, selective filters such as polaroid sunglasses, to reduce the glare of strong sunlight, or – less predictably – by chemicals, such as hallucinatory drugs, to modify the way we perceive the area of illumination.

Clearly, a natural remote sensing system such as the eye–brain is adapted best to the instantaneous assessment of environmental patterns. In the context of reflex activity this would be one of its greatest strengths. On the other hand, where conscious effort and freedom of choice are dominant this adaptation is one of its primary weaknesses. Fortunately, we may identify a number of facilities by which artificial systems are often necessary to improve our natural remote sensing performance. These include:

1. Both broader, and more selective, abilities to detect variations in environmental conditions. Spectrally our eye–brain system has a very limited performance: what is physiologically possible is restricted to the visible region of the electromagnetic spectrum.
2. A capacity for recording more permanently the patterns that are detected, to permit a more leisurely inspection of features of special interest.
3. A better recall system so that patterns observed in different areas and/or at different times might be intercompared with greater accuracy and in greater detail.

4. The ability to play back the past at different speeds if required. Many natural events may be examined best when studied by time-lapse methods, through which reality is speeded up, or in 'action replays' when recorded data are slowed down.
5. Automatic ('objective') data analysis so that personal (psychological, educational and/or neurophysical) peculiarities of the observer may be minimized.
6. Means of enhancing images to reveal or highlight selected phenomena.

In recent years great strides have been made towards the fulfilment of such desiderata through artificially improved remote sensing of the environment. Let us consider in more detail the range of opportunities presently being exploited and explored, not only in relation to remote sensing of visible light, but in relation to the exploitation of other remote sensing media also.

2.2 Technologically assisted remote sensing

Over the years our limited natural capabilities for remote sensing have been much extended by the invention and steady improvement of a variety of specialized instruments. Here we shall confine our attention to the broad fields in which such instruments are designed to operate. Later in this chapter and elsewhere in this book we will proceed to examine some of the instruments themselves, the forms and characteristics of their data, and ways in which these may be processed, analysed and interpreted.

2.2.1 Force fields

To the Earth scientist the most familiar 'force fields' are those of gravity and magnetism. Gravity is the force exerted by each body in the universe on every other body. It is directly proportional to the product of the masses of any pair of bodies, and inversely proportional to the square of the shortest distance between their centres of mass. We do not understand what causes gravity, but we can measure it, and use the results, for example, to substantiate geophysical theories of the solid Earth and to reveal facts concerning the nature and disposition of its crust. Evaluations of gravity and magnetism are important aspects of the geophysical exploration of Earth, and potentially of other bodies in the solar system, as practised by geologists.

Magnetism is the attraction which certain natural minerals (and the Earth which contains them) have for others, especially iron. The magnetic properties of such minerals permit us to prospect for commercial deposits through geomagnetic surveys. Slow changes in the magnetic field of the Earth accompany the 'wanderings' of the magnetic pole; wilder variations are caused by fluctuations in the 'solar wind' of charged particles arriving from the Sun. These are responsible for 'magnetic storms' and the associated interference to radio communications, and the aurorae; they can be monitored by magnetometers to reveal and record changes in the behaviour of the Sun. They are of interest primarily to geophysicists and astronomers.

2.2.2 Acoustical energy

Acoustic sounding of Earth's oceans is well-established as a means of profiling the sea floor. From a typical acoustic echo sounder, short pulses of audible sound are emitted in a selected direction, and the returning reflected or scattered signals are collected and interpreted as indications of ocean depth and sea floor characteristics. The atmosphere also is amenable to investigation using acoustic sounding techniques, especially aspects of its wind and thermal structures (Fig. 2.1). Indeed, the strength of interaction of acoustic waves and the atmosphere is far greater than for electromagnetic waves. Unfortunately the operational range of acoustic waves is less, and their slower speed of propagation can be a source of error with some measurements. Consequently the applications of such techniques to the atmosphere have been predominantly research. For the future, the simplicity and much lower cost of an acoustic echo sounder system should ensure the continued development of such systems, with some eventual objectives in operational meteorology.

2.2.3 Electromagnetic energy

Undoubtedly the most important medium for environmental (i.e. Earth surface/lower atmosphere) remote sensing is electromagnetic radiation. This is

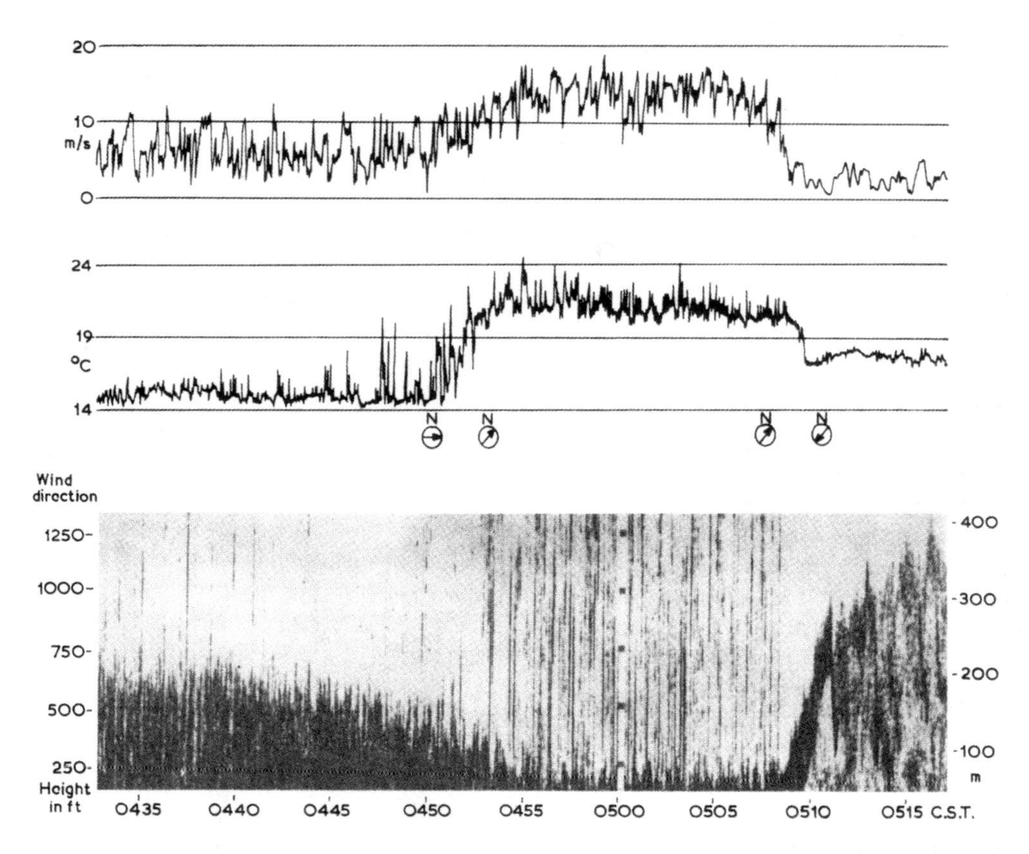

Fig. 2.1 It is possible to detect the small amount of energy back-scattered from a vertically directed beam of sound by inhomogeneities in the density structure of the lower troposphere. Figure 2.1 exemplifies the type of visual record obtained, as a function of height and time. This record was obtained during the passage of a cold front. The graphs show wind (at top) and temperature (bottom). (Source: McAllister and Pollard, 1969.)

the only form of energy transfer that can take place through free space as well as a medium. It exhibits enormous variety in its behaviour, and can be exploited by remote sensing in many different ways. So important is this field that some authorities have considered it to be the only one of significance for environmental remote sensing! More than one account of remote sensing contains a categorical statement such as this: 'In remote sensing, information transfer from an object to a sensor is accomplished by electromagnetic radiation'. From section 2.3 onwards, electromagnetism as a medium for remote sensing will be our chief focus too. This is justified by the strong preponderance of present interest and practice in this field. However, other modes of information transfer certainly do occur, as summarized above.

2.2.4 Active remote sensing

A distinction is commonly drawn between *passive* and *active* remote sensing. In passive remote sensing (exemplified earlier by the eye–brain system) the sensor detects the energy or force that emanates from the target itself, or some 'third party' source. Most environmental remote sensing systems rely upon these approaches. On the other hand, in active remote sensing there is the detection of a signal that is artificially produced. Since such signals are generated under relatively controlled conditions much can be learned from the ways in which they are affected by objects in the environment, and/or by the media through which they are transmitted. Radar (Radio Direction and Ranging) is a common example of an active system exploiting electro-

Table 2.1 Key terms associated with electromagnetic radiation

Processes and phenomena	Entities	Processes	Properties[a]	Behavioural characteristics[b]	Performances[c]
Suffixes	-er or -or	-ion	-ivity	-ance	
Terms	Emitter/Radiator Absorber Reflector Transmitter	Emission Absorption Reflection Transmission Extinction	Emissivity Absorptivity Reflectivity Transmissivity	Emittance Absorptance Reflectance Transmittance	Exitance Absorptive power

[a] Usually related to theoretical (model) maxima on 0–1 scales. Each observed value can be viewed as a natural limitation on the associated performance.
[b] Expressing actual source, medium, or target performance as a ratio of the possible maximum. Sometimes expressed as a percentage (e.g. albedo (the percentage reflectance of natural objects)).
[c] Expressed as radiant fluxes per unit area.

magnetic radiation. Sonar (Sound Navigation and Ranging) is an active system utilizing acoustical energy.

2.3 Electromagnetic energy

2.3.1 The nature of radiation

Energy is the ability to do work. It can exist in a variety of forms including chemical, electrical, heat, and mechanical energy. In the course of work being done, energy must be transferred from one body or one place to another. Such transfers are effected by:

1. *Conduction*. This involves atomic or molecular collisions.
2. *Convection*. This is a corpuscular mode of transfer in which bodies of energetic material are themselves physically moved.
3. *Radiation*. This is the only form in which electromagnetic energy may be transmitted through either a medium or a vacuum.

It is this third type of transfer with which we are primarily concerned in remote sensing studies. Table 2.1 introduces some key terms, concepts, quantities and units and related terminology.

In keeping with many other areas in the physical and environmental sciences, remote sensing uses *models* (simplified or idealized representations of reality) to describe complex phenomena, situations or interactions. In the case of electromagnetic

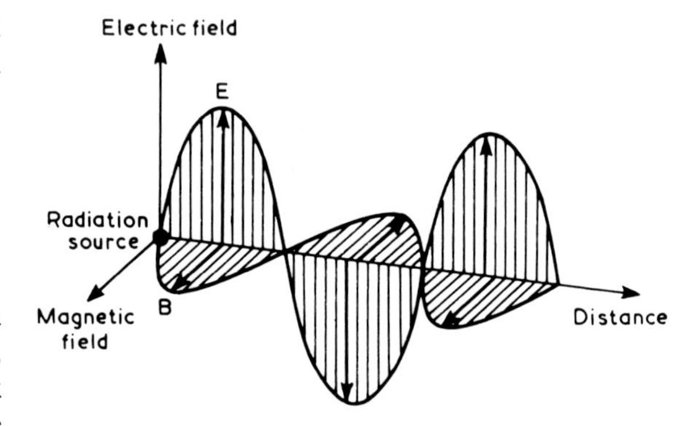

Fig. 2.2 Electric (E) and magnetic (B) vectors of an electromagnetic wave, viewed at a given instant. The intensity of radiation varies with the square of the peak amplitude of the electric field, and is proportional to the number of protons in the field. The type of radiation (Fig. 2.3) is governed by wavelength, measured in absolute units.

radiation two models are necessary to describe and elucidate its most important characteristics:

1. The wave model. Any particle with a temperature above absolute zero vibrates. This vibration sets up wave-like disturbances in the electric and magnetic fields that surround the particle, which is acting as a source of radiation. These are perpendicular to each other (Fig. 2.2); furthermore, the direction of motion of these wavy disturbances is always at right angles to them at any time. The wavy disturbances travel away from the source at a common and constant speed, the 'speed of light' (almost exactly 3 ×

$10^8 \mathrm{m\,s^{-1}}$). The spacing between successive wave crests is the wavelength of the radiation; the number of crests passing a selected point in one second is its frequency. The two types of wave disturbances vary in phase with each other (i.e. changes in the wavelength of amplitude of the one are matched by changes in the wavelength and amplitude of the other).

2. The particle model. This emphasizes aspects of the behaviour of radiation which suggest that it is comprised of many discrete units, called 'quanta' or 'photons'. These carry from the source some particle-like properties, such as energy and momentum, but differ from all other particles in having zero mass at rest. It has been hypothesized consequently that the photon is a kind of 'basic particle'.

Whilst the true nature of electromagnetic radiation remains obscure, we know that it results whenever an electrical charge is generated. In terms of the wave model, the wavelength (λ) of the resulting ray of energy is determined by the length of time that the charged particle is accelerated; the frequency (f) of the radiation waves depends upon the number of accelerations per second to which the particle is subjected. The relationship between wavelength, frequency, and the speed of light (a universal constant, c) is:

$$\lambda f = c, \text{ or } \lambda = c/f \tag{2.1}$$

This tells us that wave frequency is inversely proportional to wavelength, and directly proportional to the speed of wave advancement.

We may relate this expression of wave theory to the particle theory through the statement:

$$E = hf \tag{2.2}$$

where E is the energy of a quantum or photon, h is a constant (named after Planck, who proposed the 'Quantum Theory' in 1900), and f the frequency of the radiation waves. If we multiply $\lambda = c/f$ by h/h (which does not alter its value), and substitute E for hf, we are left with a new expression, namely:

$$E = hc/\lambda \tag{2.3}$$

This tells us that the energy of a photon varies directly with wave frequency, and inversely with radiation wavelengths. This is borne out by experiment, for it can be demonstrated that the longer the wavelength of radiation the lower the energy involved, and conversely, the higher the frequency of radiation, the greater the energy involved. These simple relationships are fundamental to the appreciation of the behaviour of electromagnetic radiation.

Probably the most familiar form of electromagnetic radiation is *visible light*. Radiation detected by the human eye ranges through the well-known sequence of component colours from red through orange, yellow, green, blue and indigo to violet. This range, the so-called 'visible spectrum', should not be confused with the much broader 'electromagnetic spectrum', of which it forms but a tiny part. Thus the eye–brain system is capable of detecting only a tiny section of the total spectrum of radiation, all of which travels at the speed of light. Some knowledge of the breakdown of the electromagnetic spectrum in regions other than visible light is essential for an adequate appreciation of many remote sensing studies.

Largely for convenience, the electromagnetic spectrum is subdivided into a number of sections. The boundaries between them are expressed in several different ways (Fig. 2.3):

1. *Wavelength*. The spectrum stretches from extremely short waves (cosmic rays) to very long ('electromagnetic') waves. The waves range from the microscopic to hundreds of kilometres in length.
2. *Frequency*. We have seen that wavelength and frequency are always inversely related, since all waves advance at a common speed, the speed of light. Therefore, longwave radiation is a low frequency propagation, whereas short wave radiation is characterized by much higher frequencies. The units of measurement are hertz (Hz), cycles per second.
3. *Photon energy*. Equation 2.3 stated that the energy of a photon is directly proportional to wave frequency. Figure 2.3 confirms that long-wave radiation has low photon energy, whereas short-wave radiation has high photon energy. This is expressed in joules, or watt-seconds.

So the whole electromagnetic spectrum ranges from almost infinitely short cosmic rays to the long waves of radio and beyond. It includes a number of familiar everyday manifestations such as gamma rays, X-rays, ultraviolet light, infrared rays and

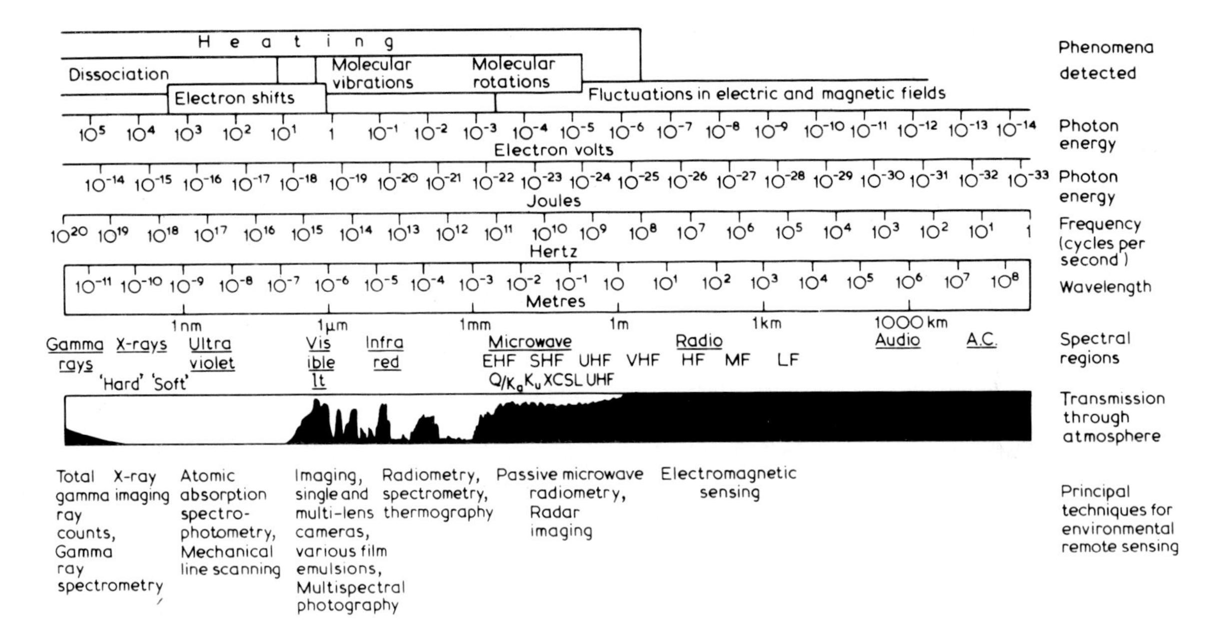

Fig. 2.3 The electromagnetic spectrum. The scales give the energy of the photons corresponding to radiation of different frequencies and wavelengths. The product of any wavelength and frequency is the speed of light. Phenomena detected at different wavelengths are shown, and the principal techniques for environmental remote sensing. The significance of the results for different environmental applications constitutes a greater part of the latter chapters of this book: the student might usefully tabulate them for themselves.

radar waves. Figure 2.3 also indicates some of their more common uses in remote sensing. However, it should be understood that the whole spectrum is a continuum, which is subdivided in a rather inconsistent way. The boundaries shown between different regions of the electromagnetic spectrum are inexact; there is considerable overlap between some neighbouring regions; no fully accepted terminology is applied to them; and different authorities prefer different boundary values.

Next we should note that radiation *amounts* may be quantified in several different ways. These are summarized in Table 2.2. Of these, the radiance (L) is the most important because it describes what is actually measured by a remote sensing instrument (section 2.7).

Lastly, we must recognize that it is not always possible to assess *quantitatively* the radiation emerging from a target (relating the actual levels of radiation involved by numbers), ideal though this practice clearly is. However, any remote sensing imagery may at least be described *qualitatively*, (in terms of its brightness and brightness variations), and for many applications this may be enough.

2.4 Radiation at source

2.4.1 The generation of radiation

An electromagnetic source, whether natural or artificial, may emit:

1. a broad continuum of wavelengths of radiation;
2. radiation within a narrow (single spectral) band; or
3. radiation of a single wavelength.

The intensity of such radiation varies with the square of the peak amplitude of the electric field, and is proportional to the number of photons in the field. Variations in the intensity, and perhaps also the wavelength or frequency, of radiation from a source may occur through time. The electrically charged particles involved in generating the emitted radiation are basic units of matter, namely atoms, electrons and ions. All are in a state of constant motion under normal circumstances, i.e. when the temperatures of the objects they constitute are above absolute zero. Since the molecules of these objects are built of electrically charged particles

Table 2.2(a) Quantities and units of special significance to remote sensing

Quantity	SI unit	Quantity	SI units
Length (l)	Metre (m) or micrometre (μm)	Area (A)	Square metre (m^2)
Time (t)	Second (s)	Volume (V)	Cubic metre (m^3)
Mass (m)	Kilogram (kg)	Frequency (v)	Hertz or cycle s^{-1} (Hz)
Temperature (T)	Kelvin (K)	Wavelength (μ)	Metre (m)
Plane angle (x)	Radian (rad)	Incidence angle: angle between line of sight and vertical (θ)	Degree (°)
Solid angle (Ω)	Steradian (sr)		
Force ($m\ lt^{-2}$)	Newton (N)	Grazing or depression angle: angle between line of sight and horizontal (γ)	Degree (°)
Energy ($m\ l^2t^{-2}$)	Joule (J)		
		Density (D)	Kilogram per cubic metre (kg m^{-3})

Table 2.2(b) Radiometric quantities, units and their meanings (After Curran, 1985)

Quantity	Defining expression	Unit	Meaning
Radiant energy (Q)	—	Joule (J)	Total energy radiated in all directions
Radiant density (W)	$\frac{dQ}{dv}$	Joule per cubic metre (J m^{-3})	Total energy radiated by a unit area in all directions
Radiant flux (#)	$\frac{dQ}{dt}$	Watt (W)	The rate of flow of radiant energy (radiant power)
Radiant exitance (M)	$\frac{d\phi}{dA}$ (out)	Watt per square metre (W m^{-2})	Total energy radiated in all directions in unit time
Irradiance (E)	$\frac{d\phi}{dA}$ (in)	Watt per square metre (W m^{-2})	Total energy radiated in all directions away from a unit area in unit time
Radiant intensity (I)	$\frac{d\phi}{d\Omega}$	Watt per steradian (W sr^{-1})	Total energy radiated on to a unit area in unit time
Radiance (L)	$\frac{d^2\phi}{d\Omega(dA \operatorname{Cos}\theta)}$	Watt per square metre per steradian (W m^{-2} sr^{-1})	Radiant flux within a selected angle of observations
Spectral radiant exitance (M)	$\frac{dM}{d\lambda}$ (out)	Watt per square metre per micrometre (W m^{-2} μm^{-1})	Total energy radiated by a unit area within a selected angle of observation
Spectral irradiance (E)	$\frac{dE}{d\lambda}$ (in)	Watt per square metre per micrometre (W m^{-2} μm^{-1})	Examples of radiometric quantities restricted by wavelength
Spectral radiance (L)	$\frac{dL}{d\lambda}$	Watt per square metre per steradian per micrometre (W m^{-2} sr^{-1} μm^{-1})	Example of radiometric quantity restricted by wavelength plus angle of observation

they possess natural resonance, either vibrational or rotational, so that they act as small oscillators which accelerate the electrical charges. The emitted photons of energy have frequencies that correspond to the resonances of the molecules. In turn, the distribution of molecules among energy states is a function of temperature, so that at higher temperatures there are proportionately more molecules in the higher energy states. This is to say that all particles do not increase their energy equally as the temperature of a radiation source rises, but the overall effect is an increase in molecular activity, accompanied by both an increase in emitted energy and a related shift in the dominant wavelength of radiation towards the higher frequencies.

2.4.2 *Emission of radiation*

All bodies with temperatures above absolute zero generate and send out, or emit, energy in radiant form. Each radiation source, or radiator (whether natural or artificial), emits a characteristic array of radiation waves which can be assessed in terms of their wavelengths and intensities. For many real-world radiators their array is very complex, comprised of a number of different contributions from the various constituents. Hence a characteristic curve or *spectral emission signature* may be sought for each type of natural remote sensing target by plotting the intensities of the emitted radiation against appropriate wavelengths in the electromagnetic spectrum. Chapter 3 provides illustrations of a number of such spectral signatures. For the present we must concern ourselves more with the principles that cause such signatures to differ.

A useful concept, widely used by physicists in radiation studies, is that of the *black body*, a model (perfect) absorber and radiator of electromagnetic radiation. A black body is conceived to be an object or substance that absorbs all the radiation incident upon it, and emits the maximum amount of radiation at all temperatures. Although there is no known substance in the natural world with such a performance, the black body concept is invaluable for the formulation of laws by comparison with which the behaviour of actual radiators may be assessed. These laws include the following:

1. Stefan's (or Stefan–Boltzmann's) law. This states that the total emissive power of a black body is proportional to the fourth power of its absolute temperature (T). This can be expressed as:

   $$M = \sigma T^4 \tag{2.4}$$

   where M is radiant exitance in watts m^{-2}, and σ is the 'Stefan–Boltzmann Constant'. This relationship applies to all wavelengths shorter than the microwave: in the microwave region radiant exitance varies as a direct function of T. The chief practical significance of Stefan–Boltzmann's law is that hot radiators emit more energy per unit area than cooler ones.
2. Kirchhoff's law. Since no real body is a perfect emitter, its exitance is less than that of a black body. Clearly it is often useful to know how the real exitance (M) of a radiator compares with that which would be anticipated from a corresponding perfect radiator (M_b). This may be established by evaluating the ratio M/M_b, which gives the emissivity (ε) of the real body. Thus, for the general case, we may say that:

   $$M = \varepsilon M_b \tag{2.5}$$

   The emissivity of a real black body would be 1, whilst the emissivity of a body absorbing none of the radiation upon it (not unexpectedly known as a 'white body') would be 0. Between these two limiting values, the 'greyness' of real radiators can be assessed, frequently to two decimal places. If we plot emission curves for real radiators against curves for corresponding black bodies, we usually find that the idealized curves are much smoother and simpler than the observed patterns. In effect the *emissivity index* is a measure of radiating efficiency across the spectrum as a whole.
3. Wien's (displacement) law. This states that the wavelength of peak radiant exitance (λ_{max}) of a black body is inversely proportional to its absolute temperature (T):

   $$\lambda_{max} = C_3/T \tag{2.6}$$

   C_3 is a constant equal to 2897 μm K. This equation tells us that, as the temperature of a black body increases, so the dominant wavelength of emitted radiation shifts towards the short wavelength end of the spectrum. This can be exemplified by reference to the Sun and Earth as radiating bodies. For the Sun, with a mean surface temperature of about 6000 K, $\lambda_{max} \simeq 0.5\,\mu$m. For the Earth, with a surface temperature of about 300 K on a warm day, $\lambda_{max} = 9.6\,\mu$m. Empirical observations have confirmed that the wavelength of maximum radiation from the Sun indeed falls within the visible waveband of the electromagnetic spectrum whilst that from the Earth falls within the infrared. We experience the former dominantly as light, and the latter as heat.
4. Planck's law. This more complex statement describes accurately the spectral relationships between the temperature and radiative prop-

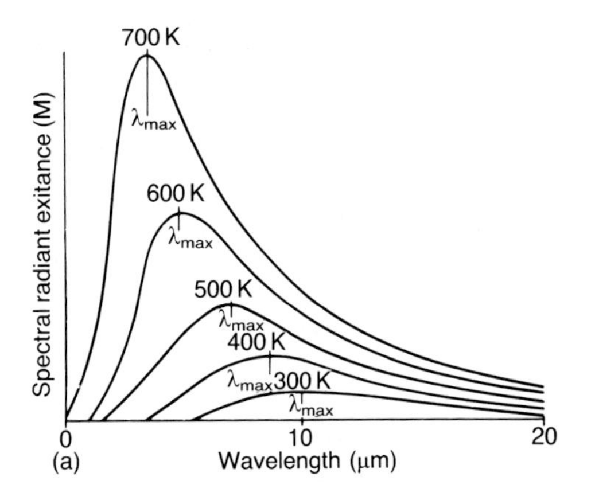

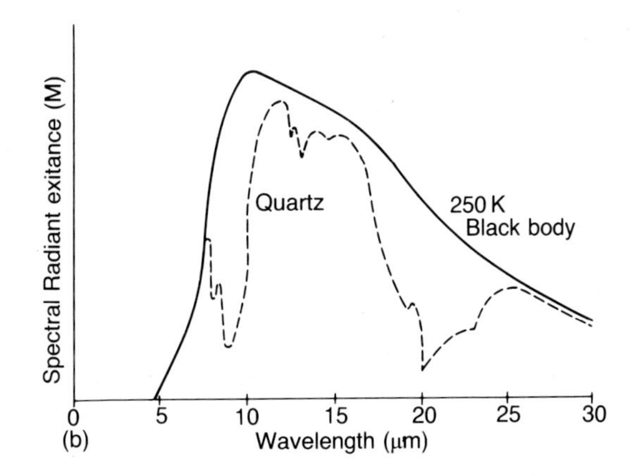

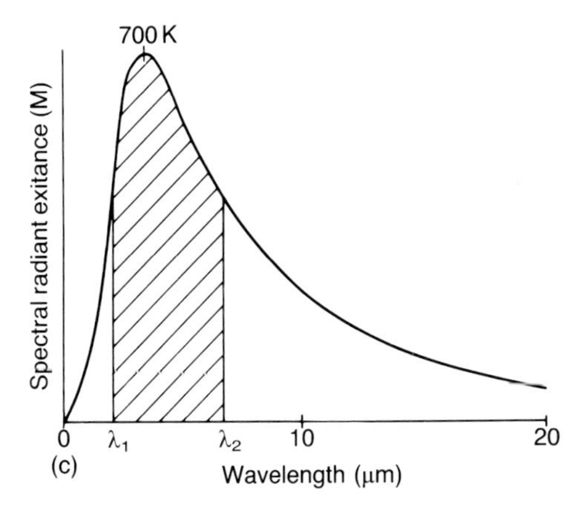

Fig. 2.4 (a) Selected black body radiation curves for various temperatures. In each case the spectral radiant exitance (M_λ) is represented by the y-axis; total radiant exitance (M) is given by the area under each curve. Note that the areas under the curves diminish as temperature decreases. Note too that λ_{max}, the wavelength of peak radiant exitance, shifts towards the longer wavelengths with decreasing black body temperature. (b) The total radiant exitance of a real body (e.g. quartz) is less than that of a black body at the same temperature. The general relationship is expressed by Kirchhoff's law. (c) One form of Planck's law of radiation permits the assessment of that portion of the total radiant exitance falling between selected wavelengths, e.g. λ_1 and λ_2.

erties of a black body. In one form, Planck's law may be expressed by

$$M_\lambda = C_1\lambda^{-5}/(e^{C_2/\lambda^T} - 1) \qquad (2.7)$$

where M is the energy emitted in unit time for our unit area within a unit range of wavelengths centred on λ, C_1 and C_2 are universal constants, and e the base of natural logarithms (2.718). A useful feature of Planck's law is that it enables us to assess the proportions of the total radiant exitance which fall between selected wavelengths. This can be useful in remote sensor design, and also in the interpretation of remote sensing observations.

In consequence of these four laws we see in Fig. 2.4 that emission curves are characteristically negatively skewed, with their peak intensities falling in the lower quartiles of their spectral bands. The wavelengths of maximum emission are progressively shorter for the hotter radiators, and the total emissions from the hotter radiators are greater than those from the cooler radiating bodies.

2.5 Radiation in propagation

2.5.1 Scattering

It is unfortunate (for remote sensing) that considerable complications arise in practice where the Earth's atmosphere intervenes between the sensor and its target. Although the speed of electromagnetic radiation is unaffected by the atmosphere, this medium may affect several of the other characteristics of this form of energy propagation. These include:

1. the *direction* of radiation;
2. the *intensity* of radiation;
3. the *wavelength and frequency* of the radiation received by a target, at the base of the atmosphere;
4. the *spectral distribution* of this radiant energy.

Frequently both the direction and intensity of radiation are altered by particles of matter borne in the atmosphere. These redirect, in a rather unpredictable way, the radiation *en route* through a turbid medium. An understanding of radiation scatter is necessary for the selection of sensors or filters where certain effects are required, and where image degradation owing to atmospheric impurities may be avoided, or at least reduced, by sensing at the most appropriate waveband(s).

The depletion or *attenuation* which results from 'scattering' (i.e. dispersion of radiation) by particles suspended in the atmosphere is related to the wavelength of radiation, the concentration and diameters of the particles, the optical density of the atmosphere (discussed under Refraction below) and its absorptivity. The common types of scatter are:

1. Rayleigh scatter. This mostly involves molecules and other tiny particles with diameters much less than the radiation wavelength in question. It is characterized by an inverse fourth power dependence on wavelength. Hence, for example, ultraviolet radiation (about one-quarter the wavelength of red light) is scattered sixteen times as much. This helps to explain the dominance of orange and red at sunset when the sun is low in the sky: the shorter wavebands of visible light are cut out by a combination of atmospheric absorption and powerful scattering.
2. Mie scatter. This occurs when the atmosphere contains essentially spherical particles whose diameters approximate to the wavelengths of radiation in question. Water vapour and particles of dust are the main agents that scatter visible light.
3. Non-selective scatter. Here particles with diameters several times the radiation wavelengths are involved. Water droplets, for example, with diameters ranging commonly from 5 to 100 μm, scatter all wavelengths of visible light (0.4–0.7 μm) with equal efficiency. As a consequence clouds and fog appear whitish, for a mixture of all colours in approximately equal quantities produces white light.

2.5.2 *Absorption*

This is the retention of radiant energy by a substance or a body. In the real world it involves the transformation of some of the incident radiation into heat, and the subsequent re-emission of that energy at a longer wavelength.

Considering the ideal case, it will be recalled that a black body is a perfect radiator; it is also a fully efficient absorber of radiant energy which is incident upon it. So, just as the *emissivity* of a real body can be defined as an inherent characteristic of its material, expressing the ease with which it gives up energy by radiation, so the *absorptivity* (α) of a real body is an expression of its ability to absorb radiant energy. Logically, for a black body, $\alpha_b = \varepsilon_b$, and both ε and α equal unity. For 'grey' bodies, ε and α are also equal (with values of less than 1) if they are opaque, in which case both absorption and emission are confined to the (physically simple) surface layer. For all other cases $\alpha_b \neq \varepsilon_b$, and modelling the processes involved can be very complicated and difficult.

In the atmosphere (which can be decidedly murky, but is never opaque) radiation absorption takes place not so much at its surface, but in transit. Three gases, namely water vapour, carbon dioxide, and ozone, are particularly efficient absorbers of radiation from the Sun. Consequently incoming solar radiation (often abbreviated to 'insolation'), which is by far the most important natural source of radiation for passive remote sensing, is attenuated significantly by its passage through the atmosphere. Often the combined effects of absorption and scattering by particulate matter are expressed in terms of an *extinction coefficient*. In turn, this enables us to evaluate the energy that reaches the surface of Earth as a ratio of the radiation incident upon the outer limits of the atmosphere. This permits us to evaluate the *transmittance* of the atmosphere. For any such medium the transmission of radiant energy through it is inversely related to the product of the thickness of the layer and its extinction coefficient. In practice, therefore, the transmittance decreases as the combined effects of absorption and scattering accumulate.

2.5.3 *Refraction*

When electromagnetic radiation passes from one medium to another, bending or 'refraction' occurs in response to the contrasting densities of the media (Fig. 2.5). A measure of this given by the Index of Refraction (n), where:

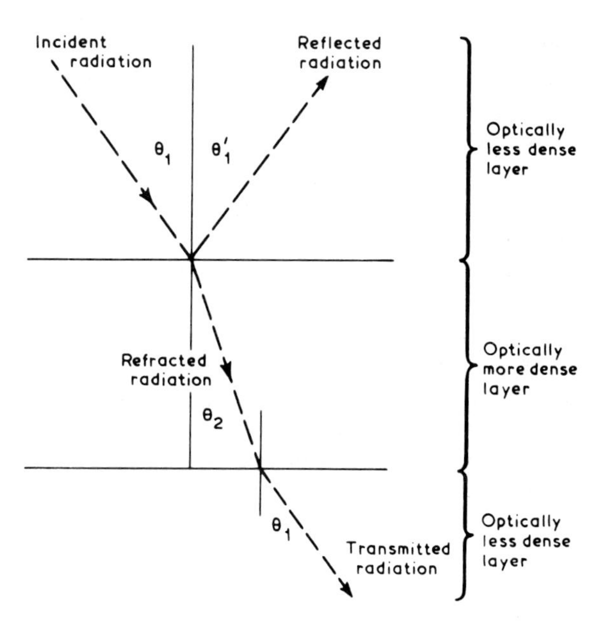

Fig. 2.5 The relationship between incident radiation, reflected radiation, refracted radiation and transmitted radiation, and a stratification of optically different layers.

$$n = c/c_n \tag{2.8}$$

The index is the ratio of the speed of light in a vacuum (c) to its speed in the substance (c_n). In a non-turbulent atmosphere (which can be conceived as a series of layers of gases each of a different density) the refraction of radiation is predictable. For, as Snell's Law states, there is a constant relationship for a given frequency of light between the index of refraction and the sine of the angle between the ray and the interface between each pair of adjacent layers. Problems do arise, however, where a turbulent atmosphere is involved. Turbulent motions are essentially random, and their effects upon radiation are consequently unpredictable.

2.6 Radiation at its target

2.6.1 Reflection

So far we have discussed the effect of particles on radiation as if its resulting redirection is another variable we cannot predict. In the real world this is not always so. The critical factors are the smoothness and orientation of the object lying in the path of an electromagnetic ray. If the object has smaller surface irregularities than the wavelength of the impinging energy, it will act as a mirror and the angle at which the energy is directed away from the object will equal the angle of its incidence upon it. This process of reflection is sometimes called *backscattering*. The angles of incidence and reflection, and a perpendicular to the reflecting surface from which those angles are measured, all lie in the same plane. Conversely scattering may be described in terms of reflection, when it is labelled *diffuse reflection* in contrast to *specular reflection*, which is the simpler ideal case. Although we might expect some reflection by particles in the atmosphere, reflection is much more important for remote sensing where continuous (e.g. land or sea) surfaces are involved.

A useful expression of the reflectivity of different terrestrial surfaces is the reflection coefficient, or *albedo* (Table 2.3). The albedo of a surface is the

Table 2.3 Values of albedos for various surfaces and types of surfaces (Source: Lockwood, 1974)

Type of surface	Surface	Albedo % of incident shortwave radiation
Soils	Fine sand	37
	Dry black soil	14
	Moist ploughed field	14
	Moist black soil	8
Water surfaces	Dense, clean and dry snow	86–95
	Woody farm, snow-covered	33–40
	Sea ice	36
	Ice sheet with water covering	26
Vegetation	Desert shrubland	20–29
	Winter wheat	16–23
	Oaks	18
	Deciduous forest	17
	Pine forest	14
	Prairie	12–13
	Swamp	10–14
	Heather	10
Geographic locations	Yuma, Arizona	20
	Winnipeg (July)	13–16
	Washington, DC (September)	12–13
	Great Salt Lake, Utah	3

percentage of the insolation incident upon it that is reflected back towards space.

We referred earlier to a white body as one that absorbs none of the radiation which impinges on it. We are now able to say that, whereas a black body is conceived as a perfect absorber of radiation, a white body is a perfect reflector. Since none of the incident radiation is absorbed, the surface temperature of a white body remains unchanged.

Because visible spectrum wavelengths are so short, most surfaces reflect light diffusely regardless of the angles at which this radiation strikes them: some of the reflected energy returns in the direction of its source. Longer wavelengths, e.g. microwaves, create more specular reflection off the same surfaces, the reflected energy generally being directed away from its source. Some of the consequences of such differences in performances may be illustrated by reference to the remote sensing of a building at night. Using visible light – torch or a searchlight – the facing wall of the building will be discernible whatever its angle to the light beam and the wall it strikes. However, since most of the reflection in this case is specular rather than diffuse, much stronger returns are obtained from a head-on, rather than an oblique, orientation.

Reflected radiation is of considerable importance to remote sensing, for many observing systems are based upon it. The natural remote sensing system of the eye and the brain observes and perceives many aspects of the natural environment through reflection of the solar radiation by which it is illuminated. The early, now well-developed, artificial technique of photography records phenomena through the light which is reflected from them, whether in the studio or out of doors. Active remote sensing systems, such as radar, often measure the reflection of specially propagated energy, so that the sizes, positions and/or densities of natural or man-made reflectors can be assessed thereby. Examples from many areas of environmental remote sensing can be found in later chapters of this book.

2.6.2 *Absorption*

Not all the radiation incident upon the surface of a target is necessarily scattered or reflected by it. Some of the incident energy may enter the target to be propagated through it as a *refracted wave front*. As we noted earlier, with a gaseous medium, such as the atmosphere, energy in transit may be attenuated or depleted, at least partially, by the process of absorption. The ability of a substance to absorb radiation (i.e. its absorptance) depends upon its composition and its thickness.

Absorptance also varies with wavelength of radiation. A target may behave very differently if exposed to radiation of different wavelengths. For example, an object or a medium may be highly absorptive in the visible range yet transparent in the infrared (like some 'semi-conductors'), or it might be transparent in the visible, yet opaque in the infrared (like glass).

Clearly some knowledge of the absorptance of a target is often invaluable in choosing the best remote sensing means to solve a specific problem, and in interpreting correctly the results of a remote sensing survey.

The net effect of absorption of radiation by most substances – including the atmosphere and terrestrial surfaces – is that the greater part of the absorbed energy is converted into heat. Thereby the temperatures of the substances are raised. This heat energy may enable the original target of radiation to become a secondary radiation source itself, emitting some of the energy absorbed. Since the peak intensity of solar radiation is in the waveband of visible light, and the Earth/atmosphere system is by no means a black body, it is not surprising that its radiation temperature – as we saw earlier – is much less than that of the Sun; indeed, the radiation peak of the Earth is in the infrared (Fig. 3.1). Fortunately for environmental remote sensing, infrared energy (a form of 'thermal radiation') is emitted from the surface of Earth both by day and by night, and is affected little by atmospheric particles such as haze and smoke. If we take care to view such remote sensing targets through atmospheric window wavebands in the infrared portion of the spectrum (where absorption by certain gaseous constituents of the atmosphere is negligible) and under cloud-free conditions, we are able to add much to our knowledge of their physical and chemical properties, and their diurnal cycles, which we could not have learnt by conventional photography. As we shall see in later chapters, thermal mapping is becoming increasingly significant in remote sensing applications today.

2.6.3 *Transmission*

The transmittance of a target (or a medium such as the atmosphere) is defined as the ratio of radiation at distance x within it to the incident radiation. Since we have already mentioned transmission in relation to other effects, in this section we may conveniently conclude our review of physical principles by revising some of the interplay between reflection, absorption and transmission at the target. Clearly the sum of the three must equal the incident radiation, much of the detail depending upon the nature of the target itself, whether it be transparent or opaque. We must not forget, that, as pointed out in section 2.6.2, a target that is relatively transparent at one wavelength may be relatively opaque at another, so that the relations between reflection, absorption and transmission generally vary across the electromagnetic spectrum. Lastly, it should be remembered that the angle of incidence of radiation may, if low, cause the proportion of reflected energy to exceed the combined proportion of absorbed and transmitted energy, whereas, if high, that angle may allow more of the energy, to enter the target providing its detailed surface characteristics permit (Fig. 2.5).

2.7 Radiation observed

We have seen that any body with a temperature above absolute zero emits electromagnetic radiation, and that this radiation is affected by many characteristics of both the emitting body itself and other bodies or media which surround it. It is of great importance in environmental remote sensing to understand exactly what it is that a sensing instrument observes, and how this can be analysed to elucidate the nature and behaviour of the radiation source. In this respect, three additional quantities must be distinguished, namely:

1. *kinetic temperature*, the 'true' or internal temperature of a body (i.e. that measurable by an *in situ* device, e.g. a thermometer);
2. *radiating (or brightness) temperature*, the apparent temperature of a body (i.e. that measurable by a remote sensing device, e.g. a radiometer);
3. *antenna temperature*, the product of the brightness temperature and the power pattern of the radiometer antenna that is making the measurement.

Of these, the third is of significance both in the design and operation of an instrument, and also in consequential uses of the data generated thereby. However, it is increasingly usual for the antenna temperatures to be transformed into brightness temperatures before data are supplied to end users; therefore they need not concern us any more at this point. For most remote sensing applications the brightness temperatures of objects are the most important, and kinetic temperatures can be estimated only from brightness temperature observations. Attempts are often made to estimate kinetic temperatures from brightness temperatures, although to do so accurately may be very difficult.

Some of the factors relating kinetic and brightness temperatures, and which complicate the estimation of the former from the latter, have been explored already, including the emissivities of the objects of interest, and the absorption and/or scattering of radiations emitted from them. Other factors that affect the relationships between the two must now be introduced. These include what we may call *thermal properties* of objects, and *patterns of heating and cooling* related principally to the succession of days and nights. The relevant thermal properties, which determine both how heat is distributed through a body and how its brightness temperature varies with time and depth, include:

1. Thermal capacity (c). This is a measure (in $J\,kg^{-1}\,K^{-1}$) of the ability of an object to store heat. For example, water has a much higher thermal capacity than either soil or vegetation.
2. Thermal conductivity (k). This is a measure (in $W\,m^{-1}\,K^{-1}$) of the rate at which heat can pass through a material. This is generally higher for artificial materials than natural materials, which are relatively poor heat conductors. Therefore diurnal temperature changes observed in rural areas are more purely functions of the upper layers of water, soil or vegetation rather than their whole depths.
3. Thermal diffusivity (K). This is a measure (in $m^2\,s^{-1}$) of the rate of change of the temperature within a body or medium. In general, dry surfaces are observed to diffuse temperature changes more slowly than wet surfaces.
4. Thermal inertia (P). This is a measure (in $W\,m^{-2}\,K^{-1}\,S^{1/2}$) of the thermal resistance of a material to changes of temperature. Some ma-

Table 2.4 Thermal properties of some surface types: high (H), high/moderate (HM), moderate/low (ML) and low (L) (After Curran, 1985)

Surface type	Thermal capacity (c)	Thermal conductivity (k)	Thermal diffusivity (K)	Thermal inertia (P)
Vegetation	H/M	L	H	H
Dry soil	M/L	M/L	M/L	H
Wet soil	M/L	M/L	M/L	H
Water	H	L	H	H
Urban areas	M/L	H	L	H/M

terials, e.g. dry sandy soils, have low thermal inertias, and exhibit wide diurnal ranges of temperature. Others, e.g. wet clay soils, have high thermal inertias, and their observed diurnal temperature ranges are relatively low.

The thermal properties of some common constituents of the environment are shown in Table 2.4.

Lastly, reference must be made to rates of heating or cooling. Surface temperature changes prompted by insolation heating during the day and outward radiation by night are important for intelligent uses of brightness temperature patterns and changes, but can be very difficult to accommodate in brightness temperature data analysis and interpretation because they are influenced by so many factors. These include angles of insolation, slope and aspect, as well as the distributions of screening bodies, such as clouds, trees and buildings. Such factors greatly complicate the task of objective classification of thermal imagery for, although many features of the natural world can be modelled, all models are simplified representations of real world conditions. For example, in remote sensing studies of vegetation the observed radiation or brightness temperatures are likely to be affected by both leaf and underlying soil temperatures, which depend in turn upon the closeness (or otherwise) of the vegetation cover, the moisture content of the leaves and the soil, the slope of the terrain, the time of day and the season of the year, plus the types and specifications of the sensors and platforms from which the remote sensing observations have been made.

2.8 General conclusions

Although it may have seemed at the beginning of this chapter that Earth scientists wishing to employ remote sensing techniques to improve their perception and appreciation of the human environment need do no more than choose the most appropriate instrument and the best platform for their purposes, it must be clear by now that many factors may influence that choice. In later chapters we shall review some of the more common sensors, sensor packages, and remote sensing platforms available for such uses. However, it is clear already that if we are to understand this rapidly expanding field of scientific activity, rather than just know of it, we must try to relate experimental and operational practice to the principles we have examined thus far. Since this might be difficult without further explanation and illustration, Chapter 3 will be more specifically concerned with the observed radiation characteristics of more real-world phenomena, so as to strengthen the bridge between the rather abstract theory that dominated in this chapter and the very technical array of practical possibilities discussed later in this book.

3 Radiation characteristics of natural phenomena

3.1 Radiation from the Sun

3.1.1 The spectrum of solar radiation

Solar radiation reaches the outer limit of Earth's atmosphere undepleted except for the effect of distance. Since the total radiation from a spherical source, such as the Sun, passes through successively larger spheres as it radiates outward, the amount passing through a unit area is inversely proportional to the square of the distance between the area and the radiation source. Although the Earth, 93 million miles from the Sun, intercepts less than one fifty millionth of the Sun's total energy output, this intercepted energy is vital not only to most remote sensing techniques (either directly or indirectly) but also to terrestrial life itself.

Observations of the radiation output from the Sun are made difficult by the attenuation effects of the Earth's atmosphere, at or near whose base most measurements of solar radiation traditionally have been made. Clearly the best vantage point for evaluating the spectrum of solar radiation is outside the atmosphere. Although some data from satellite-borne sensors are now available, probably the most complete and reliable evaluation of it is still that obtained many years ago by staff members of the Smithsonian Institute from the elevated observatory at Mount Wilson. Rocket measurements of ultraviolet radiation have been added to extend the spectrum towards the short wavelength end, where solar radiation is almost completely blocked by absorption in the upper atmosphere. Figure 3.1 summarizes the results, including a curve extrapolated to the outer limits of the atmosphere, and corrected for mean solar distance.

3.1.2 The atmospheric absorption spectrum

We remarked in Chapter 2 that solar radiation is significantly attenuated by the envelope of gases which surrounds Earth. Consequently the curve of solar radiation incident on the Earth's surface stands below that for the top of the atmosphere, as shown in Fig. 3.1. Some of the attenuated radiation is quickly lost to space by scattering processes, whilst the remainder of it is absorbed within the atmosphere itself. Almost all the solar radiation that penetrates the atmosphere and reaches the surface of Earth is of wavelengths shorter than 4.0 μm, whereas that emitted by the Earth (Fig. 3.1) is principally in the broad waveband from 4.0 to 40 μm. There is, therefore, practically no overlap between these two types of radiation, which are sometimes referred to as *shortwave* and *longwave* radiation respectively. The atmosphere has the ability to absorb not only direct solar radiation, but indirect (back-scattered) solar radiation, plus longwave (terrestrial) radiation also. In considering the absorption of radiation by the atmosphere it is more convenient to review the impact that this has upon solar and terrestrial radiation together, rather than separately.

The patterns of absorptitivity of different constituents of the atmosphere across part of the electromagnetic spectrum are portrayed in Fig. 3.2. The most striking feature of this set of curves as a whole is the great variability observed from one wavelength to another. In many cases sharp peaks of absorptitivity at some wavelengths are separated by equally abrupt troughs where the ability of the gases to absorb radiation is very slow. The most important absorbers of radiation within the mixture of gases comprising the atmosphere of Earth are:

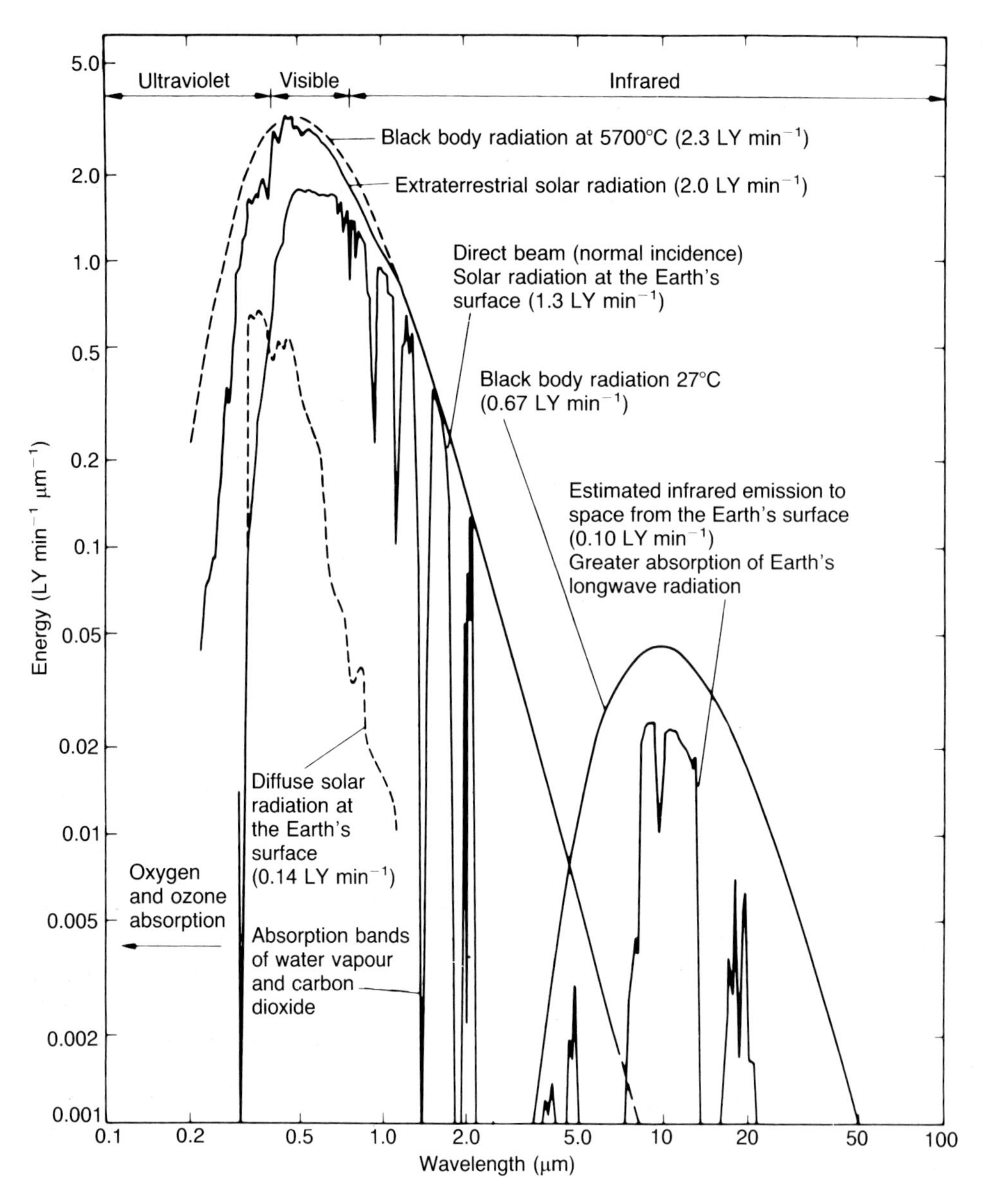

Fig. 3.1 Electromagnetic spectra of solar and terrestrial radiation. The curve for extra-terrestrial solar radiation represents that radiation which is present on the top of the Earth's atmosphere for the mean distance between the Earth and the Sun. (Source: Sellers, 1965.)

1. Oxygen and ozone. Radiation of wavelengths less than about 0.3 μm is not observed at the ground. Almost all of that reaching the top of the atmosphere is absorbed high in the atmosphere: the rest is back-scattered to space. Energy of <0.1 μm is highly absorbed by atomic and molecular oxygen (O and O_2 respectively), and also by nitrogen (N_2), in the ionosphere; energy of 0.1–0.3 μm is absorbed efficiently by ozone (O_3) in the ozonosphere. Further, but less complete, ozone absorption occurs in the 0.32–0.36 μm region, and at relatively minor levels around 0.6 μm (in the visible), and 4.75 μm, 9.6 μm and 14.1 μm (in the infrared).
2. Carbon dioxide (CO_2). This is of chief significance in the lower stratosphere. It is more evenly distributed than the other absorbing gases and is overshadowed as an absorber at

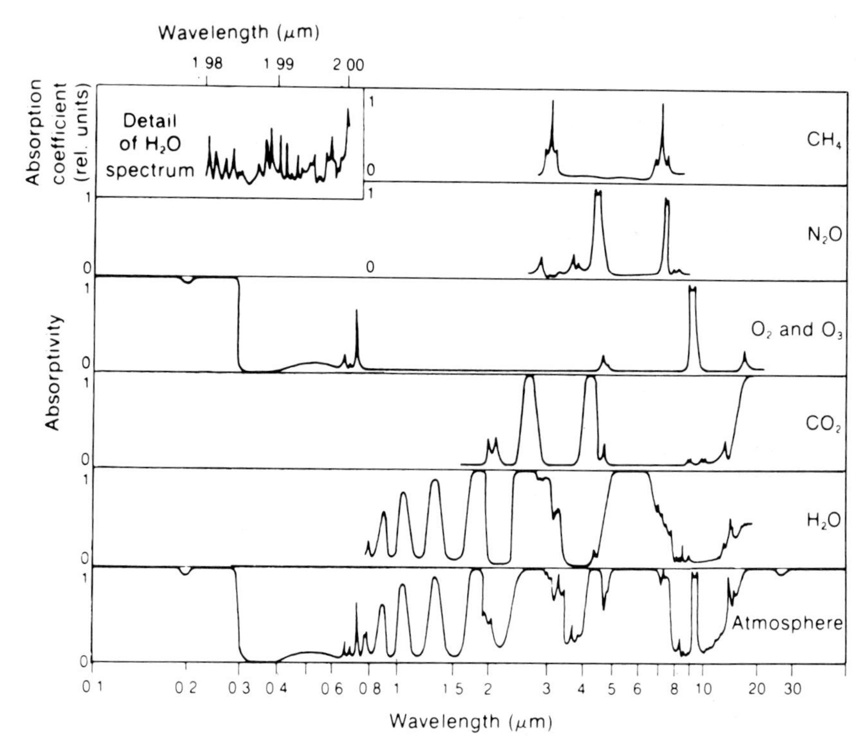

Fig. 3.2 Spectra of absorptivity by constituents of the atmosphere, and the atmosphere as a whole. (Source: Fleagle and Businger, 1963.)

higher altitudes by oxygen and ozone, and in the troposphere by water vapour, although its small tropospheric quantities are thought to be increasing, with consequential implications for global warming (Chapter 11). Carbon dioxide has weak absorption bands at about 4 μm and 10 μm, and a very strong absorption band around 15 μm, which is well known and widely exploited for atmospheric sounding.

3. Water vapour. Among the atmospheric gases this absorbs the largest amount of solar energy. Several weak absorption lines occur below 0.7 μm, while important broad bands of varying intensity have been identified between 0.7 and 8.0 μm. The strongest water vapour absorption is around 6 μm, where approaching 100% longwave radiation may be absorbed if the atmosphere is sufficiently moist.

In Fig. 3.2 we see below the curves for individual constituents of the atmosphere the total atmospheric absorption spectrum. At the short end of the atmosphere absorption spectrum the atmosphere absorbs almost all the incident radiation. However, the atmosphere is largely transparent in the visible band from 0.3 to 0.7 μm. There and thereafter in the direction of increasing wavelengths we see a sequence of more or less sharply-defined absorption bands alternating with relatively transparent regions. These transparent regions are the most useful of the so-called *atmospheric windows* in the absorption spectrum.

Broadly speaking, practitioners of environmental remote sensing whose major interests are in Earth surface features select atmospheric window wavebands to explore, and avoid those wavebands in which atmospheric absorption is strongly marked, especially if they plan to use high-level observation platforms. However, the wavebands of maximum absorption may be chosen deliberately for some kinds of meteorological and climatological studies. For example, vertical profiling or *sounding* of the atmosphere in depth by multiband radiometers focuses upon the 15 μm CO_2-absorption waveband. In studies of this kind the aim is to identify and measure the radiation emitted from a number

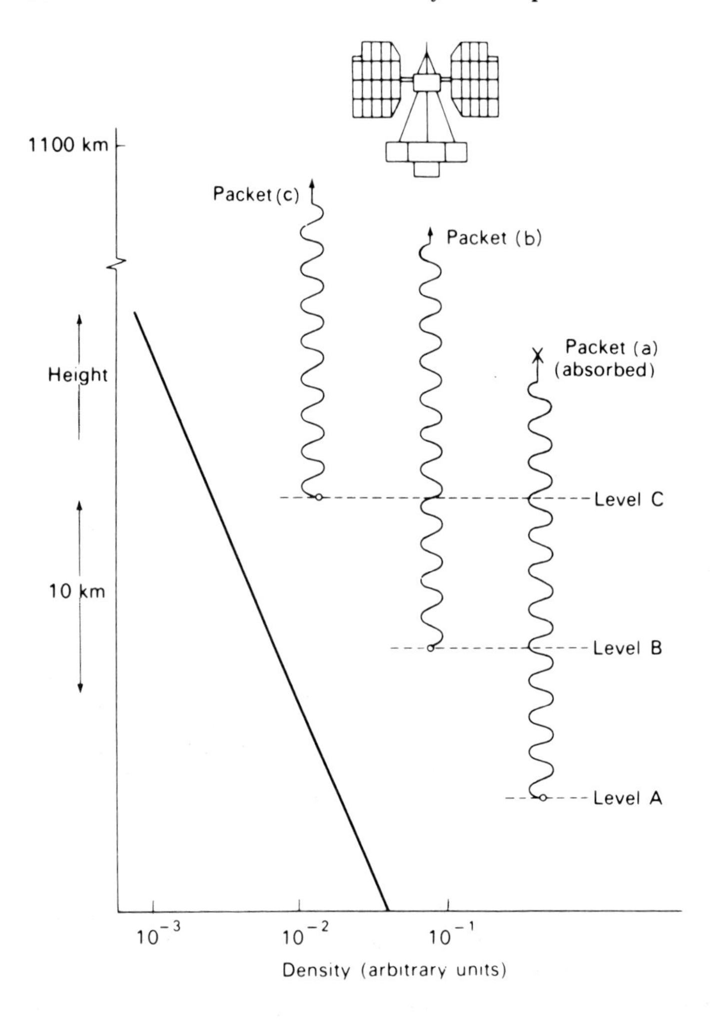

Fig. 3.3 The behaviour of packets of radiation emitted by carbon dioxide from different levels of the atmosphere. (Source: Barnett and Walshaw, 1974.)

of levels in the atmosphere, so that vertical profiles of the structure of the atmosphere may be obtained (see also Chapter 10).

The basic principle on which satellite depth sounding rests is that the molecules of a gas emit electromagnetic radiation at the frequencies at which they absorb energy. Absorption by a gas such as CO_2 is due to many different vibrational modes of the gas molecules. These modes occur in a series of more or less well-defined bands across the spectrum. So-called 'spectroscopic' sounding by satellite-borne multichannel radiometers and spectrometers involves the measurement of radiant emissions from the underlying atmosphere across a series of (often closely spaced) frequencies in each of which the atmosphere absorbs in a known way. Figure 3.3 illustrates this approach through reference to the upward emission of energy of a single frequency. This has been chosen to give a 50% transmittance from level B to the satellite, i.e. radiation packet (b) reaches the satellite half attenuated. Packet (a) is attenuated more on account of its passage through a greater depth of the atmosphere. Packet (c) arrives at the satellite virtually unattenuated, but emission from this level is small: the emission of radiation by CO_2 is proportional to the concentration of the gas, which decreases exponentially with height. Hence the radiation measured by the satellite-borne sensor is a weighted mean of that emitted from various layers, with the greatest contribution from around level B. If upwelling radiation is measured simultaneously at several frequencies, suitable weighting functions can be calculated to specify the atmospheric press-

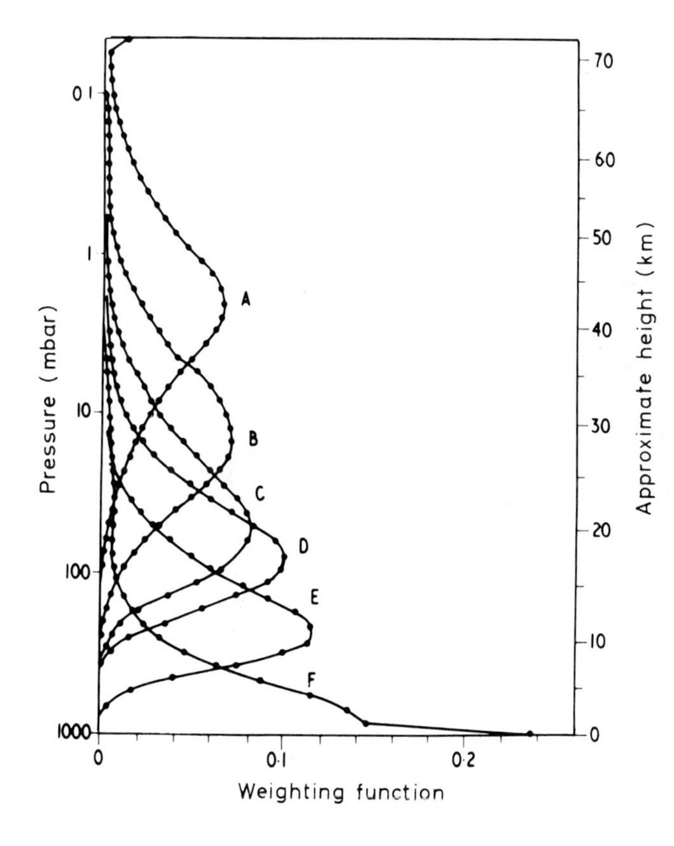

Fig. 3.4 The weighting functions of the Selective Chopper Radiometer (SCR), which was carried by Nimbus 4. The height scale is approximate, since the heights of the pressure surfaces vary by up to a few kilometres according to the temperature profile. (Source: Barnett and Walshaw, 1974.)

ure levels from which most of the measured radiations originate from the entire depth of the atmosphere (Fig. 3.4).

Radiometric soundings of the atmosphere were pioneered by experimental sensors on research and development (R and D) Nimbus satellites. Operational successors of their sounders, exploiting both atmospheric windows and the intervening 'window frames', are flying now on National Oceanic and Atmospheric Administration (NOAA) satellites (Table 3.1). Atmospheric sounding will be discussed further in Chapter 10, in relation to applications of the vertical profile data they provide.

Broad-band 'spectroscopic' sensors pioneered by the Infrared Interferometer Spectrometer (IRIS) of Nimbus have been used to retrieve thermal emission spectra for various environments on Earth, in addition to vertical profiles of temperature, humidity, and ozone concentration, as illustrated by Fig. 3.5 (see also section 4.3.3). IRIS measured planetary radiation in a broad waveband from 6.25 to 22.5 μm using a modified Michelson interferometer. Radiation from the target was split into two approximately equal beams to give an interferogram which was transmitted to the ground. Here a Fourier transform was performed to produce thermal emission spectra of regions on Earth (Fig. 3.5(a)). From the water vapour, ozone and carbon dioxide absorption bands in the spectrum, vertical profiles of the concentrations of these gases in the overlying atmosphere were then derived (Fig. 3.5(b)).

Infrared interferometer spectrometers have been used not only to investigate the atmosphere of Earth, but also the atmospheres of other planets, for example Mars. Still further spectrometers have been tested or proposed, particularly in relation to the greatly enhanced 'Earth Observation System', EOS, planned for the late 1990s onwards, as explained elsewhere in this book. As with many other remote sensing systems designed initially for investigating our terrestrial home, there is a bright future for at least some of them for use in very long range remote sensing to observe other units in the solar system. Although it is not the purpose of this book to explore such a theme, it is worth

Table 3.1 Channels of the High Resolution Infrared Sounder (HIRS-2) and associated Microwave Sounding Unit (MSU) carried by operational NOAA polar-orbiting weather satellites in the late 1980s and early 1990s (Courtesy, NOAA)

Channel	Wavelength (μm)	Wave frequency (cm^{-1})	Altitudes of peak emissions (mbar), or general sensitivity
H1	14.96	668.40	30
H2[a]	14.72	697.20	60
H3	14.47	691.10	100
H4[a]	14.21	703.60	280
H5	13.95	716.10	475
H6[b]	13.65	732.40	725
H7[b]	13.36	748.30	Surface
H8[b]	11.14	897.70	Window, sensitive to water vapour
H9	0.73	1 027.90	Window, sensitive to O_3
H10	8.22	1 217.10	Lower tropospheric water vapour
H11	7.33	1 363.70	Middle tropospheric water vapour
H12	6.74	1 484.40	Upper tropospheric water vapour
H13[abc]	4.57	2 190.40	Surface
H14[a]	4.52	2 212.60	650
H15[a]	4.46	2 240.10	340
H16	4.39	2 276.30	170
H17	4.33	2 310.70	15
H18[d]	3.98	2 512.00	Window, sensitive to solar radiation
H19[d]	3.74	2 671.80	Window, sensitive to solar radiation
H1[e]	0.596[f]	50.30[g]	Window, sensitive to surface emissivity
H2[c]	0.558[f]	53.74[g]	500
H3[a]	0.548[f]	54.96[g]	300
H4[a]	0.518[f]	57.95[g]	70

[a] Used to compute temperature profiles.
[b] Used to compute cloud fields.
[c] Used in cloud correction.
[d] Used to compute surface temperature.
[e] Used to compute surface emissivity.
[f] in cm.
[g] in GHz.

noting that already considerable progress has been made with the observation and analysis of the dynamics of atmospheres in addition to our own.

Figure 3.5 also underlines a general principle of fundamental importance in terrestrial remote sensing, namely that the best spectral regions for observing the Earth on the one hand, and its atmosphere on the other, do not coincide. We must now proceed to explore in greater detail why this is so.

3.1.3 Atmospheric transmission

In view of the almost perfect inverse relationship between the atmospheric absorption spectrum and the spectrum of atmospheric transmission the best spectral regions for observing the surface of Earth are the atmospheric windows in which absorption is low and transmission is high: these are the relative peaks in Fig. 3.5(a), where observed thermal emission spectra most closely approach related black-body spectra. Conversely, the best spectral

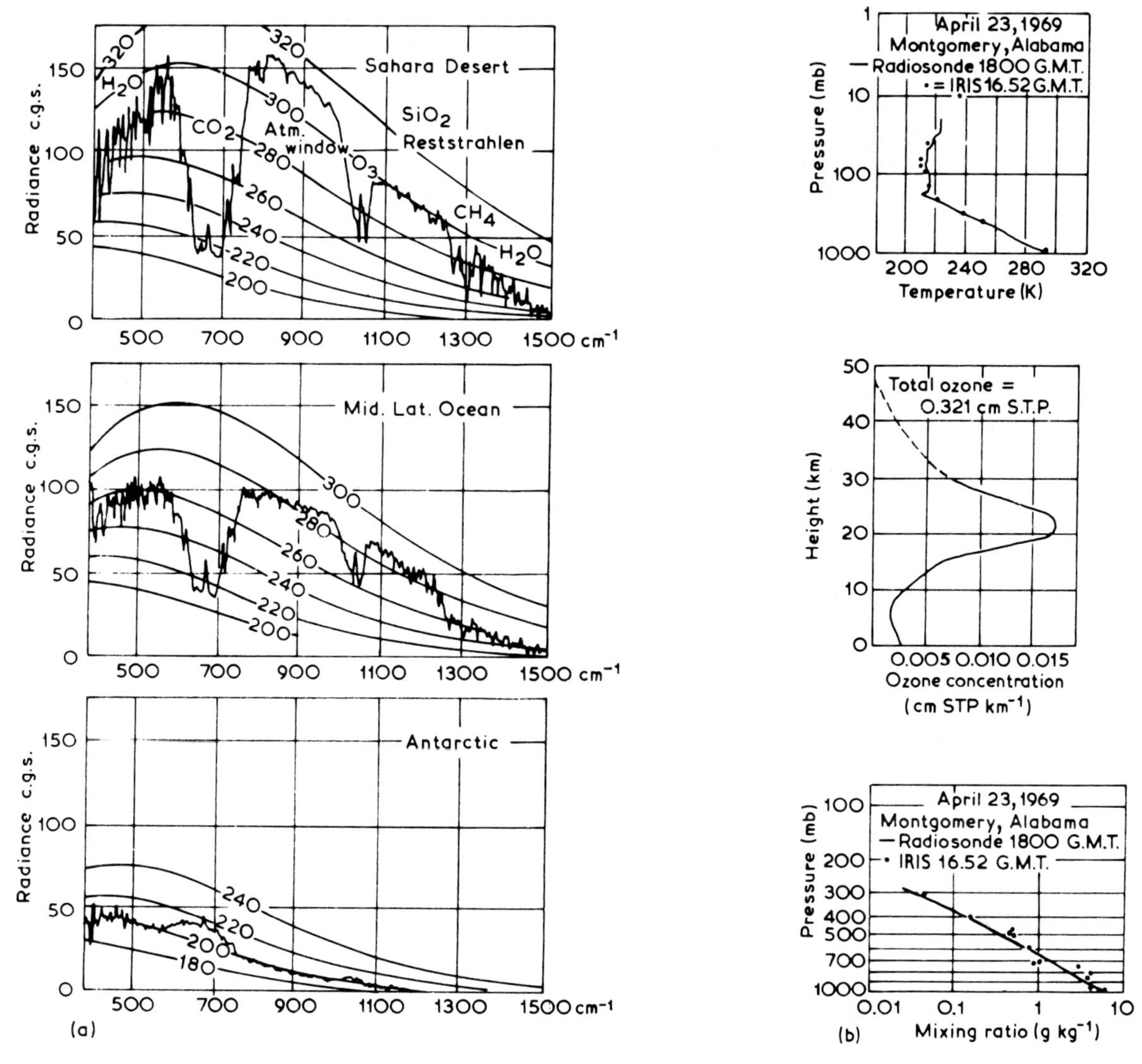

Fig. 3.5 Data received from the Nimbus Infrared Interferometer Spectrometer (IRIS) experiment were processed to give: (a) thermal emission spectra of the Earth from 5.25 to 22 μm. Here, representative spectra on one day for three regions of the world are compared with black-body curves. The departures are due to both atmospheric absorption and changes of emissivity of the source with wavelength. Within these wave numbers the primary absorption constituents of the atmosphere are water vapour (H_2O), carbon dioxide (CO_2), ozone (O_3), and methane (CH_4). In the broad atmospheric window, surface temperatures in the three graphs (from top to bottom) are 312 K, 285 K and 200 K, respectively. (b) Temperature, humidity and ozone profiles. Excellent correspondence is found between these profiles and radiosonde data and measurements from ground-based spectrometers. In sharp contradistinction to data from the scatter of surface weather stations, IRIS returns continuous strips of data practically encircling the globe. (Source: Aracon, 1971.)

regions for observing the atmosphere must be the 'window frames' or pillars of high absorption and low transmission: so the profiles in Fig. 3.5(b) were retrieved from the CO_2, O_3 and H_2O absorption bands, respectively, whose positions are indicated in Fig. 3.5(a: top). Reference back to Table 3.1 will confirm that different spectral regions must be exploited for observing the Earth's surface on the one hand and the atmosphere on the other.

Considering these facts it comes as no surprise to find that a typical atmospheric transmission curve for the atmosphere for radiation wavelengths from 0 to 14 μm (Fig. 3.6) bears a striking upside-down resemblance to the atmosphere's absorption spec-

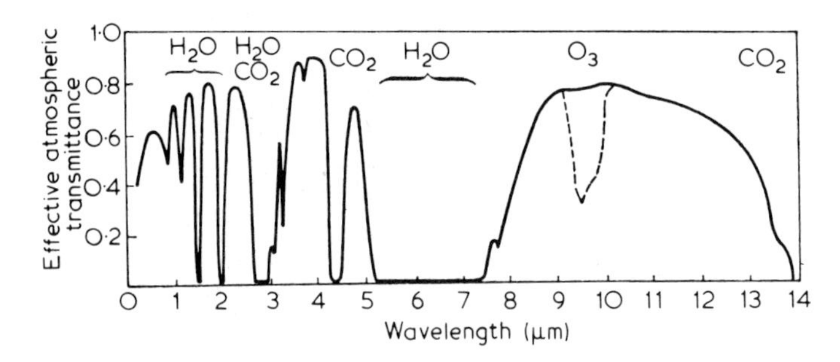

Fig. 3.6 The spectral transmittance of 650 m of sea-level atmosphere at low humidity and low haziness. The principal regions of absorption by certain gaseous constituents of the atmosphere are labelled.

trum (Fig. 3.2). However, when we come to consider remote sensing from aircraft instead of satellites, it is important to remember that the effective transmittance of any column of the atmosphere is dependent not only on absorption but on scattering too. The situation we have chosen to illustrate is one in which humidity is low, and there is little particulate matter in the column. Under hazy or foggy conditions attenuation by scattering may have a larger effect upon the spectral transmittance of the atmosphere at sea-level than attenuation by absorption. Consequently if one wishes to view the surface of Earth from aloft, especially at the short end of the spectrum (e.g. visible air photographic missions), care must be taken to choose those weather conditions under which attenuation will be as low as possible, for then transmittance will be at its peak. Often the angle of view will be critical for the decision whether or not to undertake any specific photographic mission. Under given attenuation conditions transmission is best for vertical paths, deteriorating through slant paths to the horizontal, where the view at the surface of the Earth is entirely within the layer characterized by the highest concentrations of atmospheric water and suspended particles. For this and other reasons vertical air photography is often strongly preferred to oblique air photography (Chapter 7).

3.2 Radiation from the Earth

3.2.1 Atmospheric window wavebands

We have seen why much of the planning of airborne or spaceborne missions concerned with remote sensing of the planetary surface rather than the Earth's atmospheric envelope involves the identification and exploitation of the windows between bands of peak attenuation of radiation by the atmosphere. Since it is often sufficient for the purposes of introductory studies in climatology and meteorology to refer to atmospheric window radiation in the singular it is worth stressing again that the atmosphere contains not one, but several, important windows for radiant energy transmission. Indeed, in spectral regions above radiation wavelengths of approximately 1 m, little absorption takes place excepting under extreme weather conditions, and 'window' conditions prevail.

Of particular importance to environmental remote sensing is the fact that in the infrared ('heat') region of the electromagnetic spectrum there are not one, but several, atmospheric windows (Figs 3.2 and 3.6). Of these the most valuable for use in remote sensing are those between about 3.0–4.5 μm and 8.5–14 μm. The chief advantage of the first is that it is the more sharply defined. The chief advantage of the second is that it lies wholly within the thermal emission spectrum of Earth, and embraces the wavelength of peak infrared emission from the source (approximately 10 μm). Unfortunately there is a marked atmospheric absorption band around 9.6 μm, which must be either disregarded or avoided by satellite and high-altitude rocket missions. This is why its influence on the spectral transmittance of the atmosphere is shown in Fig. 3.6 (which illustrates a sea-level case) by a broken line. As Fig. 3.6 reveals, this absorption peak is caused by ozone in the stratosphere.

Both the 3.0–4.5 μm and 8.5–14 μm window regions have been used for remote sensing studies of the surface of the Earth (and its cloud cover) from satellite altitudes. For example, weather satellites, such as the American Nimbus and NOAA, and some members of the Russian Cosmos and Meteor families, have been equipped with

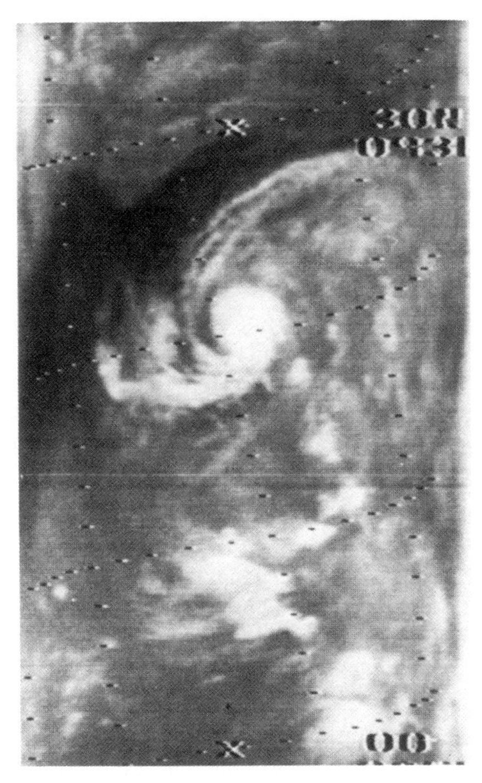

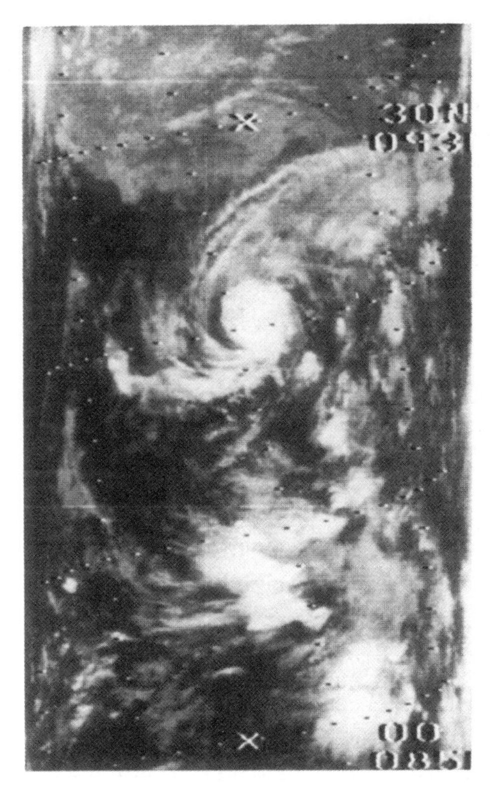

Fig. 3.7 Significant differences are evident in these simultaneous views of the Bay of Bengal obtained from the THIR (6.7 μm and 11.5 μm channels) on Nimbus 5. The 11.5 μm image (right) indicates surface and cloud-top temperatures; the 6.7 μm image (left) reveals concentrations in the moisture content of the upper troposphere and stratosphere. (Courtesy, NASA.)

infrared sensors designed in part to explore infrared window waveband radiation.

Let us consider in greater detail the relationships between choice of waveband and object of enquiry, first through reference to the Advanced Very High Resolution Radiometer (AVHRR) system on the current family of NOAA operational weather satellites (see also Chapters 10 and 11). Table 10.2 summarizes the wavebands that are involved, and the objects of their enquiry.

Differences in the transparency of the atmosphere in two commonly exploited multichannel radiometer wavebands are illustrated by Fig. 3.7. The Temperature Humidity Infrared Radiometer (THIR) on Nimbus 5 was designed for temperature evaluation of radiating surfaces of the Earth in the 10–12 μm region of the broadest infrared window, and for assessment of atmospheric moisture in the water vapour absorption waveband centred on 6.7 μm. A noteworthy complication in the first case is that, in the two principal infrared windows, clouds often obscure the surface of the Earth. Water vapour is transparent to radiation, but aggregations of water droplets reflect and scatter incident radiant energy. These attenuating effects are so strong that even quite shallow clouds reduce the effective transmittance to zero. Therefore, where clouds are present in an atmospheric column, their upper surfaces act as the effective radiating surfaces so far as sensors designed to exploit the atmospheric window wavebands are concerned. An important benefit of this is that, in cloudy areas, the observed radiation or brightness can be used to map the heights of the cloud tops, for temperatures usually decline upwards through the troposphere. By physical or statistical techniques cloud top temperatures can be translated into heights of the cloud tops above the ground.

The Surface Composition Mapping Radiometer (SCMR) of Nimbus 5 was also interesting in that it measured target radiation in spectral regions between 8.4–9.5 μm and 10.2–11.4 μm. Both of these fall within the broader infrared window, but both avoid the ozone absorption band around 9.6 μm (Fig. 3.2). Simultaneous data from the two 'split infrared window' channels yielded different brightness temperatures for the same targets, for as Chapter 2 suggested, radiant exitance varies with the wavelength of radiation, and many other factors also influence observed brightness temperatures. Intercomparisons of measurements from the two channels of SCMR gave general clues as to the types and variations of mineral surfaces as viewed from space.

Other window wavebands well worth exploiting in remote sensing include the microwave region of the electromagnetic spectrum. Within this, opaque regions are due to water vapour and atmospheric oxygen. The intervening windows have clear advantages over those in the infrared: they can be used even when weather conditions are quite severe. The relatively long wavelength rays in the microwave region (ranging from about 1 mm to 1 m) are able to penetrate thick clouds and even some rain storms (Chapters 4 and 12). Microwave sensing, whether active or passive, therefore has to a greater degree the 'all-weather' capability that other systems lack. At these longer wavelengths, scattering is relatively insignificant, and for normal atmospheric conditions, absorption can be read for attenuation with little resultant error.

Before leaving, for the present, the question of atmospheric windows it remains to be stressed that the first to be exploited was in the visible region of the spectrum, and this is still of great importance today, although it is true to say that conventional photography, both colour and black and white, is often hindered much more by the common attenuating agencies of atmospheric water in droplet form and particulate matter than absorption by gases in the atmosphere. We shall discuss conventional photography in more detail later, but will spend most time considering data from outside the visible portion of the spectrum.

3.2.2 Recording reflected and/or emitted radiation

Forgetting neither the significance of the altitude of remote sensing platforms, nor the special opportunities for remote sensing within atmospheric window wavebands, it is useful to note that phenomena on the surface of the Earth can be investigated by several passive remote sensing means:

1. Broad waveband sensing. Broad-band, non-specific, sensors can be used to integrate the energy from many wavelengths into a composite image. An example is the common ('panchromatic', literally 'all colours') camera–film combination. This records radiation across the visible portion of the spectrum within the upper and lower wavelength limits of the film emulsion.
2. Narrow waveband sensing. Here radiation from the target is recorded only in a single selected waveband of the electromagnetic spectrum. We have seen how most objects reflect or emit energy over a broad range of individual wavelengths. There is, however, normally a peak wavelength at which the maximum amount of energy is being reflected or emitted. Under such circumstances objects can best be differentiated from their backgrounds by measurements attuned to their radiation peaks.
3. Bispectral sensing. Sometimes the location and identification of environmental phenomena is made easier by simultaneously recording radiation from the target in two non-adjacent wavebands. The data are then used comparatively. The Nimbus THIR and SCMR sensors performed such a type of operation.
4. Multispectral sensing. A series of sensors arranged to operate at several very narrow bandwidths (often equally spaced across a selected region of the radiation spectrum) permit the compilation of three or more simultaneous images of the target area (Fig. 3.8). Much current work in remote sensing is now concerned with the compilation of *spectral signatures* derived from multispectral data, and their interpretation by comparison with 'fingerprint banks' of signatures known to be associated with particular objects or environmental phenomena. Unfortunately too little is known of the spectral responses of many targets in the natural environment, and spectral signature interpretation is often less easy and/or less definitive than some end-users would wish. Many factors affect spectral signatures, including temporal variations such as time of day and season of the year. New multispectral scanners are using many separate channels, and the

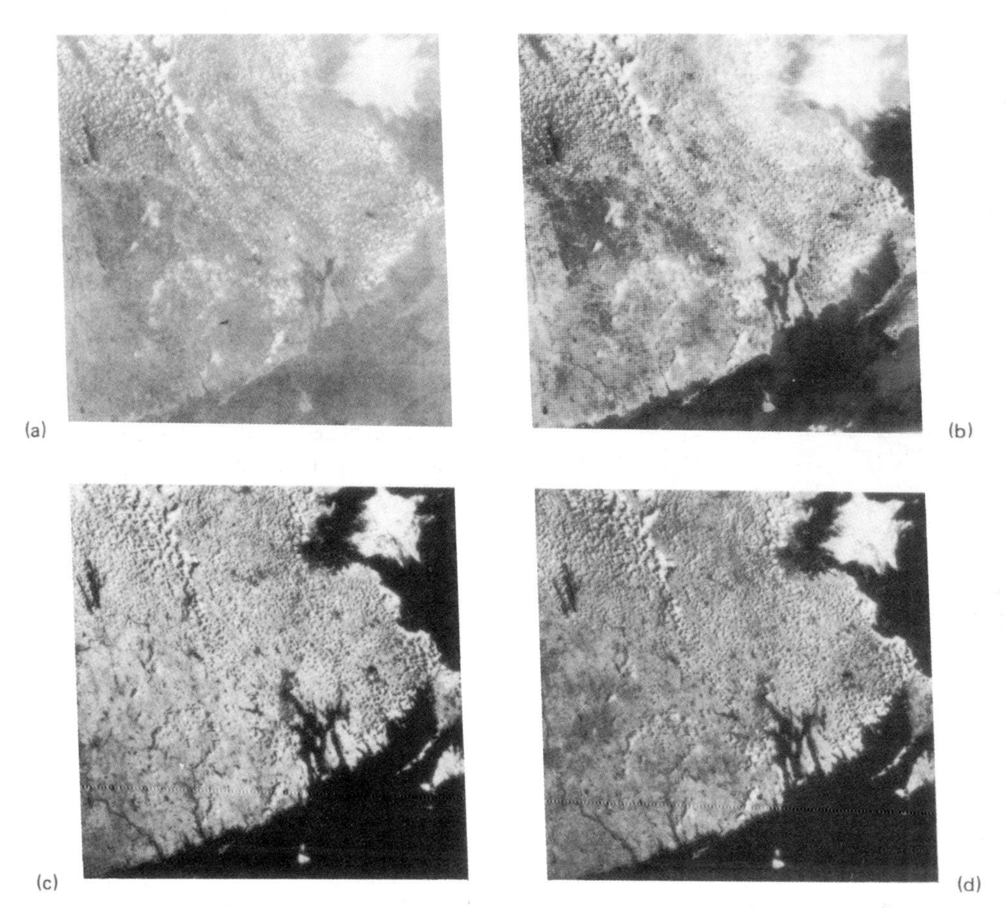

Fig. 3.8 A set of Landsat 1 multispectral scanner film transparencies for New England, 28 July 1972. (a) Band 4 (0.5–0.6 μm); (b) Band 5 (0.6–0.7 μm); (c) Band 6 (0.7–0.8 μm); (d) Band 7 (0.8–1.1 μm). More recently Landsat has carried a seven-channel 'Thematic Mapper' providing even more information than is illustrated here (Chapter 4). (Courtesy, NASA.)

analysis of the signatures based on large numbers of target responses is almost impossible without the aid of some mechanical or electrical back-up system. For manual analyses, sensors with as few as four carefully selected channels have to suffice to provide maximum contrast for ready qualitative intercomparison.

Since we have already considered introductory examples of remote sensing by the first three of the four means listed above we may fruitfully turn our attention finally to examples of the fourth. As the analysis of multispectral signatures from the environment is hedged about by many difficulties and uncertainties we should consider first the nature of the responses from some individual constituents of that environment.

3.3 Spectral signatures of the Earth's surface

3.3.1 Signatures of selected features

Rocks, the most fundamental constituents of the solid surface of Earth, can be distinguished from each other under ideal conditions by their spectral signatures in the thermal emission region of the spectrum (Fig. 3.9). We noted earlier that the emission spectrum from the Sun deviates in some details from the spectrum for a black body with the same mean surface temperature. Similarly many elements of the surface of the Earth have emissivities that vary both in temperature and in frequency, and act more like grey bodies than black bodies. In Fig. 2.4(b), illustrating the distribution of energy emitted by quartz (SiO_2), there were wide

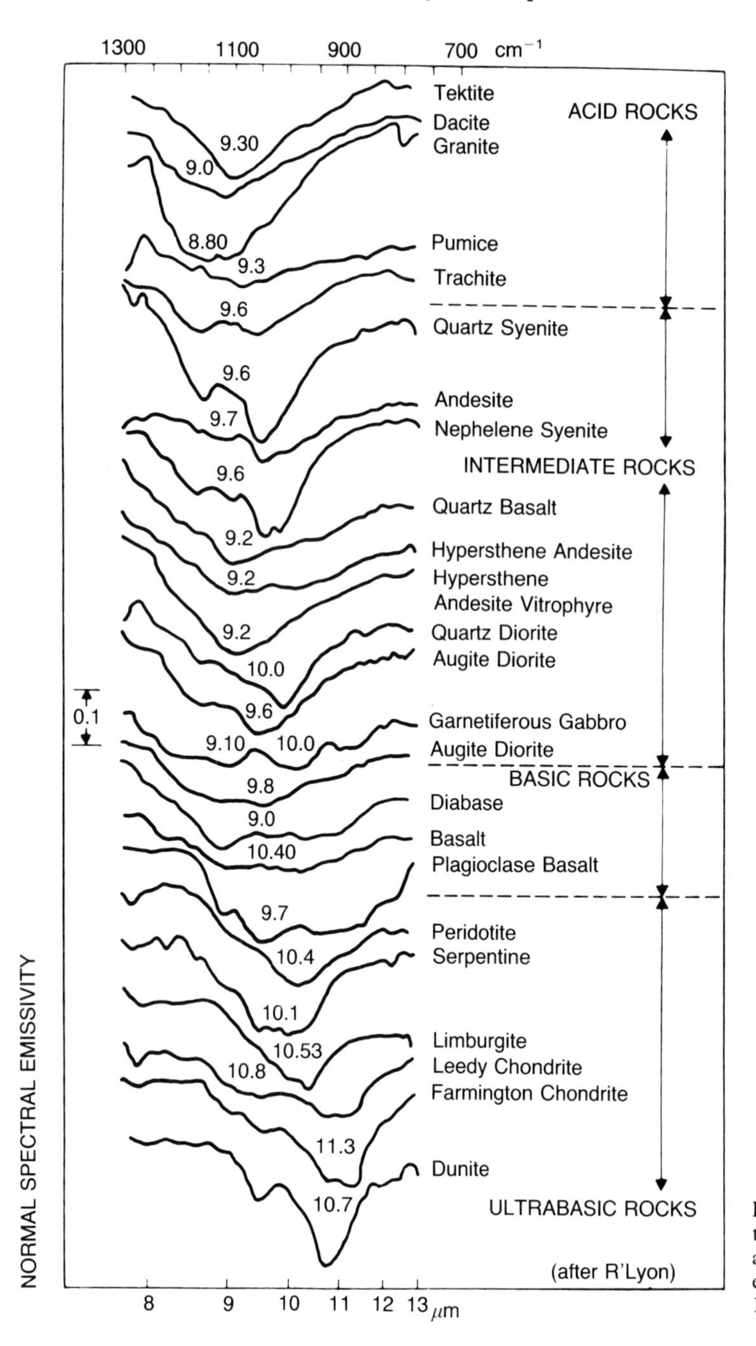

Fig. 3.9 Spectral signatures of rocks. Many types of rocks are differentiated from one another by the spectra of their radiation emissions in the infrared. (Source: Laing, 1971.)

differences between the two curves, especially around 9 μm and 20 μm. At these wavelengths incident energy is absorbed sharply by quartz, producing lower rates of emission here. Natural substances often behave more like perfect absorbers and radiators at some frequencies than others, revealing the so-called 'reststrahlen' or residual ray effect, when the actual emission curve is compared with the ideal. In the remote sensing of geological constituents of the environment basic rocks can be distinguished generally from acidic rocks by their signatures in the infrared, since the exact wavelength of the absorption peak varies from the one to the other (Fig. 3.9). Fortunately

Table 3.2 Aircraft and laboratory studies reveal differences in the percentage reflectances of different types of surfaces and crops. The four bands indicated here are those covered by the Multispectral Sensor (MSS) on the Landsat series of satellites

	Reflectance (%)			
	Band 1 (0.5–0.6 μm)	Band 2 (0.6–0.7 μm)	Band 3 (0.7–0.8 μm)	Band 4 (0.8–1.1 μm)
Rock and soil materials and covers				
Sand	5.19	4.32	3.46	6.71
Loam 1% H_2O	6.70	6.79	6.10	14.01
Loam 20% H_2O	4.21	4.02	3.38	7.57
Ice	18.30	16.10	12.20	11.00
Snow	19.10	15.00	10.90	9.20
Cultivated land	3.27	2.39	1.58	(not given)
Clay	14.34	14.40	11.99	(not given)
Gneiss	7.02	6.54	5.37	10.70
Loose soil	7.40	6.91	5.68	(not given)
Vegetation				
Wheat (low fertilizers)	3.44	2.27	3.56	8.95
Wheat (high fertilizers)	3.69	2.58	3.67	9.29
Water	3.75	2.24	1.20	1.89
Barley (healthy)	3.96	4.07	4.47	9.29
Barley (mildewed)	4.42	4.07	5.16	11.60
Oats	4.02	2.25	3.50	9.64
Oats	3.21	2.20	3.27	9.46
Soybean (high H_2O)	3.29	2.78	4.11	8.67
Soybean (low H_2O)	3.35	2.60	3.92	11.01

such radiation differences can be observed even from satellite altitudes, since they fall within the broad atmospheric window waveband from about 8 to 13 μm. It has become possible to differentiate rough rock specimens in the laboratory by sole use of the spectrum of infrared energies emitted by the samples. The same method has been found equally applicable in the field and from the air.

Similar differences between spectral signatures of different rocks have been described for the ultraviolet, visible and photographic infrared, based on reflection, rather than emission, spectra. One key source of variability in the observed spectra from particular types of rocks is the water content of the samples. Another is their carbon dioxide content. Figure 3.10 illustrates reflectance spectra of red sandstone surfaces under different moisture conditions. It is notable that the percentage reflection from the wet surfaces is much less than that from the dry surfaces, especially around 1.4 and 1.9 μm, which are strong water absorption bands. Clearly detailed 'ground truth' data must be obtained if such remote sensing data are to be interpreted correctly (Chapter 6).

Broadening our discussion from rocks and weathered rocks to a wide variety of other environmental constituents, we see from Table 3.2 that many types of soils, crops, and other active surfaces, such as ice and snow, are amenable to multispectral identification from airborne or spaceborne sensor systems. Further results from Landsat satellites will be examined in later chapters.

3.3.2 Signatures of complex environments

A general feature of multispectral images, such as those from the Landsat satellites (e.g. Fig. 3.8(a–d)), is that they represent the Earth's surface in great spatial and spectral detail. Although the spectral reflectance properties of some surface

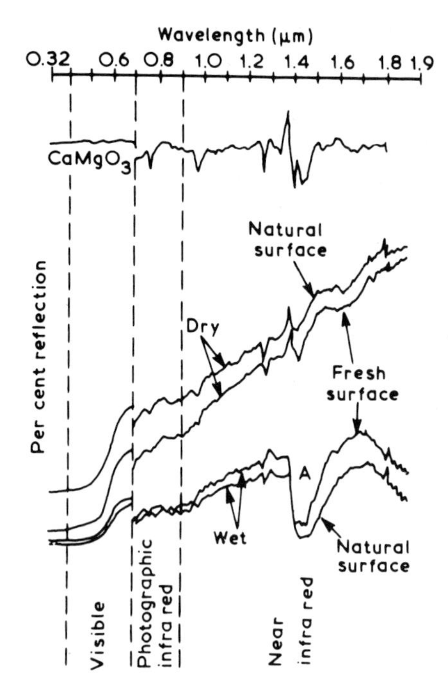

Fig. 3.10 Reflectance spectra of wet and dry, fresh and weathered (salt encrusted) surfaces of a red sandstone.

Table 3.3 Sources of variation in multispectral signatures of vegetation (Source: Polcyn *et al.*, 1969)

Illumination conditions
Illumination geometry (sun angle, cloud distribution)
Spectral distribution of radiation

Site environmental conditions
Meteorologic
Micrometeorologic
Hydrologic
Edaphic
Geomorphologic

Reflective and emissive properties
Spatial properties (geometrical form, density of plants, and pattern of distribution)
Spectral properties (e.g. reflectance or colour)
Thermal properties (emittance and temperature)

Plant conditions
Maturity
Variety
Physiological condition
  Turgidity
  Nutrient levels
  Disease
  Heat-exchange processes

Atmospheric conditions
Water vapour, aerosols, etc. (absorption, scattering, emission)

Viewing conditions
Observation geometry (scan angle, heading relative to Sun)
Time of observation
Altitude

Multichannel sensor parameters
Electronic noise, drift, gain change
Accuracy and precision of measurements on calibration references and standards
Differences in spectral responses of systems

constituents can be identified readily by eye, and reported in qualitative terms (e.g. the greater propensity for water to show differences in reflection from MSS Bands 4 to Bands 5 and 7 than is found over land), more detailed and precise statements are required for many practical purposes. A wide range of automatic and semi-automatic procedures has been developed to meet the resulting need for more objective image analysis, comparison and interpretation. Whilst later chapters will be more concerned with these methods, we remember that many involve some element of spectral signature recognition. However, as Table 3.3 indicates, many influences have a bearing on the spectral signatures which may be derived from multispectral aircraft or satellite data. Often it is not possible to evaluate all the influences acting on a given set of data. It is rarely (if ever) possible to assess the effects of such influences on each other. Not surprisingly, therefore, we must note yet again that, contrary to earlier over-optimistic expectations, the real world is so complex and variable that satellite mapping of environmental patterns is always subject to some doubt and error: many early promises concerning automatic feature classification were at best premature. Objective feature recognition and mapping is often more accurate when multispectral data are available, but even multispectral feature recognition and delimitation is often inexact. A further problem which has been recognized relatively recently is that satellite data often lend themselves best to classifications which do not precisely match those established through large periods of surface ob-

Table 3.4 Characteristics of the radiometers used in a programme of multispectral sensing over Bear Lake, Utah/Idaho, March 1971 (Source: Schmugge *et al.*, 1973)

Frequency (GHz)	Wavelength (cm)	Pointing relative to nadir (deg.)	3dB beam width (deg.)	RMS temperature sensitivity (K)
1.42	21	0	15	5
2.69	11	0	27	0.5
4.99	6.0	0	5	15
10.69	2.8	0	7	1.5
19.35 H	1.55	Scanner	2.8	1.5
37 V	0.81	45	5	3.5
37 H	0.81	45	5	3.5
Infrared	1.0×10^{-3}	14	<1	<1

servations. Thus, for example, a distinction is now drawn commonly between (conventional) *land use* data and (satellite derived) *land cover* analyses. One considerable challenge for the future will continue to be how best to relate *in situ* and remotely sensed data; a second will be to improve methods of combining the two different types of data for best possible use of both.

The complexity of the real world, and multispectral investigations of it, can be well illustrated by reference to a programme carried out over Bear Lake on the border between Utah and Idaho. Plate 1(a) (colour section) shows the flight path of the instrument platform, on that occasion a NASA Convair 990 Airborne Observatory. In all, eight sensors were used simultaneously to view the frozen lake and its surrounding snowfields and bare rock. The sensors are summarized in Table 3.4. One viewed in the infrared at 10 μm through the broad atmospheric window. The remainder were all microwave sensors, of which the 1.55 cm sensor was different from the rest in that it was a scanning device, whereas the others all viewed at fixed angles relative to the nadir angle of the aircraft.

Stripchart results from all the sensors are shown in Fig. 3.11. The plot for the 1.55 cm scanner is the average of the five central beam positions across the flight path of the aircraft. Considerable differences are evident from one curve to another. Some of these concern their general forms. For example, the curves of the data from channels 2 and 3 are approximately the inverse of those from channels 5 and 6. Other differences relate to detail embroidered on the basic shapes.

One of the conclusions of this study was that at the longer wavelengths (especially 21 cm) the observed variability in brightness temperatures, especially over the lake, was related to thickness variations in the ice. Results indicated that the transparency of ice and snow is a function of wavelength, and that ice and snow depth may be assessed by these longer microwave emissions.

Plate 1(b) (colour section) is the 1.55 cm microwave image of the pass over Bear Lake at an altitude of 3400 m. The area it covers corresponds with part of Plate 1(a). The outline of the frozen variably snow-covered lake can be clearly seen. The low brightness temperatures of the frozen surface contrast sharply with the higher temperatures along the steep slopes of the eastern edge of the lake. Although a considerable amount of cloud was present over the target area when the flight was made, useful results were still obtained since microwaves, as we noted earlier, penetrate most clouds.

More recently, multispectral scanning and analytical science have developed very rapidly, involving complex imaging spectrometers (or 'spectroradiometers') and advanced processing techniques: investigating and evaluating the radiation characteristics of specific phenomena and environments simultaneously through a large number of wavebands is an increasingly attractive and effective form of remote sensing. For the most part

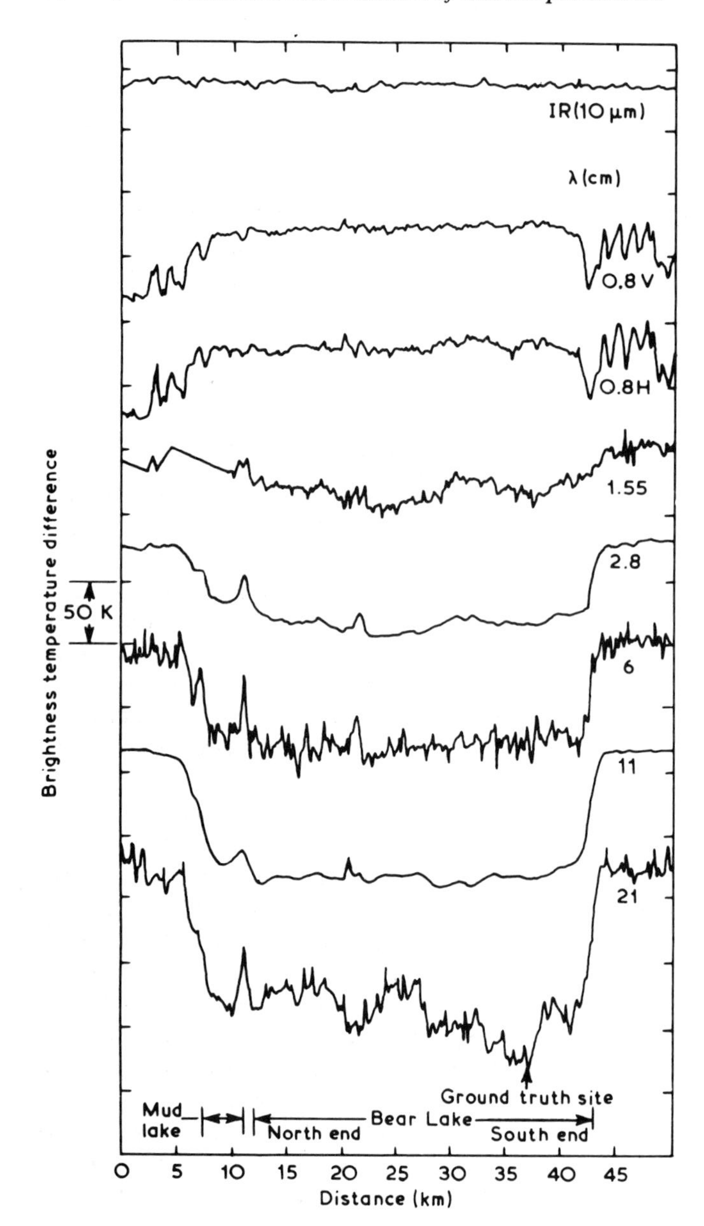

Fig. 3.11 Multispectral data obtained over Bear Lake, Utah/Idaho, 3 March 1971. H and V refer to the horizontal and vertical channels of the 0.8 cm radiometer, which viewed the surface at an angle of 45°. The remaining radiometers were nadir viewing. Each graph is related to the average temperature value at the frequency in question. (Source: Schmugge *et al.*, 1973.)

operational environmental mapping and monitoring to date have been concerned with the relatively simple – but often only moderately successful – task of analysing integrated radiation or radiation patterns obtained through relatively small numbers of wavebands at one time. Thus there is a new challenge for the future: we know much about the basic physics of radiation reflection and emission, but we still need to learn much more about the often highly complex reflection and emission patterns that are associated with real phenomena and situations.

A good example of a situation that should be clarified by spectroradiometry from space is afforded by phytoplankton, sudden upsurges (or 'blooms') of which occur from time to time, especially in lakes, coastal waters and enclosed seas. Such blooms usually occupy quite small areas (from a few tens of square kilometres to hundreds), last for 10–100 days, and are often nearly mono-

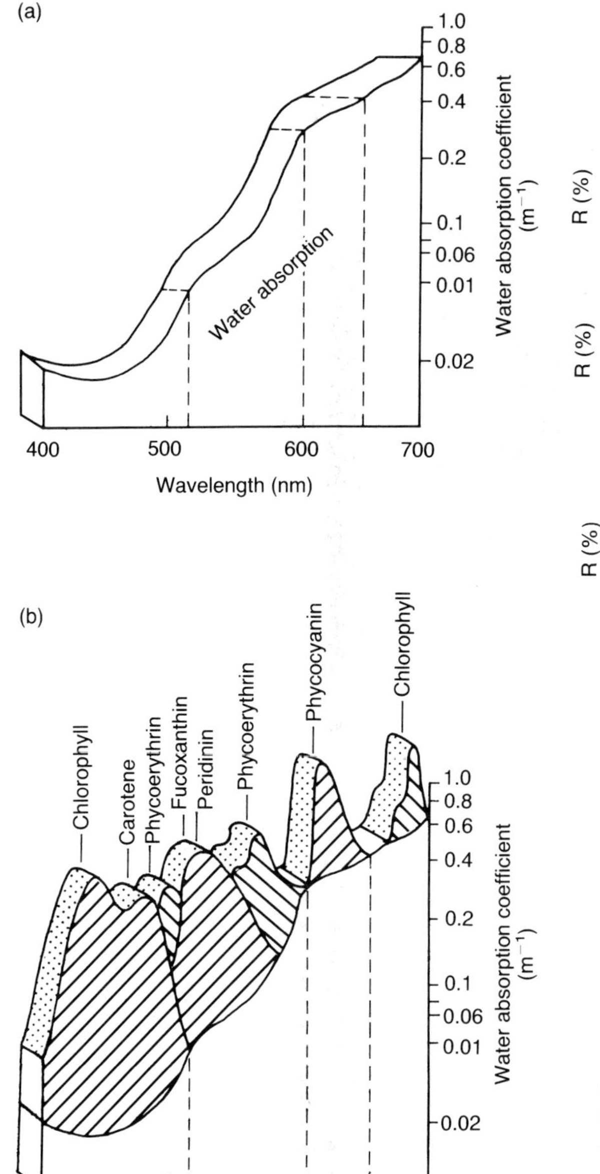

Fig. 3.12 Absorption of light (a) by water and (b) by different algal pigments in the windows of 'clarity' of water; the spectra for the pigments approximate those measured *in vivo*; fucoxanthin and peridin are superimposed. (After Yentsch and Yentsch, 1984.)

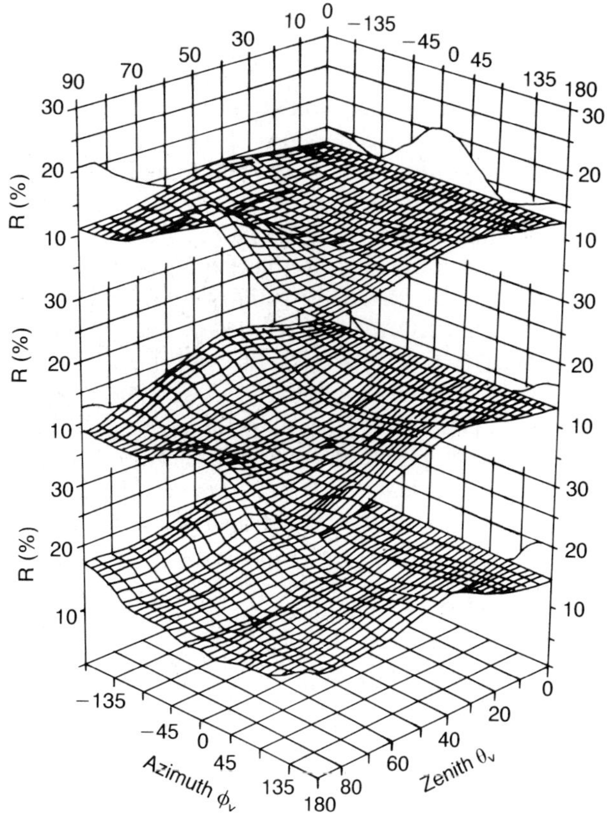

Fig. 3.13 Angular reflectance distributions above savannah at the surface (bottom data) for a clear atmosphere (aerosol free), at the top of a mildly aerosol-loaded atmosphere with surface visual range V_0 = 50 km (middle data), and at the top of a heavier aerosol-loaded atmosphere with V_0 = 10 km (top data). Sun direction is θ = 30°, ϕ = ±180°, λ = 0.85 μm. (Courtesy, NASA.)

specific. Since they are known to encourage significant fluxes of carbon to deeper waters as they decay, their creation and dissipation is now being urgently sought in relation to primary oceanic productivity, the global carbon cycle, and the possibility of continued global atmospheric warming. Figure 3.12(a) illustrates the common profile of sensible light absorption by water in the absence of phytoplankton; Figure 3.12(b) shows the increased absorptions associated with common phytoplankton pigments. For example algae containing chlorophyll (chlorophytes) usually transmit light in the green region, those containing phycocyanin (cyanophyceans) transmit mostly in the blue region, while algae containing chlorophyll a, chlorophyll c (diatoms, dinoflagellates, chrysophytes) and phycoerythrin (cyanophyceans) transmit mostly in the yellow. Sensors such as the Landsat MSS and TM, and the Nimbus-7 CZCS provide only enough information to yield spectral signatures with the level of detail in Fig. 3.12(a); to generate the much more complex signatures exemplified by Fig. 3.12(b) spectrometers providing for the simultaneous collection of image in a hundred or more continuous spectral bands are required, yielding continuous reflectance spectra for each pixel in the scene.

One such system being developed for the EOS programme is the High Resolution Imaging Spectrometer (HIRIS). Along with planned developments of the NOAA-AVHRR sensor (section 5.9) and MODIS-N (Table 14.2), this will provide data with the spectral resolution necessary to assess plankton populations, as well as other recognition/classification problems of a physically similar nature, e.g. detailed rock and mineral identification, plant classification, and analyses of organic particles suspended in water. The development testing and use of such systems will become a major thrust of satellite remote sensing of natural phenomena in the 1990s and beyond. However, interesting challenges may well remain. For example, it is generally recognized that improved corrections are required to adjust for atmospheric effects on upwelling radiances measured by satellites, for whatever purpose. Figure 3.13 emphasizes that, in order to observe the very complex effects of the atmosphere on such radiances, account must be taken of the sensor viewing angle (for different responses will be observed from different viewing angles) – and for greatest accuracy simultaneous observations from several viewing angles are required. As yet these are still beyond the capabilities of the sensor systems now being built.

4 Sensors for environmental monitoring

4.1 Introduction

In this chapter the sensors available for remote sensing studies will be examined in more detail in order to explain the advantages and disadvantages attached to each sensing system. It will be convenient to discuss the sensors in relation to the wavebands in which they can be applied, beginning with passive sensing systems and concluding with active sensing systems.

Two broad categories of passive sensors can be identified namely *photographic* and *non-photographic* (scanning dependent) systems (Fig. 4.1). Photographic systems operate in the visible and near infrared parts of the spectrum (0.36–0.9 μm) only, whereas non-photographic sensors can range from X-ray to radio wavelengths (Fig. 4.1).

4.2 Passive sensors in the visible wavelengths

4.2.1 Photographic cameras

The photographic camera is a well-known remote sensing system in which the focused image is usually recorded by a photographic emulsion on a flexible film base. There are three basic types of camera: *framing*, *panoramic* and *strip* cameras. The framing camera usually provides a square image with an angle of view up to 70°. The panoramic camera gives a very wide field of view which can extend to the horizons on either side, with consequent distortion of the image. There are various types of panoramic camera systems and the commonest type uses a reciprocating, or continuously rotating, narrow field lens. The strip camera consists of a stationary lens and slit together with a moving film of width equal to, or greater than, the slit length. The film is moved across the slit at a speed which ensures no blurring of the image at the height flown (i.e. it is *image-motion compensated*). The exposure is controlled by the slit width. Such cameras have been used mostly in low level sorties by high speed aircraft in military reconnaissance flights.

In the framing and panoramic cameras the exposure interval can be chosen so that a sequence of overlapping images can be obtained (Fig. 4.2(a) and (b)). This overlap is essential if stereoscopic (i.e. three dimensional) viewing is desired. In order to avoid gaps in the stereoscopic cover it is necessary to ensure that the camera is oriented in the direction of track of the aircraft or space platform (Fig. 4.2(c)). The scale of the photograph obtained by a framing camera is dependent on the focal length of the lens and the height of the camera above ground. Mapping and reconnaissance cameras commonly use 6 inch lenses but 12, 3½, 3 and 1½ inch lenses are used in special camera systems. Where the 6 inch (150 mm) lens is used the calculation of scale of the photograph is a simple operation. For example, if the camera used was a Wild RC8 Universal Aviogon with a focal length of 6 inches and the flying height was 10 000 feet above ground the photographic scale can be calculated as follows:

$$\text{Photographic scale} = \frac{\text{Focal length}}{\text{Height of the camera above ground}} = \frac{6}{10\,000 \times 12} = \frac{1}{20\,000} \tag{4.1}$$

The principal factors that limit resolution of the camera system are:

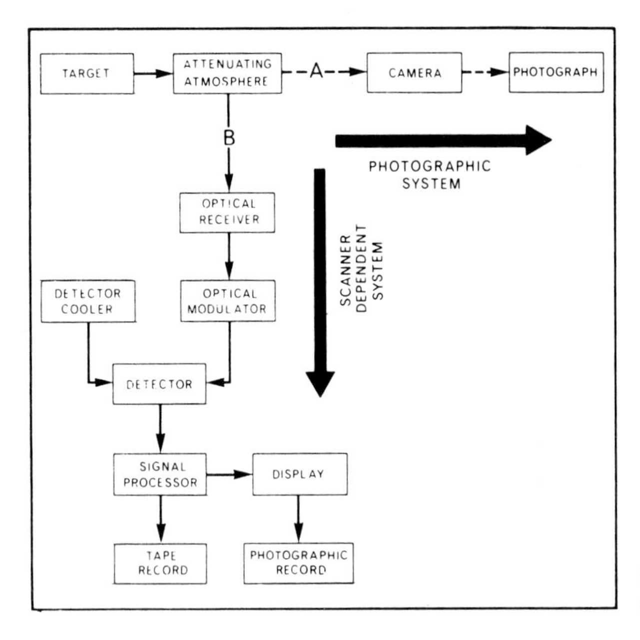

Fig. 4.1 Photographic and non-photographic (scanner dependent) systems. (Source: Curtis, 1973.)

1. lens resolution;
2. film resolution;
3. film flatness and focal plane location;
4. accuracy of image-motion compensation (IMC);
5. control of roll, pitch, yaw, vibration;
6. optical quality of any filter or window placed in front of the lens.

The definition of photographic image quality has been quoted conventionally in terms of 'resolving power'. This usually is measured by imaging a standard target pattern (Fig. 4.3) and determining the spatial frequency (in lines mm^{-1}) at which the image is no longer distinguishable. Resolving power also can be determined by examining the contrast and distortion of the image of a sinusoidal grating object. This measurement, as a function of the spatial frequency of the grating gives what is termed the *modulation transfer function* (MTF), which is commonly used to express the resolution characteristics of different films. Examples of the sensitivity and resolving power of Kodak aerial films are given in Table 4.1.

A summary of the characteristics of the different types of film available is given in Table 4.2 from which it will be apparent that the versatility of the camera sensing system is increased by the availability of different films with different spectral ranges, sensitivities and resolving powers.

Table 4.1 Sensitivity and resolution for Kodak aerial films (Source: EMI, Department of Trade and Industry, 1973)

Type	Sensitivity (aerial film speed)[a]	Resolving power (lines mm^{-1}) at test object contrast	
		1000 : 1	1.6 : 1
High-definition Aerial 3414	8	630	250
Panatomic X 3400	64	160	63
Plus X Aerographic 2402	200	100	50
Double X Aerographic 2405	320	80	40
Tri X Aerographic 2403	640	80	20
Infrared Aerographic 2424	200	80	32
Aerochrome Infrared 2443	40	63	32
Ektachrome Aerographic 2448	32	80	40

[a] Aerial film speed for monochrome negative material is defined as $3/2E$, where E is the exposure (in metre-candela-seconds) at the point on the characteristic curve where the density is 0.3 above base plus fog density, under strictly defined conditions given in ANSI Standard PH2.34–1969. A doubling of aerial film speed number denotes a doubling of sensitivity.

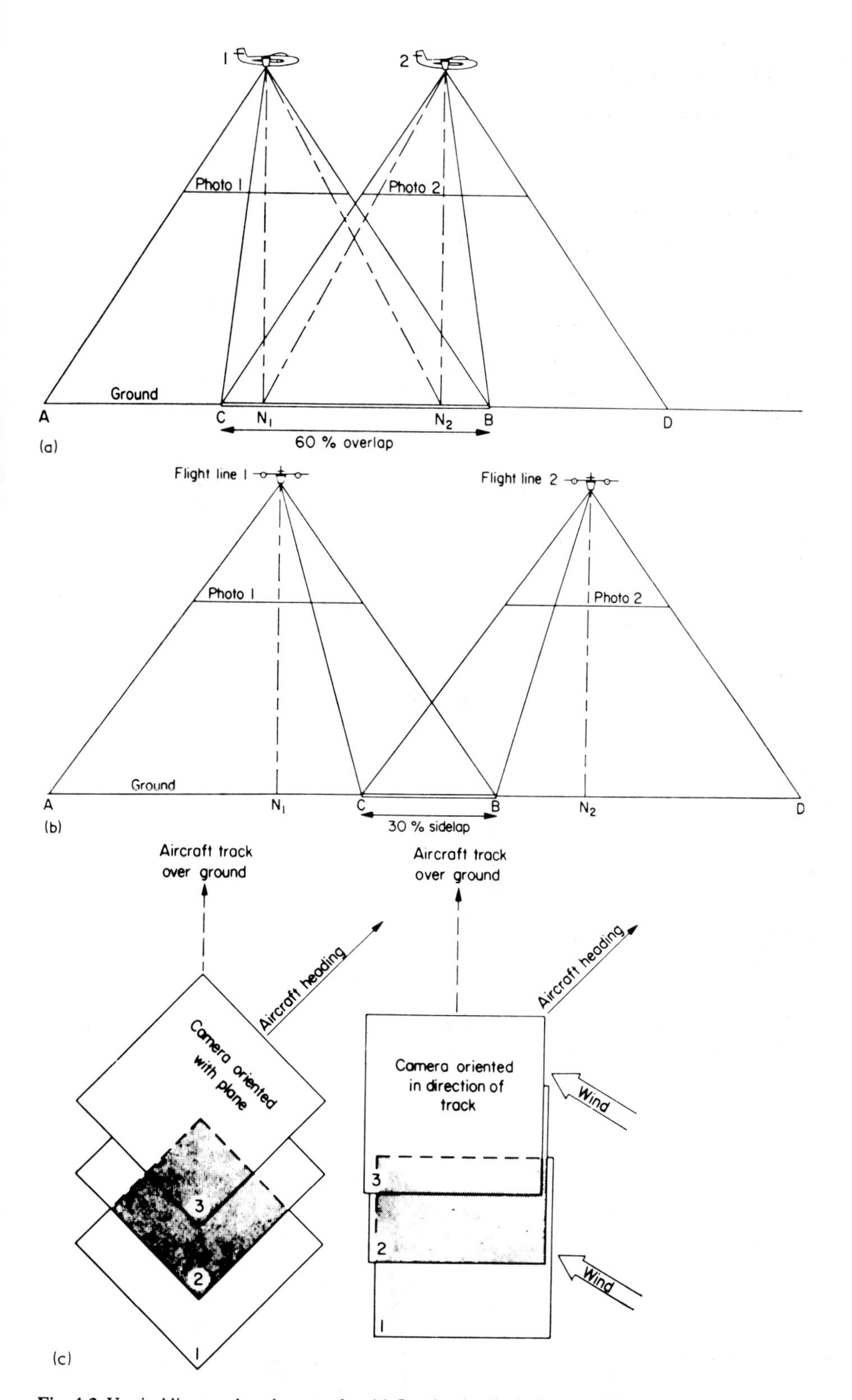

Fig. 4.2 Vertical line overlap photography. (a) Overlap (endlap) photographic coverage is obtained by photography at time intervals which provide coverage AB at position 1 and CD at 2. Overlap BC is 60% and the nadir points N_1 and N_2 are within the overlap area. (b) Sidelap (lateral) coverage is obtained by flight lines overlapping by 30%. (c) The effect of crabbing. Overlap areas are shaded to show reduced overlap when camera is oriented with plane and the correction when camera is rotated to the direction of the aircraft track. (Source: Curtis, 1973.)

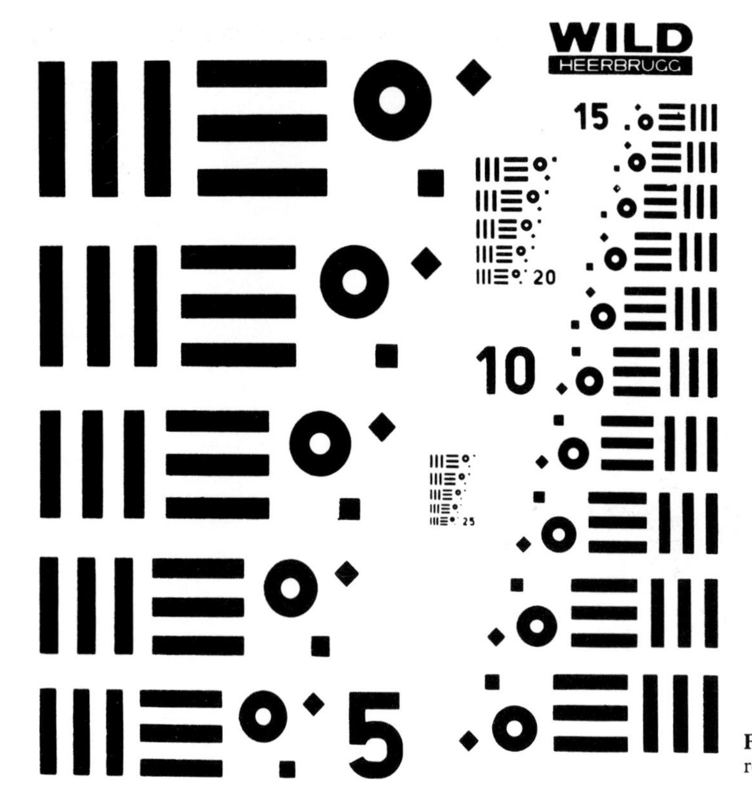

Fig. 4.3 Three-line resolution test to determine resolving power. (Source: Smith, 1968.)

Table 4.2 Summary of film characteristics (Source: Curtis, 1973)

Film	Disadvantages	Advantages
Panchromatic	Limited tonal range	Sharp definition Good contrast Good exposure latitude Inexpensive
Infrared	Limited tonal range Very high contrast Slight resolution fall-off (without correction for focal distance Difficult to determine correct exposure Loss of shadow detail	Vegetation evident Tracing of water courses facilitated Inexpensive
Colour	Expensive Special processing facilities necessary Diffusion of image under high magnification and slightly less definition than panchromatic	Excellent all-round interpretation properties owing to good contrast and great tonal range Good exposure latitude when used as a negative film Good-quality black and white prints can be produced from negatives
False colour	Expensive Special processing facilities necessary Diffusion of image under high magnification Critical exposure Low ASA rating Less tonal range than colour Duplicates expensive and difficult from positive film	Sharp resolution Superior rendering of vegetation and moisture

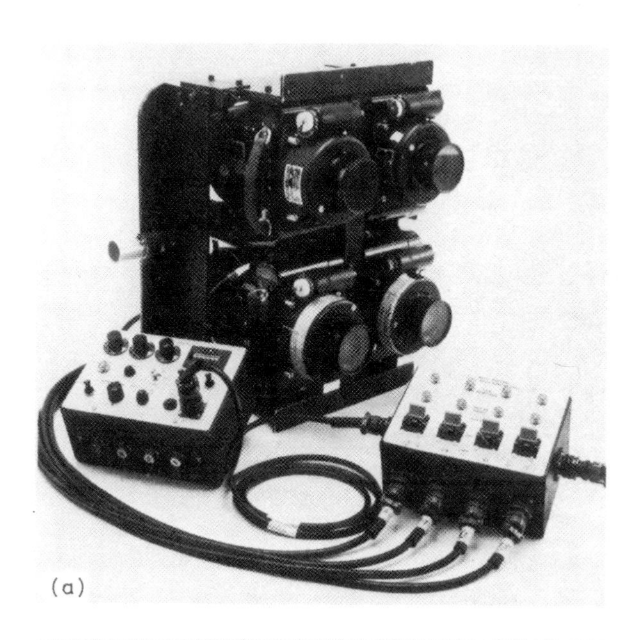

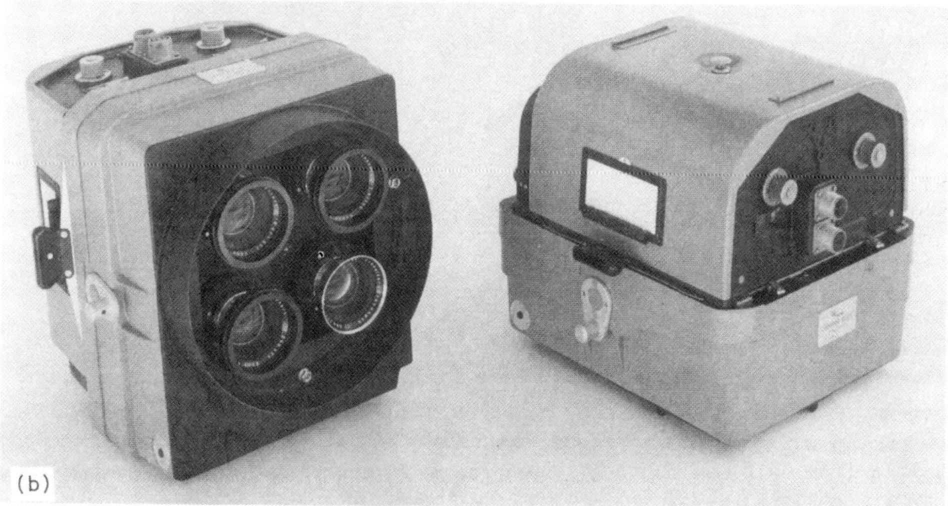

Fig. 4.4 (a) Fairey multispectral camera system consisting of four Vinten cameras with lens focal length of 102 mm (4 in) loaded with 70 mm film. (Courtesy, Clyde Surveys Ltd, Maidenhead, Berks.) (b) International Imaging Systems (I^2S) multispectral aerial camera. Four Schneider Xenotar 150 mm or 100 mm lens. Film capacity 242 mm in width by 76.2 m long. (Courtesy, John Hadland (PI) Ltd, Bovington, Herts.)

Where images are required at different wavelengths, multispectral photographs may be taken either by mounting several identical cameras with different film emulsions or using a special multispectral camera (Fig. 4.4(a) and (b)). In the latter, several identical lenses record their separate images on a common film. The four lens system is the most common arrangement, but the Itek camera has nine lenses. An example of multispectral photography by the I^2S camera system is given in Fig. 4.5. This photograph was obtained with Infrared Aerograph 2424 film using four filters, which provided images at 0.4–0.5 μm (1), 0.5–0.6 μm (2), 0.6–0.7 μm (3) and 0.7–0.9 μm (4). These separate images can be recombined and analysed as discussed in Chapter 7.

There is constant research aimed at improving the sensitivity and resolution of films, whilst reduc-

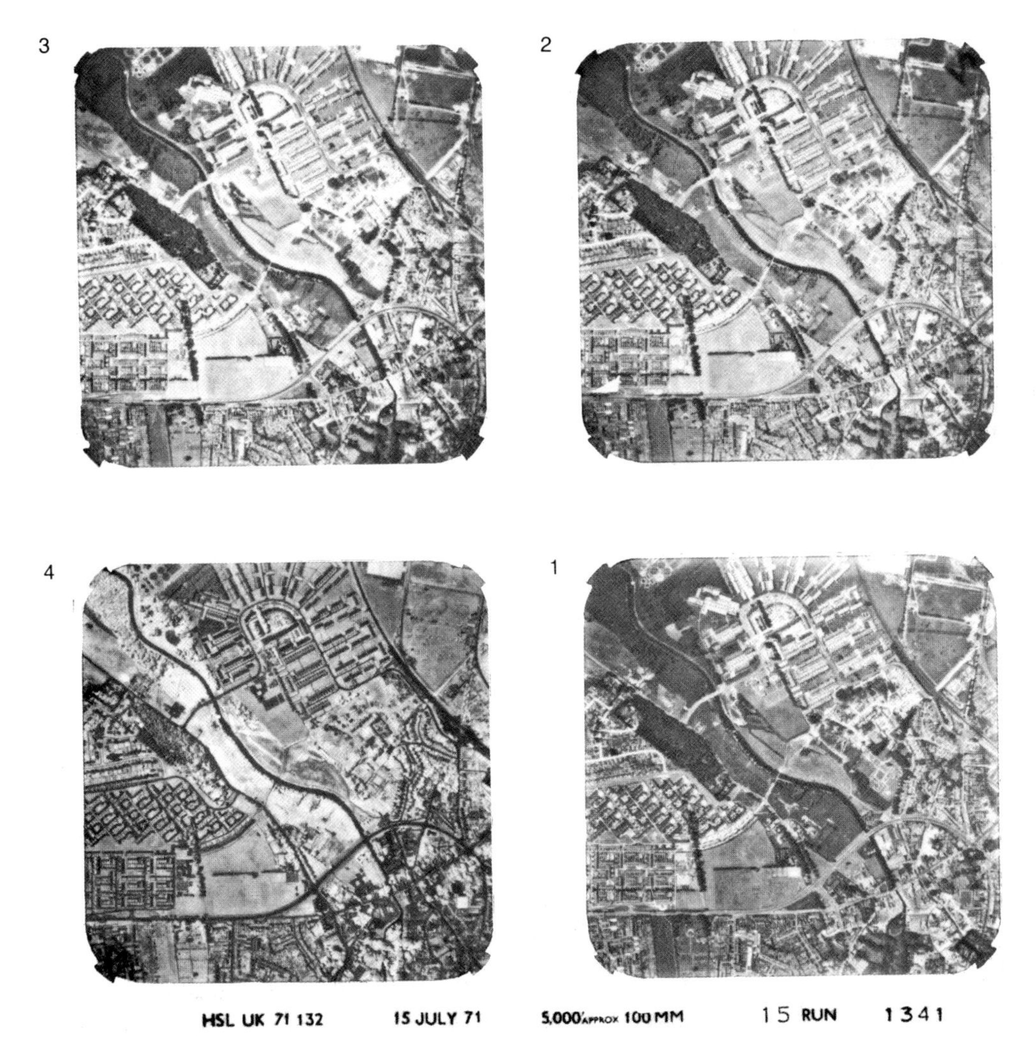

Fig. 4.5 Multispectral photograph of part of Thetford, Norfolk. Apart from the differentiation of deciduous and coniferous trees, Band 4 (infrared) also provides considerable additional detail concerning roof structures. Road details are better portrayed in Band 3 (red) at top left. A combination of Bands 3 and 4 offers advantages in studies of urban landscapes. (Courtesy, NERC.)

ing their thickness (weight). Sensitivity has been approximately doubled each decade since 1850.

The principal advantages of the photographic camera are its large information storage capacity, high ground resolution, relatively high sensitivity, and high reliability.

4.2.2 Space-borne photographic cameras

The main shortcomings of photographic surveys from spacecraft are that exposures can be made only in daylight, cloud obscures ground detail, and the photographic film cannot be reused. It has been estimated that, if scanning systems had not been used on Landsat, the film weight required to cover one year of filming operations to obtain 60 m ground resolution would be 300 kg. On board processing would have nearly doubled this weight, and improving the resolution to 20 m (the approximate resolution of Landsat TM data) would have increased the film weight to 3000 kg. Such weight factors must be considered alongside the difficulty

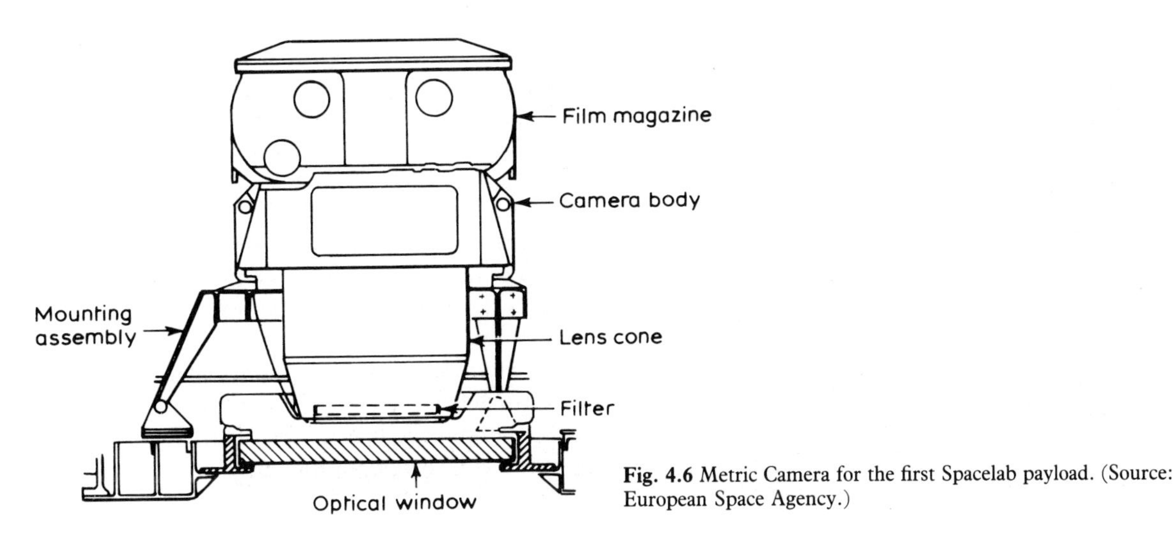

Fig. 4.6 Metric Camera for the first Spacelab payload. (Source: European Space Agency.)

Table 4.3 Operational characteristics for the Large Format Camera (Source: European Space Agency)

Parameter	Specification
Lens	
Focal length	30.5 cm
Aperture	f/6.0
Spectral range	400–900 nm
Image format	22.8 × 45.7 cm
Field of view	
Along track: degrees	73.7
altitude ratio	1.5 × H
Across track: degrees	41.1
altitude ratio	0.75 × H
Forward overlap modes	80, 70 or 60%

in transferring the information on the film to Earth. One method of transfer involves processing the film on the satellite and electronically scanning it for picture transmission to ground stations by broadband telemetry. The other method is by ejecting the film from the satellite for recovery in the air or on the ground. Capsules of 39 kg (air retrieval) and 136 kg (ground retrieval) have been used in this way with military reconnaissance satellites.

The Metric Camera, developed in Europe, together with the American Large Format Camera provide examples of recent sensors in the civilian domain which afford high resolution space photography. The Metric Camera configuration is shown in Fig. 4.6 and was first tested in the 'Metric Camera Experiment' (Chapter 5). Excellent images of the Middle East, North Africa, Pakistan, Afghanistan and West China were obtained by this camera. A catalogue together with a set of mocrofiches covering all acquired images are available at the UK National Remote Sensing Centre, Farnborough (see also section 5.7).

The Large Format Camera (LFC) was designed in the USA and made its maiden voyage in a Space Shuttle Challenger mission in 1984 (Chapter 5, Table 4.3 and Fig. 4.7). Although designed primarily for mapping, individual photographs have proved useful for many types of qualitative studies for Earth resources studies and environmental monitoring. For example, studies of the uses of the Large Format Camera photographs for coastal mapping showed that high resolution space photography has a role to play in change detection of coastal features. As the effects of global warming become apparent through raised sea-levels such observations may be important in planning sea defences. Recent cutbacks in funding of space observations have led to the suggestion that high-resolution cameras might be mothballed. However, such actions may be motivated partly by a desire to avoid too much high resolution imagery to be available to civilian as against military users. Undoubtedly there is continued interest in high resolution photography for cartographic purposes, especially for those territories that are large in extent and relatively inaccessible.

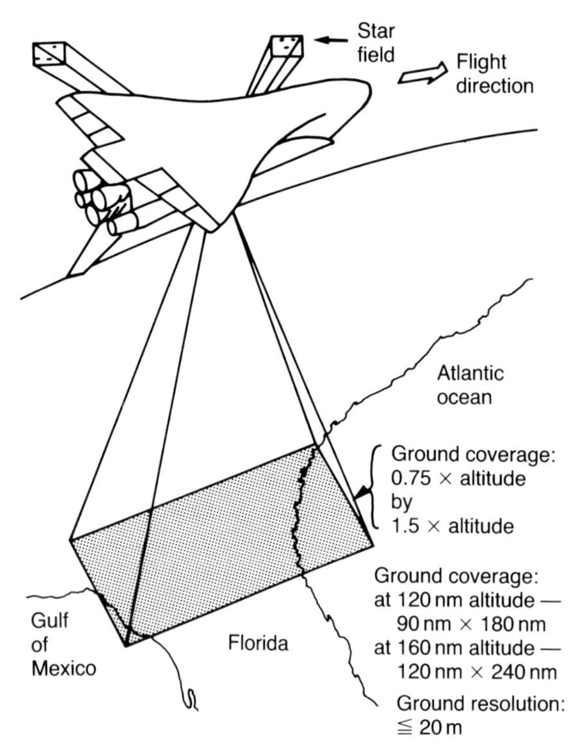

Fig. 4.7 Large Format Camera (LFC) payload ground coverage. (Source: NASA.)

The highest resolution photography that is commercially available is now provided for sale by Soyuzcarta, the Moscow based agency for Soviet low Earth-orbit imagery. Until recently it has been obtained by the Russian KFA-1000 camera, now superseded by the MK-4: this gives lower resolution (about 63% of that of the KFA-1000) but with greater ground coverage per frame. The ground resolution of the KFA-1000 images is 5–6 m using Spectrozonal film at 270–280 km altitude from Cosmos satellites. Examination of Earth photography shows that there has been progressive improvement in the resolution of the spaceborne photographic systems used, from about 125 m in respect of early Gemini photography in the mid-1960s to the 5–6 m of the recent Soyuz photographs.

4.2.3 *Vidicon cameras*

The development of the vidicon camera represented a major advance in the evolution of remote sensing systems used on both airborne and space platforms. Vidicon cameras can obtain very high resolution pictures in the wavebands 0.35–1.1 μm without the need for film replacement or converting a film image into a video signal (Chapter 7). The principle of operation of a vidicon is shown in Fig. 4.8. The optical image is focused on the photoconductive target and a charge pattern, which is a replica of the optical image, is built up on the target surface. This positive pattern is retained until the electron beam again scans the surface and restores it to equilibrium by depositing electrons. An analogue television signal is thereby created which is then amplified. In the case of the Return Beam Vidicon (RBV) camera the signal is derived from the unused portion of the electron beam. This returns along the same path as the forward beam and is amplified in an electron multiplier. The RBV tube provides greater sensitivity at low light levels and higher resolution than the ordinary vidicon tube.

Unlike the photographic film the slow scan vidicon photoconductive target may retain the previous image in vestigial form. Thus a ‘priming’ cycle is needed to allow complete discharge of the image and restoration of the target for the next exposure. In the Landsat 1 RBV camera this needed about 7 s and the complete cycle time could not be less than 11 s.

The vidicon camera can provide multispectral data either with a series of band pass filters on the optical system or with a number of vidicon sensors operating simultaneously to cover each band. This latter approach was used in Landsat 1, where three vidicon cameras covered different spectral ranges between 0.475 and 0.830 μm. The data transmission rate from this camera assembly was about 35 megabits s^{-1}. A ground resolution of 80 m was achieved with this system with a field of view of 185 × 185 km.

Landsat 2 carried a similar RBV system to Landsat 1, but Landsat 3 provided much improved resolution of c. 30 m. It also provided for one spectral band (stereo) in the range 0.5–0.75 μm with two cameras aligned to view adjacent ground scenes with a 14 km overlap giving 183 × 98 km per scene pair.

Vidicon cameras have proved extremely valuable for producing real time and near real time imaging of the Earth’s surface, especially in weather satellites. Experience with satellite borne television

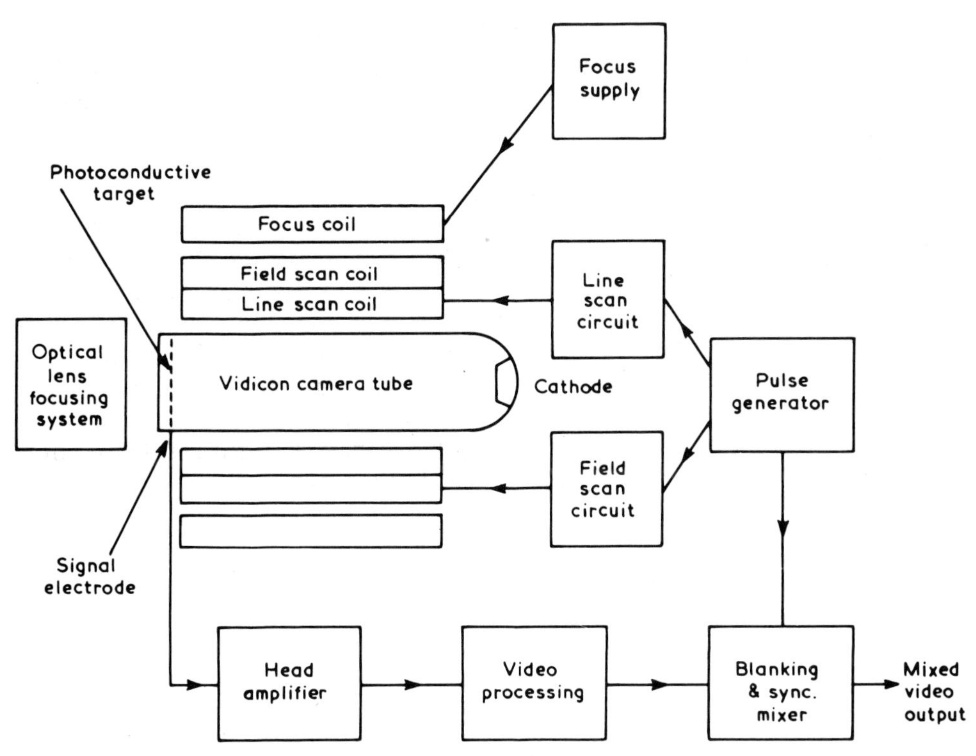

Fig. 4.8 Schematic diagram of a vidicon camera. The incoming radiation is focused on to the photoconductive target on which a replica of the optical image is formed. (Source: EMI, 1973.)

cameras carried in TIROS, Nimbus 1 and 2, the Ranger Moon shots, Apollo 8–16 and Mariners 1–9 led to great improvements in vidicon systems. In the Mariner 6 and 7 systems good quality images were obtained of the surface of Mars. One camera (A) gave a ground resolution of 1 km and coverage of 900 × 700 km whereas a second camera (B) gave a resolution of about 100 m in an area of 70 × 90 km.

4.3 Passive sensors outside the visible wavelengths

4.3.1 Infrared sensors

Sensors in the infrared can be classified broadly as 'non-imaging' or 'image-forming'. The non-imaging sensors include *radiometers*, *spectrometers* and may be considered first.

An infrared radiometer is a radiation measuring instrument with substantially equal response to a relatively wide band of wavelengths in the infrared region. It measures the difference between the source radiation incident on the radiometer detector and a radiant energy reference level. The source reflection results from reflected and scattered solar radiation and self emission from the Earth's surface and atmosphere.

A spectrometer permits selection and isolation of a desired wavelength (or band of wavelengths) in the infrared spectrum. This can be achieved by dispersing the incoming radiation by means of prisms, gratings, dichroic mirrors or filter. The dispersed radiation then can be measured in various wavebands using a number of detectors (see further discussion in section 6.4).

Detectors for use in scanning systems can be classed as thermal or photon detectors, but Earth resource scanners invariably use photon detectors because of their greater sensitivity. A selection of photon detectors can be used to cover all the high atmospheric transmission wavebands from 0.4 to 14 μm) (Table 4.4). Detector cooling is necessary to achieve satisfactory results for all detectors

Table 4.4 Optimum detector choice and waveband detectivities (Source: EMI, Department of Trade and Industry)

Cooling	Thermoelectric →			
	Radiative	→		
	Closed cycle engines and solid cryogens		→	
Temperature (K)	300	195	77	35
Waveband (μm)	*Detectivities* (W^{-1} cm $Hz^{1/2}$)			
1.5–1.8	(1) InAs 4.5×10^{10} (2) HgCdTe (3) InSb 1.5×10^{8}	InAs 7.5×10^{10} HgCdTe	InAs 1.5×10^{11}	
2.0–2.5	InAs 6×10^{9} HgCdTe InSb 2×10^{8}	InAs 2×10^{11} HgCdTe 1.5×10^{10} InSb 3.5×10^{9}	InAs 3×10^{11} InSb 3.5×10^{10} HgCdTe 2×10^{10}	
3.5–4.1	HgCdTe InSb 3×10^{8}	HgCdTe 2.5×10^{10} InSb 6×10^{9}	InSb 6.5×10^{10} (PV) InSb 4.5×10^{10} (PC) HgCdTe 4×10^{10}	
8–10	(TGS 6×10^{8}) (Thermistor 4×10^{8}		HgCdTe 3×10^{10} PbSnTe 7×10^{9}	GeHg 2.5×10^{10}
10–12	(Thermal detectors)		HgCdTe 2×10^{10} PbSnTe 9×10^{9}	GeHg 3×10^{10}

except for silicon (0.4–1.1 μm) and germanium (1.1–1.75 μm). Since long-term stability of the detector cannot be assumed it is necessary also to provide some built-in calibration system. Calibrating signals normally are obtained using tungsten lamp, diffused solar radiation, or controlled black body temperature sources.

In Table 4.4 the relative cooling ability obtainable by thermoelectric, radiative, closed cycle engines and solid cryogens are shown. Also the range of chemical elements that can be used for detectors in the range 1.5–12 μm are shown, together with the approximate cooling required for particular sensors, and for given detectivities.

Since the maximum achievable radiance resolution can be limited by the degree of available detector cooling, considerable attention has been given to cooling systems. Cooling down to 100–70 K is required for adequate resolution within the 8–13 μm waveband. Detectors used in aircraft usually use liquid coolant (liquid nitrogen), but for satellite multispectral scanners solid cryogen, closed cycle refrigerators or passive 'radiation to space' methods are used. With the solid cryogen a pre-cooled solidified gas is stored in a high performance dewar system and vented to space. Closed cycle refrigerators (e.g. Stirling) provide higher cooling capacity/weight/size ratios than solid cryogens but are demanding in terms of power requirements. Passive radiation cooling is an attractive proposition because of its lack of power requirements and long life. However, the cooling capacity of this system may be too low for some large detector arrays.

The very wide range of environmental uses of satellite radiometers is well demonstrated by the Advanced Very High Resolution Radiometer (AVHRR) carried on NOAA Polar Orbiting satellites, as discussed especially in later chapters of this book. However, it must be noted that infrared radiometers also are used extensively for local studies from aircraft and helicopter platforms, whether alone or in support of satellite investigations.

The image forming sensors include Infrared Linescan, Thermal Imager and Imaging Spectro-

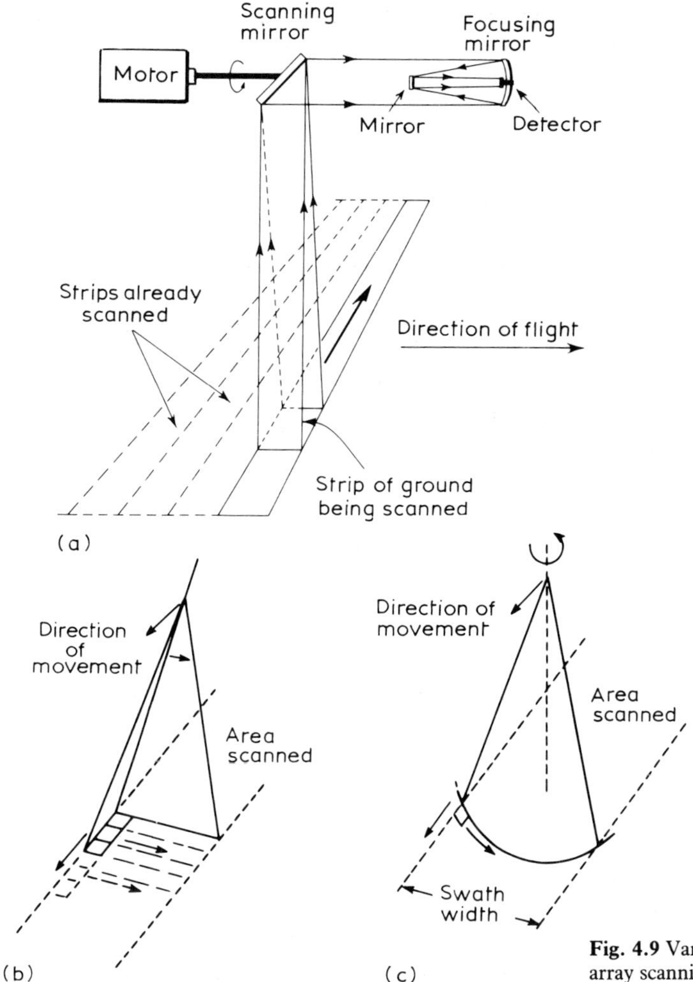

Fig. 4.9 Various forms of scanning: (a) the linescan system; (b) array scanning – side and forward motion; (c) conical scanning.

meter (section 4.3.3). The simplest form of imager is one that carries a single detector and scans a series of lines or narrow strips across the scene so that an image can be built up progressively as the platform carrying the instrument advances along its flight path or track. Special arrangements are required for systems of this type flown on geostationary satellites: these will be described in sections 5.5.3 and 5.10.

The infrared linescanner (IRLS) is a simple form of imager (Fig. 4.9(a)). Some systems allow for a set of, or an array of 'solid-state' sensors to be swept across the area (Fig. 4.9(b)) in 'pushbroom' fashion, and this can be advantageous in that it eliminates moving parts from the sensor, reducing the risk of serious instrument malfunction. In such systems each minisensor in the array functions as a separate unit, providing the information for one series of pixels in the resulting scene. The disadvantage of such a system is that each minisensor requires individual calibration. A third linescan type is conical scanning (Fig. 4.9(c)). This is advantageous because all the data therefrom are obtained the same distance from the satellite, and from the same viewing angle. This ensures that all pixels have the same size on the ground, and eliminates *limb-darkening* effects (the reductions in radiation observed from a target owing to atmospheric effects, which increase with increasing obliqueness of view from the satellite). It will be

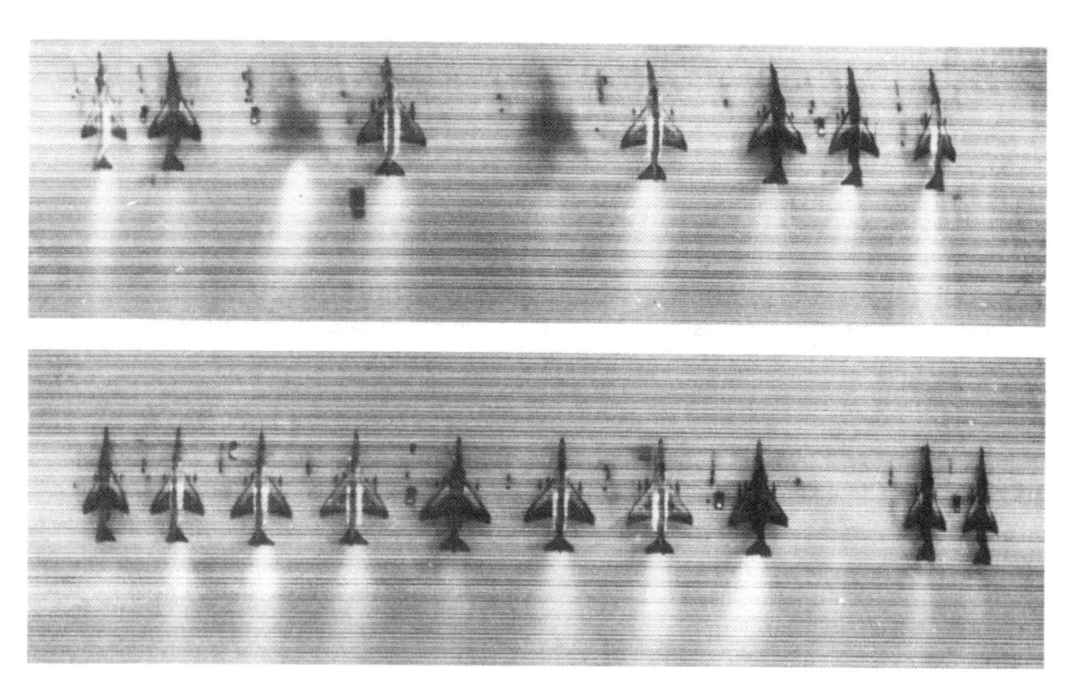

Fig. 4.10 Infrared linescan image of aircraft on an airfield. The heat from engines is shown in light tones. Note also the thermal shadows of recently departed aircraft at the top of the picture. (Courtesy, Hawker Siddely Dynamics Ltd, Hatfield, Herts.)

Fig. 4.11 Infrared linescan imagery of land near Mark Yeo, Somerset. Note shelter effect of hedges showing in light tones of area with higher temperature. Grazing animals are easily seen by reason of their high body temperature. (Courtesy, Royal Signals and Radar Establishment, Malvern. Crown Copyright.)

apparent in each case that scan rate must be related to the altitude and velocity of the sensor when assessing the performance of the system.

Infrared line scanning systems are used extensively from aircraft and helicopter platforms. For example the Infrared Linescan Type 212 manufactured by Hawker Siddeley Dynamics has a temperature resolution of 0.25°C at 0°C background temperature, thus surface temperatures can be detected to an accuracy of less than 1°C. When detector signals are processed from an image the resultant thermal maps can provide great detail concerning temperature variations. Two examples are shown in Figs 4.10 and 4.11 where the heat from aircraft engines and the warmer areas in the shelter of hedgerows can be easily recognized.

Typically, such airborne infrared line scanning systems generate data which indicate *relative* thermal energy differences between objects scanned. An experienced observer, therefore, can look at such data and say that Point A has a greater apparent temperature than Point B. However, the *absolute* difference in temperature can only be estimated in such cases.

To achieve absolute temperature relationships from the scanner data several additional constraints must be met. First, the video electronics must have DC (direct current) response to follow faithfully each thermal level variation from the detector. Secondly, radiation received by the detector must come predominantly from the area of interest being sensed. Finally, the complete electronic chain from detector through to final recording must be 'drift stabilized' so that the system responsivity, V/W (volts output/watts input), is constant.

The Daedalus DS-1230 is a quantitative scanner fulfilling the absolute temperature requirements set out above. This scanner provides high spatial resolution using a detector size of 1.7 milliradians combined with temperature resolution of 0.2°C.

The linear array or 'pushbroom' system now deployed on SPOT satellites is described in Section 5.8. Suffice to say here that this has provided higher resolution imagery than any other civilian satellite-borne radiometer or spectrometer down to about 10 m on the ground (see Fig. 4.13) and this system will be used increasingly in the future.

Thermal imagers produce images very similar to those obtained by linescanners, but with sensors in this case providing scanning in both dimensions. A high quality display can be obtained by using an array of detector elements matched to a suitable scanning system, as shown in Fig. 4.12. This is the so-called *banded scan* system in which the master scan is generated by a rotating drum with progressively angled facets. This type of sensor is essential for geostationary satellites where there is no vehicle motion relative to the Earth (see Chapter 5). In any case, as scanning height increases the thermal imager may become preferable to the simple line scan in that it can be matched more efficiently to the scan angle to be covered, and its scanning rate can be controlled to provide the desired compromise between sensitivity, resolution and scanning frame time. This type of sensor lends itself readily to the use of a range of detector elements, thereby increasing sensitivity for particular conditions of optics and scan rate. The HCMM and CZCS radiometers (sections 5.11.1 and 5.11.3) are examples of space systems using infrared.

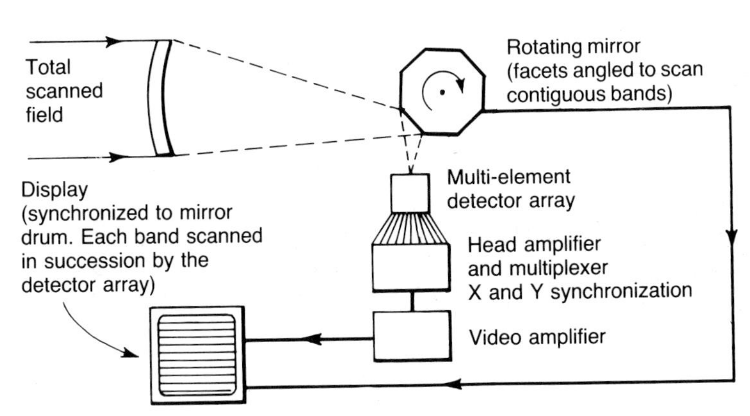

Fig. 4.12 Thermal imager system – block schematic. (Source: Laird, 1977.)

Fig. 4.13 SPOT HRV (panchromatic) scene of the east London docklands and neighbouring areas, 30 June 1986. (Copyright, CNES.)

4.3.2 Microwave radiometers

Beyond the infrared wavelengths, wave energy in the range 1–100 mm forms the basis of remote sensing by microwave radiometry. Microwave measurements have been part of the remote sensing programme in space since its beginning, though interest in their value has risen rapidly in recent years. This is illustrated by Table 4.5, which gives a history of microwave sensors in space. The instruments listed in this table can be separated into two categories, namely: *imagers* of surface features, and *sounders* of atmospheric properties such as temperature or water vapour. For example, the Nimbus-5 Microwave Spectrometer (NEMS), Scanning Microwave Spectrometer (SCAMS) and Microwave Sounding Unit (MSU) were microwave sounders designed to obtain profiles of atmospheric temperatures, although the Nimbus-5 NEMS and SCAMS also obtained ancillary information on features such as rain, snow and ice coverage.

Among the earliest of the imaging passive microwave radiometers was the Electronically Scanned Microwave Radiometer (ESMR), a single fre-

the single frequency sensors was the Scanning Multichannel Microwave Radiometer (SMMR). The SMMR was a multifrequency imaging device, with five frequencies each dual polarized, scanning at a fixed angle of about 50° at the Earth. Today's successor is the Special Sensor Microwave Imager (SSM/I) on USA military satellites (see Chapter 10).

Table 4.6 illustrates the fact that for a large number of applications it is essential to make measurements in multiple frequencies. Multiple-frequency microwave imaging of the Earth's surface has proved successful because, by analogy with multispectral photography, the increased information provided by simultaneous measurements at different frequencies makes it possible to detect additional phenomena and to isolate relevant data from other surface or atmospheric radiation effects by such means.

Although the naturally emitted microwave radiation intensities are much lower than those in the infrared (Fig. 3.1), resulting in poorer brightness

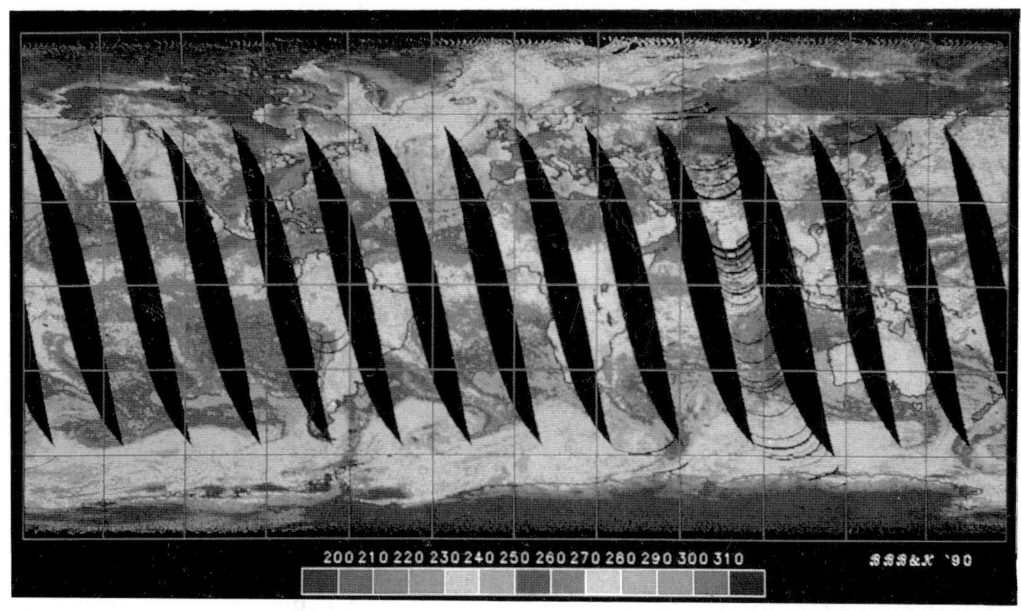

(a)

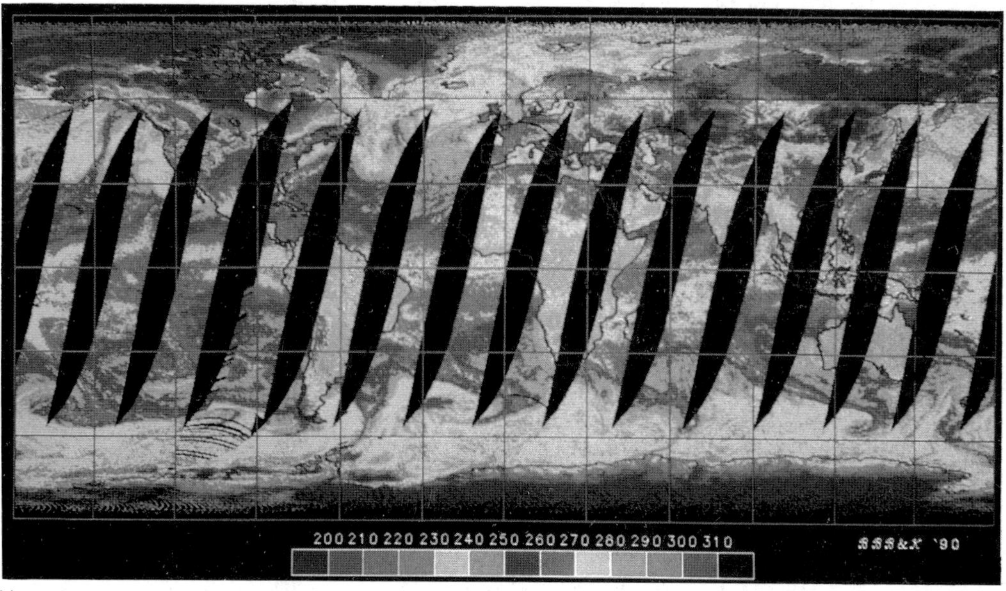

(b)

Fig. 4.14 Global mosaics of passive microwave imagery from the 85 GHz (vertical polarization) channel on SSM/I instrument for 20 February 1988. This instrument was first flown in 1987 on the DMSP F8 satellite. DMSP is a near-polar orbiting satellite family permitting night-time imaging along its *ascending* tracks (Fig. 4.14a), and day-time imaging along its *descending* tracks (Fig. 4.14b). Note the different swath patterns in these two cases, and the higher emissions from land areas especially in the southern (summer) hemisphere by day, and from sea areas in the tropics where atmospheric moisture contents are highest.

temperature resolution, the longer wavelengths have the advantage that they allow sensing through cloud cover.

The microwave radiometer is made up of a directional aerial, a receiver (for selection and amplification) and a detector. A block diagram of a typical microwave system is shown in Fig. 4.15. The ground resolution achieved by microwave radio-

Table 4.5 History of microwave sensors in space (Courtesy, NASA)

Year	Spacecraft/ instrument	Spectral regions										Resolution nadir (km)[b]	Frequencies (GHz)
		1.4	6	10	18	21	37	50–60[a]	90	160	183		
1962	Mariner				X	X						1 300	15.8, 22.2
1968	Cosmos 243	X		X								37	3.5, 8.8
1970	Cosmos 384					X	X					13	
1972	Nimbus-5												
	ESMR				X							25	19.35
	NEMS					X	X	X(3)				180	22, 31.4
1973	Skylab												
	S-193			X								16	13.9
	S-194	X										115	
1974	Meteor						X						
1975	Nimbus-6												
	ESMR						X					20 × 43	
	SCAMS					X	X	X(3)				150	22, 31
1978	DMSP												
	SSM/T							X(7)				175	
1978	Tiros-N												
	MSU							X(4)				110	
1978	Nimbus-7												
	SMMR		X	X	X	X	X					18 × 27	(Data at 50 km)
	Seasat												
	SMMR		X	X	X	X	X					22 × 35	
1987	DMSP												
	SSM/I				X	X	X		X			16 × 14	19.3, 22, 85.5
1991	SSM/T-2								X	X	X	50	150
and	NROSS												
after	LFMR		X	X								15 × 25	6
	NOAA												
	AMSU-A					X	X	X(12)	X			50	24, 31
	AMSU-B								X	X	X(3)	15	
	Eos												
	ESTAR	X										10	
	AMSR		X	X	X	X	X		X			1.5	

[a] Superscript indicates number of channels.
[b] At highest frequency.

meters depends on the aerial size and the orbiting altitude (Fig. 4.16). Since it is important that aerial size should be limited in order to avoid undue weight or distortion of the aerial array the lowest altitude orbit should be used for maximum resolution.

The principal characteristics of passive microwave radiometer (PMR) systems are, therefore:

1. relatively low spatial resolution capability of the order of 10 km or more;
2. wide range in the observed surface emissivity;
3. nearly all weather capability.

The emissivity of an observed body changes with observation angle, polarization, frequency and surface roughness. An example of an object with low emissivity is a smooth water surface for which $\varepsilon = 0.4$ when observed at normal incidence. Rough soil on the other hand has an emissivity close to unity. As in other situations, it is convenient when considering the measurements made by microwave radiometers to use the concept of *brightness temperature*, T_B (p. 27).

Both natural and man-made surfaces present a large range of microwave brightness temperatures.

Table 4.6 Parameters measured with microwave sensors and the frequencies at which the measurements should be made (Courtesy, NASA)

Parameter	Frequency of observation (GHz)									
	1.4	6	10	18	21	37	50–60	90	160	183
Soil moisture	●	○								
Snow		○	○	●		●		◒		
Precipitation										
Ocean			◒	●	○	◒				
Land				◒		●		●		◒
Sea-surface temperature		●	◒	◒	◒	○				
Sea ice										
Extent				●		●		○		
Type		○	◒	●		●		◒		
Wind speed (sea surface)			●	◒	○	○				
Water vapour										
Total (over ocean)				●	●	◒				
Profile					◒	○	◒	○	◒	●
Cloud water (over ocean)					◒	●		◒		
Temperature profile					○	○	●	◒		

Key: ● necessary, ◒ important, ○ helpful.

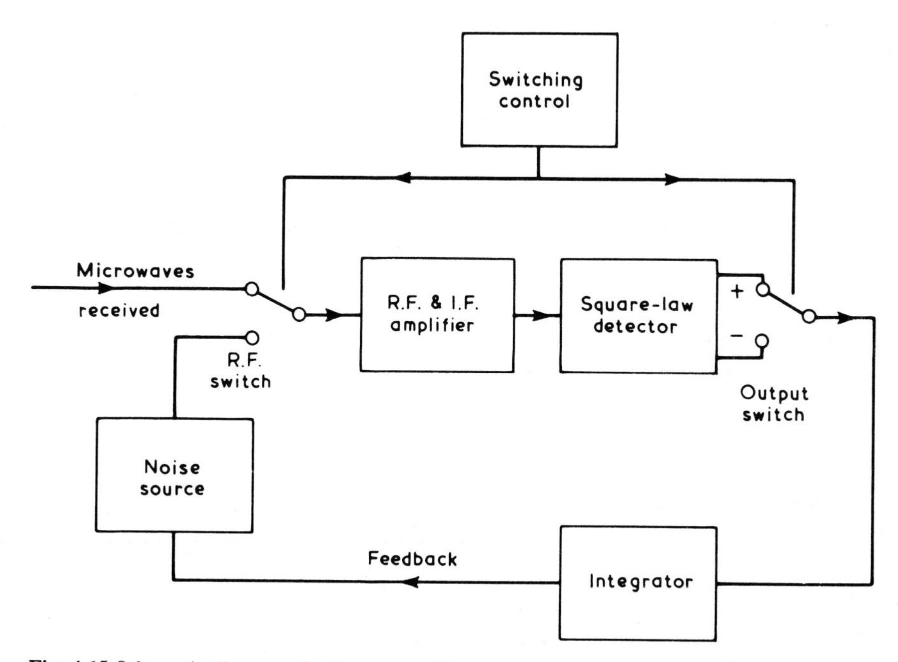

Fig. 4.15 Schematic diagram of microwave radiometer system.

Whereas for a black body the brightness temperature is equal to the physical temperature and ε is unity for all incidence angles, natural objects (grey bodies) show variations in emissivity depending on incidence angle, wavelength and the degree of surface roughness. Where surfaces are very rough the brightness temperature is independent of angle of incidence, but if roughness is limited T_B will become lower as the angle of incidence increases.

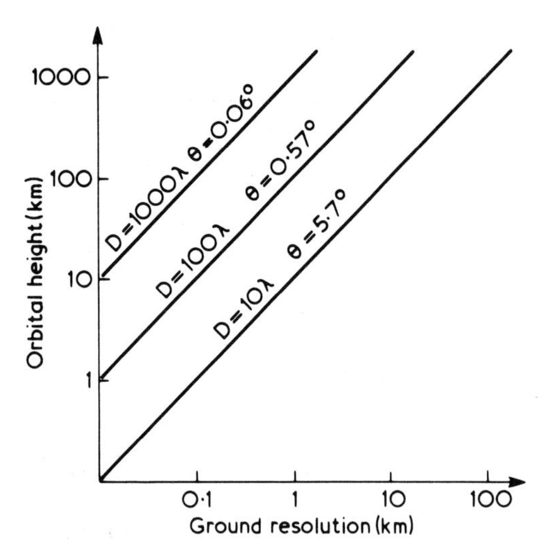

Fig. 4.16 The relationship between ground resolution, orbital height and effective aerial diameter, *D*, for microwave radiometers. The aerial beamwidth is given by θ. (Source: EMI, 1973.)

radiometry concerns the depth to which ground properties contribute to the observed brightness temperature (i.e. the 'penetration depth'). The penetration depth depends on the wavelength used and the dielectric properties of the material. Typical values of the penetration depth are 20 wavelengths for asphalt and sand and only 0.5–0.1 wavelengths for water. Thus wet materials provide little penetration whereas dry substances, e.g. desert sands may show substantial (approximately 1 m) penetration. The wavelength dependence of the penetration depth makes profiling of the surface layer possible by using multifrequency radiometry.

Another important parameter characterizing the radiation properties of an object is the *polarization*, i.e. the distribution of the electrical field in the plane normal to the propagation direction. As a rule the radiometer is only sensitive to the field in one direction. Black body radiation is completely unpolarized, but the emission of many natural features shows pronounced polarization effects which can be useful for identifying the nature of the feature.

In consequence of the above the measured radiation of a surface feature depends on the following characteristics of the microwave system:

1. polarization direction of the radiometer;
2. observation angle related to the surface plane of the object;
3. frequency (wavelengths) of bands used.

The radiation received by the microwave system is also influenced by characteristics of the material that is being sensed. The important characteristics are:

1. electrical and thermal properties of the material;
2. surface roughness and size of the object;
3. temperature and its distribution in the body.

Notwithstanding such problems and complications, passive microwave systems can tell us much about the environment which cannot be deduced from other types of passive remote sensing data. Table 4.7 lists some of the uses to which SSM/I data are now being put. This confirms the great flexibility of such systems – whose future use will be enhanced greatly as and when new radiation collection methods are developed enabling the spatial resolutions of the data to be improved significantly.

Indeed, the chief limitation in the use of passive microwave sensing from space for some environmental applications is the low resolution of the system from orbital altitudes. This is perhaps the chief area in which progress has yet to be made in the evolution of microwave remote sensing.

4.3.3 Absorption spectrometers

Spectroscopic techniques can be applied to various types of passive sensors leading to instruments commonly referred to as spectrometers. Essentially, spectrometers are instruments used to determine the *wavelength distribution* of radiation (Fig. 3.5). This can be done by dispersing the radiation spatially according to wavelength. The prism spectrometer relies on the dependence of the index of refraction of various prism materials on the wavelength of radiation entering the prism. The grating spectrometer achieves dispersion by diffraction and interference. One of the most popular grating spectrometer systems used today was devised by Ebert some 60 years ago. Scanning of the spectrum is achieved by applying a sawtooth type of oscillatory motion to the grating.

A rotating, circularly variable filter may replace the prism or grating and thereby gain an increase

Table 4.7 Primary environmental products prepared from SSM/I-1 data by the US Navy (Courtesy, US Naval Research Laboratory)

Parameter	Geometric resolution (km)	Range of values	Quantization levels	Absolute accuracy
Ocean surface wind speed	25	3–25	1	$\pm 2\,m\,s^{-1}$
Ice				
Area covered	25	0–100	5	±12%
Age	50	1st year, multiyear	1 year, >2 years	None
Edge location	25	N/A	N/A	±12.5 km
Precipitation over land areas	25	0–25	0, 5, 10, 15 20, ⩾25	$\pm 5\,mm\,h^{-1}$
Cloud water	25	0–1	0.05	$\pm 0.1\,kg\,m^{-2}$
Integrated water vapour	25	0–80	0.10	$\pm 2.0\,kg\,m^{-2}$
Precipitation over water	25	0–25	0, 5, 10, 15, 20, ⩾25	$\pm 5\,mm\,h^{-1}$
Soil moisture	50	0–60	1	None
Land surface temperature	25	180–340	1	None
Snow water content	25	0–50	1	±3 cm
Surface type	25	12 types	N/A	N/A
Cloud amount	25	0–100	1	±20%

in the rate at which incoming radiation can be scanned. When a circular filter of this kind is used with a rotating beam modulator (chopper) and an internal black-body reference source, calibration of the spectrometer measurements can be achieved (Fig. 4.17). Commercial circularly variable spectrometers are available with the capability of interchanging detectors. These instruments allow coverage of waveband frequencies in the range from 0.35 to 23 μm and can accommodate both cooled and uncooled detectors. Infrared spectrometers have been used to obtain atmospheric profiles – generally values of temperature and water vapour as a function of height. The first such infrared sensor that demonstrated the feasibility of remote profiling of the atmosphere was the Satellite Infrared Spectrometer (SIRS), which used an Ebert type grating spectrometer. Today's much more advanced vertical sounders cover many infrared bands, with high temperature accuracies (better than 1 K), useful relative humidity accuracies (better than 10%), and very good to total ozone content sensitivities (about 0.01 cm).

In the context of airborne remote sensing, Airborne Imaging Spectrometer (AIS) data are now being used in mineral exploration. The AIS has enabled geologists to collect data in a continuous spectrum and carry out spectral analysis to determine individual absorption band characteristics.

The use of absorption spectrometry in remote gas sensing depends on the unique spectral signature of each gas (Fig. 4.18). At present, spaceborne applications of remote gas sensing have been mainly limited to meteorological studies and to Martian atmospheric studies. However, the principles and instruments described can also be applied to many features, e.g. the detection of pollutants and emitted gases and vapours from Earth sources. The development of space instruments specifically for the detection of Earth pollutants or Earth resources is now being actively pursued. New instruments are being built, and further technological development, should enable more of the applications listed in Table 4.8 to be investigated (see also Chapters 3 and 19).

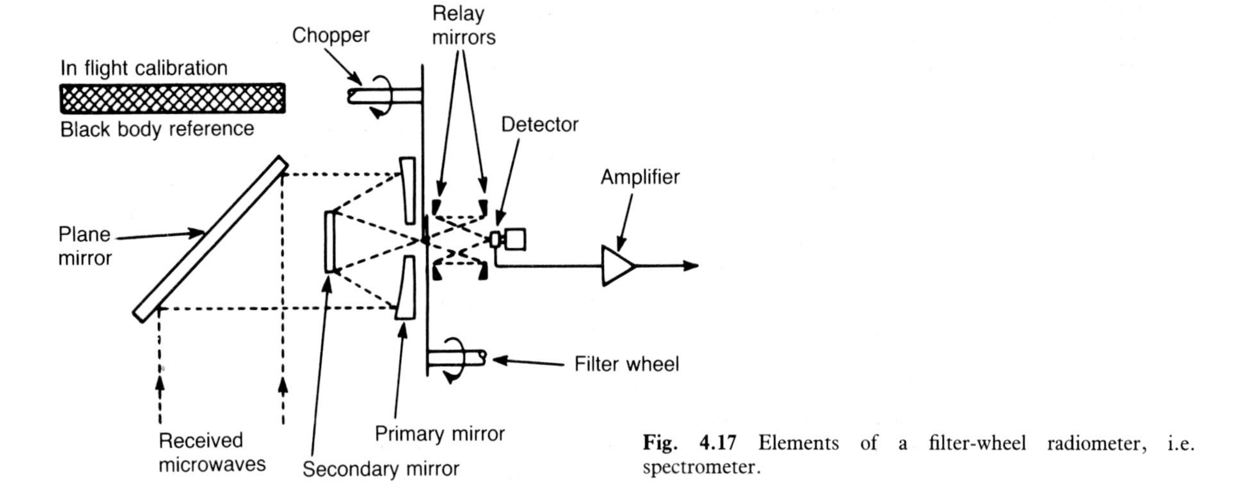

Fig. 4.17 Elements of a filter-wheel radiometer, i.e. spectrometer.

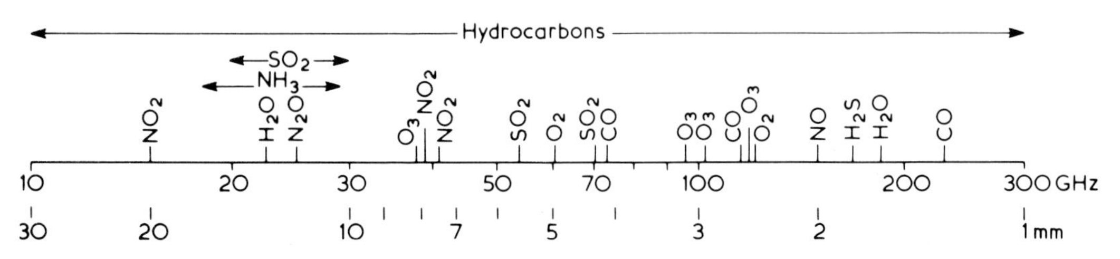

Fig. 4.18 Principal absorption frequencies of relevant gases in the microwave region. (Source: EMI, 1973.)

4.4 Active sensors outside the visible wavelengths

4.4.1 Microwave radar

The microwave systems discussed so far have been *passive* systems, that is those receiving natural emissions. It is possible, however, to devise an *active* sensing system in which waves are propagated near the sensor and are bounced off the Earth surface to be recorded on their return. This is the essence of the Radar (radio direction and ranging) system. It was first operated in the VHF radio band (30–300 MHz) but later improvements have led to the use of microwave radar (300 MHz–100 GHz).

Airborne imaging radars fall into two general categories, namely *real aperture* radar (RAR) and *synthetic aperture* radar (SAR). Real aperture radars are commonly associated with the Side-Looking Airborne Radar (SLAR) which is an airborne sensor for displaying the back-scatter characteristics of the Earth's surface in the form of a strip image of a selected area. Ground-based radars are described in Chapter 12.

High resolution RARs are obtained by means of a fixed antenna. The length of the antenna is conditioned by the size of the aircraft or spacecraft and its effect on the flight performance of the craft. Real aperture radar requires a high power output, hence RARs are sometimes referred to as 'brute force' systems. The requirements usually can be met with airborne systems but they are difficult to satisfy for satellite systems. Thus synthetic aperture radar systems (SAR) have been designed as a means of avoiding this problem.

Images of landscape derived from airborne side-looking radar resemble air photographs with low angle sun illumination in that shadow effects are produced (Fig. 4.19). This enhances the im-

Table 4.8 Applications of remote gas sensing (Source: EMI, Department of Trade and Industry, 1973)

Discipline	Applications	Ground resolution desirable[a]
Pollution	Determination of total vertical amount and vertical distribution of atmospheric gases (both man-made and natural). Can then determine Global pollution background Regional pollution dispersion and circulating patterns Spatial and temporal variations of pollutants over urban, rural and ocean areas Sink mechanisms Atmospheric chemical mechanisms Pollution flows across international boundaries	L
	Mapping of small area pollutant sources, e.g. paper/sand/pulp mills, iron and steel plants, petroleum refineries, smelters and chemical plants, housing estates (where fuel is burnt for domestic purposes), busy road junctions (CO)	H
	Monitoring residues from large scale spraying of pesticides	M
	Mapping of small area pollutant sources, e.g. paper/sand/pulp mills	M
Detection of Earth resources	Detection of metals from vapours emitted (e.g. arsenic, cadmium, zinc, sodium, mercury)	H
	Detection of micro gas seeps (especially NH_3) associated with natural gas fields	H
	Detection of trace gases emitted from oxidizing ore deposits	H
	Detection of iodine (iodine occurs in high concentrations in water in petroleum bearing strata, and is also associated with marine life which concentrates in primary fish food areas in oceans)	M
	Detection of chlorine (chlorine is indicative of bioproductivity in the oceans)	M
Other disciplines	Detection of volcanic activity by detection of SO_2	H
	Determination of the composition of volcanic gases (enables predictions to be made of the composition of the molten mantle)	H
	Monitoring of volcanic emissions of SO_2 and H_2S to give warnings of eruptions	H

[a] L, low ground resolutions (10 km); M, medium ground resolution (2 km); H, high ground resolution (0.1–0.5 km).

pression of morphological relief in the imagery but also leads to some obscure areas where the radar shadows occur. The geometry of radar pictures is entirely different from that of conventional air photographs. This is because radar is essentially a distance ranging device which consequently produces lateral distortion of elevated objects, as shown in Fig. 4.20(a).

The basic elements of a real aperture side-looking airborne radar system (SLAR) are shown in Fig. 4.20(b). A long aerial mounted on the airborne platforms scans the terrain by means of a radar beam (pulse). The radar echoes are recorded by a receiver and processor so that a cathode ray tube (CRT) glows in response to the strength of each echo. The resultant CRT display is then recorded on the film. These radars can be subdivided into six variants:

1. Monofrequency. Monopolarized, transmitting a radar carrier polarized in one plane, usually horizontal.

Fig. 4.19 Radar imagery of the Malvern Hills. (Courtesy, Royal Signals and Radar Establishment, Malvern. Crown Copyright.)

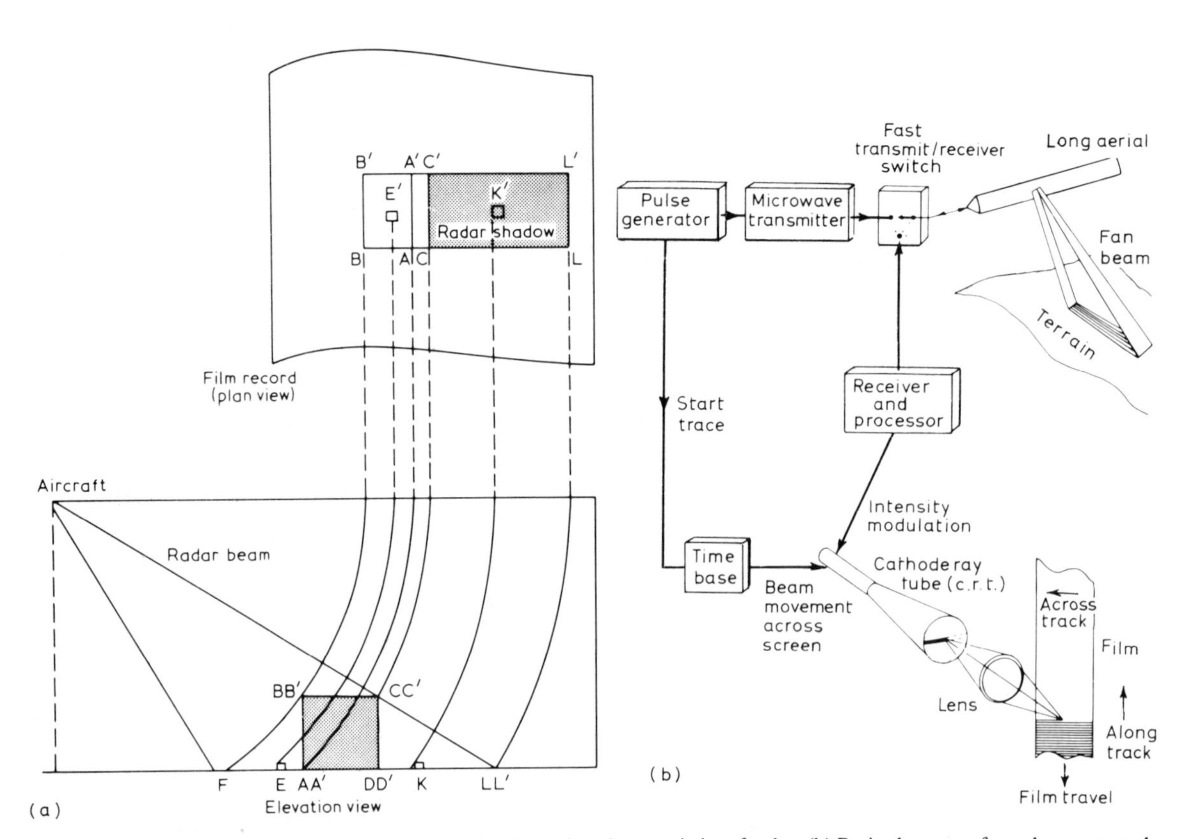

Fig. 4.20 (a) Image distortions due to the directional and ranging characteristics of radar. (b) Basic elements of a real aperture radar system (side-looking airborne radar). (Source: Grant, 1974.)

2. Multipolarized. This normally transmits a horizontally polarized electromagnetic wave but can receive horizontally and vertically polarized waves separately.
3. Circular polarized. This transmits a circularly polarized wave and can receive right-hand or left-hand circularly polarized ground reflections.
4. Multifrequency. This transmits two or more modulated radar carriers at widely differing wavelengths to show up differences in penetration and scattering from various surfaces.
5. Panchromatic. This transmits a broad-band carrier in order to minimize diffraction effects (speckling).
6. Polypanchromatic (a multifrequency panchromatic radar).

In most real aperture side-looking radar (SLAR) systems the resolution across the film record is up to ten times better than that achieved in the along track direction. The latter can be improved only by increasing the size of the antenna used. There is a limit, however, to the size of the antenna that can be mounted on aircraft and spacecraft.

In fact, further improvements in real aperture resolution became increasingly hampered by limitations in:

1. millimetric radar technology, where it becomes difficult to acquire components for engineered radars at wavelengths less than 8 mm.
2. mechanical engineering tolerances on antenna manufacture. Even at 8 mm the demanded tolerances are approximately 0.1 mm over a 5 m aperture. Furthermore, these must not only be achieved in manufacture but also maintained in flight in the face of temperature and vibration effects.

So it was that attention was drawn to the technique of *aperture synthesis* whereby the movement of the aircraft (or spacecraft) along its track can be used to form a long synthetic aerial (Fig. 4.21). This system depends on a particular method of processing the Doppler information associated with the radar returns, and is termed the 'synthetic aperture' system.

4.4.2 *Synthetic aperture radar*

The synthetic aperture system is most easily understood by analogy with more familiar optical

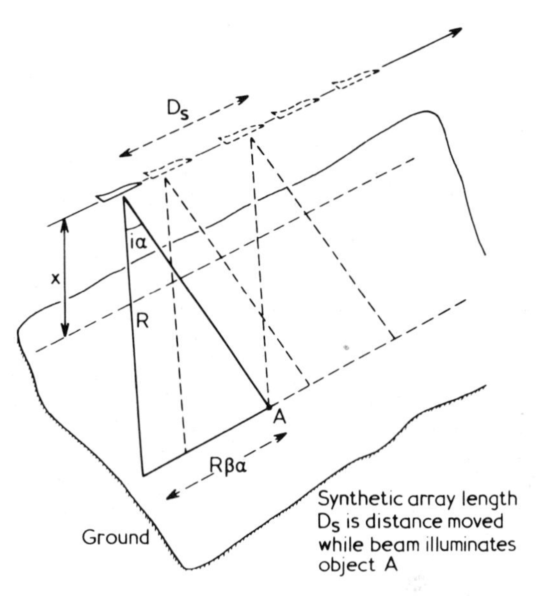

Fig. 4.21 Schematic diagram to show elements of a synthetic aperture radar system. (Source: Grant, 1974.)

concepts. If one imagines that a plane wave of monochromatic light falls on a narrow slit the wave on the far side of the slit will diverge to form the familiar Fresnel zone pattern for a slit (Fig. 4.22). The Fresnel zones on the screen are alternately dark and light in tone and are caused by wave interference patterns. The light wave amplitude that causes this observed variation can be shown to be both positive and negative. This represents the variation in the *phase* of the wave fronts at these points relative to the *phase* of the wave arriving at the zero point, i.e. the centre of the pattern on the screen.

It would be possible to place a film at the screen B and record the complete intensity image. Alternatively the same photograph could be obtained by traversing a narrow slit across the film (Fig. 4.22(a)) to expose it a piece at a time. If the phase of the waves could be added back it would then be possible to reconstruct an image of the slit A by shining a monochromatic (laser) light through the film record.

Now if one compares this example with the synthetic aperture system radar the slit A can be regarded as a small part of the ground terrain. The slit C (Fig. 4.22(b)) is the recording system (CRT) for one item of information obtained by the radar

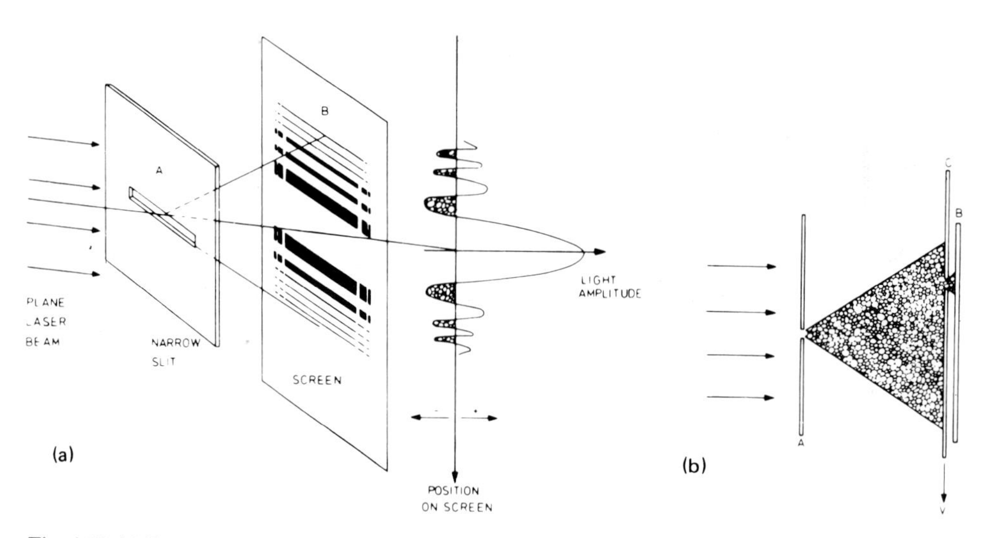

Fig. 4.22 (a) Fresnel zones appearing on screen and diagrammatic appearance of the phases of the waves. (b) Slit (C) moved across the face of the film (B). (Source: Grant, 1974.)

receiver. The screen B (Fig. 4.22(b)) is the complete film record obtained from the CRT display for the pulse striking the ground element represented by slit A.

If the image of the element (A) is to be reconstructed it is necessary to record both the phase and the amplitude of the signals received. This is achieved by carrying a very stable reference oscillator in the aircraft or spacecraft and comparing the phase of the return at each position of the sampling slit C with the phase of the stable oscillator. A radar with this type of stable oscillator is known as a 'coherent' radar and it is an essential part of the synthetic radar system.

The SLAR system acquires an oblique view of the terrain. Subsequent processing provides the interpreter with an image in which the radar geometry has its effect. The images are presented in one of two modes, either slant range or ground range. Slant range and ground range characteristics are shown in Fig. 4.23. Slant ranged imagery is related to a timed pulse to and from a target, whereas ground range imagery is heavily dependent on the height of the platform above the ground. If the terrain were absolutely flat a constant factor could be applied, but in practice relief is often uneven and variations in aircraft altitude result in changes in height above ground. The occurrence of relief in the target area introduces a distortion known as *foreshortening*. In such circumstance slopes facing the radar will be shortened relative to their true projected distance (Fig. 4.24). The amount of foreshortening is related to the incidence angle of the radar and so slopes in the near range will be foreshortened more than those in the far range. Thus radar images, even when corrected to give a pseudo-vertical view, have geometric variations across the image dependent on both the incidence angle and the shape of the terrain.

The Seasat mission launched in 1978 carried a Synthetic Aperture Radar (SAR) with a frequency of 1.37 GHz providing a ground resolution of 25 m and a swath width of 100 km (see section 5.11.2).

The *advantages* of side-looking radar compared with conventional aerial photography may be summarized as follows:

1. it has all weather and day/night capability;
2. it is capable of measuring surface roughness and dielectric properties;
3. it provides active illumination, with ability to select wavelength.

The *disadvantages* of this radar system can be:

1. its poor resolution, although at satellite altitudes the potential resolution of a synthetic aperture system is superior to that of an optical system;

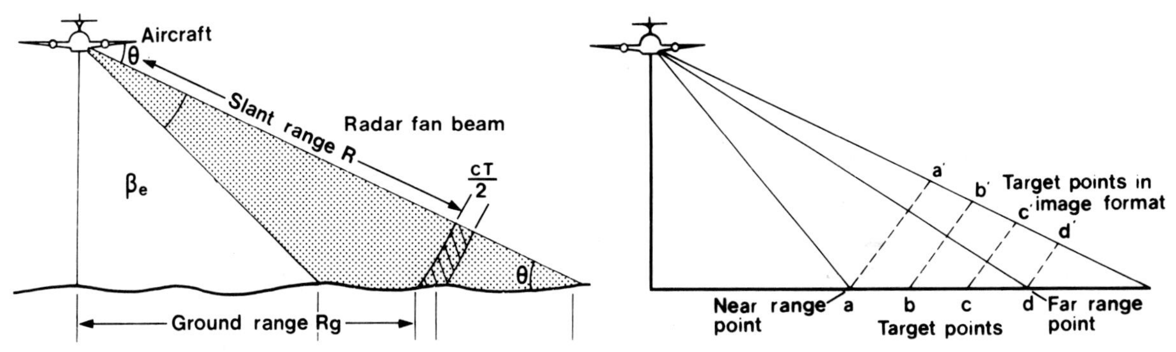

Fig. 4.23 Slant range and ground range. (Source: Trevett, 1986.)

Fig. 4.24 Foreshortening. (Source: Trevett, 1986.)

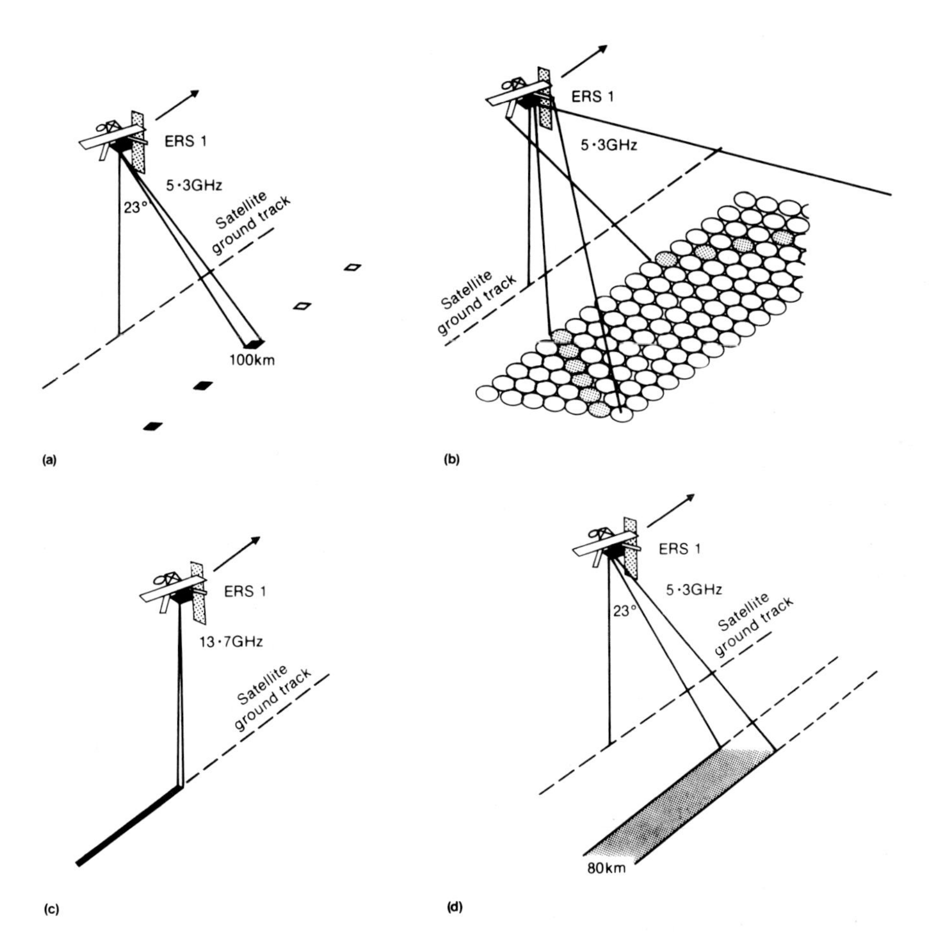

Fig. 4.25 ERS-1: the microwave payload instruments. (Source: Trevett, 1986.) (a) Wave scatterometer. The SAR will be used on a sampling basis as a wave scatterometer obtaining data for the determination of directional wave image spectra. (b) Wind scatterometer. Designed to obtain all-weather day and night measurements of surface wind speed and direction. (c) Radar altimeter. Measures the distance between the satellite and the ocean surface and extracts the significant wave height. (d) Synthetic aperture radar mode.

2. its small scale;
3. its image distortion;
4. shadowing in areas of pronounced relief.

4.4.3 Scatterometers

A particular type of radar system designed to obtain a measure of back-scattered energy for various incident angles of the radar beam is termed the *Scatterometer*. The data obtained for each resolution element is presented in a graph of reflected energy versus incident angle. The variation of radar returns can be used by the scientist to determine roughness, texture, and orientation of the terrain. Different surface materials also may be identified.

Scattering signatures have been investigated experimentally for many materials and theory has been developed for statistical roughness parameters and electromagnetic parameters of various materials. Operational scatterometers are used to gather data along a strip underneath the spacecraft or aircraft. Resolution capability is of the order of 1 m. On Skylab a narrow round beam was used which discretely shifted from the maximum angle to the nadir of the spacecraft as it moved over a given area.

A recent major European satellite operation is that of the launch of the ERS-1 satellite. ERS-1 is designed primarily as a sea and ice observation satellite and for this purpose it carries a radar altimeter and wind and wave scatterometers as well as an imaging radar in C band. The general characteristics of the entire suite of ERS-1 microwave radars are shown in Fig. 4.25.

4.4.4 Lidar (or laser radar)

The lidar is an active system similar to a microwave radar, but operating in that part of the spectrum comprising ultraviolet to near infrared regions. It consists of a laser which emits radiation in pulse or continuous mode through a collimating system. A second optical system collects the radiation returned and focuses it on to a detector. Three types of lidar are at present available: an altimeter type, which can plot a terrain profile; a scanning type, which can be used as a mapping instrument; and a third type using spectroscopic techniques, which may be used for mapping air pollutants.

Several techniques evolved in the 1970s for measuring three-dimensional wind velocities with continuous wave Doppler lidar systems. A compact, remotely operable airborne system was built at the Royal Signals and Radar Establishment, UK. An outstanding feature has been its long-term reliability and as a result pulsed CO_2 lidars have set the scene for the development of spaceborne lidar wind measuring systems to measure global wind profiles. The ongoing programme of development of the Laser Atmospheric Wind Sounder (LAWS) represents an important application of lidar technology.

5 Sensor platforms, sensor packages and satellite data distribution

5.1 Introduction

Remote sensing techniques can be applied to data from different types of observation platform and each platform, mobile or stationary, has its own characteristics. Generally speaking, three types of platforms are of interest for remote sensing: ground, airborne and spaceborne observation platforms. A classification of remote sensing platforms based on these three categories is given in Fig. 5.1.

Each type of platform has its own particular advantages and disadvantages. For example, the satellite platform offers great synoptic viewing potential, stability of the platform and orbital movements of predictable character. Yet the space platform is difficult to reach and instrumental failure may be hard to correct. Airborne platforms offer easier maintenance of sensors, lower operating costs and good accessibility for manned flights. Yet airborne platforms may suffer from instability and flight lines may be difficult to maintain. Therefore, it is not uncommon for remote sensing monitoring programmes to involve *multiplatform*

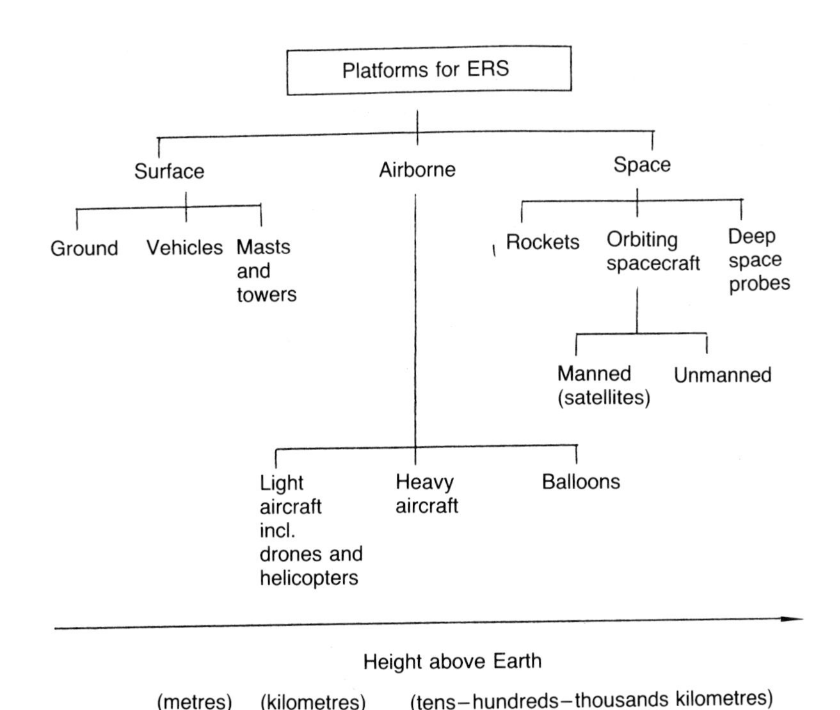

Fig. 5.1 A classification of remote sensing platforms by type and altitude.

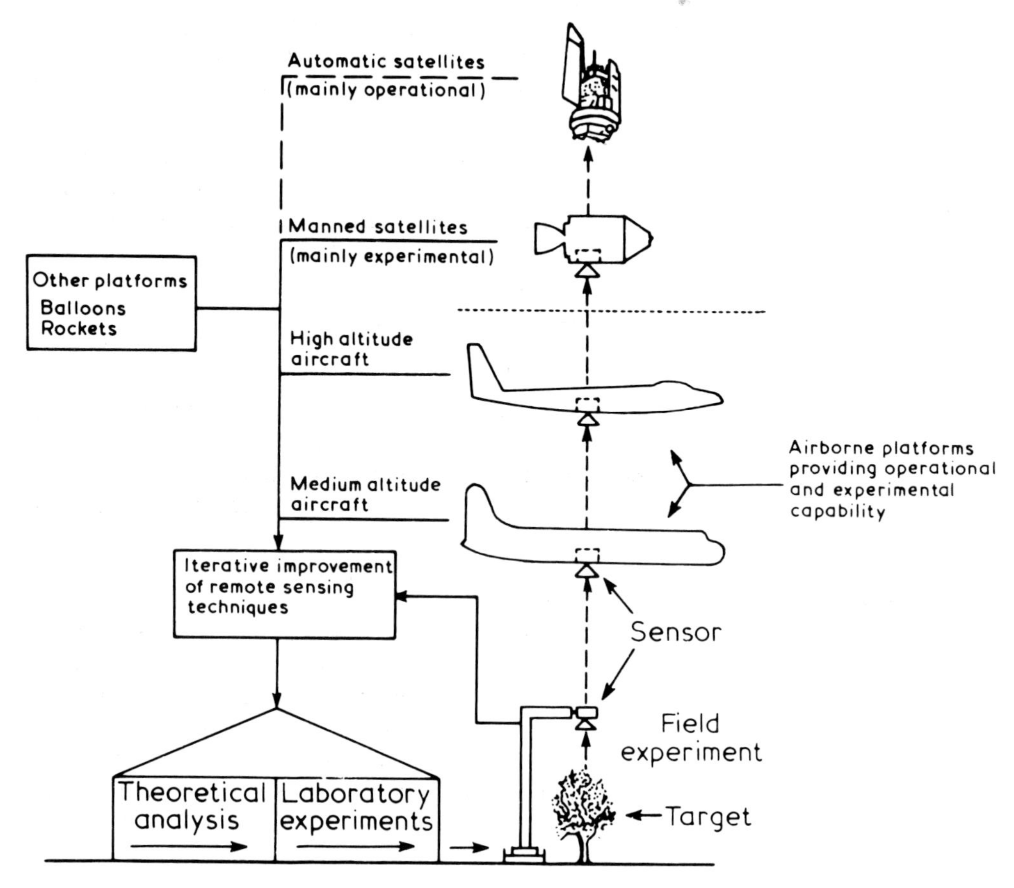

Fig. 5.2 Multiplatform remote sensing operations.

operations, as illustrated by Fig. 5.2. The following sections in this chapter discuss each type of platform with these issues in mind. However, at the same time, it must be remembered that, to make effective use of any remote sensing system, one must consider a number of other factors, such as:

1. The time and space scales of the environmental features that are essential to the fulfilment of the objectives of the programme.
2. The time of day or the season when the features of concern can best be observed (e.g. related to the phenology of a crop or natural vegetation cover, the seasonal/diurnal behaviour of animal species, or weather feature development such as thunderstorms (diurnal growth cycles) or hurricanes (seasonally distributed).
3. The effects of key physical influences, such as solar illumination, surface temperature and surface moisture, on the imagery obtained.
4. Atmospheric attenuation of the energy collected by the sensor.
5. The repetition rate of observations required to resolve the questions it is hoped to solve by remotely sensed means.

5.2 Ground observation platforms

In many areas of study, a scientific field and laboratory programme is necessary to provide a good understanding of the basic physics of the object/sensor interaction. Measurements of physical characteristics, such as the spectral reflectances and emissivities of different natural phenomena, are

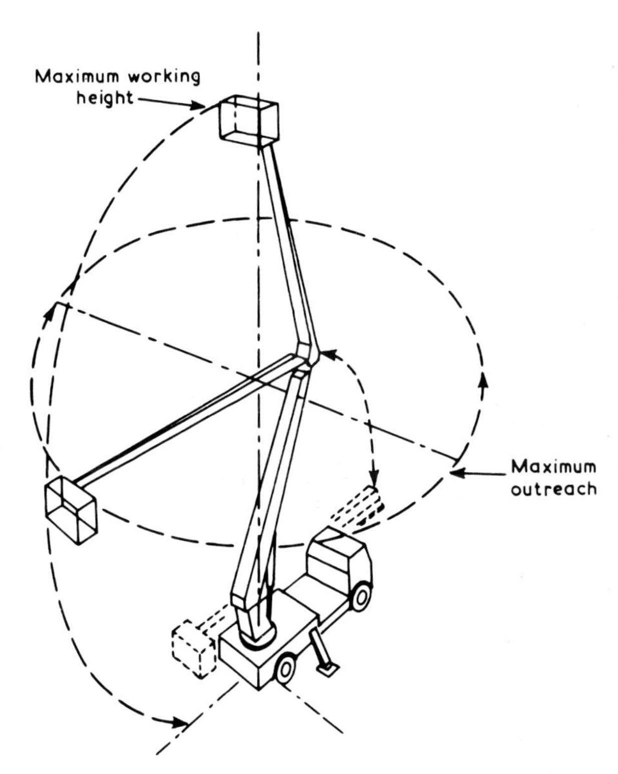

Fig. 5.3 Mobile hydraulic platform of the 'Cherry Picker' type.

necessary in order to design and develop sensors. Normally such studies are of two kinds:

1. Laboratory studies where soils, plants, building materials, water bodies, etc., are subjected to an external source of radiation and detectors of various kinds are used to measure reflection and emission spectra.
2. Field investigations in which the spectral characteristics of surface phenomena or crops are investigated in different atmospheric conditions.

Many studies of spectral response and emissivity are carried out by means of hand-held radiometers. These are, however, limited by such factors as field of view, weight of sensor, periodicity of observations and continuity of data collection. These have led to the use of various types of ground-based platforms for the collection of highly detailed data by remote sensing means.

In field investigations some sensors actually may be located on the ground itself, or platforms at or very near ground level, e.g. cameras on one hillslope to record overland flow on an opposite hillslope. Amongst mobile platforms, the most popular to date has been the 'Cherry Picker' platform (Fig. 5.3). These platforms can be extended to approximately 15 m. They have been used by various American and European research organizations to carry sensors such as spectral reflectance meters, photographic systems and scanners in the infrared or radar wavelengths.

Frequently these platforms are linked to automatic recording apparatus in field vehicles; wherever possible cross-country vehicles not limited to roads.

Portable masts also are available in various forms and can be used to support cameras and sensors for testing. For example, a Land Rover fitted with an extending mast was used successfully in conjunction with airborne studies (Fig. 5.4(a)). However, such masts can be very unstable in windy conditions.

Towers that can be dismantled and moved from one place to another (Fig. 5.4(b)) also are available. These offer greater rigidity than masts, but are less mobile and require more time to erect. Where heavy instruments, such as radar sensors, are undergoing tests it may be economic to place towers on a wheeled platform so that they can be moved on rails.

5.3 Airborne observation platforms

5.3.1 Balloon platforms

Balloon platforms can be considered under two headings: tethered and free flying. Tethered balloons were used for remote sensing observations (air photography) as long ago as the American Civil War (1862) and very recently for nature conservancy studies on beaches in Britain. Modern studies are, however, often carried out by free-flying balloons, which can offer reasonably stable platforms up to very considerable altitudes. For example the balloons developed by the Société Européenne de Propulsion can carry a photographic nacelle (Fig. 5.5(a) and (b)) to altitudes of 30 km. This nacelle consists of a rigid circular base plate supporting the whole equipment to which is nested a tight casing, externally protected by an insulating and shock-proof coating. It is roll-stabilized and the inner temperature is kept at 20°C and at ground atmospheric pressure. Standard equipment

Fig. 5.4 (a) An extending mast fitted to a Land Rover. The mast can be extended to 10 m height and is equipped with camera and photodiode sensors for studies of crop reflectance. (Courtesy, University of Bristol.) (b) Tower for supporting sensors. (Courtesy, Hawker Siddely Dynamics Ltd, Hatfield, Herts.)

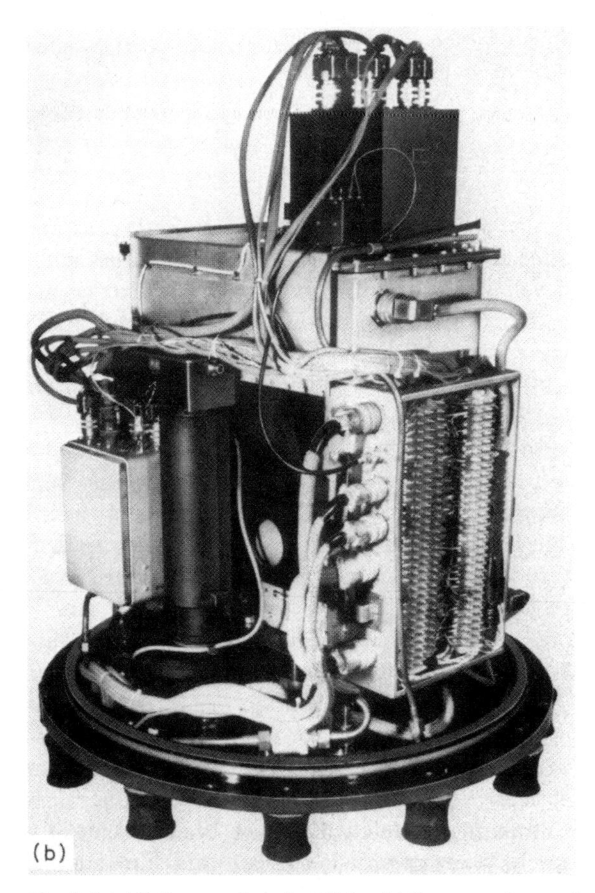

Fig. 5.5 (a) Ballon nacelle before flight. (b) Sensor equipment inside balloon nacelle. Photometer (black) and cameras in base plate. (Courtesy, Société Européenne de Propulsion.)

Fig. 5.6 Remotely Piloted Vehicle (drone), which is launched from a trailer. (Courtesy, AERO Electronics (AEL) Ltd, Horley, Surrey.)

includes two cameras, multispectral photometer, power supply units and remote control apparatus. Its return to Earth is achieved by parachute, after remote controlled tearing of the carrying balloon.

5.3.2 Powered aircraft

Commercially orientated remote sensing from aircraft has been carried out for more than 60 years using fixed-wing aircraft. Increasingly in recent years helicopters have been used widely for such purposes also. The fixed-wing aircraft have been used mainly for obtaining stereoscopic black and white photography for subsequent interpolation and photogrammetric mapping (Chapter 7). Many resource development and planning programmes use aerial photography from altitudes of approximately 1500–3000 m. These missions at low or medium altitudes are appropriate for surveys of local or limited regional interest. Although they are used widely for periodic surveys it must be recognized that flights at specified times at short intervals or for defined repetition rates over long periods are rarely achieved. Nevertheless, aircraft platforms offer an economic method of testing sensors under development. Thus photographic cameras and various scanners in the infrared, radar and microwave wavelengths have been flown over ground truth sites in many countries, especially in the NASA and ESA programmes, and the UK.

In order to obtain a greater areal coverage and to test atmospheric effects on sensors special high altitude aircraft have been used as platforms. These have often been developed from military reconnaissance aircraft, such as the American U2 plane. The altitudes attained by such aircraft are normally about 15 km, from which ground coverage of 50–400 km^2 per frame can be obtained.

In some cases the use of normal aircraft may be deemed too expensive, or it may be that there is a demand for flights at very low levels and at low speeds. In such circumstances the microlight (ultralight) aircraft which have been developed in recent years are promising platforms. For example, a microlight equipped with two radiometers, vidicon camera, portable stereo videorecorder and

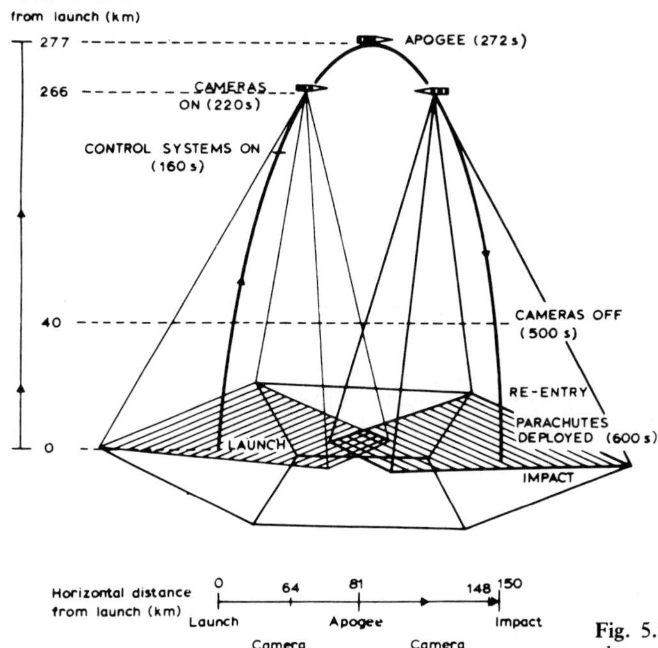

Fig. 5.7 Schematic diagram of Skylark rocket trajectory and photographic coverage. (Source: Savigear *et al.*, 1974.)

a laser range finder has been used for low level flights at 5–10 m altitude over specific test sites in the UK.

Today helicopters are also widely used for environmental studies. In studies of inaccessible areas, such as swamps and bogs, they can carry out quadrat-like studies speedily and economically. Helicopters are capable of carrying quite heavy loads over substantial distances and are now used widely as photographic camera, vidicon camera and thermal heat sensor platforms.

A new type of airborne platform has emerged recently as a result of military requirements for target systems. This is usually termed the 'remotely piloted vehicle' (RPV), or more commonly the 'drone'. A drone is a radio-controlled miniature aircraft. Since it is controlled from the ground its relatively short control range of 5 km is dictated by the visual range obtainable. However, aerial photography of local, inaccessible, areas is highly practicable. For example, surveys of oil pipelines, fence lines, power transmission lines or forest areas are possible using an underwing robot camera which is triggered by remote control signal from the ground.

An example of such a system is the SNIPE manufactured by Aero Electronics Ltd (Fig. 5.6). This drone has a guaranteed flight time of 45 min, and a maximum speed of 177 km h^{-1}. It is launched by a catapult system and is retrievable by parachute. Such systems can obtain high-definition photographs from altitudes up to 900 m. Typically the 35 mm camera has automatic wind system, f 2.8 lens and variable, high-speed, shutter.

5.4 High-altitude sounding rockets

In the late 1960s and early 1970s some experiments in the use of sensors from space altitudes were carried out using modified sounding rockets as the platforms. One limitation upon the use of such a system was that there was a need to ensure that the descending rocket did not cause damage. As a result its use was limited to sparsely populated areas e.g. in surveys over Australia and Argentina.

The Skylark Earth Resource Rocket (Fig. 5.7) was an example of such a platform. The rocket was fired from a mobile launcher to altitudes between 90 and 400 km. During the flight the rocket motor and payload separate and its sensors were held in a stable attitude by an automatic control system. By stepping the payload around six times the sensors can scan a 360° field of view. The payload and the spent motor returned to ground slowly by parachute enabling speedy recovery of the photographic records. The sensor payload consists of two cameras, normally of the Hasselblad 500 EL/70M mm type.

Unfortunately the imagery was obtained from a range of different altitudes and look-angles, posing problems for subsequent use of the data. Fitted with a digital scanning device, such a cheap platform could have had a brighter future owing to the rapid advances that have been made on the handling of numerical rather than photographic image data but the improvements made in the spatial resolutions of data obtained from satellites seem likely to close the Earth resources rocket chapter entirely.

5.5 Satellite platforms

5.5.1 Orbital considerations

It is now generally recognized that most environmental monitoring can be achieved best by a combination of satellite missions embracing a variety of sensors and sensor packages. As well as providing the broadest views of regional relationships satellite platforms can be put into orbits that ensure repeated coverage of the whole of the Earth's surface. The interval between satellite observations at a particular location can, to some extent, be selected by choice of an appropriate orbit. Furthermore, such orbits can be so defined that observations can be made under comparable or identical conditions of illumination over comparatively long periods. Therefore, some description and explanation of orbital configurations is appropriate before considering details of the platforms and sensor packages themselves.

The orbital relationships of satellites are affected by many factors but one may note first that there is a basic relationship between the *heights* of satellite orbits and their lives before re-entry to the Earth's atmosphere. Satellites in low orbits are affected much more by upper atmospheric drag and resulting increases in the effects of gravitational pull (Table 5.1).

Table 5.1 Relationships between average altitudes of satellites in near-circular orbits, and their expected lives before re-entry

Height above Earth surface (km)	Life before re-entry
250	12 days
500	10 years
600	50 years
1 000	1000 years
10 000	Indefinite

Thus there are lower limits below which Earth observation satellites will not normally be operated. Since the design lives of their payloads are usually within the span from 1 to 5 years, a minimum operational altitude of about 450 km is indicated. Actual orbital altitudes of unmanned remote sensing satellites are characteristically well above that level; whilst most operate between 800 and 1500 km some are at approximately 36 000 km altitude. The first of these groups is composed mostly of polar or near-polar orbiting satellites occupying so-called 'sun-synchronous' orbits (crossing the Equator at the same sun time each day); the second group are mostly 'geostationary' satellites that appear to 'hover' over preselected points on the Equator.

Although all satellite orbits are elliptical, for most Earth observation purposes it is desirable to achieve as nearly a circular orbit as possible. However, this is not always the case. Consider two satellites launched on the same day, 27 June 1979. Of these, NOAA-6 was placed into a nearly circular orbit. The apogee (top) of the NOAA-6 orbit was 826 km, and the perigee (bottom) was 810 km. A Russian satellite launched on the same day was Cosmos 1109. Its orbit was highly elliptical with an apogee of 40 058 km and a perigee of 613 km: such orbits may be preferred for satellites whose viewing missions differ from one region of the world to another. (See also Fig. 5.9).

5.5.2 Low-level satellites

Low-level (800–1500 km) Earth observation satellites encircle the Earth whilst the Earth itself revolves on its polar axis within their orbits. These satellites fall into three broad groups on the basis of detailed orbital differences. These are (Fig. 5.8):

1. Equatorial-orbiting satellites, whose orbits are wholly within the plane of the Equator (Fig. 5.8(a)).

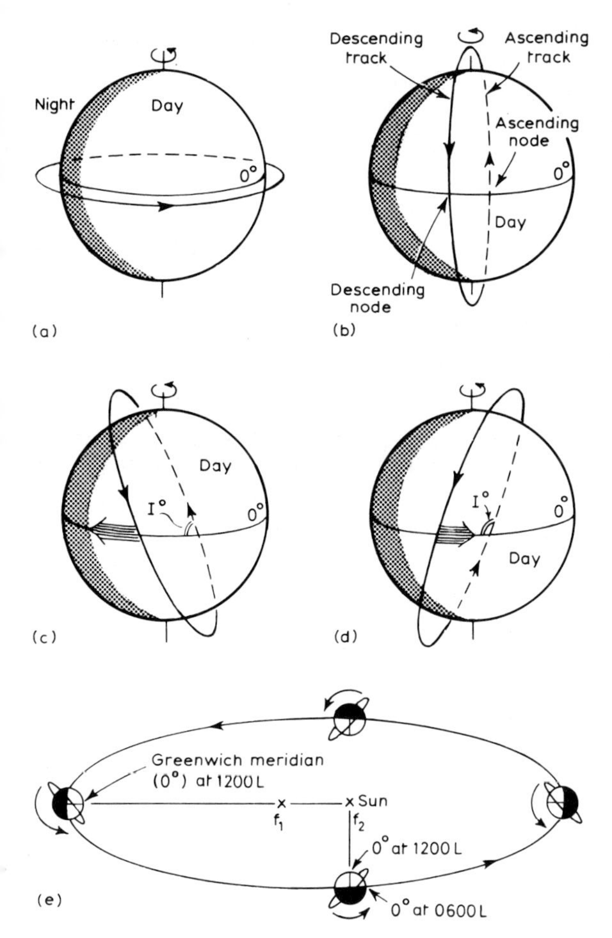

Fig. 5.8 Orbital characteristics of polar-orbiting satellites, and the Earth as a satellite of the Sun. (a) Equatorial orbit. (b) True polar orbit. (c) Prograde oblique orbit (launched eastwards, regresses westwards). (d) Retrograde oblique orbit (launched westwards, precesses eastwards). (e) Perpendicular from Sun to Earth, precesses eastwards throughout the year.

2. Polar-orbiting satellites, whose orbits are in the plane of the Earth's polar axis; each successive orbit crosses the Equator at a different sun time (Fig. 5.8(b)). True polar orbits are preferred for missions whose aim is to view longitudinal zones under the full range of illumination conditions.
3. Oblique-orbiting (near polar orbit) satellites (by far the largest group) whose orbital planes cross the plane of the Equator at an angle other than 90°. Because the range of possibilities here is much greater than in (a) and (b), further discussion of oblique orbits is required (Fig. 5.8(c and d)).

Oblique-orbiting satellites may be launched eastwards into direct, or *prograde*, orbits (so called because the movement of such satellites is in the same direction as the rotation of the Earth), or westwards into *retrograde* orbits. The inclination of any orbit is specified in terms of the angle between its ascending track and the Equator, i.e. at the ascending node. Because the Earth is not a perfect sphere it exercises a gyroscopic influence on satellites in oblique orbits such that those in prograde orbits (Fig. 5.8(c)) *regress* whilst retrograde orbits *advance* or *precess* (Fig. 5.8(d)) with respect to the planes of their initial orbits.

Figure 5.8(e) shows that a perpendicular from the Sun to the Earth also advances around the globe in the direction of Earth rotation on its polar axis, because of the behaviour of the Earth itself as a satellite travelling around the Sun. It is clear that a special type of Earth satellite orbit can be achieved if the rate of precession of an orbit is geared to the orbiting of the Earth around the Sun: this is the *sun-synchronous* configuration (a retrograde oblique orbit) which ensures that satellites view the same point on the surface at the *same local time* at predetermined intervals (twice each day with near polar-orbiting weather satellites, once every 18 days with Landsat). Thus the altitude of the sun in the sky is approximately the same *within similar seasons* at every crossing of the same parallel. However, it should be noted that differences in the elevation of the sun affect each local area through the annual cycle of the seasons as a result of the apparent motion of the Sun relative to Earth.

It is in this context that the observation of the surface from a satellite must be seen to be a function not only of the satellite altitude and orbital configuration, but also of the 'field of view' (FOV) of its sensor system, i.e. the angle through which a sensor can collect radiation from its target. One very desirable sun-synchronous orbit is a retrograde orbit inclined at 99.1° to the Equator, at an altitude of 1100 km. This orbit is completed once every 100 min. From this information the number of orbits that will be completed in 24 h can be calculated approximately (14.5), as also can the breadth of the FOV required to ensure complete global imaging cover even in equatorial latitudes. In the case of NOAA satellites, once-daily coverage is achieved in the visible by a scanning radiometer with a viewing angle of 112°; twice-daily coverage

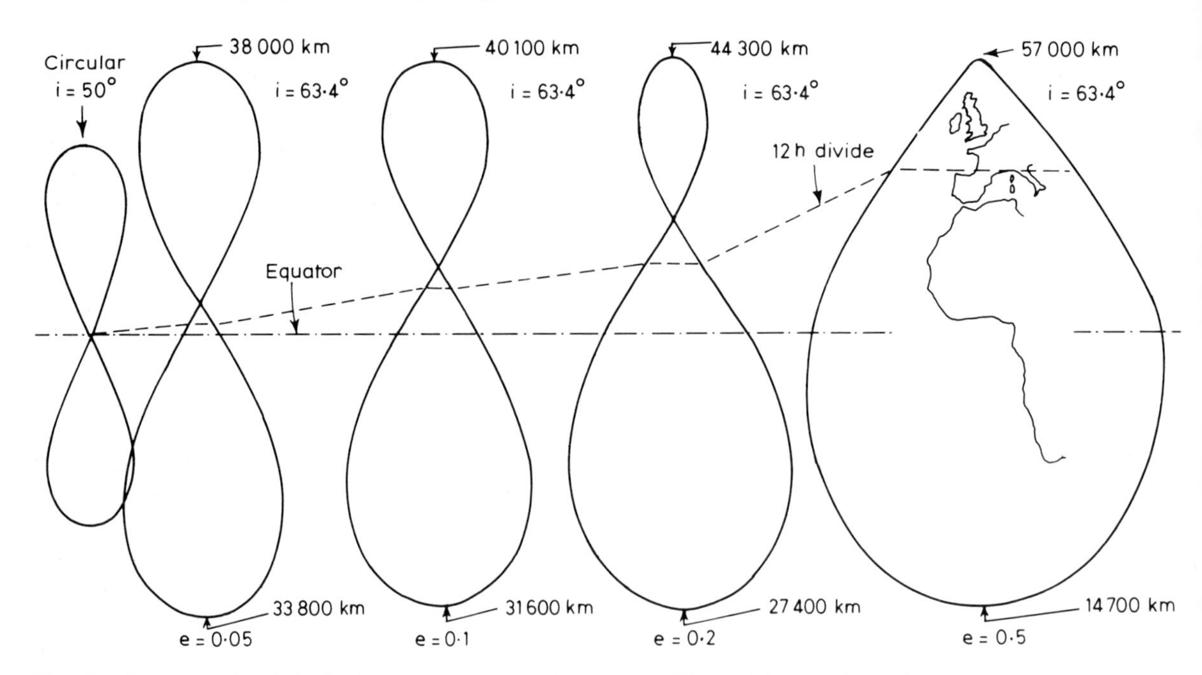

Fig. 5.9 Ground tracks of inclined geosynchronous orbits to exemplify possible variations of eccentricity, inclination and orientation. Associated (possibly advantageous) variations would result in imaging resolutions but also in image geometries (generally disadvantageous). (Source: Velten, 1976.)

is achieved in the infrared by the same sensor system because heat radiation is emitted from both the daytime (descending track) and night-time (ascending track) hemispheres. In the case of Landsat, narrower swaths of imagery have been obtained from scanning radiometers, such as the Multispectral Scanner (MSS) of Landsat 3, which had a viewing angle of only 11.6° and the satellite occupied a lower orbit, at about 900 km. In this way more detailed target information is acquired, but at the expense of breadth and frequency of coverage. Such trade-offs are commonplace, for communication links between satellites and ground stations constrain the volumes of data that can be transferred from the one to the other whilst they are intervisible.

5.5.3 High-level satellites

Whilst the sun-synchronous orbit is the most popular low orbit, *geosynchronous* orbits are virtually the only useful high orbits. In these the satellites precess around the Earth at rates related to the rotation of the Earth on its polar axis. The most useful single orbit is at approximately 36 000 km, at which altitude a prograde satellite orbiting in the plane of the Equator 'hovers': i.e. it is apparently 'fixed' above a given point on the surface. This special type of geosynchronous orbit is known as *geostationary*. It is exploited by many communications satellites, and some Earth observation satellites, especially within the programme of the World Weather Watch (WWW) of the World Meteorological Organization (WMO) (Chapter 10). These have included ATS, SMS, GOES, Meteosat, Insat and GMS satellites. The key advantage of the geostationary orbit is that it permits satellites to view the visible disc (limited by the curvature of the Earth) much more frequently than any low orbit satellite can do. Other geosynchronous possibilities exist but have yet to be exploited for Earth observation missions. Figure 5.9 exemplifies some of these, drawn from design studies carried out by the European Space Agency.

5.6 The evolution of Earth resources satellites

The first photographs of the Earth from space were taken from the NASA spacecraft named Mercury, Gemini and Apollo in the 1960s. The interest shown in the 70 mm format colour photographs first taken from a Mercury satellite in 1961 was to grow rapidly as the relief and geological features of many unmapped areas of the world were seen in detail for the first time.

Subsequently some 2400 photographs were taken in colour and infrared colour from the Gemini spacecraft in the mid-1960s. These pictures created considerable interest and the Apollo mission which followed built upon this work, eventually leading to the acquisition of multiband photography using a multicamera array. It was this imagery that laid the foundations for the subsequent series of Landsat sensor packages. NASA's Goddard Space Flight Center (GSFC) began a conceptual study of Earth resources satellites in 1967 as part of this evolutionary stage and the Landsat family of satellites was the result.

5.6.1 The Landsat system

Perhaps the most important of all the Earth resources satellite families flown to date has been the Landsat family.

The Landsat system was initially designed to make automatic observations using a payload consisting of a return beam vidicon (RBV) camera system and multispectral scanner (MSS). The RBV system on Landsats 1 and 2 operated by shuttering three independent cameras simultaneously (Fig. 5.10(a)), each sensing a different spectral band in the range 0.48–0.83 μm. On Landsat 3 the RBV system was changed to two RBV panchromatic cameras operating in the range 0.51–0.75 μm (Fig. 5.10(b)). These cameras produced two side-by-side images, each covering a ground scene of approximately 99 $\times$ 99 km. By using a focal length of 25 cm the ground resolution obtained was approximately 30 m. Digitization of the analogue signal gave an incoming data stream of 45 megabits s^{-1}.

The MSS system is a line scanning device using an oscillating mirror to scan at right angles to the space craft flight direction (Fig. 5.10(c)). Optical energy is sensed simultaneously by an array of detectors in four spectral bands from 0.5 to 1.1 μm. The area of coverage was approximately equal to the of the RBV images.

The early Landsat satellite platforms operated in near circular, sun-synchronous, near-polar orbits at an altitude of 915 km, circling the Earth every 103 min, completing 14 orbits per day and viewing the entire Earth every 18 days. The orbit was selected so that the satellite ground trace repeated its Earth coverage at the same local time every 18-day period to within 37 km of its first orbit. Each day the paths shifted 160 km westward (Fig. 5.11). The amount of overlap between successive passes varied from 14% sidelap at the Equator to 70% at polar latitudes. For example, at 40°N the sidelap of successive images is 62 km (Fig. 5.11(b)). Images were acquired between 0900 hours and 1000 hours local sun time, except at high latitudes.

A major improvement in the Landsat system was introduced in 1982 with the launch of Landsat 4 and continued with subsequent satellites of this series. Whilst an MSS of the same design as on Landsats 1, 2 and 3 is carried, the RBV has been replaced by a new instrument termed the Thematic Mapper, so called because of its intended use for the mapping of different surface types, categories or 'themes'. The choice of spectral bands in the Thematic Mapper (TM) was made in the framework of the needs of scientists concerned with environmental monitoring (Fig. 5.10(d)). The different bands investigated by the Landsats are listed in Table 5.2.

The improvements achieved by the TM included both the location and number of spectral bands available, together with better spatial resolution and geometric fidelity of the data.

Alongside changes in instrumentation, the orbital characteristics of Landsats 4, 5 and 6 were modified in relation to Landsats 1–3. The orbit was reduced to 705 km altitude giving a faster repeat cycle of 16 days for Earth coverage. This lower orbit resulted in less overlap between images obtained on adjacent orbit paths than occurred in the case of Landsats 1–3: Landsats 4, 5 and 6 have areas near the sub-satellite point where only single images are obtained every 16 days. The pattern of overpasses is now such that adjacent swaths are imaged at 8-day intervals instead of on successive days as in Landsats 1–3.

The spatial resolution of the Thematic Mapper as measured by the Instantaneous Field of View (IFOV) is 30 m compared with a value of 79 m for the MSS sensor of the earlier Landsat missions. This spatial improvement provides for greatly improved visual analysis of TM images. However, the higher spatial resolution together with an increased number of bands and finer radiometric resolution has resulted in much increased data

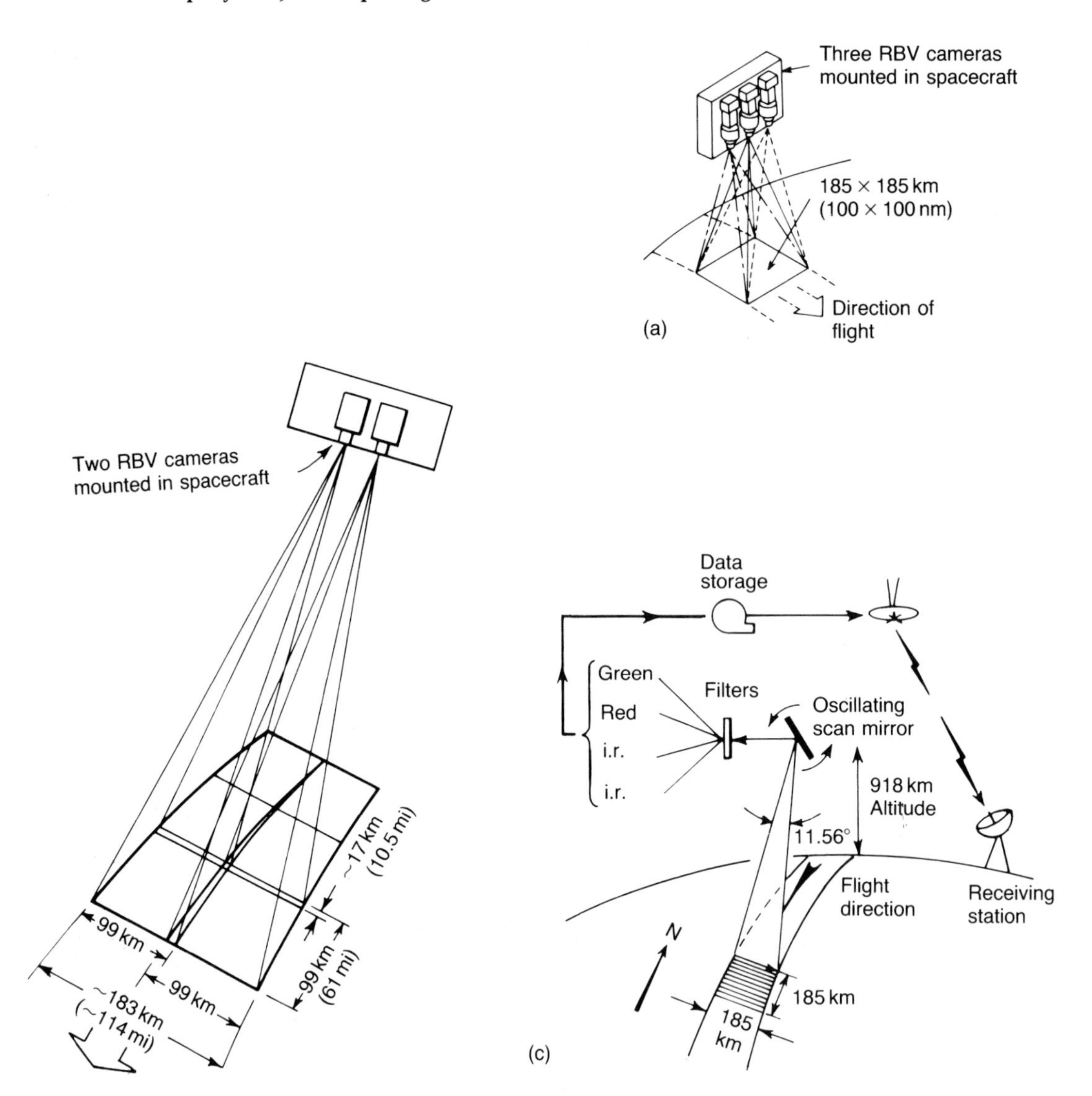

Fig. 5.10 Imaging systems and Landsat satellites. (a) Return Beam Vidicon system for Landsats 1 and 2. (b) Return Beam vidicon system for Landsat 3. (c) Landsat Multispectral Scanner. (d) Thematic Mapper for Landsat 4 and above. For each terrain scene, four images are transmitted to a receiving station. (Source: NASA.)

rates. There are consequential implications for computing time. One early MSS scene of 185 × 185 km was comprised of 33 megabytes of data, which could be contained on a single 1600 bpi tape. For the Thematic Mapper the same scene contains 300 megabytes of data and requires seven 1600 bpi tapes.

The geometric characteristics of the Thematic Mapper data are superior to those of the earlier MSS systems owing to improved sensor pointing and finer spatial resolution (Fig. 5.12). As a result Thematic Mapper data requires less geometric correction in order to achieve mapping tasks and the data can be used for mapping at much finer scales than MSS data.

The Landsat Data Collection System consists of a combination of ground stations, operations control centre and NASA Data Processing Facility. The principal receiving stations for Landsat are shown in Fig. 5.13.

In parallel with the Landsat programme NASA promoted the idea of developing an Orbital Research

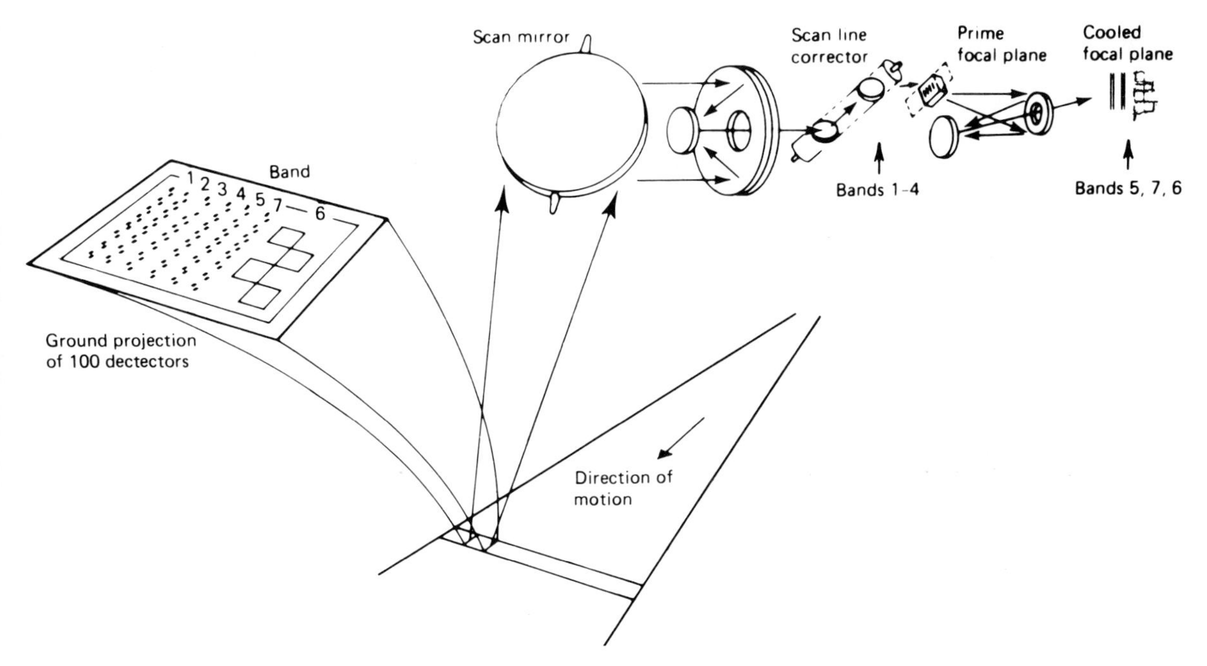

Fig. 5.10 (d) Continued

Laboratory and this concept was built into the Apollo programme. Named the Skylab project, the Skylab mission of 1973 carried a three-man crew and the observations were coordinated with the collection of ground data on test sites and aircraft underflights. Skylab can be regarded as the forerunner of the Space Shuttle, which is currently one of the principal sensor platforms for Earth environment studies.

It is noteworthy that in the former USSR a similar progression using both manned space stations (Salyut/Mir) launched from Soyuz spacecraft and unmanned (Cosmos) series satellites took place. Much of the Russian effort has been devoted to a wide range of applications, ranging from weather forecasting to military reconnaissance. The meteorological satellites included the Molniya series which were used for the collection of meteorological data in the 1960s. Later the Molniya craft were supplanted by the more advanced Meteor satellites in 1969. Since then an increasing number of other nations have followed the USA and Soviet lead. The first Indian remote sensing satellite, IRS-1A, was launched in 1988 using a Soviet launcher. This carried multispectral scanning systems (Linear Self Scanning (LISS)) cameras in a sun-synchronous orbit of 904 km altitude with a 22-day repeat cycle. A Japanese Marine Observation satellite, MOS-1, was launched in 1990, carrying a Multispectral Electronic Self-synchronous orbit at a 909 km altitude.

In the remainder of this chapter either Earth resources satellite systems or some of the key systems from which environmental satellite data sets of special importance and wide accessibility are described. Systems planned for the future will be summarized in Chapter 19.

5.7 The Space Shuttle

The Space Shuttle concept arose from the need for prosecution of two principal objectives associated with any experimental Earth-observation programme using advanced sensing methods:

1. To understand the capabilities and limitations of the various sensor systems proposed and to develop measurement techniques that can be applied to the different areas of application under investigation.
2. To deploy proven remote sensing instruments and methods for experiments in Earth observation disciplines aimed either at purely scientific objectives or for perfecting measure-

Table 5.2 Wavebands and applications of (a) the Multispectral Scanner (MSS) on Landsats 1–3, and (b) the MSS and Thematic Mapper (TM) on Landsats 4–6 (Source: Harris, 1988)

(a) Band	Wavelength range (μm)	Applications
4	0.5–0.6	Sediment loads, shallow water
5	0.6–0.7	Vegetation, cultural features
6	0.7–0.8	Land/water separation
7	0.8–1.1 (1.0[a])	Vegetation and geological studies

[a] The upper limit of Band 7 is normally quoted as 1.1 μm, but also has been quoted as 1.0 μm.

(b) Band	MSS wavelength range (μm)	TM wavelength range (μm)	Principal applications for TM data
1	0.5–0.6	0.45–0.52	Coastal water mapping Soil/vegetation differentiation
2	0.6–0.7	0.52–0.60	Green reflectance by healthy vegetation
3	0.7–0.8	0.63–0.69	Chlorophyll absorption for plant species differentiation
4	0.8–1.1 (1.0)	0.76–0.90	Biomass surveys
5		1.55–1.75	Vegetation moisture Snow/cloud discrimination
6		10.4–11.7[a]	Thermal mapping including plant stress
7		2.08–2.35	Vegetation moisture and geological mapping
Ground pixel size	82 m	30 m (Bands 1–5, 7) 120 m (Band 6)	
Quantization levels	64	256	
Data rate	15 Mbps	85 Mbps	
Weight	68 kg	258 kg	
Size	0.35 × 0.4 × 0.9 m	1.1 × 0.7 × 2 m	
Power	50 W	332 W	

[a] Prelaunch calibration showed this wavelength range. The design range was 10.4–12.5 μm.

ment methods for later applications-oriented missions.

The Shuttle System, as its name suggests, consists of four NASA spacecraft which can be used to maintain a frequent pattern of spaceflights. As one spacecraft returns to Earth, another is in preparation for the next experiment. The first flights began in 1981. The Shuttle consists of three major components – two rocket boosters, liquid propellant tank and orbiter vehicle. The boosters return to Earth by parachute and the fuel burns from the propellant tank until the orbiter reaches its desired altitude. The propellant tank is then jettisoned too, leaving the orbiter vehicle to return from space as a glider.

The orbiter vehicles have carried many civilian and military satellites into space. In the context of environmental monitoring we may especially note the use of the Shuttle for the launch of Shuttle Imaging Radar experiments. The Shuttle Imaging Radars (SIR-A and SIR-B) operated at a wavelength of 23 cm (L Band) and were horizontally polarized. They were launched in November 1981 and October 1984, as part of the scientific payloads of Shuttle missions. The objectives of these missions were to explore areas that were difficult to study using visible and thermal sensors. SIR-B was

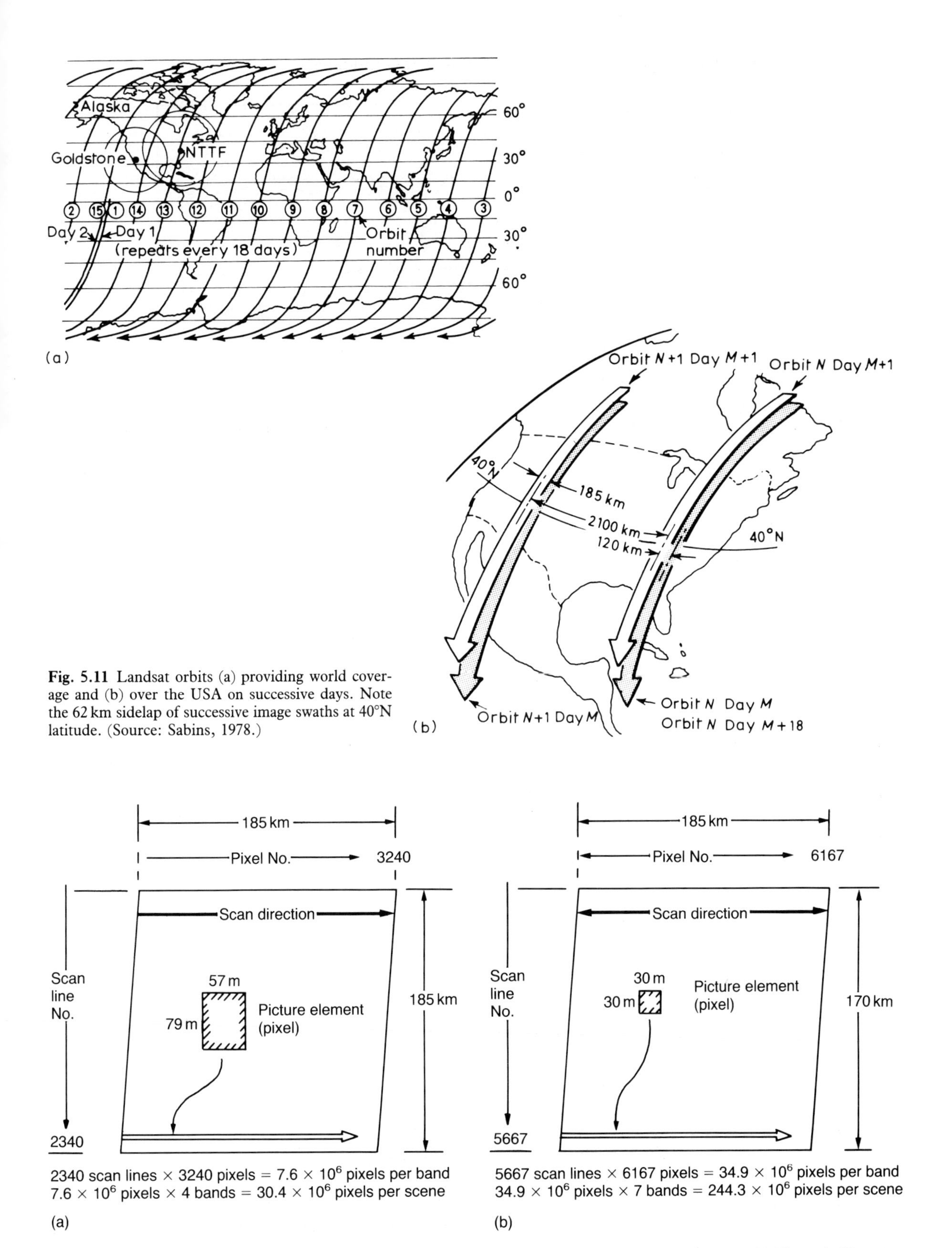

Fig. 5.11 Landsat orbits (a) providing world coverage and (b) over the USA on successive days. Note the 62 km sidelap of successive image swaths at 40°N latitude. (Source: Sabins, 1978.)

Fig. 5.12 Arrangement of scan lines and pixels in Landsat MSS and TM images. (a) Multispectral Scanner. (b) Thematic Mapper. (Courtesy, NASA.)

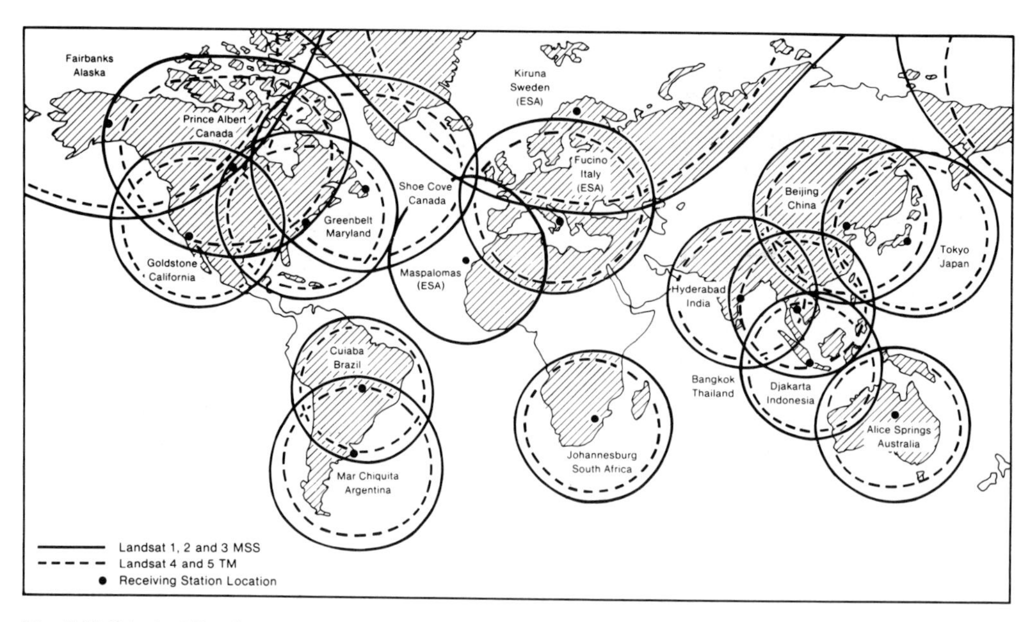

Fig. 5.13 Principal Landsat receiving stations.

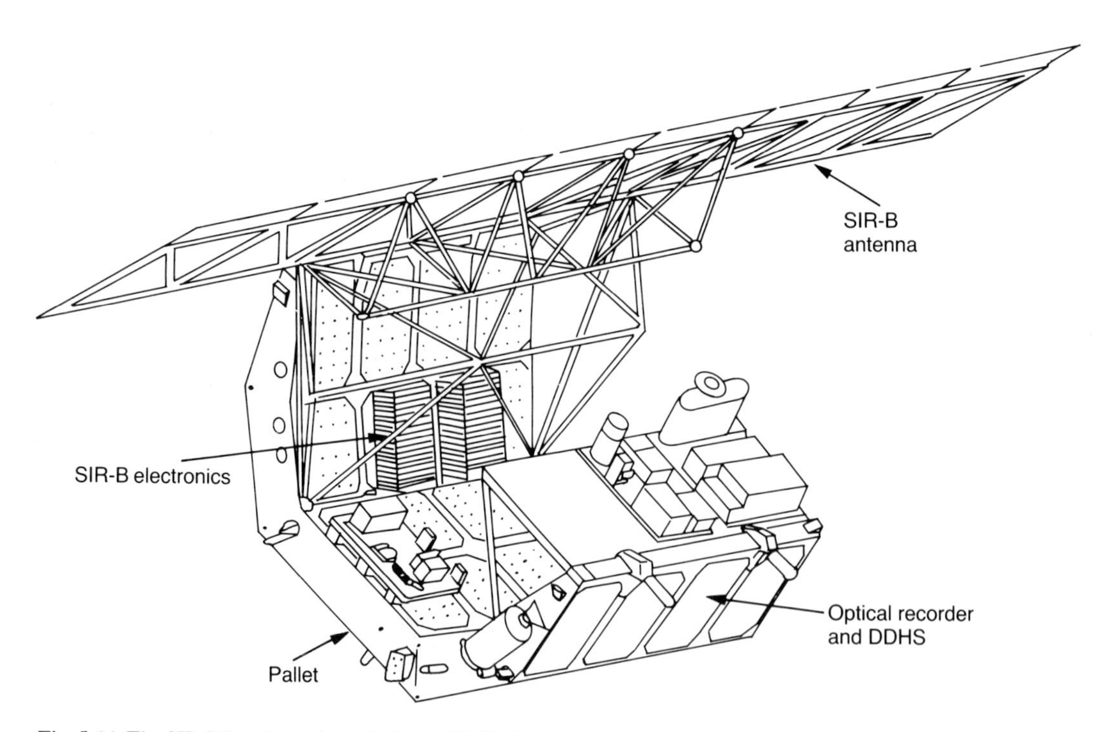

Fig. 5.14 The SIR-B imaging radar, which provided L Band (23 cm) imagery with a resolution of 30 m from swaths 20–50 km wide.

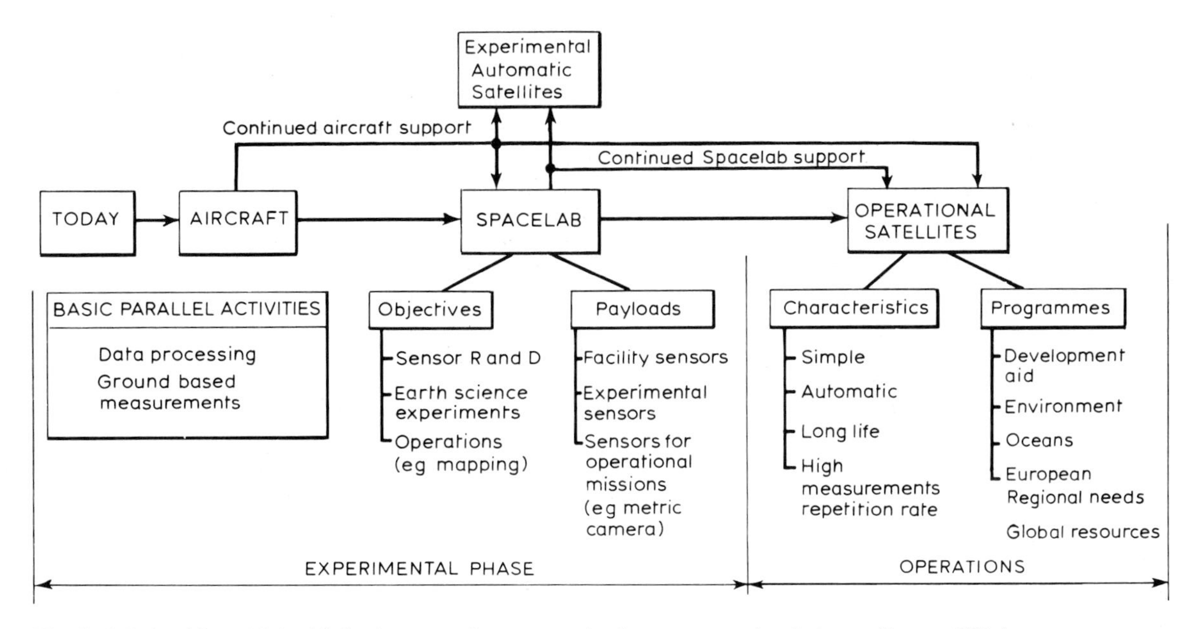

Fig. 5.15 Role of Spacelab in the development of remote sensing instruments and techniques. (Source: ESA.)

designed to acquire digitally processed imagery at selectable incidence angles between 15 and 60°. In essence the central objective of SIR-B was to study the effect of different illumination geometries on the data: topography is revealed best at low incidence angles, whereas high angles give more information on surface roughness. The main applications studied were:

1. penetration of dry sand to discover fossil drainage patterns;
2. geological studies, especially in heavily forested areas;
3. geomorphological mapping in desert and glacial environments;
4. topographic and land use mapping in cloudy tropical areas;
5. irrigation (soil moisture and salinity) studies.

Only 20% of the planned coverage was obtained, but this corresponded to an area of more than 1 million square miles. The principal features of SIR-B are shown in Fig. 5.14.

The Large Format Camera (LFC), see section 4.2.2, also made its maiden space flight aboard the Shuttle in 1984. This camera, with a focal length of 305 mm, was designed for map making purposes and several investigators prepared 1 : 25 000 scale topographic maps with much smaller contour intervals than previously considered possible. Its maiden voyage took place in 1984 on Challenger mission 41-g. Although dogged by poor weather conditions it provided some remarkable imagery on four film types (colour, colour infrared, and two black and white negative films).

A major project designed for the Space Shuttle was the Space Laboratory (Spacelab), a European-constructed manned space laboratory. The roles for Spacelab are summarized in Fig. 5.15. This was planned to act as a bridge between ground and airborne measurement systems and long-life, automatic satellites.

Considerable flexibility was built into the modular approach for the Spacelab design. The sensor platform consisted essentially of two parts. First, a pressurized laboratory (module) providing a 'shirt sleeve' environment for the space crew, scientists and equipment. Second, a pallet area outside the laboratory that was unpressurized but could be used for certain items of equipment which could be remotely controlled. The varieties of Spacelab configurations on offer are shown in Table 5.3. The orbiter and payload bay are shown in Fig. 5.16. The first Spacelab mission took place between 28 November and 8 December 1983, when the principal experiment carried out was a cartographic investigation termed the 'Metric

Table 5.3 Spacelab facilities (Source: ESA)

Available to users	Spacelab configuration			
	Short module +9 m pallet	Long module	15 m pallet	Independently suspended pallet
Payload weight (kg)	5500	5500	8000	9100
Volume for experiment equipment				
Inside module (m^3)	8	22	—	—
On pallet (m^3)	100	—	160	100
Pallet mounting area (m^2)	51	—	85	51
Electrical power (28 V DC 115/200 V at 400 Hz AC)				
Average (kW)	3–4	3–4	4–5	4–5
Peak (kW)	8	8	9	9
Energy[a] (kW h)	300	300	500	500
Experiment support computer with central processing unit	← 64 K core memory of 16-bit words →			
and data acquisition system	← 320 000 operations s^{-1} →			
Data handling				
Transmission through orbiter	← Up to 50 Megabits s^{-1} →			
Storage digital data	← Up to 30 Megabits s^{-1} →			
Instrument pointing subsystem IPS	Mounted on pallet, will provide arc second pointing for payloads up to 5000 kg			

[a] Energy can be increased by the addition of payload-chargeable kits, each providing 840 kW h and weighing approximately 350 kg (at landing).

Camera Experiment' (section 4.2.2). The objective was to test the mapping capability of high-resolution space photography using a modified Zeiss RMK-A 30.23 survey camera. It was designed to provide very high spatial resolution, geometric integrity and large area coverage. It produced imagery at ground resolution of about 20 m from an altitude of 250 km.

The Space Shuttle also was used as the platform on which the Modular Opto-electronic Multi-spectral Scanner (MOMS) was launched in 1983 and 1984, orbiting at 289–300 km. The MOMS uses the principle of electronic scanning across the entire flight path by a dual lens system (Fig. 5.17) with a large number of discrete photosensitive elements. The principal characteristics of the system include the following:

1. spatial resolution of 10–20 m (i.e. similar to SPOT);
2. narrow bandwidths optimized for applications;
3. stereoscopic capabilities;
4. flexibility, allowing for extension of spectral bands towards the middle infrared (1.6–2.3 μm) and thermal infrared (8–12 μm).

5.8 The SPOT satellite system

The first French satellite 'Systeme Probatoire de l'Observation de la Terre' (SPOT) was launched by the Ariane rocket in February 1986. It marked a departure from the conventional multispectral scanners of the Landsat series. Whereas Landsat used the mechanical systems of the scanning mirror, SPOT was designed to adopt a 'pushbroom' sensor (section 4.3.1).

The pushbroom scanner (or 'multispectral solid state linear array') consists of a line of detectors on a fixed assembly. With the pushbroom device the scanning of an individual line of a scene is performed electronically by successively measuring the current generated by each detector within the linear array. Each spectral band uses four linear arrays with each array consisting of 1728 detectors. The characteristics of the SPOT-1 Satellite, including its High Resolution Visible (HRV) imaging

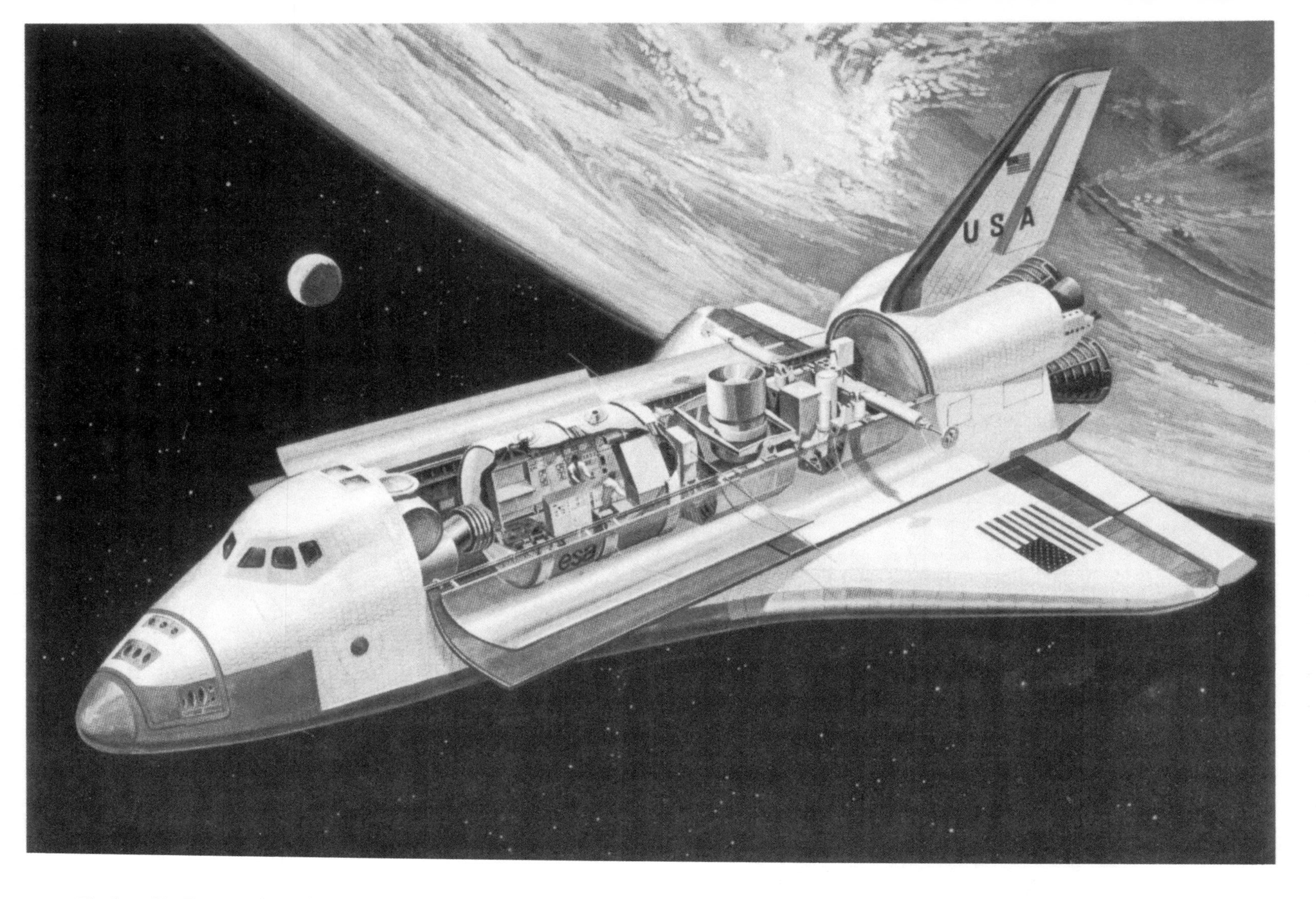

Fig. 5.16 The European Space Agency (ESA) builds Spacelab, a habitable laboratory that will be carried in the Space Shuttle orbiter cargo bay for Earth-orbital periods up to 30 days. Spacelab will be used to conduct experiments in the physical sciences, health sciences, and manufacturing processes. (Source: ESA.)

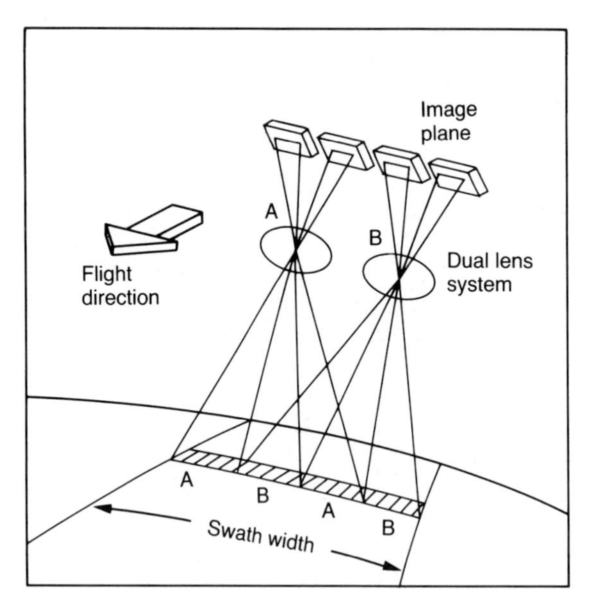

Fig. 5.17 Schematic diagram of MOMS imagery system in operation.

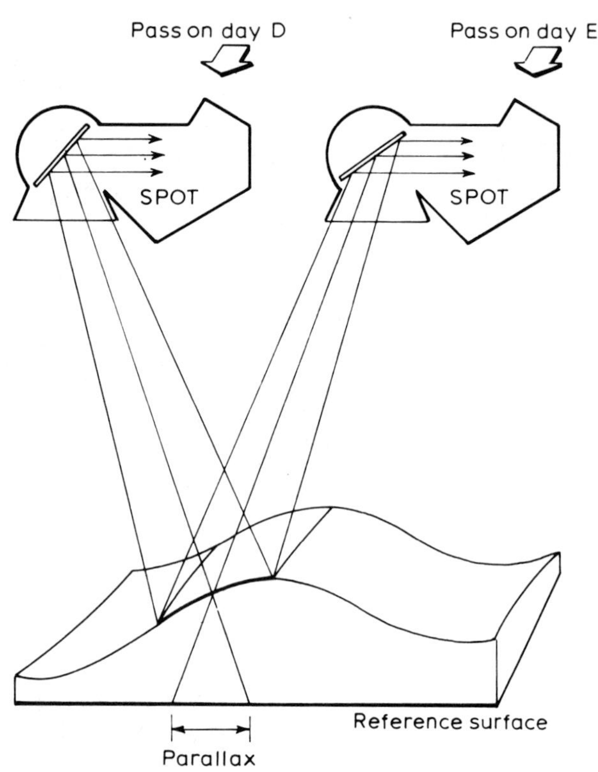

Fig. 5.18 The French satellite SPOT carries two High Resolution Visible (HRV) imaging systems designed to permit stereoscopic viewing of a given scene by making two observations on successive days such that the two images correspond to pointing angles on either side of the vertical. Stereoscopic imagery from SPOT is very useful in photo-interpretation and photogrammetry for a wide range of practical applications.

Table 5.4 Orbital and sensor characteristics of the SPOT-1 and SPOT-2 satellites (Source: NRSC, Farnborough)

Orbital parameters
Orbit: near polar, sun-synchronous
Altitude: 832 km
Inclination: 98.7°
Timing: crossing Equator at 10.30 hours
Local sun time (from N to S)
Repeat cycle: 26 days

Sensors: High Resolution Visible (HRV)

	Wavelength (μm)	Resolution (m)
Multi-spectral mode:		
Band 1	0.50–0.59	20
Band 2	0.61–0.68	20
Band 3	0.79–0.89	20
Pancromatic mode:	0.51–0.73	10

instruments and orbital parameters are shown in Table 5.4. It will be apparent that the ground resolution is higher than that obtained in Landsat TM data, particularly in the panchromatic (black and white) mode which offers 10 m resolution.

SPOT is able to record images of the same area acquired at different viewing angles during successive satellite passes, and by this means stereoscopic pairs of images of a given scene can be obtained (Fig. 5.18). The main applications of stereoscopic imagery obtained in this way are as follows:

1. compilation of topographic maps with uniform contour intervals of 20–50 m;
2. direct compilation of digital terrain models;
3. improved perception and interpretation of large-scale vegetation and man-made features.

We may note that the SPOT satellite cannot sense wavelengths longer than the near infrared. In this respect it contrasts with the Landsat TM data which extends into the thermal infrared bands.

SPOT characteristics also include:

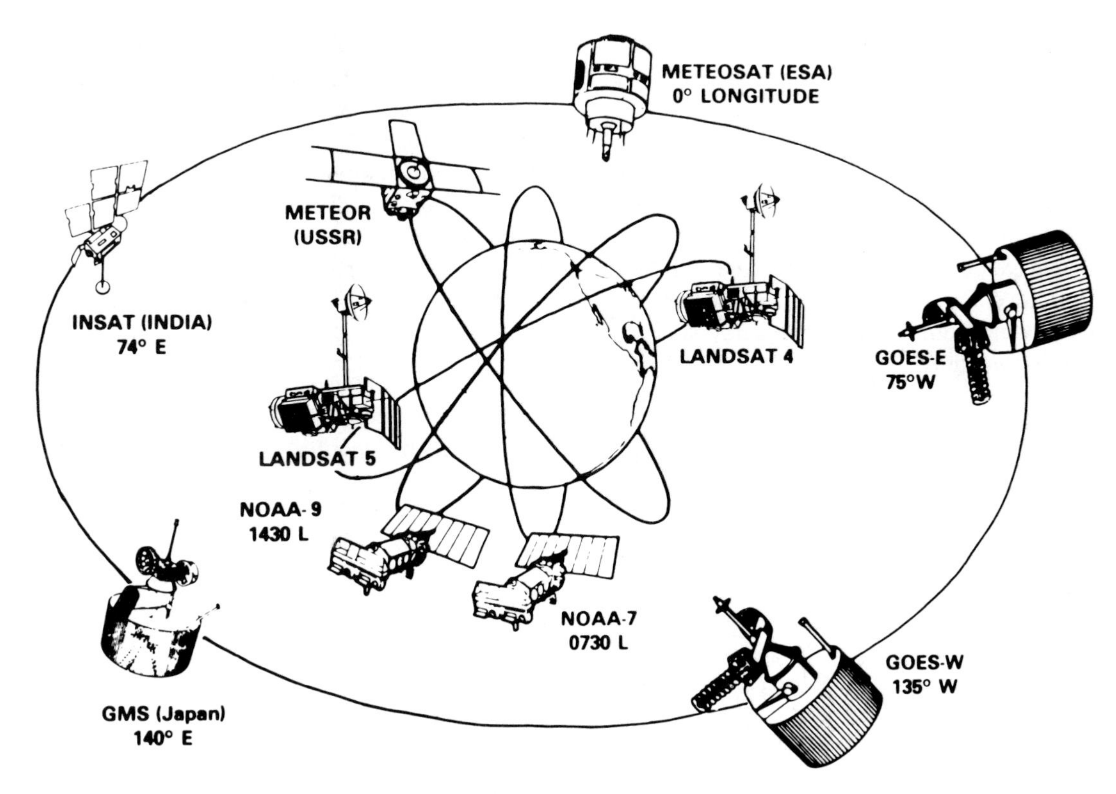

Fig. 5.19 The civilian operational satellite system of the late 1980s included satellites from five different countries and the European Space Agency. The geostationary satellites view Earth from a distance of over 35 000 km; the polar-orbiting satellites are at a distance of about 850 km. (The French and Japanese Earth resources satellites are not shown.) (Courtesy, NASA.)

1. nadir viewing (near-vertical) providing a swath width of 117 km;
2. off-nadir viewing (oblique pointing facility) up to 27° from the vertical;
3. revisit possibilities to increase the frequency of imaging individual sites;
4. stereoscopic imaging (Fig. 5.18).

If the satellite's instruments were only capable of nadir viewing the interval between repeat observations for any given point would be 26 days. However, during the 26-day period the programmable steering capability allows more frequent cover of any areas of interest, within a strip 950 km wide. This strip is centred on the satellite ground track, and by steering the sensors, the SPOT satellite can observe a point on the Equator seven times during the 26-day period. Further north, for example in northern England, it would be possible to cover a point 13 times in the same period – although competition for areas to be covered complicates the management equations.

SPOT-1 data has been used in many applications but following the launch of SPOT-2 it was decided that the SPOT-1 data would no longer be available. After that date civilian users were solely dependent on SPOT-2.

5.9 The meteorological satellites

Detailed assessments of the role of meteorological satellites will be made in later chapters (Chapters 10 and 11). However, one may note here that the widespread need for meteorological data has led to the development of an operational Earth-observation satellite system in which Earth resources satellites (e.g. Landsat and SPOT) are supplemented by a wide-ranging global network of environmental satellites including low-level, near-polar, sun-synchronous platforms (e.g. NOAA,

Meteor) together with a separate system of high altitude, geostationary meteorological satellites, as shown in Fig. 5.19.

Weather and climate are vital factors affecting environmental management in all parts of the world. The well-equipped NOAA polar-orbiting satellites are of particular interest in environmental management on a global scale. NOAA currently aims to maintain two polar-orbiting satellites viewing both the night and day sides of the Earth for two global views each day. For Earth surface and lower atmosphere monitoring the primary instruments are the Advanced Very High Resolution Radiometer (AVHRR) and the Television and Infrared Observation Satellite (TIROS) Operational Vertical Sounder (TOVS).

The AVHRR instrument has five channels in the visible, near infrared and thermal infrared, each with a 1.1 km ground resolution, increased to six channels with the NOAA K, L, M and N satellites launched from 1994. From 1997 to 2000 it is expected that a further development of the AVHRR will be flown, namely a seven-channel VIRSR (Visible and Infra-red Scanning Radiometer). Although the spatial resolution of the AVHRR is considerably coarser than that of Landsat its data are very useful for some aspects of applications. The repetitive coverage (four times per day) improves the chances of obtaining cloud-free imagery and allows the monitoring of large-scale and regional features. In particular global *vegetation index* maps can be produced from the visible and near infrared data depicting the general health of vegetation (Chapters 13 and 16). Table 5.5 shows the characteristics and evolution of AVHRR instruments.

The vegetation index mapping reduces the effects of cloud cover by temporarily compositing geographically registered data sets over weekly imaging periods. Each day the normalized difference vegetation index (NDVI) is calculated. The NDVI can be expressed as:

$$\frac{\text{IR band} - \text{red band}}{\text{IR band} + \text{red band}} \tag{5.1}$$

Its sensitivity to green vegetation is based on the radiance reflected in the red band being related to the amount of chlorophyll, and the radiance in the IR band being related to the density of green leaves (see also p. 272). The infrared channel itself also has been found to be of value in locating forest fires. The main advantage that NOAA AVHRR data provides is that they can support multi-temporal approaches to vegetation and terrain classification much more readily and cost effectively than with Landsat or SPOT data.

Other elements of the NOAA payload are summarized in Chapter 10.

5.10 Geostationary meteorological satellites

When complete, the geostationary satellite system of the World Meteorological Organization's World Weather Watch includes the US Geostationary Operational Environmental Satellites (GOES) West and East, ESA's Meteosat, Japan's Himawari, and India's Insat. The GOES are operated by NOAA. GOES West covers the Western Americas and the Pacific, whereas GOES East covers the eastern Americas and the Atlantic region. The GOES sensors record visible (0.66–0.77 μm) and thermal infrared (10.5–12.6 μm) wavelengths. GOES images can be produced every 30 min. These are essentially aimed at regional or large-scale investigations, although their use has been investigated for assessments of thermal regimes of soils and fire risks in forests.

The Meteosat satellites are operated by the European Space Agency (ESA) and are in geostationary orbit located on the Greenwich meridian over the Equator. Meteosat records images every 30 min in three wavebands. The main orbit and sensor characteristics are shown in Table 5.6.

In addition to their use in connection with the production of 4–5-day weather forecasts, Meteosats also have been used for a wide range of terrestrial studies, including:

1. monitoring regional vegetation changes in response to drought;
2. mapping soil and vegetation boundaries particularly in savannah and semi-arid African regions;
3. monitoring the coverage of swamps and salt pans in Africa;
4. evaluating biomass and crop production;
5. estimating surface and plant canopy temperatures;
6. hazard monitoring;
7. modelling drainage basin characteristics;
8. monitoring ocean features, including currents, areas of up-welling, and sea ice.

Table 5.5 Characteristics of the evolving family of multichannel radiometers on NOAA satellites, 1972–2000 (Courtesy, NASA)

	VHRR	AVHRR/1	AVHRR/2	AVHRR/3	VIRSR
Date first flown	1972	1979	1981	1994	1997–2000
Spatial resolution (m)	900	1100	1100	1100	1100
Thermal resolution (°C)	0.5	0.12	0.12	0.12	0.10
Radiometric resolution (bits)	8	10	10	10	12
Number of channels	2	4	5	6	7
IR calibration	Yes	Yes	Yes	Yes	Yes
Visible calibration	Yes	No	No	No	Yes

Table 5.6 Orbital and sensor characteristics of the Meteosat satellites (Source: NRSC, Farnborough)

Launch	Wavebands (mm)
Series initiated by	0.4–1.1
Meteosat-1:	5.7–7.1
November 1977	10.5–12.5
Orbital parameters	Resolution (at nadir)
Orbit: geostationary	visible: 2.4 km
Altitude: 35 900 km	infrared: 5.0 km
Position: 0° latitude, 0° longitude	Data acquisition every 30 min

An importance feature of Meteosat is its ability to collect and relay information from various Data Collection Platforms (DCP), such as land stations, ocean buoys, ships, aircraft, balloons and other satellites. The specification for the standard Meteosat version of the data collection platform (DCP) provides for operation on three channels and a data storage capacity of 5192 bits.

The Indian and Japanese geostationary meteorological satellites have been generally similar to GOES and Meteosat. As with all other operational satellite families these are evolving, with major changes anticipated in the 'GOES-Next' series from about 1994, and the Meteosat Second Generation satellites also due in the mid-1990s.

5.11 Special purpose sensors and satellites

In addition to the satellites outlined above there have been an increasing number of satellites designed to undertake specific tasks. Several of these missions are of short duration so that particular sensor systems and sensor packages can be tested. The following brief review aims to provide the reader with an outline of the more successful of such systems and their relevance to environmental monitoring.

5.11.1 The Heat Capacity Mapping Mission (HCMM)

The Heat Capacity Mapping Mission was launched by NASA in April 1978 and operated until September 1980. The HCMM occupied a circular, sun-synchronous orbit at an average of 620 km, and observed in the visible/near infrared (0.5–1.1 μm) and thermal (10.5–12.5 μm) regions of the spectrum. The data observed at about 500 m and 600 m respectively. The main objectives of the HCMM programme were to conduct research into the feasibility of using day/night thermal infrared remote sensing data for the following:

1. rock type discrimination and mineral resource location;
2. measurement of plant canopy temperature at frequent intervals to determine evapotranspiration and water stress;
3. measurement and monitoring of soil moisture change;
4. mapping natural and man-made thermal effluents;
5. detection of thermal gradients in water bodies;
6. mapping and monitoring snow fields for water run-off prediction;
7. measuring the effects of urban heat centres;
8. monitoring marine oil pollution.

5.11.2 Seasat

This satellite, specially developed for ocean monitoring, completed 106 days of successful operation before a sudden failure in October 1978. Although Seasat was designed primarily for oceanographic mission objectives carrying a radar altimeter, radar scatterometer, and microwave imaging radiometer, its SAR data were possibly of the greatest environmental interest, being used to study a variety of both sea and land phenomena. High-quality SAR images enabled the following to be carried out:

1. oceanographic monitoring;
2. polar ice mapping;
3. geological mapping to detect lineaments, folds, faults, fractures;
4. drainage network analysis;
5. crop type and growth stage;
6. vegetation mapping and soil studies;

The most important feature of Seasat for environmental monitoring and management was that it demonstrated comprehensively for the first time the use of an active microwave sensor, which was an all-weather capability. There is little doubt that had Seasat not failed the data would have found wide applications. It is interesting to note that the first Japanese MOS (Marine Observation Satellites) (section 5.6) and the Earth resources satellite launched by the European Space Agency, namely ERS-1, have similar radars as their prime sensors, as detailed below.

5.11.3 The Nimbus-7 Coastal Zone Colour Scanner (CZCS)

The CZCS was launched on Nimbus-7 in 1978, and continued to provide good data until the late 1980s. It was the first satellite sensor dedicated to monitoring coastal zone and ocean environments. This was a five-channel instrument resolving at c. 800 m, with four visible waveband channels, plus one in the infrared. Its repeat cycle was nominally 6 days. In particular the main objectives of CZCS included:

1. mapping and measurements of suspended materials and other phenomena over large areas of water;
2. improving scientific knowledge of marine ecosystems;
3. assessment of existing fisheries and potential fishing grounds;
4. experimenting with real-time data acquisition and the rapid production of sea parameter maps;
5. defining requirements for future ocean monitoring instruments.

The new SeaWifs (Sea-viewing, Wide Field-of-View Sensor) satellite planned for launch by the USA in 1993 is expected to be a worthy and much-needed successor to the CZCS for ocean colour monitoring and analysis.

5.11.4 The European Remote Sensing Satellites (ERS)

In order to achieve full benefit from remote sensing it is necessary to look towards all-weather systems based on microwave/radar techniques. The first European Remote Sensing Satellite (ERS-1) is representative of a new generation of space missions seeking for global measurements irrespective of cloud and sunlight conditions. The ERS-1 satellite is shown in Fig. 5.20. Significant advances in our knowledge of the oceans, ice bodies, sea-surface winds, and coastal processes are sought from ERS-1 data. The satellite payload (partly illustrated by Fig. 4.25) is made up as follows:

1. An Active Microwave Instrument (AMI), which operates in three different modes.
2. A Radar Altimeter, which provides accurate measurements of sea-surface elevation, significant wave heights, sea-surface wind speeds and various ice parameters.
3. An Along-track Scanning Radiometer (ATSR) and Microwave Sounder combining infrared and microwave sensors for the measurement of sea-surface temperature, cloud-top temperature, cloud cover, and atmospheric water vapour.
4. Precise Range and Range-Rate Equipment (PRARE) for the accurate determination of satellite position and orbit characteristics and for geodetic 'fixing' of ground station.
5. A Laser Retro-reflector (LRR) for the measure-

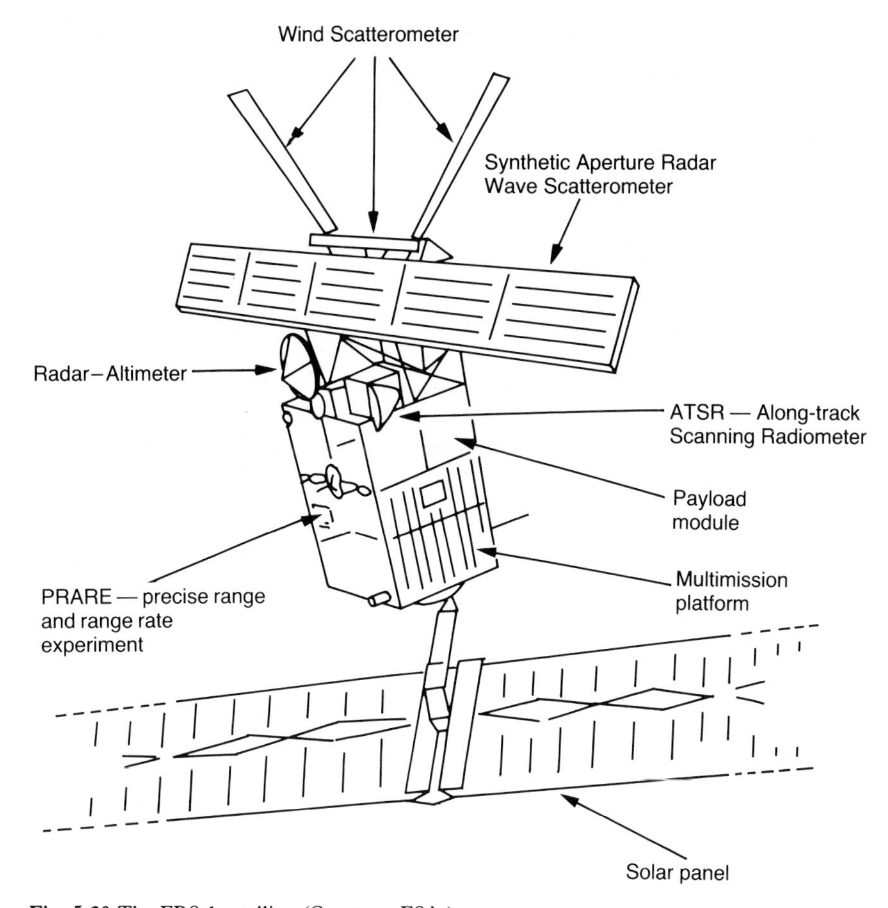

Fig. 5.20 The ERS-1 satellite. (Courtesy, ESA.)

ment of satellite position and orbit using laser ranging stations on the ground.

ERS-1 has a sun-synchronous circular orbit (quasi-polar) with a mean altitude of 785 km and an inclination of 98.5°. The satellite is designed to provide various repeat cycles between 3 and 176 days. The 3-day repeat cycle provides for frequent revisiting of dedicated calibration sites under constant geometrical and illumination conditions. The 3-day cycle also provides for slightly different phases to give highly repetitive coverage of ice zones during the Arctic winter. The main limitations of the 3-day cycle are restricted coverage of the imaging SAR and wide separation of the Radar Altimeter tracks.

Meanwhile, a 35-day cycle provides for SAR imaging of every part of the Earth's surface and density of the Altimeter tracks increases to give a separation of just 39 km at 60° latitude. A 176-day cycle provides measurement of the mean sea-surface and ocean geoid. However, the conflicts with other requirements means that this cycle can be used only sparingly late in the mission programme.

The ERS-1 ground segment and user interfaces involve four processing and archiving facilities, in Germany (Oberpfaffenhofen), France (Brest), UK (Farnborough), and Italy (Matera), which are the main centres for the generation of off-line precision products and the archiving of ERS-1 data and products.

5.12 The ground segment and satellite data distribution

It will be evident that a great deal of data from different sensors and platforms has been collected in the 1970s and 1980s, necessitating increasingly careful attention to all aspects of the satellite remote sensing 'ground segment' whereby satellites

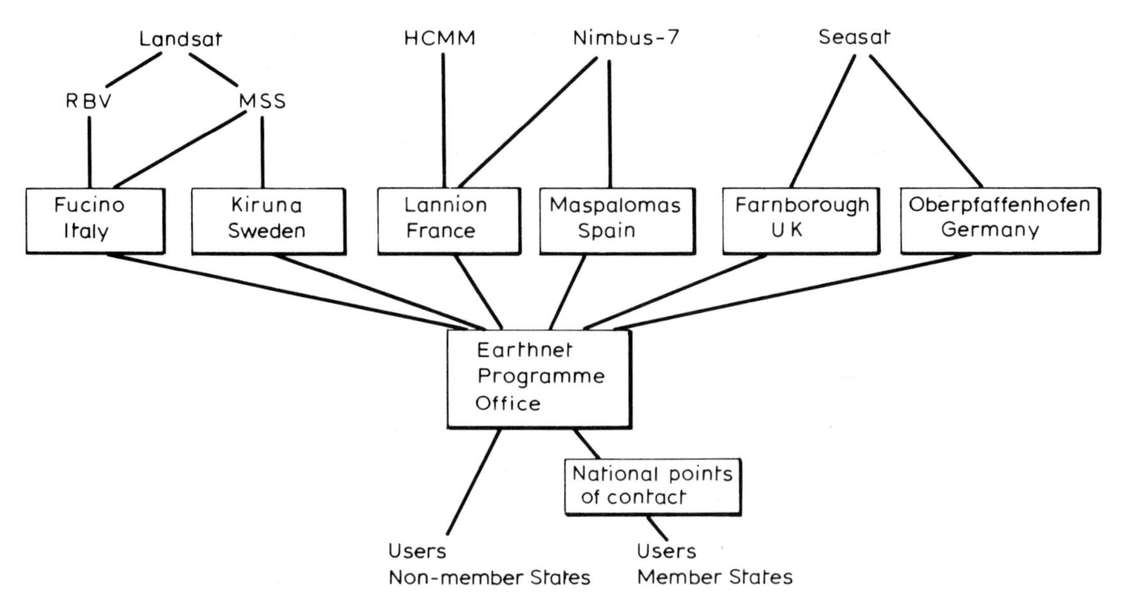

Fig. 5.21 The original Earthnet programme. (Source: ESA.)

are controlled, data received, and in which use is made of the data received. Even greater volumes are expected in the 1990s calling for more highly organized systems of data reception, archiving, and dissemination. In Europe the need for a central organization to handle data from satellites not designed for direct read-out has led to the creation of Earthnet as a focal point for European users of remote sensing products. The Earthnet Programme Office is sited at the European Space Agency establishment near Frascati in Italy. It is responsible for programme management, monitoring of the network, user interface, data distribution and preparation and data requirements in future expansions of European activities. A schematic diagram of the early Earthnet programme is given in Fig. 5.21.

Each Member State of the European Space Agency has a National Point of Contact (NPOC) through which the tasks of Earthnet are carried out. Users in Europe should apply to their National Point of Contact for remote sensing data and these are listed in Table 5.7. The United Kingdom NPOC is located at National Remote Sensing Centre, Space Department, Farnborough, whence further information on available imagery can be obtained.

Similar data distribution centres have been established in other countries and regions of the world, some being government-sponsored, others (increasingly) being commercial or quasi-commercial organizations which have been licensed to function between the satellite operator and the user/customer. In the USA, for example, the Earth Resources Observation System (EROS) Data Center, Sioux Falls, South Dakota 57198 is the primary source of Landsat data, though the partial commercialization of the Landsat programme has led to the establishment of a new Landsat data marketing company, Eosat, with offices and representatives in several countries of the world. The headquarters of Eosat are located at 4300 Forbes Boulevard, Lanham, Maryland 20706.

The most important source of data from American meteorological satellites is NOAA/NESDIS through its Satellite Data Supply Division (SDSD), Room 100, Princeton Executive Center, Washington, DC 20233. Comparable sources have been set up to supply or market data from most of the other major meteorological satellite operators. Of special interest to research users of environmental satellite data will be the special 'Pathfinder' data sets now being prepared from US satellite data archives paying special attention to commonly required degrees of preprocessing, and quality control. These will provide the best possible long-term data sets for global change research, and will be available through SDSD and other outlets.

Table 5.7 National points of contact in Europe for the Earthnet system

Belgium
Services de Programmation
de la Politique Scientifique
Rue de la Science 8
1040 Bruxelles
Tel 02-2304100 Twx 24501 PROSCI B

Denmark
National Technological Library
Centre for Documentation
Anker Engelundsvej 1
2800 Lyngby
Tel 02-883088 Twx 37148 DTBC

Netherlands
National Aerospace Laboratory
NLR
Anthony Fokkerweg 2
1059 CM Amsterdam
Tel 020-5113113 Twx 11118 NLRAA NL

France
GDTA, Centre Spatial de Toulouse
18 Avenue Edouard Belin
31055 Toulouse
Tel 61-531112 Twx 531081 F CNEST

Germany
DFVLR Hauptabteilung Raumflugbetrieb
8031 Oberpfaffenhofen
Post Wessling
Tel 08153-28740 Twx 526401

Ireland
National Board for Science & Technology
Shelbourne House
Shelbourne Road
Dublin 4
Tel Dublin 683311 Twx 30327 NBST EI

Italy
Telespazio
Corso d'Italia 43
Roma
Tel 06-8497306 Twx 610654
TELEDIRO

Spain
CONIE
Pintor Rosales 34
Madrid 8
Tel 34 1 2479800 Twx 23495
INVES E

Sweden
Swedish Space Corporation
Tritonvagen 27
S-17154 Solna
Tel 08 980200 Twx 17128
SPACECO S

Switzerland
Bundesamt fur
Landestopographie
Seftigenstrasse 264
CH-3084 Wabern
Tel 031-541331 Twx 32498
UEM CH

UK
National Remote Sensing
Centre
Space Department
Royal Aircraft Establishment
Farnborough
Hants GU14 6TD
Tel 252 24461 Twx 858442
PE MOD G

Because of the very rapid rate of growth of satellite data types, products, uses, and dissemination channels, the reader must be directed elsewhere for more detailed and comprehensive lists of satellite data sources. Indeed, these are areas of remote sensing which have grown more rapidly than perhaps any other since the second edition of this book was published in 1982. One of the best overall summaries of global sources of remotely sensed imagery is the *Keyguide to Information Sources in Remote Sensing* edited by Edward Hyatt, and published by Mansell in London and New York in 1988. This gives details of many remote sensing data catalogues, and data sources throughout the world. However, the problem of unravelling the route to be followed between a specific remote sensing data need and actual acquisition of the data is very like the proverbial ball of string: quite easy once a suitable lead has been identified. In cases of difficulty, most national remote sensing centres or points of contact will be happy to provide advice, and even practical help if necessary.

6 *Calibration, evaluation and validation in remote sensing*

6.1 Introduction

In the early years of remote sensing the extent of ground checking was usually somewhat limited and statistical tests for accuracy were less demanding than are commonplace today. In the case of aerial photography effort was expended on the ground surveying necessary to establish ground height control points. These were then used to set up stereoscopic models for photogrammetric mapping. This work is still necessary but the extension of remote sensing techniques into other wavelengths and into spectral modelling of surface features has resulted in a new range of *in situ* data collection procedures which, more recently, has embraced surface remote sensing technology also. The use of remote sensing techniques by airborne and satellite platforms for most present-day purposes demands that there should be some method of *calibrating* and *evaluating* the sensors in use. In addition there is a need for checking the accuracy of interpretations made from the data, i.e. for *verifying* or *validating* remote sensing programme results.

In the case of calibration and evaluation there are examples where measurement of a parameter, such as temperature, is necessary. For instance there are airborne thermal sensors designed to detect relative differences in temperature and display them to advantage, i.e. they are self-regulating systems that alter the scale over which a measurement is made according to the dynamic range in the target area. In such cases it is essential to measure temperature at points within the target area in order to obtain some absolute values from the sensor data.

In other cases it is helpful to make detailed measurements of the spectral characteristics of surface features using remote sensors very close to the target so that its spectral responses can be modelled and the interpretation of airborne or satellite data can be facilitated. One such study involves *field spectroscopy*, which is discussed below as an example of this increasingly important means of bridging the gap between conventional evaluations of the environment and those based on aircraft or satellite remote sensing.

It has been commonplace to use the term 'ground truth' for observations made on the surface of the Earth in contrast to remote sensing data. This term is now being abandoned in favour of *in situ* data because the measurements are often made in water ('sea truth'), air ('air truth'), etc. rather than at the ground surface and, even more important, there is a growing realization that the data so collected are often subject to error, i.e. may not be 'true' at all. A further term which may be preferable is 'collateral data' (section 6.3), although this is a broader term which more properly covers other types of information bearing on the phenomena to be measured by remote sensing, but of a different, perhaps secondary or supplementary kind, e.g. soil maps which may help in the interpretation of satellite vegetation algorithm outputs.

Mostly such data are based on samples taken at arbitrary points and at intervals of time, so there are errors attached to such observations which need to be assessed. For example, if sensor data are being used to identify agricultural crops it is necessary to know the condition of a sample population of fields before assessing the accuracy of interpretation of the remote sensing data set itself. It is now recognized that there are many factors which may introduce errors into such field surveys. These include boundary/locational problems, ambiguity in crop classification, inconsistency between surveyors and time lapses between the surveys and the acquisition of the related remotely sensed data. Further, the range of observations and so called 'surfaces' observed or observable is potentially very broad. Sometimes it is difficult to define the measured 'surface' very precisely, for example it may be necessary to monitor meteorological factors, such as wind-speed, solar radiation, rainfall, atmospheric humidity and cloud cover, in order to assess the effects of atmospheric conditions on sensor performance. Such measurements may be made close to the ground (microclimatic observations), others may be obtained from standard meteorological screens (mesoclimatic observations), but some may be recorded at considerable heights in the atmosphere by radiosonde techniques.

Similarly a wide variety of methods are employed for measuring *in situ* conditions in water bodies. Oceanographic observations may include measurements of sea temperature, salinity, wave motion (height and wavelength), and biological content as well as meteorological conditions. These measurements are made from a variety of platforms including weather ships, coastal protection vessels, automatic data collection buoys and coastguard stations.

In estuaries and rivers additional factors, such as suspended sediment load, biological oxygen demand, water reaction and pollution, are often recorded. These observations may be made from boats or by automatic samplers used from the banks.

In this chapter we shall draw examples mainly from ground and water surface studies, but the general problems of data collection, such as sampling and selection of sites for observations, are relevant to other types of surfaces also. The data transmission methods are always essentially the same in each case.

It should be noted that large-area atmospheric, hydrologic and oceanographic studies commonly depend on data from extant stations or station networks. Special efforts to collect *in situ* data in these contexts are often necessary.

Finally, a word about the concept of 'conventional observations'. Sometimes this is taken to be a synonym for *in situ* data. However, the concept should be used with care, for some remote sensing methods (e.g. weather radar methods in meteorology) have become so well-established as to be viewed by some at least as being 'conventional' *vis-a-vis* satellite techniques. Therefore, like some other terms and concepts discussed in this section the concepts of 'conventional data' or 'conventional observations' are generally best avoided.

6.2 Selection of ground data sites

The location of areas for ground data collection in support of particular aircraft and/or satellite remote sensing 'campaigns' may be decided on the basis of a number of criteria. These include study objectives, sample size satisfactory for statistical purposes, repeatability and continuity of the experimental study, access to the study area, availability of existing *in situ* data for the area, personnel, equipment resources and the orbit characteristics of the space platform.

The shape of the ground data collection area is dependent on statistical requirements and speed of access. The allocation of sample plots in a data collection area is made more statistically efficient if the area can be stratified into relatively homogeneous areas. For stratification to be useful, strata boundaries should separate areas where within-class variance is less than between-class variance. The number of survey plots then can be estimated by the method of proportional allocation. Thus, the shapes of homogeneous areas may influence the shapes of ground survey areas. In addition, the sampling technique used in the ground observations (grid, area, line) may influence the shape of the area, for example, line sampling along existing road networks may be preferred to block area sampling.

The size of the ground data collection area will be affected by study objectives, statistical considerations, scale of the ground phenomena, angle of view of the sensors and time factors. Where sensor testing is the objective, it is desirable to select the smallest ground area that allows detailed ground monitoring by equipment and personnel over a wide range of ground/atmospheric conditions. Many small sites may be necessary if it is desired to check on the validity of a spectral signature for a particular surface condition. An example of such a site is that used by one of the authors at Long Ashton near Bristol (Fig. 6.1). The total size of the site is 120 ha within which smaller areas of approximately 2 ha were repeatedly monitored.

Where relationships are sought between the sensor response and particular surface conditions

Fig. 6.1 Ground truth site, Long Ashton. Note the soil moisture measuring equipment in the foreground, including neutron moisture probe and tensiometers. (Photo: L.F. Curtis.)

(e.g. a particular crop such as grass) it is necessary to obtain sufficient samples of the crop (generally more than 30) to allow statistical tests to be carried out. The size of area which will provide sufficient samples must be determined.

Time constraints affect the size of ground data collection areas through:

1. quantity of data required;
2. resources available for collection;
3. the rate of change in ground environmental conditions.

For example, where evaporation rates are high it may be necessary to monitor changes in soil moisture content frequently (several times a day) whereas in conditions of low evaporation loss the soil moisture change may be slow and fewer observations are required in a given time.

The prime objective of ground data collection is to provide a *contemporaneous record* of ground conditions at the time of imagery. In practice it is difficult to obtain synchronous data for more than a small area or selected sample sites. The aim, however, is to obtain sample ground truth data

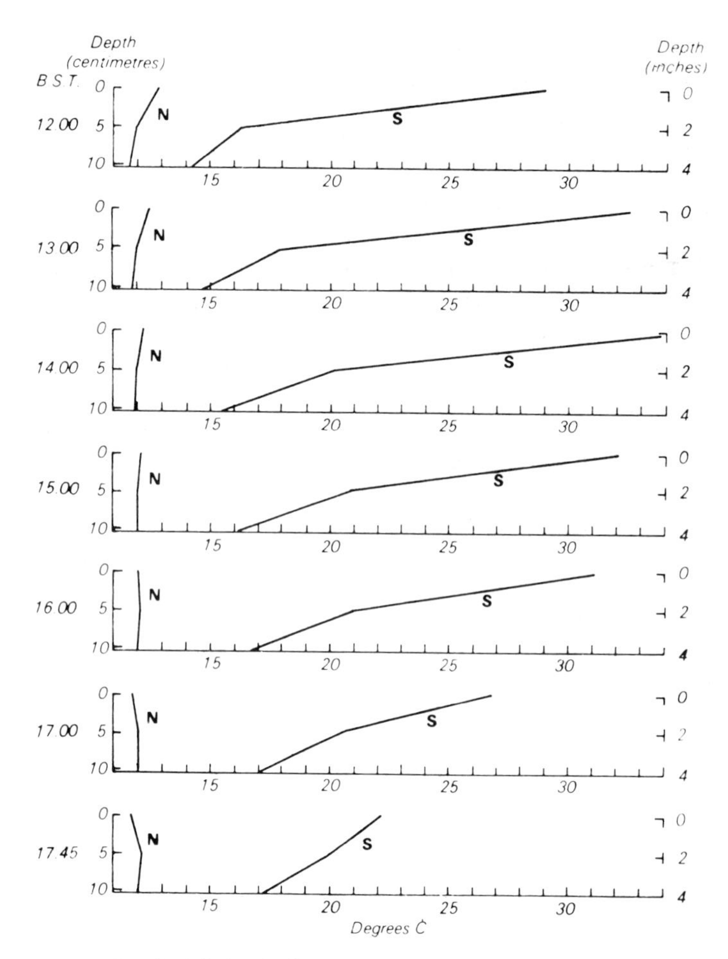

Fig. 6.2 Contrasts in soil temperatures on north and south facing slopes on Exmoor, England. (Source: Curtis, 1971.)

Fig. 6.3 Field examination of rock and soil samples in relation to air survey. Adequate labelling and field description are essential elements of camp work. (Courtesy, Hunting Technical Services Ltd, Elstree.)

within a short time of the acquisition sensor data. In planning ground data collection, special attention should be given to the rate of change of the variables to be observed. These variables can be categorized as *transient* or *non-transient*. Data recording of transient features (e.g. crop stage, leaf cover, wind speed, surface moisture) must be near synchronous. Recording of non-transient features (e.g. slope, aspect, soil texture) can be carried out prior to, or after, the sensing mission.

As an example of rate change in a transient surface condition one may note that data for soil temperature variation on slopes of an Exmoor Valley (Fig. 6.2) show that frequent observations are necessary on south facing slopes, but the rate of change on north facing slopes is much less.

The nature of the ground data required varies according to the type of investigation being made. Geological surveys often demand that rock and soil samples be taken for analysis and description (Fig. 6.3). Hydrological studies require a range of information including stream gauging, suspended sediment contents, water-table measurements and local climatic data. In studies of soil conditions the ground truth data normally include records of soil phase (or soil series), soil moisture, soil temperature, soil mixture, structure, stoniness, organic matter content, soil colour and bulk density.

Some data can only be obtained a short time before the remote sensing mission takes place because the phenomena under investigation are continually changing. For example, in crop studies it is necessary to observe transient features such as the type, stage height and colour of crop, disease types, weed species, effects of husbandry (ploughed, harrowed, drilled, rolled, wheeling marks), the grazing method, number of livestock and per cent crop cover.

Other types of data are more permanent and features such as the morphology of the terrain normally can be recorded by field survey and analysis of contour maps before the sensing missions take place. The morphology of the ground over which the mission is carried out is an important element in respect of data interpretation. Gradient, slope form and aspect often have significant effects on sensor data and this is particularly the case where radar data is being used. Terrain classification and evaluation techniques are now well developed and they can be used for recording site morphology at various levels of detail.

6.3 Collateral data for crop studies

In this section selected aspects of supplementary or collateral data collection for crop and sea monitoring must be reviewed. It must be emphasized, however, that each combination of sensors and user applications requires careful consideration before any particular mission takes place: frequently, the nature of the data to be collected will be affected by the geographic location of the area of investigation.

The range of collateral data required varies according to the farming region and its soil, relief and climatic conditions. Furthermore, the cost of acquisition of remote sensing imagery often determines that it be used for more than one purpose, e.g. in the agricultural context for crop recognition, soil drainage mapping and land quality evaluation. Generally, four categories of data are required for multipurpose land use studies in rural areas as follows (Fig. 6.4):

1. site morphology;
2. crop/vegetation cover characteristics;
3. cultivation/husbandry features;
4. soil surface conditions.

The range of data to be collected also should be related to the organizational structure and personnel resources. A summary table (Table 6.1) gives manpower requirements for ground data collection in respect of remote sensing studies in three English counties. The prime objective of ground data is to provide a contemporaneous record of ground conditions at the time of imagery. In practice it is rarely possible to obtain detailed synchronous agricultural data for more than a small area or selected sample sites. In planning ground data collection, special attention should be given to the rate of change of the variables to be observed. These variables can be categorized as transient or non-transient. Data recording of transient features (e.g. crop stage, leaf cover) must be nearly synchronous. For example, data for spring barley in Nottinghamshire showed that mean percentage leaf cover increased from 18 to 40 in a period of 8–10 days in the first half of May. These changes are of sufficient magnitude to necessitate repetition of

DATE .... LAND USE * .... FIELD REF ....

CROP CONDITIONS

Stage *

Height: average
range
pattern *
extent
comment

Colour: general *
pattern *
extent *
comment

Disease: type
extent *
comment

Weeds: species
density *
comment

Husbandry

Ploughed
Harrowed
Drilled
Rolled
Wheelings

Grazing method
Livestock

SOIL CONDITIONS

% Soil exposed

Surface colour: general
pattern *
extent *
comment

Roughness: furrowed
normal tilth
cloddy
panned

Surface stones: abundance %
size
type

Surface moisture *

Site Morphology:

Gradient *
Slope type *

Microrelief
type
extent *
Field Boundary *

* Codes available (e.g. extent 1: < 5%; 2: 5-50%; 3: > 50%)

General Comments

Fig. 6.4 Sample data collection form for use in agricultural studies. (Source: Curtis and Hooper, 1974.)

ground data collection since the proportion of bare soil exposed beneath a growing crop will have a major effect on image response. Quantitative observations of soil exposure using quadrat sample methods are time consuming. A single observer measuring crop cover using a 50 × 50 cm, 100 point quadrat for 500 observations per field could cover approximately 25 fields per day. The most

Table 6.1(a) Assessments of manpower requirements for ground data collection (Source: Curtis and Hooper, 1974)

Area observed (km^2)	Total fields	Total observers	Date	Sampling method	Sampling density	Prior training	Progress ($km\,h^{-1}$)	Fields per hour per person
635	933	10 (Working in pairs)	June/July	Line traverse (road)	One field in four along traverse	Agricultural officers familiar with area	8	22
70	341	1	May	Line traverse (road)	Continuous – all fields on traverse	Agricultural officers familiar with area	4.4	15
70	341	1	August	Line traverse (road)	Continuous – all fields on traverse	Agricultural officers familiar with area	3.8	14

Table 6.1(b) Sample study of the time allocation in ground truth data collection

Task	Percentage of total time	
	May (45 h)	August (46 h)
Ancillary data collection – soil samples and farming operations	12	5
Data collection and recognition on traverses	48	52
Data checking and office compilation of data interpretation of imagery	40	43

successful method of estimating leaf cover in a cereal crop such as barley may be to establish a relationship between crop stage and leaf cover for a sample of fields by quadrat measurements. Data for crop stage/leaf cover relationships are shown in Fig. 6.5, leaf cover in each field having been determined by 500 point observations.

Where large quantities of ground data are to be collected and handled it is necessary to develop a system of computer storage. In these circumstances coding of data in a computer compatible form becomes desirable. Such coding is relatively straightforward where a limited range of data is to be recorded and where class intervals are known or can be predicted. In our experience, however, it is difficult to provide unambiguous codes which will cover all, or even most, land use conditions of possible significance to imagery evaluation. There is a real risk that different ground surveyors will code the same conditions differently. Attempts to devise comprehensive coding systems can, therefore, lead to complex systems which impair speed and efficiency in field surveys. Thus, where coding is employed, experimentation is necessary to test the coding system and to train field staff.

6.4 Field measurements of spectral reflectance

We commented in section 6.1 about the growing need in some application areas to make very low-level remote sensing observations in order to better model and monitor phenomena from higher altitude platforms. Field spectroscopy is one such activity.

Measurement of the spectral reflectance of different surfaces in the field environment is difficult. Yet it is preferable to make field rather than laboratory measurements since the surfaces and the conditions of illumination constructed in the laboratory cannot fully reproduce the outdoor states of reflectance as sensed by airborne or satellite systems.

Targets (from the entire bowl of the sky, e.g. when cloudy) may be illuminated either *hemispherically* as from a cloudy sky or *directionally* (for one direction only) as from direct sunlight on a sunny day, although natural targets are normally

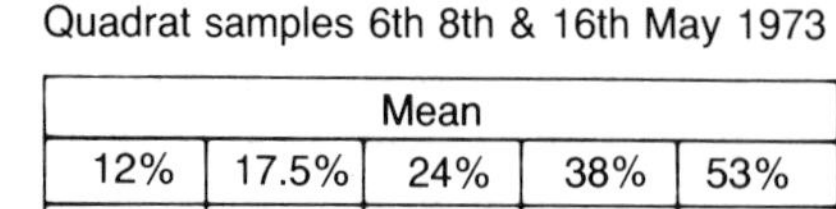

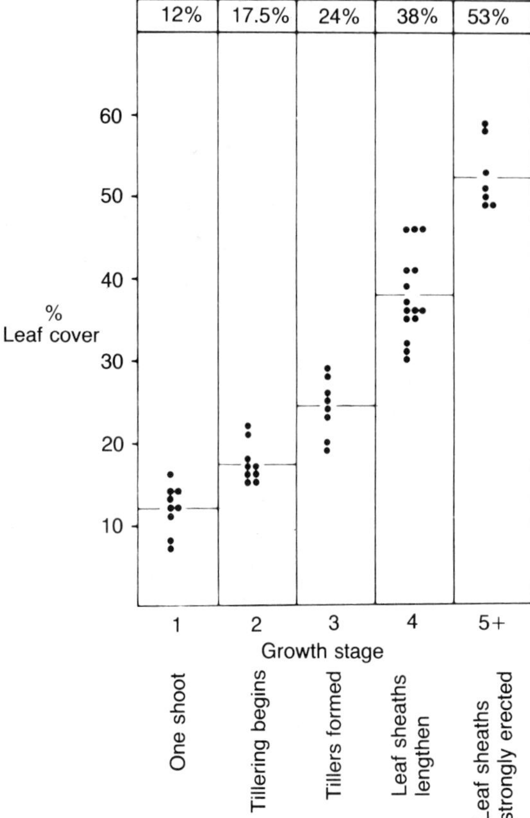

Fig. 6.5 Relationship between growth stage and leaf cover in spring barley. (Source: Curtis and Hooper, 1974.)

illuminated by the entire hemisphere of the sky. The radiation environment is composed of two distributions of electromagnetic radiation – one incoming, *irradiance*, and the other outgoing, *radiance*. The radiation geometry of the field environment in these circumstances is shown in Fig. 6.6. In practice the positions of the source of irradiation and the sensor are both defined by two angles, the *zenith* angle (angle from the vertical) and the *azimuth* angle (measured in the horizontal plane), as shown in Fig. 6.7. In order to specify the reflectance from a target completely it would be necessary to measure reflectance *at all possible* positions of the source radiation and the sensor, i.e. when the 'bidirectional reflectance distribution function' (BRDF) would be defined. Such detailed measurements are not possible in the field environment, so an alternative to the BDRF has been sought. The alternative often adopted rests on the use of a standard reflectance panel. Since a perfect reflecting panel does not exist, a correction is made for the spectral reflectance characteristics of the panel. When this is done a 'bidirectional reflectance factor' (BRF) can be determined which can be related to the BDRF.

The reflectance of a field target can be represented by the following functional equation:

$$f(\theta_i, \phi_i, \theta_\tau, \phi_\tau) = \frac{dL(\theta_r, \phi_\tau)}{dE(\theta_i, \phi_i)} \tag{6.1}$$

where dL is the radiance per unit solid angle, dE the irradiance per unit angle, and i,r the incident and reflected rays, respectively.

An alternative to the use of standard reflectance panels is to use an upward-looking spectral sensor, which shows a dependence on the zenith or azimuth angle of the incident irradiation. Both methods can be used to obtain reflectance values in the sensor viewing elevation and azimuth. The term 'spectral indicatrix' is sometimes used to describe the reflectance characteristics as measured either over a limited range of wavelengths, or over a limited set of irradiation source positions, or over a limited set of sensor positions. However, highly specialized radiometers such as the 'parabola' instrument can scan radiance and irradiance under the control of a dedicated microprocessor to provide a very large number of measurements. It may well be that such instrumentation will be used extensively in the next few decades.

Field spectroscopy is, therefore, a technique used for the measurement of spectral reflectance under field conditions and it is widely recognized that such measurements are fundamental to quantitative studies of vegetation using remote sensing. The data obtained are used in three areas of remote sensing: *calibration of data* from different platforms and sensors (or from the same sensor at different times), *prediction of best conditions* for observation and recording of data, and *modelling the reflection* from different surface structures.

The instruments used can be divided into those that allow the wavelength to be varied in a continuous fashion across a wide range (spectroradiometers or spectrometers) and those that sense a limited number of preset spectral bands (radiometers). However, recent advances in detector

INCIDENCE

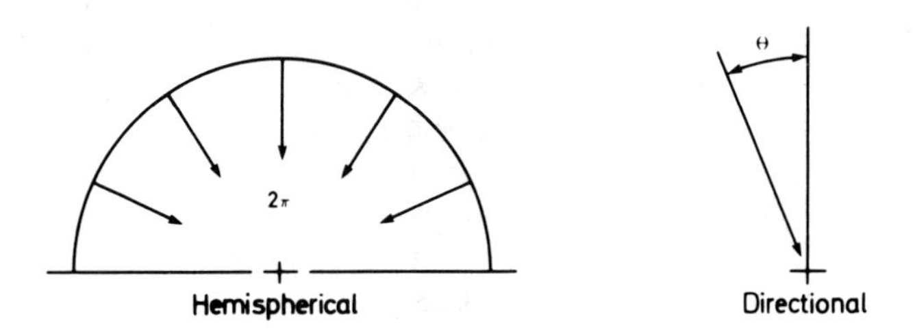

COLLECTION

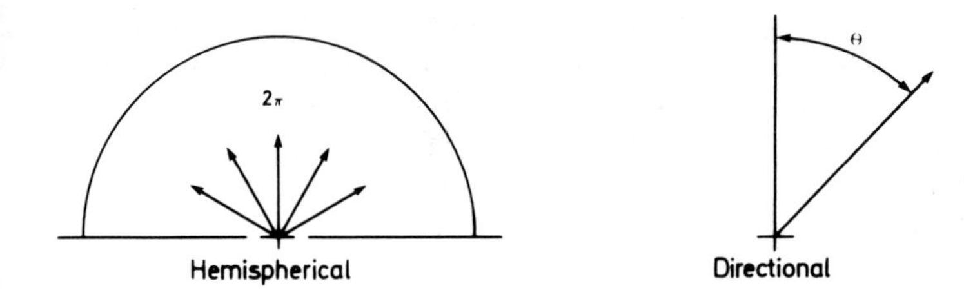

Fig. 6.6 Angular nature of reflectance measurements. A graphical description of hemispherical and directional radiation and collection. (Source: Curran, 1985.)

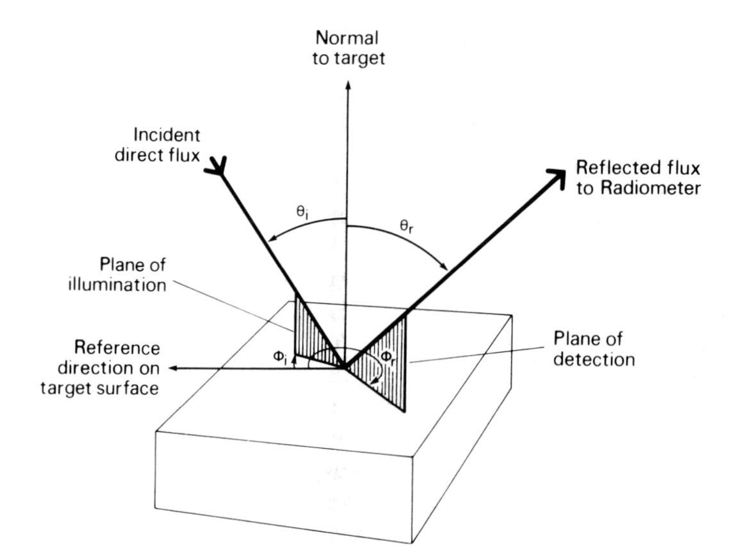

Fig. 6.7 Radiation geometry of the field environment amenable to measurement in small-site experiments. (Source: Milton, 1987.)

technology have blurred this distinction because it is now possible to use a linear array of several hundred detectors located behind an interference filter (*multispectral linear array*).

If field spectroscopic data is to have long-term value it must be collected with care and the conditions of observation must be documented in a consistent fashion. General rules for these are now emerging so that research can be standardized between different research groups. The salient points to be observed in spectral data collection are as follows:

1. Use a mast or tripod to ensure a fixed geometry between sensor and target.
2. Ensure that the sensor is at least 1 m (preferably 2 m) above the surface of the target.
3. Be consistent in always orientating the sensor

support (and positioning other equipment, and people) in the same positions relative to the Sun.
4. Check that the standard (reference) panel fills the field of view of all bands of the sensor.
5. Make contemporaneous measurements with a continuously recording solarimeter.
6. Operators should wear dark clothing and kneel some distance away during measurements.
7. Keep vehicles at least 3 m from the target.
8. Be aware that gusts of wind can have large effects on measurement (e.g. up to 60% variation in cereal barley in red wavelengths).

In order to promote the collection of field spectroscopic data in the United Kingdom the Natural Environment Research Council has set up an Equipment Pool for Field Spectroscopy in the Department of Geography at the University of Southampton. Supporting advice is given by the administrators of the Pool, which maintains a range of laboratory facilities and calibration standards. Portable spectroradiometers are available together with information masts and tripods suitable for land and water measurements.

The collection of spectral data must be accompanied by close observation and recording of data concerning the natural or man-made features which are the subject of study. In this way an archive of the spectral properties associated with different surfaces can be built up. There are, however, considerable demands to be met in describing accurately the canopy features of natural cover. Components such as green leaves, twigs and shadows require close attention in order to make accurate definitions of the structural nature of vegetation covers.

On the basis of careful field studies together with theoretical studies the modelling role of field spectroscopy is growing steadily in importance. For example, the relationship between green leaf area measurements and the green/red reflectance ratio has been tested. Likewise various 'structural models' of the reflectance from different types of vegetation canopy following general studies have been tested by field spectroscopy.

6.5 Fully integrated local area studies

It will be evident from the earlier sections of this chapter that, in addition to detailed *in situ* studies of specific sites there is a growing requirement for comprehensive studies of the nature of surface phenomena across a wide range of scales as a means of relating satellite image characteristics to different surface materials. Increasingly, comprehensive studies are now being undertaken which will further add to our knowledge and understanding of the relations between *in situ* data, low-level remote sensing data and data from aircraft and satellites, so benefiting the models and monitoring of vegetation and crops from space.

Of course there are many interactions between surface vegetation, weather and climate. In particular we need to know how land surface vegetation regulates the rate of soil moisture return to the atmosphere, ultimately influencing local weather and regional climate. Some wide-ranging and well-coordinated studies have been inaugurated to address these complex issues encountered in the real world. For example, as part of the International Satellite Land Surface Climatology Project (ISLSCP) NASA namely First ISLSCP Field Experiment, beginning with FIFE 89. This was carried out over a selected area of prairie in central Kansas using a pattern of observations tested two years earlier (Fig. 6.8). The data for the FIFE study came from:

1. automatic meteorological stations reporting on temperature, humidity, wind speed, several components of radiation flux, soil temperature and precipitation at 15 min intervals;
2. aircraft deployed to measure fluxes of water and energy to the atmosphere, and to remotely sense active and passive microwave energy reflected from the Earth's surface at different wavelengths through the spectrum;
3. satellites in the form of data from GOES, NOAA, Landsat, and SPOT sensors.

Similar experiments were undertaken for an area in the Soviet Union in the locality of Kursk in 1991, and more will follow, of increasing sophistication.

It is experiments such as these, which combine detailed data from ground, air and satellite observations, that will form the basis not only for much improved understanding of the physical system being investigated, but also for better evaluations of the strengths and weaknesses of remote sensing techniques.

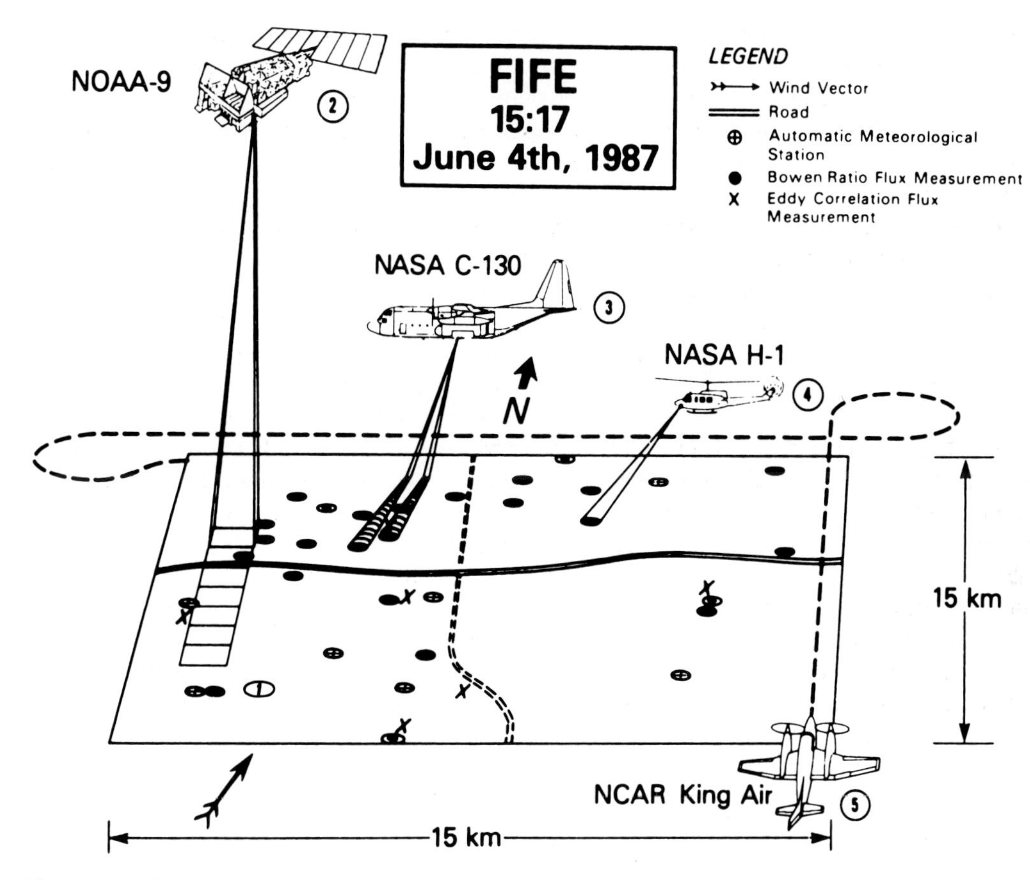

Fig. 6.8 Proof-of-concept data collection for FIFE 89: the observing network in place on 4 June 1987. (Source: Becker *et al.*, 1988; courtesy, NASA.)

6.5.1 Sea-state observations

By way of contrast to the data collection needs and problems outlined above, briefer reference may be made to the collection of sea-state information in support of oceanographic remote sensing (see also Chapter 12).

Sea observations may be obtained from a wide variety of seaborne platforms. These may include weather ships, oil rig platforms, data collection buoys and specially commissioned ships. As in the case of crop studies, the monitoring of sea conditions must be carried out at the time of the remote sensing mission. One of the chief problems facing the mission planner is that of co-ordinating and positioning ships and buoys so that the transient state of the sea can be most accurately assessed. Frequently, the most satisfactory method is to combine continuous sampling along a ship's traverse with discrete samples at selected points along the traverse.

A good example of this approach is provided by the successful collection of data for the EURASEP programme organized by the Joint Research Centre, Ispra, in 1977. In this investigation sea truth was required for analysis of the performance of the Ocean Colour Scanner (OCS). A number of ships were used to obtain coastal data off Belgium, France and Holland. In a ship's traverse, continuous samples were taken of chlorophyll *a*, turbidity, salinity, and temperatures from the top 10–15 cm of water. At particular positions discrete samples were taken from the surface and at depths of 1 m and 5 m in a vertical profile. These samples provided a 5 l quantity sample which was then subdivided while still being agitated into subsamples of 2 l each for chlorophyll and sediment analysis, 250 ml for plankton analysis and 100 ml

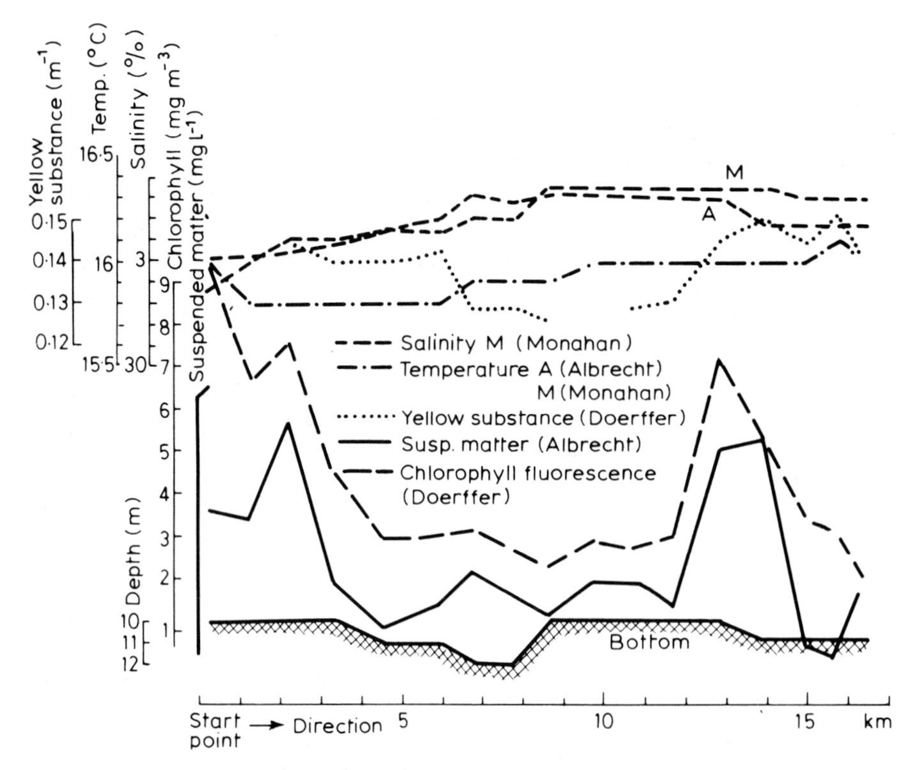

Fig. 6.9 An example of traverse data obtained in the Eurosep Programme on Track Z, 29 June 1977, off the coast of Belgium. EURASEP (European Association of Scientists in Environmental Pollution) was jointly programmed by the EEC (General Directorate XII) and the Joint Research Centre, Ispra. (Source: Sorensen, 1979.)

for yellow substance ('gelbstoffe') analysis (see Fig. 6.9). A standard format for sea station identification and sample numbering is essential in such work. Also positioning must be accurate and this is often achieved by a DECCA navigation system. A traverse speed of 5 knots has been found to be acceptable in such sea truth observations.

6.6 Data sampling and tests for accuracy

Remote sensing data from aerial or satellite platforms can provide a synoptic view of a large area of the Earth's surface. On the other hand an *in situ* survey can only be spatially selective no matter how well organized it may be. *In situ* surveys obtain data for particular sample areas. It is, therefore, important to make some assessment of the *number* of sample observations necessary in order to characterize the surface(s) being studied, and the surface data samples must be *representative* of the conditions concerned.

If the *in situ* data set is representative at a given confidence level it can then be used to check the accuracy of estimates based on the relationships between the remote sensing data (e.g. spectral data) and the target phenomena of interest. In many cases the optimum sample size is too large to collect in the limited time that exists between collection of the remotely sensed data and changes in the state (e.g. overland, the moisture) of the surface being observed. Clearly, this problem of change of state is most acute in the case of phenomena of constant flux (e.g. sea state). It is less of a problem in situations where change is slow in the feature under observation (e.g. with natural vegetation cover).

On the whole the problems of collection of accurate *in situ* data have been touched on only lightly in other remote sensing texts. Furthermore, there has been a tendency to underestimate the number of samples needed in particular studies, so that some early ground data may have included errors of around 25% of the mean. Also, there has been a lack of standardization of methods of data collection both within individual studies and between separate studies of selected phenomena, e.g. land cover. Above all there has been some reluctance to face up to the very considerable costs of obtaining valid *in situ* data.

In the case of vegetation surveys, improvements have arisen largely from the development of *ground radiometry* in order to characterize sample sites in relation to aerial and satellite radiometric observations. It is often suggested that at least 50 sample points per class (e.g. heather areas) is a ball-park figure for phenomena that change little over time. However, experience has shown that it is best to make an estimate of the sample number using the formula below.

$$\mathrm{SN} = 100\,(\mathrm{CV})^2 \tag{6.2}$$

where SN is the sample number, and CV the coefficient of variation (standard deviation divided by the mean).

Radiometric variability is often less than that of ground data for mixed land cover sites. Thus a sample of some 60–70 may be required for heterogeneous sites, whereas homogeneous sites may be characterized by some 16 samples only.

Whether the researcher is dealing with the measurement of ground data or radiometric data it is often desirable to consider a pilot study to determine the SN appropriate, bearing in mind the acceptable error of the user. This can be done using the following:

$$\mathrm{SN} = (\mathrm{CV}\ t/e)^2 \tag{6.3}$$

where SN is the sample number, CV the coefficient of variation, t the Students 't' value for $n - 1$ at the 95% level of confidence, and e the the acceptable error.

Examples of such estimates are now found in the literature, particularly in relation to green leaf area indices (GLAI) and reflectance for grassland areas. However, the researcher may be faced with the problem of deciding on the SN for a wide range of surface types. In such circumstances it may not be possible to conduct a pilot study for each one. One may then choose to generalize by characterizing three types of surface, e.g. homogeneous, transitional and heterogeneous.

It is desirable that the area of a sample site should be a function of the spatial resolution and geometrical accuracy of the remote sensing data. For a sample site (S) and a ground diameter of a pixed (P_{d}), and where the geometric accuracy of a pixel is determined in pixel units (P_{g}), the minimum area of a sample site can be expressed as follows:

$$S = (P_{\mathrm{d}}\,(1 + 2P_{\mathrm{g}}))^2 \tag{6.4}$$

Tests for accuracy applied to satellite interpretations may involve a large number of points for which collateral data has been obtained from maps, airphotos, ground survey, etc. In such cases the standard error can be used to estimate the accuracy of classification. If examination of the data shows that for a large number of points (N) the correct data are represented as (X) and the incorrect data by (Y), then the standard error (SE) can be represented as:

$$\mathrm{SE} = \frac{x\%\ y\%}{N} \tag{6.5}$$

Alternatively the accuracy of classification for a number of classification classes can be estimated by use of Contingency Tables ('Confusion Matrices'), examples of which are presented as Tables 6.2(a) and (b) for 'training sites' over which the satellite data collection is developed, and 'check sites' over which it is then tested, respectively. A measure of the discrepancy is given by the statistic χ^2 (chi-square). The larger the value of χ^2 the greater is the discrepancy between observed and expected frequencies. This test can be used to determine the goodness of fit between two sets of data. For additional references the reader is directed to the Bibliography for this chapter at the end of the book.

6.7 Data collection systems for transmissions to satellites

In modern satellite surveys a data collection system (DCS) provides the capability to collect, transmit, and disseminate data from Earth-based sensors. Normally such a system involves data collection platforms, satellite relay equipment, gound receiving site equipment and a ground data handling system. The data collection platform (DCP) collects, encodes and transmits ground sensor data to the space platform (e.g. Meteosat). The general characteristics of one data collection platform are shown in Fig. 6.10(a) and (b). Such a platform will accept analog, serial-digital, or parallel-digital input data as well as combinations of those.

In the case of the Landsat system the spacecraft

Table 6.2 Accuracy of satellite-based land use classifications: examples from a Nature Conservancy Council study in the UK
(a) For 'training sites' (method calibration)

Observed classes (in field)	Predicted classes (on image)									
	UNIMP	MG	BRACK	BOG	HEATH	ROCK	I	WATER	CONIFW	Totals
UNIMP	11	1	2	—	—	—	—	—	—	14
MG	—	4	—	1	—	—	—	—	—	5
BRACK	—	—	8	1	1	—	—	—	—	10
BOG	—	2	—	6	1	—	—	—	—	9
HEATH	—	3	1	—	6	1	—	—	—	11
ROCK	—	—	—	—	1	5	—	—	—	6
I	1	—	—	—	—	—	4	—	—	5
WATER	—	—	—	—	—	—	—	5	—	5
CONIFW	—	—	—	—	—	—	—	—	5	5
Totals	12	10	11	8	9	6	4	5	5	70

Classification accuracy $= \frac{54}{70} \times \frac{100}{1} = 77\%$

(b) For 'check sites' (method verification)

Observed classes (in field)	Predicted classes (on image)									
	UNIMP	MG	BRACK	BOG	HEATH	ROCK	I	WATER	CONIFW	Totals
UNIMP	63	2	13	9	10	1	—	—	—	98
MARSHYG	4	7	—	3	—	—	—	—	—	14
BRACK	13	2	66	5	3	—	—	—	—	89
BOG	9	5	3	12	4	—	—	—	—	33
HEATH	4	3	5	2	8	—	—	—	—	22
ROCK	—	—	1	—	4	10	—	3	—	18
I	—	—	4	—	—	—	6	—	—	10
WATER	—	—	—	—	—	—	—	11	—	11
CONIFW	—	1	2	—	—	—	—	—	7	10
Totals	93	20	94	31	29	11	6	14	7	305

Classification accuracy $= \frac{190}{305} \times \frac{100}{1} = 62\%$

acts as a simple relay unit. It receives, translates the frequency and then retransmits the burst messages from the DCPs. No onboard recording, processing or decoding of the data is performed in the Landsat system. In the design of Spacelab, however, it is anticipated that some onboard processing of DCP data from Earth sensors may be carried out.

When the DCP data is retransmitted from the spacecraft it is put on a subcarrier of the Unified S-band (USB) which allows for narrow band telemetry to three primary receiving sites (Fig. 6.10(c)).

The role of *in situ* data collection platforms can be well illustrated by reference to the Meteosat programme (Fig. 6.11). In addition to the transmission of WEFAX (Weather Facsimile) compatible data the satellites can interrogate observational platforms on ships, buoys, hydrological stations and ground stations. An Automatic Collection and Telemetry (ACT) series has been developed for operation with the worldwide meteorological satellite network. The ACT unit samples data from the sensors installed on the ground/sea truth station

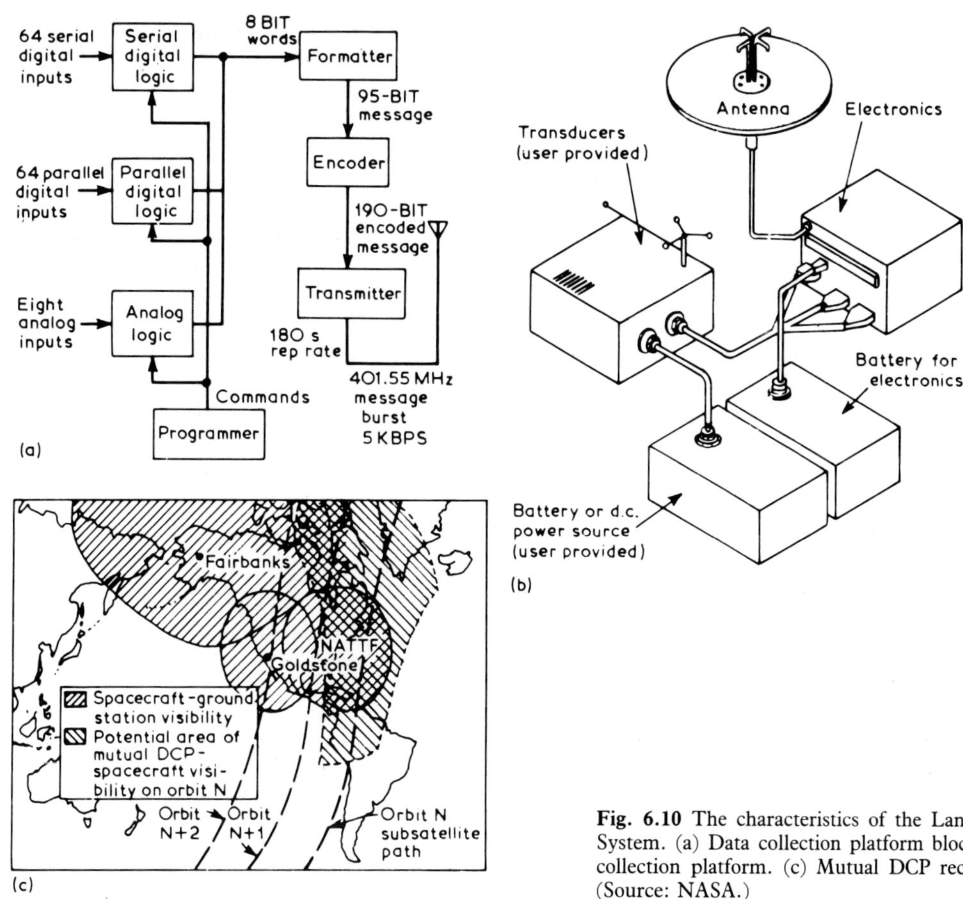

Fig. 6.10 The characteristics of the Landsat Data Collection System. (a) Data collection platform block diagram. (b) Data collection platform. (c) Mutual DCP receiving site visibility. (Source: NASA.)

at preset times, checks data limits, and correctly formats the data, which it then stores (up to 5 K bits) until the allotted transmission time. If the data is outside the preset limits (e.g. above a dangerous flood level) the unit initiates an 'alert' transmission, giving an immediate warning of dangerous conditions. The specification for the standard Meteosat version of the data collection platform provides for operation on three channels (402.0–402.2 MHz) and a data storage capacity of 5192 bits.

Data collection platforms also have been used very successfully as aids in trans-Atlantic yacht races. For example, five yachts participating in the trans-Atlantic race for two-man crews held in 1979 carried Argos meteorological keypads. These keypads were used for twice daily transmission of data at synoptic time 0 and 12 hours for the following: atmospheric pressure, water temperature, wind strength and direction, and wave height and direction. The remainder of the yachts in the race were equipped with Argos system position tracking transmitters which also carried data from a sea-surface temperature probe. Between 26 May and 15 August 1979, 13 400 location calculations were made using data gathered from 22 300 passes. The winning yacht was named 'VSD' and its position was determined 213 times over 34 days using data from just one satellite (TIROS-N). The track of the winning yacht is shown in Fig. 6.12 but it may be even more important to note that accurate location of four yachts in distress and their resultant rescue was one of the results of the Argos facility.

It will be evident that a monitoring system using surface-based sensors installed at a number of sites on ice, sea and land could be very effective if linked by satellite transmissions. Such ground-based systems would provide not only calibration data for the overhead space and airborne sensors but would also allow overhead sensors to be directed to points

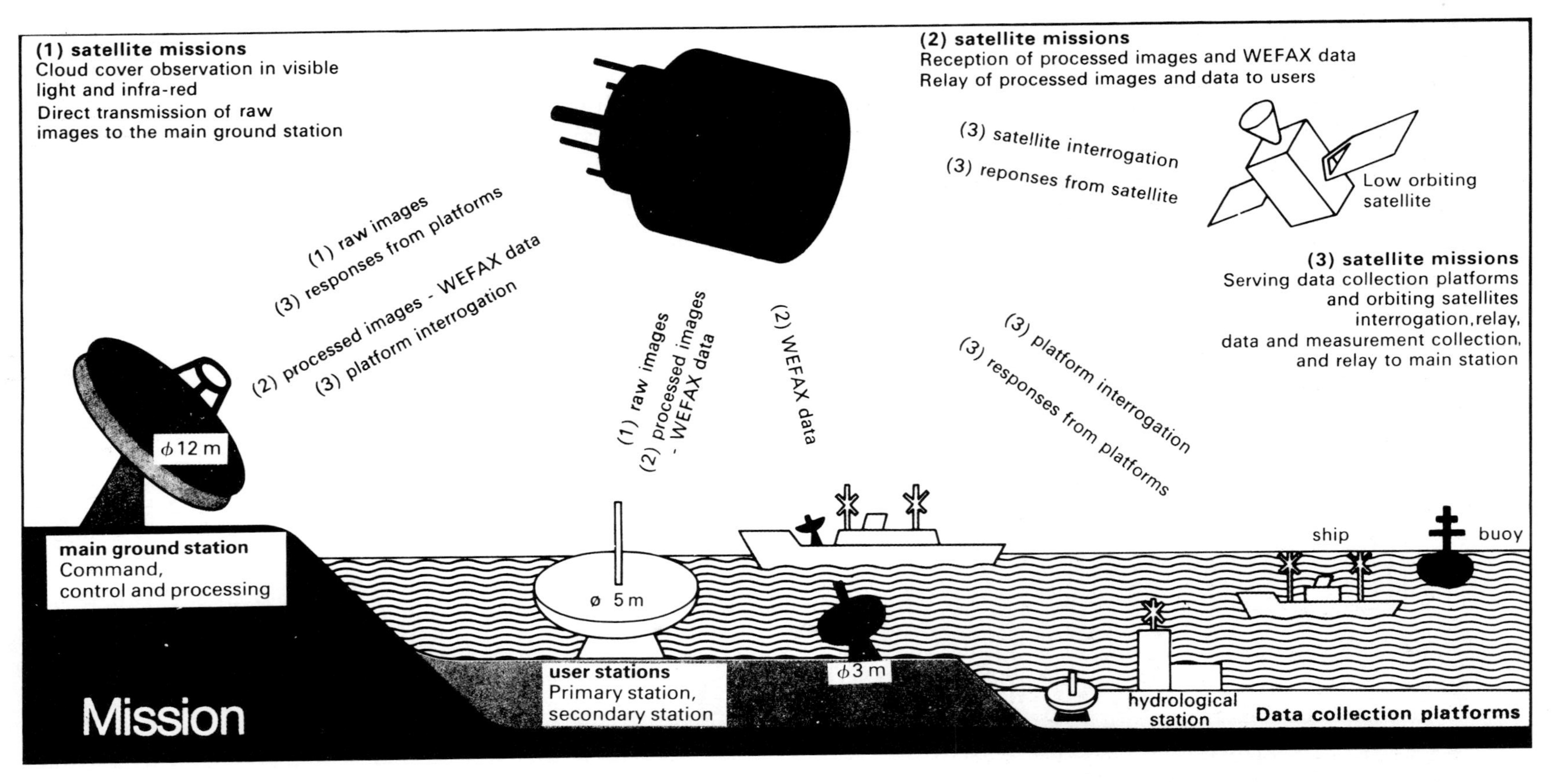

Fig. 6.11 Meteosat data collecting system. (Source: ESA.)

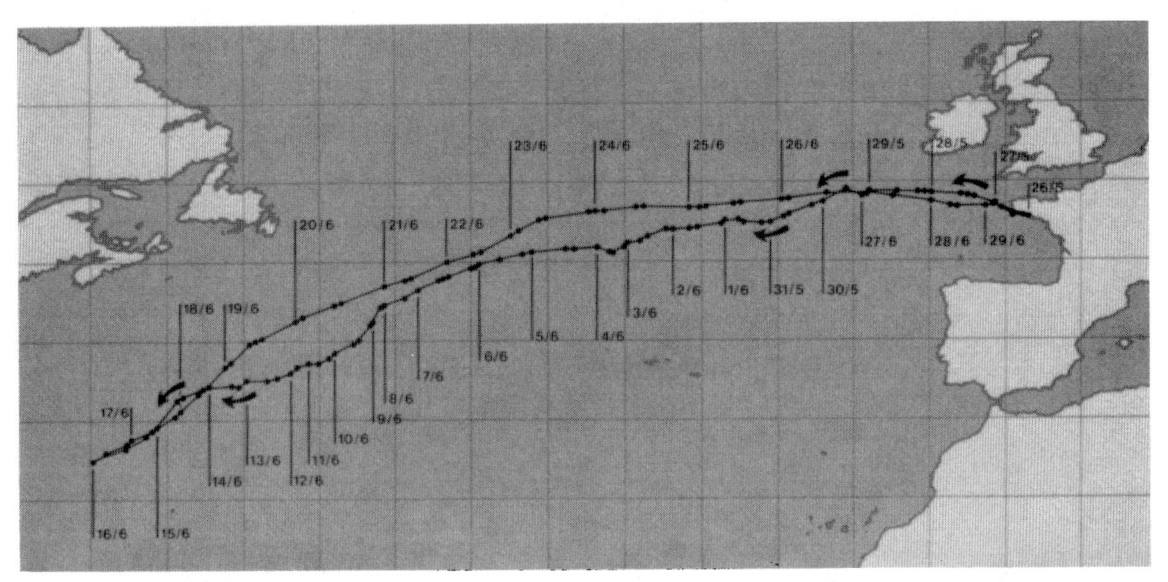

Fig. 6.12 Track of the winning yacht 'VSD' as recorded by the Argos system. (Source: ESA.)

of interest. Many of the future sensing systems on satellites will be capable of being steered to observe targets of opportunity by means of ground control. One may envisage that in the future thermal sensors will be activated and steered to monitor forest fires, radar sensors will be directed to monitor floods and camera systems will capture records of crop damage. The control of these overhead sensors will depend upon information from *in situ* data sites. In many respects one can expect the ground stations of the future to provide the datum points around which the space and airborne sensors will draw the contours of the environmental conditions at particular moments of time.

7 *Remote sensing data interchangeability, preprocessing and processing*

7.1 Introduction

We have reviewed the needs for environmental remote sensing, the physical bases of remote sensing, and data collection both from remote sensing systems and from longer established sources. We must proceed next to consider the ways and means which are being developed to present the hard-won remotely sensed data to the user in the form he or she finds most convenient, so that they can be exploited as efficiently, effectively and economically as possible in environmental research or operations. This necessitates attention to an important 'eternal triangle' linking three basic and quite different forms of remote sensing data, and to a commonly recognized quartet of stages through which most remote sensing data pass from initial acquisition to final, and hopefully profitable, deployment in the solution of environmental problems.

7.2 The interchangeability of basic remote sensing data types

Original remote sensing data are obtained in one or other of the three following forms:

1. latent photographic images;
2. analog data;
3. digital data.

Each deserves some individual comments, before the relationship between the three are examined.

Latent photographic images are obtained by cameras, through which scenes are recorded on a film base that is developed to give photographs either as photographic transparencies (directly), or photographic prints (indirectly). The former can be viewed by projection, as light is passed through the transparency on to a screen. The latter can be viewed directly, usually on a paper base.

Analog data are essentially (one-dimensional) graphs of radiation variations obtained by scanning a target (Fig. 7.1), or as we will see below, a photograph of it, the graphs representing the continuously variable patterns of radiation across the scenes. Two-dimensional images of targets may be constructed from analog data if successive analog signals relate to adjacent strips or transects of the target, e.g. successive scan lines from a suitable airborne or spaceborne sensor system. Most of the remote sensing products that appear in the media or adorn the walls of our departments and laboratories are either photographs or images. Both types of products are often referred to as 'hardcopy', to distinguish them from analog data, which are originally recorded electronically. So too are digital data, which we must review next.

Digital data are numerical representations of radiation received from small areas of the target, usually comprising some kind of digital array. For example, each SPOT scene is built up from large numbers of picture elements (or 'pixels'), each line of which is obtained from instantaneous measurements made by the 1728 miniature sensors arranged along the SPOT multispectral sensor's 'pushbroom' or multilinear array (see Fig. 7.2).

It is very important to understand that images, analog records and digital data sets are *all completely interchangeable*, as Fig. 7.3 illustrates. The processes and means by which the conversions from one data type to another are effected are as follows:

1. optical (e.g. photograph) to analog, i.e. O–A conversion, by a scanning microdensitometer;

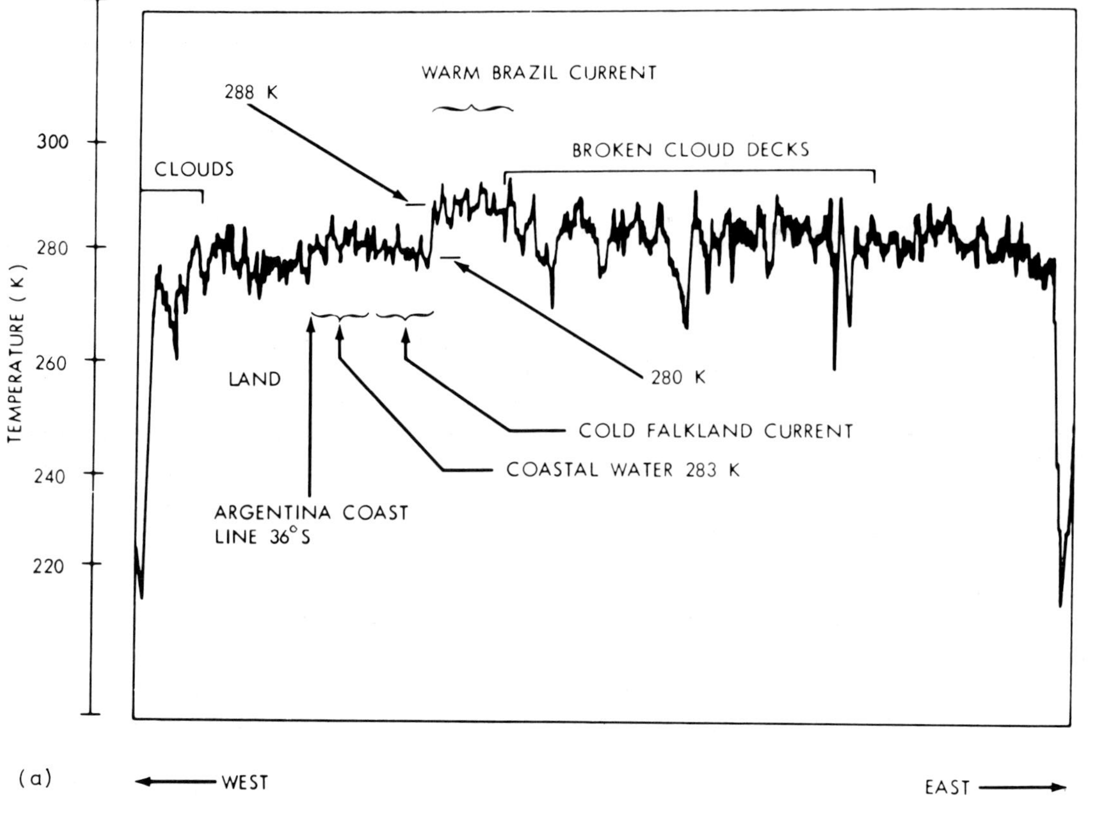

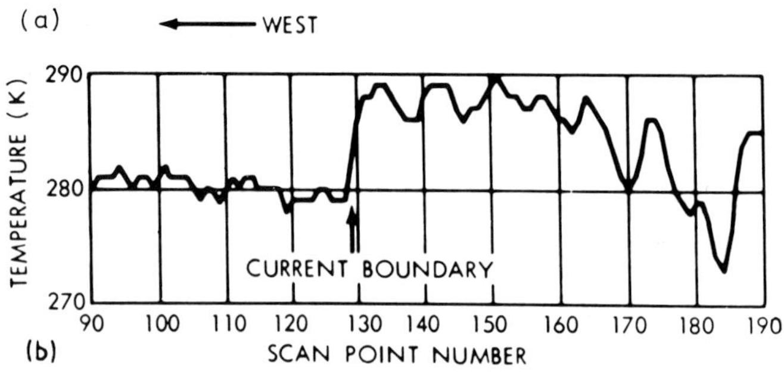

Fig. 7.1 Nimbus 2 observations of currents in the South Atlantic. (a) An analog record of an individual HRIR night-time scan line taken on 17 August 1966. (b) A section of the analog record in (a) after digitization and application of a numerical filter to remove oscillatory noise. (Source: Warnecke *et al.*, 1969.)

2. optical (e.g. photograph) to digital, i.e. O–D conversion, by a flying spot scanner;
3. analog to optical, i.e., A–O conversion, by an image processing system;
4. analog to digital, i.e. A–D conversion, by a sampling digitizer;
5. digital to analog, i.e. D–A conversion, by a graphing system;
6. digital to optical, i.e. D–O conversion, by an image processing system.

The most fundamental and common requirement in remote sensing is to be able to convert data between the optical and digital domains, for these dominate most remote sensing science and applications. However, the analog domain has been much involved in these processes, for the commonest types of instruments have been those which have followed an O–A–D route, or the reverse D–A–O route. These may now be examined in more detail.

7.2.1 Optical to digital conversion

One of the most familiar types of instrument for the conversion of film data into analog, and thence computer, data is the microdensitometer. This instrument allows the film to be traversed by a

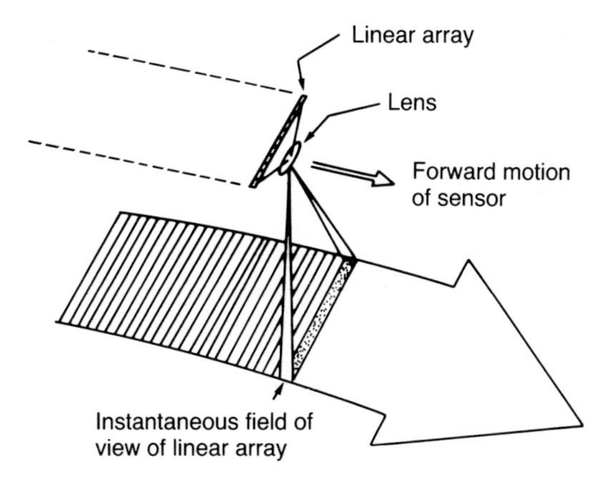

Fig. 7.2 Pushbroom scanning by the SPOT multilinear array sensor system.

beam of transmitted light. The light source is accurately controlled so that the difference between the light beam received and the light beam emitted by the film can be measured by means of a photomultiplier. The spot size of the transmitted light can be varied, so the amount of detail in the original film can be sampled according to the nature of the study to be undertaken. For example, in order to examine variations in the density of reflectance from a field containing different crops the spot size might be 50 μm. On the other hand scanning an

Table 7.1 Scanning times for the P-1000 Photoscan

Drum speed (rev s^{-1})	Raster (μm)	Data rate (Hz)	Scan time (min)
2	50	14.4	20
4	100	14.4	5
8	200	14.4	1.25

image for highly contrasting and coarse detail, such as rivers, may be carried out with a raster (sample frame) size of 200 μm.

Some microdensitometers use a flat bed system in which the film is laid flat on a moving table mechanism and light is transmitted vertically through the image. However, the most common type of densitometer in use in remote sensing studies is the drum microdensitometer. Drum type machines are usually faster than flat bed machines and are high resolution microdensitometers where the film is fastened on to a rotary drum and scanned by a light source when in rotary motion. A typical high-speed digital microdensitometer that has been widely used is the System P-1000 Photoscan, which provides the scanning times for a 12.5 × 12.5 cm film listed in Table 7.1. A schematic diagram

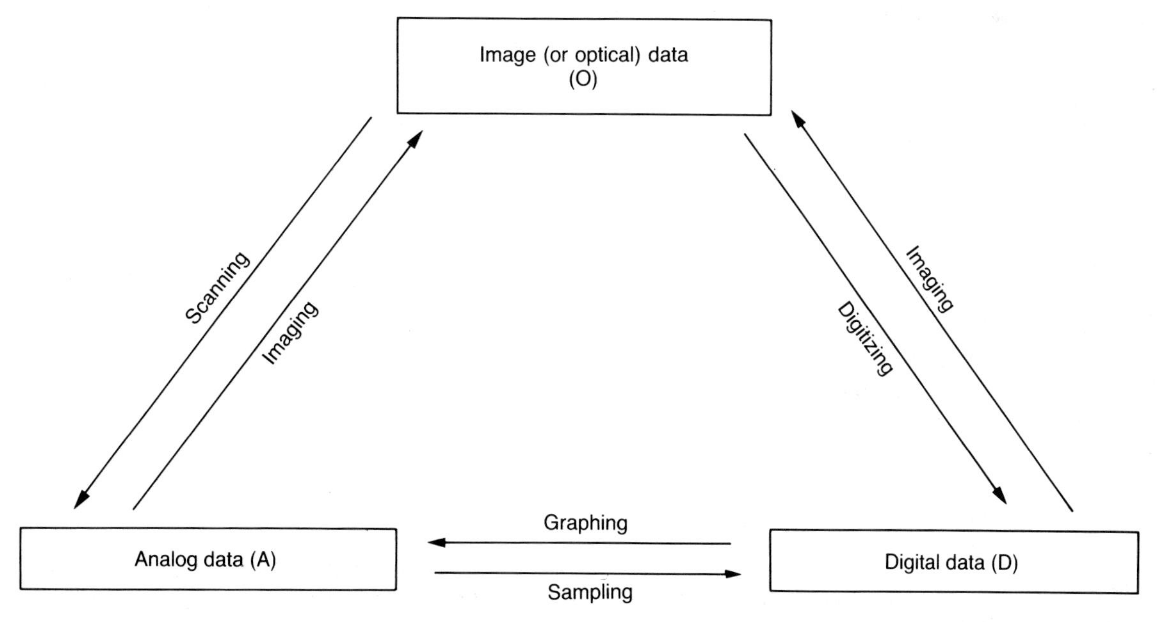

Fig. 7.3 The remote sensing data 'eternal triangle', showing the process by which conversions from one type of data to another can be effected.

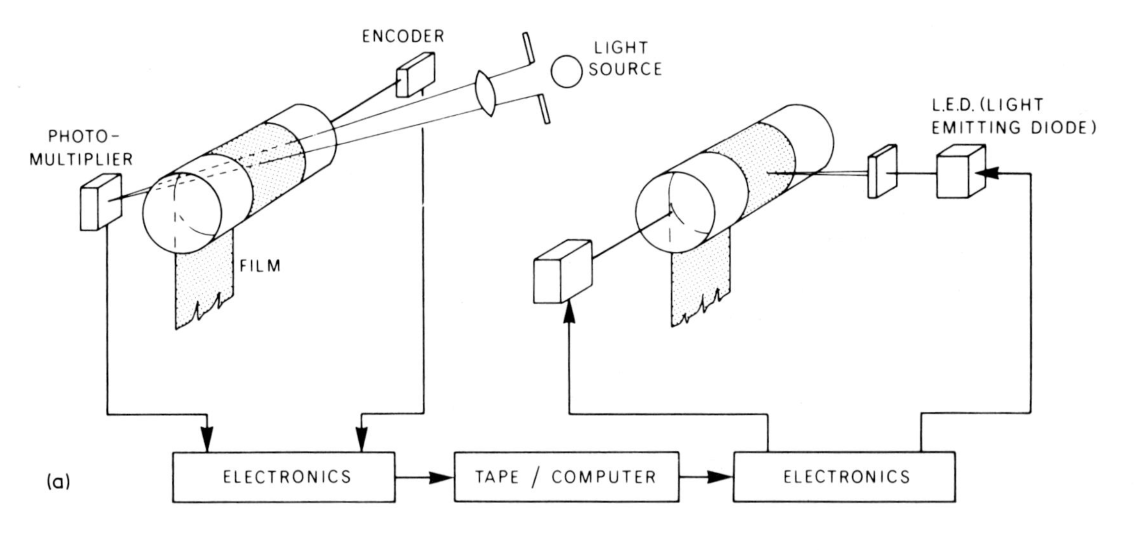

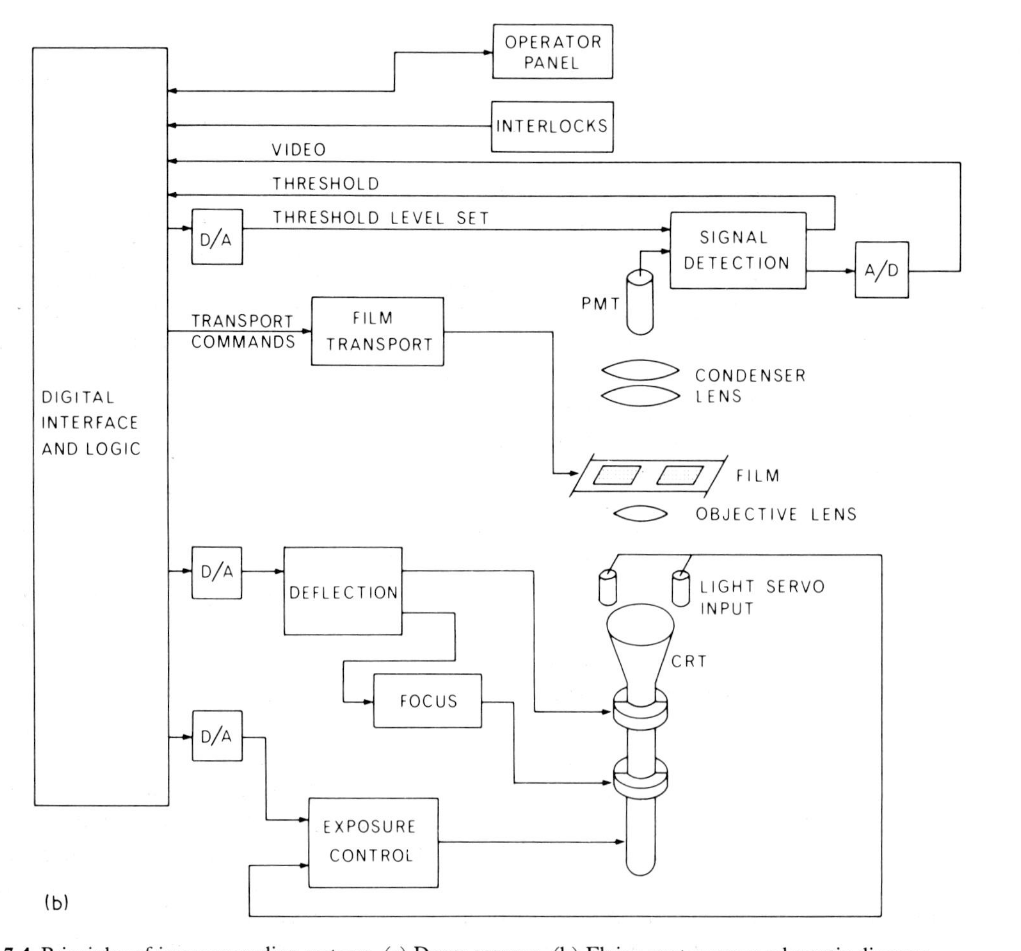

Fig. 7.4 Principles of image recording systems. (a) Drum scanner. (b) Flying spot scanner schematic diagram.

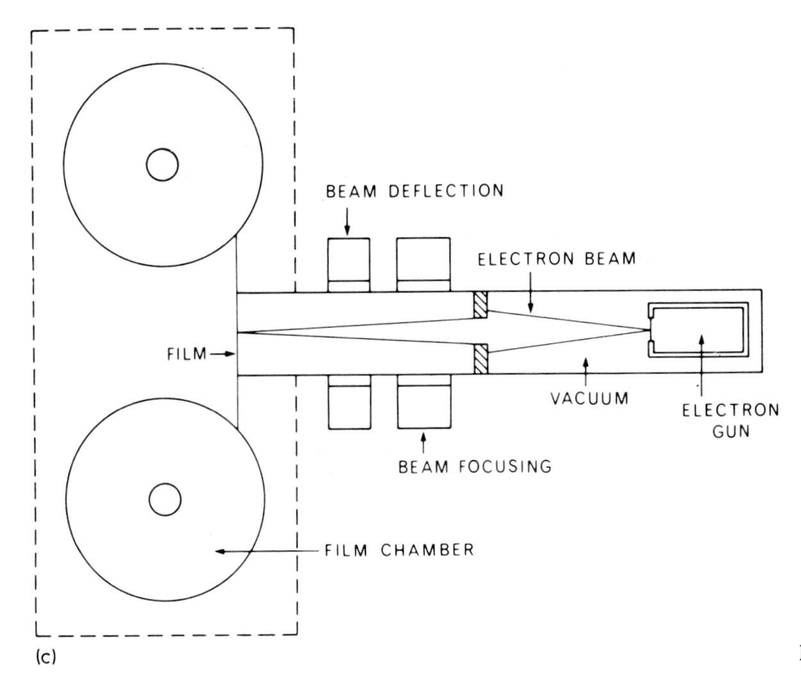

Fig. 7.4 cont. (c) Electron beam recorder.

showing the basic characteristics of the drum scanning microdensitometer is shown in Fig. 7.4(a).

Another method of converting film data into digital data is by scanning the film by means of flying spot CRT (Cathode Ray Tube) and vidicon systems (Fig. 7.4(b)). This machine offers high speed of conversion, easy interaction with computers, possibility for data reduction by selective access to images for spectral separation, and for filtering and image improvement techniques. Flying spot scanners can read 64 levels on a grey scale and record 60 levels. They require only 20 μs for reading and 27 μs or less for recording. Spot sizes of 0.03–0.05 mm can be selected for scanning.

Traditional light sources are normally used for these scanning machines. However, exceptionally fast machines have been developed using laser beams. Laser scanners consist essentially of three components, namely an optical system, a film transport system and a rotating scanner system. The laser provides a high power, collimated beam of monochromatic light which scans the film image.

Another form of converter is the electron beam recorder (EBR). The basic EBR consists of a high-resolution electron gun, an electron optical system for controlling the electron beam, a film transportation mechanism, an automatic vacuum system and regulators and electronic circuits that operate the recorder. The electron gun provides an electron spot (3–10 μm diameter) which is focused by coils on the sides of the vacuum tube. A schematic diagram of an electron beam recorder is given in Fig. 7.4(c).

7.2.2 Digital to optical conversion

With the development of multispectral scanning devices it is now necessary to have a means whereby digital (D) data can be converted into analog (A), then optical (O) data. Such a system enables photographic imagery to be generated from digital data.

Generally D–A conversion is achieved by reading digital data in the computer from computer compatible tape, transforming it into a form acceptable for digital-analog hardware equipment, and then passing the transformed data into the D–A device. Since analog tape recorders are not able to stop and start rapidly (unlike digital recorders) it is desirable to maintain a continuous flow of data to the analog hardware. This is often achieved by setting up an input and output queue. The system is arranged so that there is always a backlog of input data waiting for processing and also a reserve (buffer) amount of data in the output to the analog hardware. This enables corrections to the input data to be made without interruption of flow to the analog device. Once the analog form of data has been generated an (optical) image can be constructed, e.g. in a visual display unit such as a domestic TV screen or computer monitor. If prints of the images are required, these can be produced either by photographing the screen or through the use of a plotting machine.

A further logical step towards the provision of digital satellite data to the user has been taken by the designers of the SPOT satellite, for, as noted earlier, the pushbroom or multilinear array sensor system generates digital data directly, obviating the need for any data conversion steps to be undertaken before the observations can be processed by a computer. This approach may become the dominant one in the future.

7.3 Stages in the handling and use of remote sensing data

From the above discussion the reader will have begun to realize that not only are there several different ways in which remote sensing data may be presented, manipulated and inspected, but also that there may be different stages in the inspection and utilization of the various data types. Indeed, these stages can be listed in the form of a sequence, leading from the original observations through to the generation of the final outputs or products required by the end user. The following stages are those most widely recognized by the remote sensing community, although it may be noted that in some circumstances other stages maybe interpolated, e.g. in the form of feedback loops or refinements.

7.3.1 Data preprocessing

This includes any process that has to take place before remote sensing data can be processed and analysed by the scientific user. Ideally, therefore, it should be undertaken before the environmental scientist acquires the data from a central reception facility or archive; increasingly plans are being made to ensure that in future at least some of these types of numerical treatments are carried out on the satellite itself. Data preprocessing commonly includes data *navigation* and *registration*, *rectification*, *cleaning* and *primary calibration*. Most of these procedures are carried out at central reception and/or archiving facilities.

7.3.2 Data processing

This includes any data manipulation necessary or helpful for later scientific analysis and interpretation, but which the analyst or interpreter does not want to have to perform. Increasingly data processing is being undertaken or offered by central facilities, although in the simpler cases it is usually carried out by the user. Its chief object is to present remote sensing data in manageable forms and quantities. It often involves some element of *selection* from, and/or *compression* of the preprocessed data, and/or *enhancement* of selected features of particular significance to the customer.

7.3.3 Data analysis

This entails inspections of the information content of the data set in its own terms, i.e. searches for important features and repetitive patterns inherent in the remote sensing data themselves. Thus, the key element is *feature recognition*, whether based on instantaneous (single time) imagery, multitemporal or multidate imagery, or time series data, of unispectral, bispectral or multispectral types. The results may or may not be immediately comparable with classifications of *in situ* data.

7.3.4 Data interpretation

This covers the assessment of the contents of remote sensing data sets through comparison with the nature and condition of the target, wherever possible assessed also by *in situ* observational information. Appropriate relationships, models and algorithms are often required for meaningful intercomparisons of these two different types of data. Results of interrelations observed in selected test areas for which both remote sensing and ground data are available (i.e. 'training sets') are often prepared as the bases for dependent *supervised classifications* of remote sensing data from conventional data-sparse areas. In the absence of analyst-specified training data, *unsupervised classifi-*

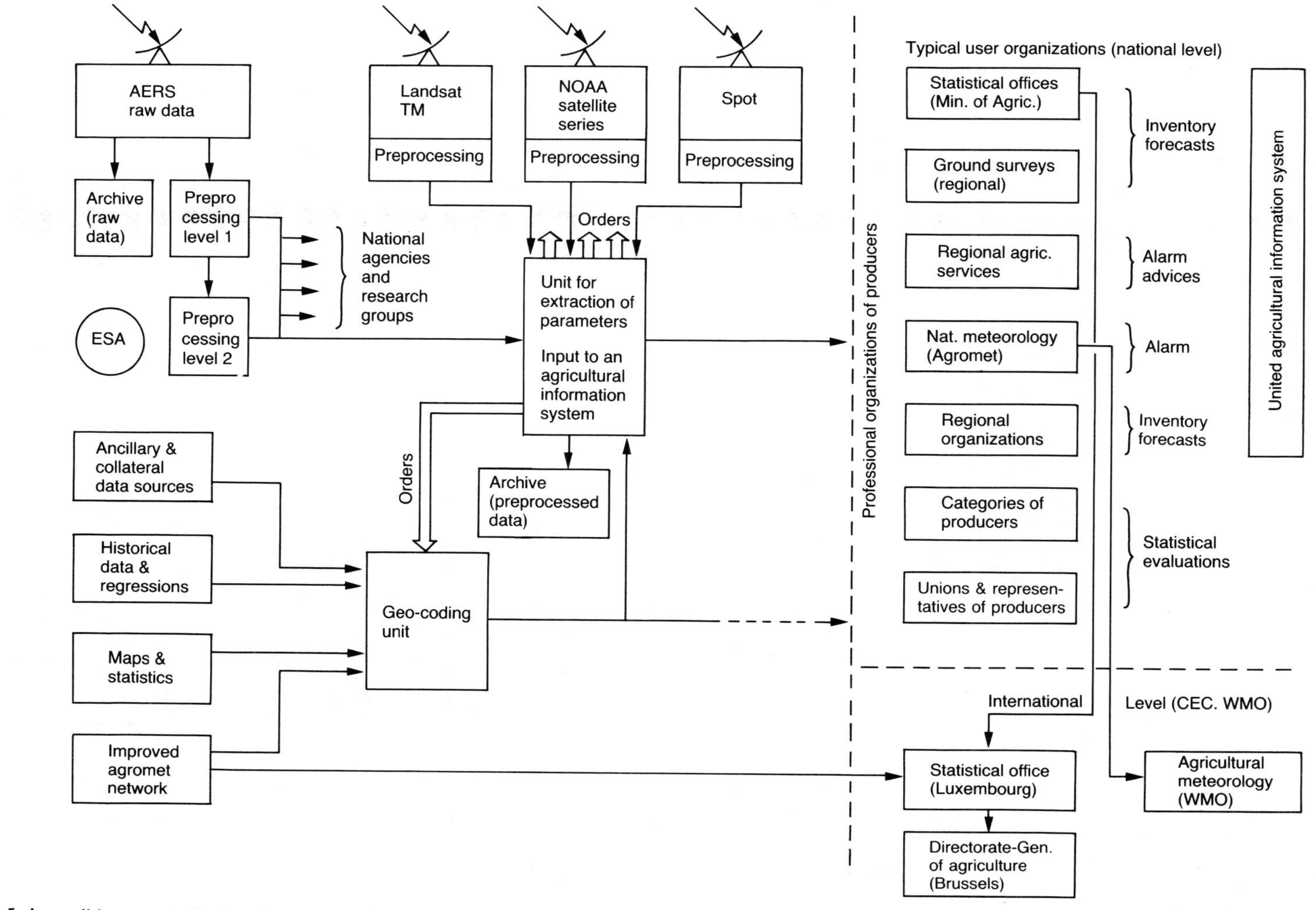

Fig. 7.5 A possible scenario for data flow from a central facility to an (agricultural) end-user community. (Courtesy, ESA.)

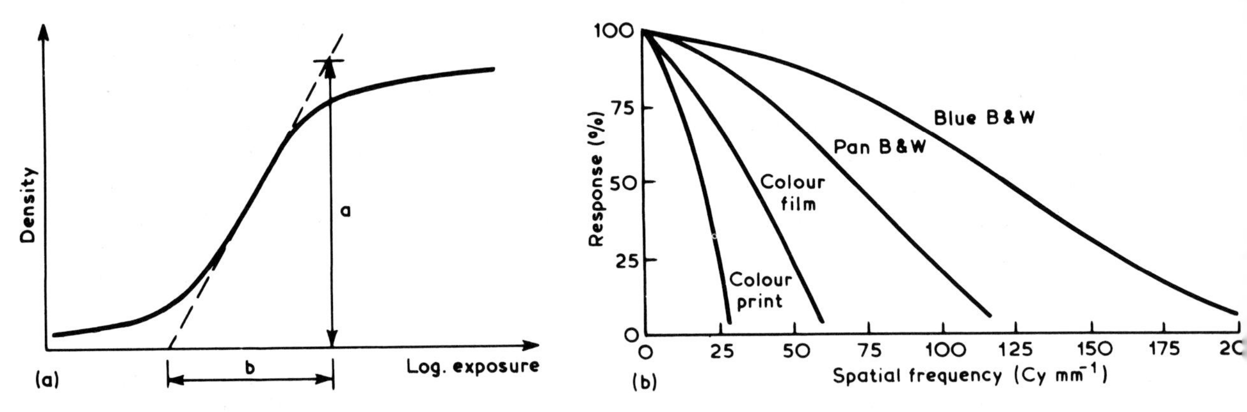

Fig. 7.6 (a) Shape of typical characteristic curve for photographic film. (b) Film transfer function curves.

cations are used instead. Here the search is for natural groupings inherent in the remote sensing data themselves, which then may be interpretable in terms of recognizable features or classes of features in the target area(s).

Lastly it must be emphasized that more and more data preprocessing is being undertaken on the satellites themselves, a point well illustrated by the case of the onboard conversion of Landsat TM data from its raw analog form to its transmitted digital form as described earlier in this chapter – and that national and international data centres are becoming increasingly well-equipped to process data to meet a variety of customer requirements (Fig. 7.5). Thus the scientist is able to focus more of his or her attention on data analysis and interpretation, and the final end-user to obtain better quality products as a result of improved quality control of the data on which they are based.

7.4 Preprocessing and processing examples

Before advancing in subsequent chapters to review in considerable breadth and depth the methods and results of remote sensing data analysis and interpretation activities it is necessary to establish some of the key practices and problems in data preprocessing and processing. This we will do by reference to two selected fields, first the preparation of airborne photographic imagery both as a background to the subject matter of Chapter 8 ('Analysis and interpretation of aerial photography') – and to all other areas of remote sensing in which photographic products are required – and second the provision of digital information from the Landsat system, which has been the most important single source of Earth resource satellite data since the early 1970s.

7.4.1 *The preparation of photographs*

Photographs can be prepared by either black and white technology or colour technology. In the black and white case attention has to be given to a number of variables which affect the quality of the product:

1. The *characteristic curve* shows density as a function of log exposure. Density is a measure of the degree of blackening of the exposed film, plate or paper after development and it can be seen in Fig. 7.6(a) that density is nearly proportional to the log of exposure in the central part of the S-shaped curve. The *proportionality factor* as measured by the ratio a/b is referred to as the film *gamma*(γ). Where density measurements are to be made on a number of films it is necessary to process each film to the same gamma.
2. *Spectral sensitivity* of the film describes the sensitivity of the film in a given region of the spectrum. Panchromatic films, as their name suggests, are sensitive to all parts of the visible spectrum up to wavelengths of 700 nm and must be developed in complete darkness, whereas certain duplicating films are sensitive only up to 500 nm and can be developed in red light.
3. The *modulation transfer function* (MTF) is the ratio of intensity variations in the image to those occurring in the original. In other words it is an expression of the resolving power of the film,

and the achievement of maximum resolution depends on the chemistry and duration of the photographic process.

4. The *dimensional stability* of the film used in the photographic process is important in order to limit distortion in the image. Also, some information concerning the *granularity* of the film is desirable since it affects the amount of detail that is registered.

Stability improves with thickness of the film, but in space applications films must be as thin as possible in order to conserve weight. It is necessary also to control the film environment if film stability is to be maintained – especially the temperature and humidity. Polyester base material is usually superior to other materials in dimensional stability. For example its expansion coefficient is normally 0.001–0.01% $°F^{-1}$, whereas cellulose triacetate is less good by a factor of 2–3 times. The *granularity* of a film becomes important because it represents unevenness in the emulsion which creates 'noise' in the photographic image. A microdensitometer trace on a uniformly exposed and processed film can be used to detect irregularities or discontinuities and thus assess its granularity. Root mean square (r.m.s.) graininess can be determined from the standard deviation of the density measurements and from the diameter of the scanning aperture:

$$G = K\sigma_K(D) \tag{7.1}$$

where G is the r.m.s. graininess, K the diameter of the scanning aperture used, and $\sigma_K(D)$ the standard deviation of density measurements.

In processing photographic film it is usually necessary to control the density of the product. This is done by a method which removes silver from the negative and is termed *reduction*. Ammonia and potassium persulphates are reducers that are commonly used but their action is liable to be uneven and hard to control. Other methods include 'dodging' which may be carried out manually by inserting tissue paper or a mask to decrease the density range. Automatic dodging printers are available and are reliable and efficient. The *printing papers* used in the photographic process can accommodate a wide range of density ranges and chloride, chlorobromide and bromide papers are the major types used.

In colour technology the chief areas of interest are in dyeing and reversal processes, resolution, colorimetry, colour balance and reproducibility.

The colour films used are primarily natural colour and infrared colour. The use of aerial colour only became significant in the 1960s following the breakthrough achieved by film manufacturers who succeeded in producing a range of colour emulsions with speed and resolution characteristics comparable to panchromatic films. Aerial colour films are available as either reversal colour films or negative colour films. Generally, the reversal films are used for medium to high altitude photography and negative colour films for work at low altitudes.

It has been appreciated for some time that the visual contrast presented by colour to the human eye is much greater than that of panchromatic photography. The eye will distinguish about 200 gradations on a neutral or grey scale, whereas it is capable of differentiating some 20 000 different spectral hues and chromas (intensities of colour). Therefore colour photography immediately offers the opportunity of more detailed study of surface objects.

Infrared colour is a false-colour reversal film. It differs from ordinary colour film in that the three sensitized layers are sensitive to green, red and infrared radiation instead of having the usual blue, green and red sensitivities. The green sensitive layer is developed to a yellow positive image, the red sensitive to a magenta image and the infrared to a cyan positive image; thus the colours are false for most natural objects.

False colour film is highly sensitive to the green–red wavelengths of light as well as the near infrared. As a result it has particular characteristics, such as that water and wet surfaces image in blue and blue–grey tones. An important feature of this type of film is that it records healthy green vegetation in various shades of red in the positive images (Tables 4.1 and 4.2).

Colour films are generally characterized by lower spatial resolution and higher contrast than black and white (Fig. 7.6(b)). Colour balance is particularly important in treatment of false colour films because unlike in true colour photography the user has no reference available for comparison. The main aim of colour balancing is to ensure that brightness differences occurring in the target area are reproduced faithfully in the film image without affecting hues and colour response. In order to

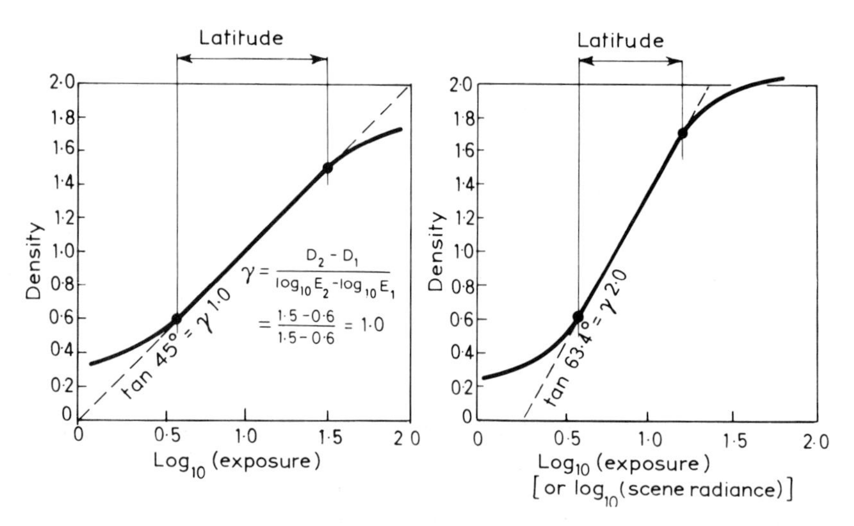

Fig. 7.7 Density (D), gamma (γ), and exposure (E) relationships. Two films of different gammas (contrasts) are compared. At $\gamma 1.0$, scene brightness ($\log_{10}E$) on the straight part of the response curve produces densities at the same contrast ratios as viewed by the sensor; $\Delta\log_{10}E = \Delta D$. When $\gamma 2.0$ film is used the apparent contrast is amplified by a factor of 2: i.e. $\log_{10}E = 2(\Delta D)$. Note that increased gamma leads to reduction in exposure latitude. (Source: Ross, 1976.)

achieve this, adjustments of the exposure for each layer are required. Sequential printers are available that allow independent layer corrections. Problems sometimes arise, however, owing to variations in the characteristics of stored false colour film. These can be largely overcome by storing the film at −20°C and making some densitometric tests before processing.

The photographic equipment necessary for photographic processing includes enlargers, contact printers, film and paper print processors, mixers, drier/cutters, microfilm processors and light tables.

7.4.2 *Photographic enhancement techniques*

If the analyst is to recognize information in the image, it must be reproduced in a way that permits visual identification. Unfortunately, in some environmental scenes important information is characterized by very low spectral reflectance levels, or by low inherent contrast between the subject and its background, or both. There are also atmospheric effects which further complicate the situation. Every photo-interpreter is familiar with the feeling that more information is buried in the image than he or she is able to see, or interpret with confidence and accuracy. As a result various enhancement techniques have been used to emphasize the tonal differences between objects.

The enhancement techniques most commonly employed consist of contrast stretching, photographic masking and density slicing. Increasing or manipulating the contrast of data is a powerful means of raising such data to levels where it can be recognized by visual interpretation. Gamma (γ) and contrast are often used as interchangeable terms for describing an image characteristic, but gamma is a measurable value whereas assessment of 'contrast' is highly subjective and varies among observers. Nevertheless, in photographic image enhancement processes, films of different gammas from 1.0 to over 6.0 may be used to achieve higher contrast. For example, in Fig. 7.7 two films of different gammas are compared where the apparent contrast of the scene is amplified by a factor of 2.

In the case of photographic masking a positive and negative image of the scene are registered together and a new image, the mask, is printed from the combination. The mask may be either positive or negative, and will add to, or subtract from, the density range of the image being masked. For example, where the data used consists of multispectral photography or scanning images three frames might be selected from a nine-frame set because they show the most marked differences.

Table 7.2 Enhancement processing

Camera negatives	Film positives	Duplicate negatives	Intermediates	Additive printing filters	Integral colour print
N1	→ P1	→ N1′	l_1 (N1′ + P6)	→ G	→ Colour
N6	→ P6		l_2 (N7′ + P1)	→ R	→ derivative
N7	→ P7	→ N7′	l_3 (N1′ + P7)	→ B	→

These frames N1, N6, N7 then can be treated as shown below (Table 7.2).

The positives and negatives are printed subtractively (superimposed) in registration on to prepunched film to form intermediates. Each intermediate is made from superimposition of one positive and one negative each from a different band. Each intermediate is then printed additively with a filter so that the image of each one is transferred to only one of the colour layers in the colour film (i.e. I_1 = magenta layer; I_2 = yellow layer; I_3 = cyan layer). The resulting images when developed from a single tri-pack film provide the colour derivative transparency for viewing. The intermediates represent differences of reflectance between bands. Where no differences occur, the negatives and positives cancel out each other, so that no colour shows on the final enhancement. It has been found that the relative brightness and colour differences of objects appearing in the colour derivatives can be used to detect such changes as moisture content, soil density and vegetative conditions.

From the above description it will be evident that the making of colour enhancement prints is complex and time-consuming. It is, therefore, somewhat expensive and it is not always easy to ascertain the best selection of negatives which will give the maximum amount of information at the interpretation phase. In these circumstances, it is an advantage to have the capability of experimenting with different combinations of the negative frames by means of a visual display. In order to achieve this a method is required for projecting spectral positive transparencies, one superimposed upon the other in accurate registration while illuminating each with a different coloured light. If the transparencies are illuminated by the primary colours, a full colour reproduction of the scene is produced. It is also possible to make reconditioned false colour (infrared colour) and artificial derivatives of various kinds by suitable combinations of transparencies and colour addition. Additive colour viewers (Fig. 7.8) are important, low-cost aids to photo-interpretation. The brightness and saturation controls of each spectral transparency permit alterations of the final composite screen presentations. In this way subtle differences in the scene can be detected which would not be readily distinguished by examination of normal panchromatic or colour prints.

Photographic density slicing converts the analog, continuous-tone image into one displaying a series of steps, each representing a separate, different increment of density, completely isolated from lesser or greater densities. It is also called 'equidensitometry' or 'isodensity contouring'. Agfa-Gevaert makes Contour Film, a special emulsion for this purpose. Very precise and narrow density slices can be made with conventional lithographic films and developers. For example, lithographic films, such as Kodak Ortho Film 2556, can reach very high gammas when processed as recommended and can prove to be ideal for density slicing.

7.4.3 Preprocessing and processing of digital data from satellites

As digital analysis systems have become ever more easily available and more capable, yet also affordable, so increasing efforts have been made to provide the user with remote sensing data basically in a digital form. Thus steps have been taken to ensure that key data type conversion processes can be carried out on board satellites rather than on the ground. Good examples of this are given by the Landsat RBV, MSS and TM systems, which have been designed to record data in an analog form, but to transmit only digital data.

The Landsat data production facilities have had

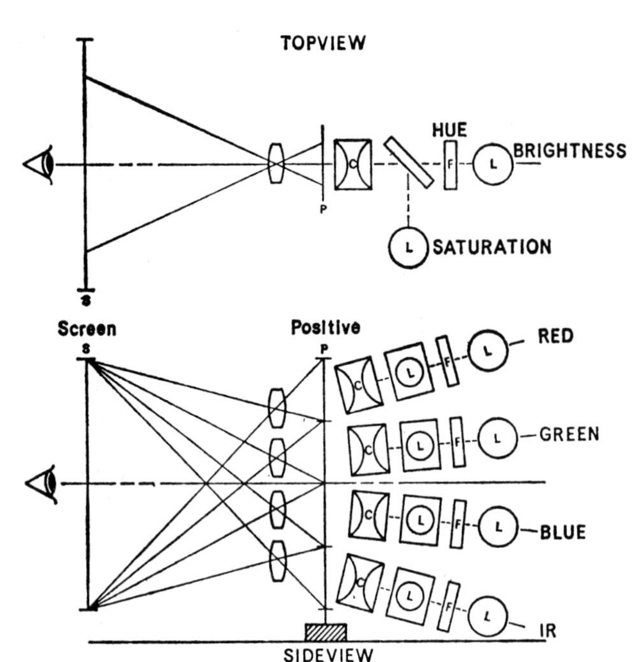

Fig. 7.8 Additive colour viewing. The positive transparencies (P) are illuminated by light from lamps (L) passing through filters (F), inner saturation and condensing lenses (C). They are then viewed on the screen (S). (Source: Curtis, 1973.)

Table 7.3 Data processing facility requirements for Landsat Return Beam Vidicon observations (Source: NASA)

RBV input (3 bands)	Scenes per week (%)	Scenes per week	For each scene[a]	Items per week
1316 scenes/wk	100% bulk	1316	3 B–W masters	3948
3948 images/wk			30+	39480
			30−	39480
			30 +prints	39480
	20% bulk colour	263	1 C−	263
			10 C prints	2630
	5% precision	66	3 B – W masters	198
			30−	1980
			30+	1980
			30 +prints	1980
			1 C−	66
			10 C prints	660
			Ability to digitize	—
	1% digitized	13.2	3 copies computer readable	≈158 tapes

[a] B–W, black and white; C, colour; +, positive transparency; −, negative transparency; +prints, positive paper prints; C prints, colour positive paper prints.

to be capable of handling very large quantities of data, as evidenced by Tables 7.3 and 7.4. These provide the statistics on which Landsat ground station design criteria have been based. The structure of the wreathed image processing subsystems for the early Landsats is shown in Fig. 7.9, showing the four main processing modes set up to produce photographs, video tapes, and (of greatest significance in the present context), computer compatible tapes (CCTs) containing data in digital form.

More recently, with the advent of the TM sensor

Table 7.4 Data processing facility requirements for Landsat Multispectral Scanner system observations (Source: NASA)

RBV input (4 bands)	Scenes per week (%)	Scenes per week	For each scene[a]	Items per week
1316 scenes/wk	100% bulk	1316	4 B–W masters	5 264
5264 images/wk			40+	52 640
			40–	52 640
			40 +prints	52 640
	20% bulk colour	263	2 C–	526
		66	20 C prints	5 260
	5% precision		4 B–W masters	264
			40+	2 640
			40–	2 640
			40 +prints	2 640
			2 C–	132
			20 C prints	1 320
			Ability to digitize	—
	5% computer readable	66	3 copies computer readable	≈713 tapes

[a] B–W, black and white; C, colour; +, positive transparency; –, negative transparency; +prints, positive paper prints; C prints, colour positive paper prints.

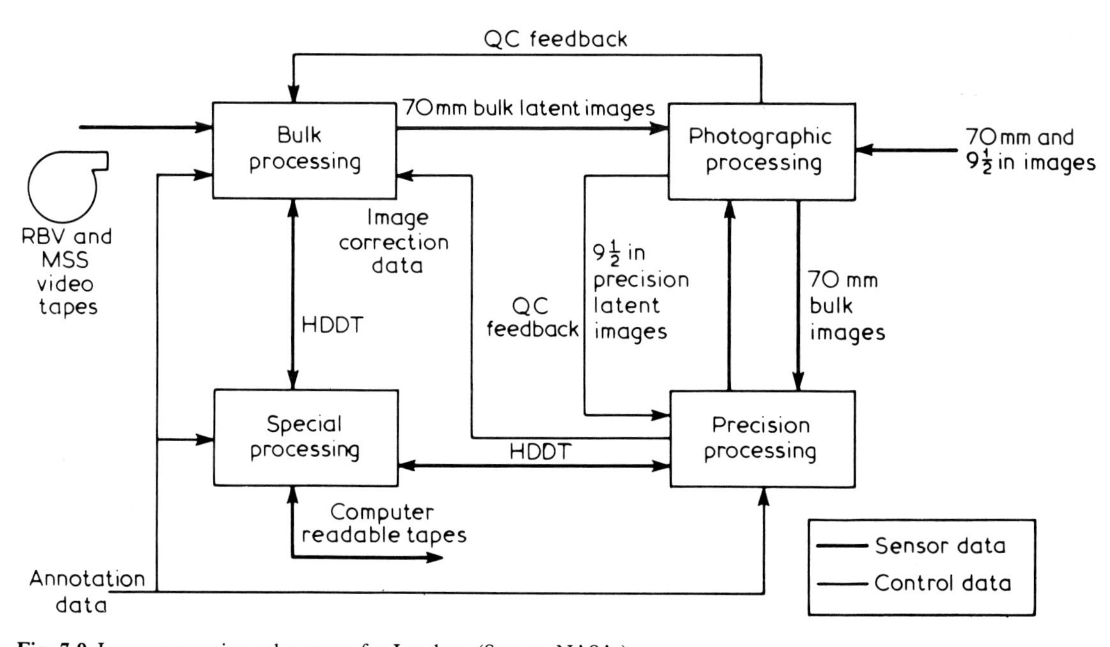

Fig. 7.9 Image processing subsystems for Landsat. (Source: NASA.)

system, with its even higher spatial and spectral resolutions, substantial modifications to the ground station facilities have been necessary. However, the process has been greatly aided by the fact that the analog-to-digital conversion is, in this case, carried out on the spacecraft itself.

Radiometrically, the TM performs its onboard analog-to-digital signal conversion over a quantization range of 256 digital numbers (8 bits). This corresponds to a fourfold increase in the grey scale range relative to the 64 digital numbers (6 bits) used by the MSS. This finer radiometric precision permits observation of smaller changes in radiometric magnitudes in a given band and provides greater sensitivity to changes in relationships between bands. Thus, differences in radiometric values that are lost in one digital number in MSS data may now be distinguished.

Geometrically, TM data are collected using a 30 m IFOV (for all but the thermal band which has a 120 m IFOV). Compared with the Landsat MSS this represents a decrease in the lineal dimensions of the IFOV of approximately 2.6 times, or a reduction in the area of the IFOV of approximately seven times. At the same time, several design changes have been incorporated within the TM to improve the accuracy of the geodetic positioning of the data. Most geometrically corrected TM data are supplied using 28.5 × 28.5 m pixels registered to the Space Oblique Mercator (SOM) cartographic projection. The data also may be fit to the Universal Transverse Mercator (UTM) or Polar Stereographic projections.

Because of the very large amount of data generated by the Landsat system, it is necessary to use a computerized data base to record essential data concerning each scene. The primary data base, located at the EROS Data Center (EDC) in Sioux Falls, South Dakota, includes all necessary information concerning date, location, and quality of all Landsat imagery. The public has access to this data base through requests mailed or telephoned to EDC as described in the information distributed by EDC, the US Geological Survey, and other organizations (e.g. EOSAT). In general, a user also can obtain, free of charge, information regarding availability of aircraft imagery, a service specially valuable to all who wish to use remote sensing imagery synergistically.

The computer search is based upon user-supplied information regarding the area of interest, desirable dates of coverage, and the minimum quality of coverage. The response is a computer tabulation of all coverage meeting the constraints specified by the user. Landsat products are identified by scene identification numbers, listed in the annotation block of each image.

The computer listing will vary in length according to the nature of the user's request. A conservative, restrictive request closely tailored to the user's needs will result in a much shorter listing than will a loosely defined, very broad request. Unless the user suspects that there is very little coverage of an area, it is usually best to define the request to match his or her needs closely, as it can be very tedious to search through a long list to identify the best images.

8 Analysis and interpretation of aerial photography

8.1 Introduction

Although progress is being made continually in all areas of environmental remote sensing to reduce dependence on the human analyst or interpreter, data analysis and interpretation by manual processes are still of basic importance and great practical value in many areas of applications. Therefore, it is right that we should review some of the most important of these areas of application before proceeding to consider elements of numerical data analysis in Chapter 9.

Whilst – as we will go on to see later – manual analysis and interpretation procedures are useful in relation to many programmes utilizing remote sensing data from satellites, it is in the use of air photographs for land surface mapping and topographic surveying that such procedures are most dominant today. Thus *air photo interpretation* (API) and *photogrammetry* will be the foci of our attention in this chapter, providing a timely reminder that much practical and commercial use of remote sensing data today follows very well-established, even quite traditional lines and supplies some important markets. However, it will become evident by the end of this chapter that even here methods of data usage are advancing, and, as in many other areas of remote sensing today, techniques developed first for use with airborne data are being applied to spaceborne data also (and all the more as the spatial resolution of the satellite imagery improves), and these different types of data are being more and more frequently exploited side by side in today's research and operations.

Therefore, whilst in the following sections we deal with some relatively traditional remote sensing practices, the reader must bear in mind both that other types of images can be treated similarly, and that increasingly the relatively traditional skills of API and photogrammetry are being supplemented by mechanical and computer aids – many of which have been developed primarily to handle the enormous quantities of satellite images to which we now have access.

8.2 Methods of photo-interpretation

Once the objectives of an eyeball interpretation exercise have been established, the completeness and accuracy of the results depend upon the interpreter's ability to disaggregate the contents of the imagery either consciously or subconsciously. Not surprisingly, therefore, many consider image interpretation (of whatever form, or for whatever purpose) to be as much an *art* as a *science*. Certainly, in the case of interpretation of aerial photographs we find a process that is highly deductive, and proceeds best in stages. These include:

1. A *general examination* of the image to take account of the general patterns of relief, vegetation and cultural development.
2. Concentration on specific areas or features, usually through a systemmatic approach to the *identification* of selected image contents.
3. *Classification* of the phenomena in the scene.

It has been suggested that it is the *level of reference* that governs the performance in photo-interpretation. The level of reference can be regarded as the amount of knowledge that is stored in the mind of any person or group of persons interpreting photographs. Three levels of reference can be distinguished – *general*, *local* and *specific*. The general level is the interpreter's general knowledge of the phenomena and processes to be interpreted.

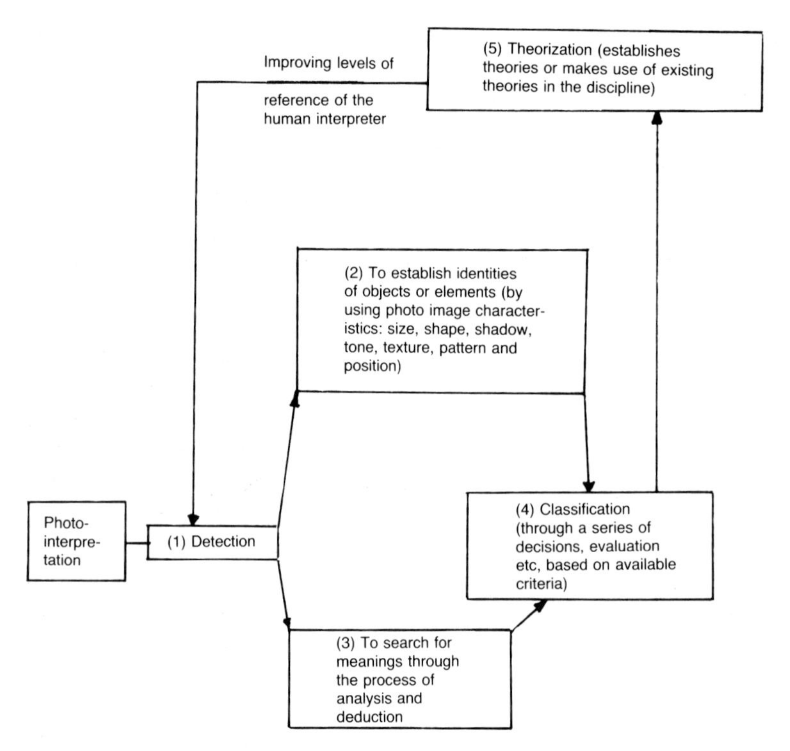

Fig. 8.1 Processes of photo-interpretation. (Source: Lo, 1976.)

Extra knowledge is added by the interpreter's intimacy with the relevant local environment. Finally, there is the specific level. At this level the interpreter employs highly specialized knowledge involving a much deeper understanding of the processes and phenomena to be interpreted.

One can identify two basic tasks in photo-interpretation as follows:

1. a process of establishing the identities of the objects and elements detected in the image;
2. a process of searching for their meanings.

The first process uses photo-image characteristics such as shape, size, tone, shadow, pattern, texture, situation and resolution to identify the objects. The second process uses more sophisticated types of analysis and deduction to discover meaningful relationships. It is the second process which seeks to give order to the qualitative information obtained and thereby provide some form of classification. Classification in itself demands some form of theorization. As a result the skill of the interpreter is raised and a feed-back of skill raises the level of reference of the human interpreter (Fig. 8.1).

Although we all 'interpret' photographs reproduced on television or in newspapers, special training is required for image or photo-interpretation. This is partly because of the unfamiliar viewpoint of the imagery and partly because of the special types of information which are usually demanded as the end products of their analysis. Nine *elements of photographic interpretation* (related to the image characteristics mentioned above) are regarded as being of general significance, largely irrespective of the precise nature of the imagery and the features it portrays. Selected examples are given in the list that follows. These should be compared with image characteristics appropriate to satellite imagery, e.g. those listed in the context of cloud type recognition from meteorological satellite imagery in Chapter 10. Thus, most of these elements apply to imagery obtained by other sensors also, even to those of the multispectral scanner type.

1. Shape. Numerous components of the environment can be identified with reasonable certainty merely by their shapes or forms. This is true of both natural features (e.g. geologic structures) and man-made objects (e.g. different types of industrial plant).
2. Size. In many cases the lengths, breadths, heights, areas, and/or volumes of imaged objects are significant, whether these are surface features (e.g. different tree species) or atmospheric phenomena (e.g. cumulus vs. cumulonimbus clouds). The approximate scale of many objects can be judged by comparisons with familiar features (e.g. roads) in the scene.
3. Tone. We have seen how different objects emit or reflect different wavelengths and intensities of radiant energy. Such differences may be recorded as variations of picture tone, colour, or density. They permit the discrimination of many spatial variables, for example on land (different crop types) or at sea (water bodies of contrasting depths or temperatures). The terms light, medium and dark are used to describe variations in tone.
4. Shadow. Hidden profiles may be revealed in silhouette (e.g. the shapes of buildings or the forms of field boundaries). Shadows are especially useful in geomorphological studies where microrelief features may be easier to detect under conditions of low-angled solar illumination than when the sun is high in the sky. Unfortunately, deep shadows in areas of complex detail may obscure significant features, for example the volume and distribution of traffic in a city street.
5. Pattern. Repetitive arrangements of both natural and cultural features are quite common, which is fortunate because much photo-interpretation is aimed at the mapping and analysis of relatively complex features, rather than the more basic units of which they may be comprised. Such features include agricultural complexes (e.g. farms and orchards), and terrain features (e.g. alluvial river valleys and coastal plains).
6. Texture. This is an important characteristic closely associated with tone in the sense that it is a quality which permits two areas of the same overall tone to be differentiated on the basis of microtonal patterns. Common photographic textures include smooth, rippled, mottled, lineated and irregular. Unfortunately, texture analysis tends to be rather subjective since different interpreters may use the same terms in slightly different ways. Texture is rarely the only criterion of identification or correlation used in interpretational procedures. More often it is invoked as the basis for a subdivision of categories already established using more fundamental criteria. For example two rock units may have the same tone, but different textures.
7. Site. At an advanced stage in a photo-interpretation procedure the location of objects with respect to terrain features or other objects may be helpful in refining the identification and classification of certain picture contents. For example, some tree species are found more commonly in one topographic situation than in others, whilst in industrial areas the association of several clustered, identifiable structures may help us determine the precise nature of the local enterprise. For example, the combination of one or two tall chimneys, a large central building, conveyors, cooling towers, and solid fuel piles point to a correct identification of an installation as a thermal power station.
8. Resolution. More than most other picture characteristics, resolution depends upon aspects of the remote sensing system itself, including its nature, design and performance, as well as the ambient conditions during the sensing programme, and subsequent processing of the acquired data. Resolution always limits the size and therefore in many cases, the nature, of features which might be recognized. Some objects will always be too small to be resolved (e.g. fair weather cumulus clouds on the average low-orbiting weather satellite photograph), while others lack sharpness or clarity of outline (e.g. the exact position of a shoreline is often difficult to deduce from an air-photo of scale, say 1 : 10 000).
9. Stereoscopic appearance. When the same feature is photographed from two different positions with overlap between successive images, an apparently solid model of the feature can be seen under a stereoscope (Fig. 8.2). Such a model is termed a 'stereo model' and the three-dimensional view it provides can aid interpretation. This valuable information cannot be obtained from a single print and is usually less

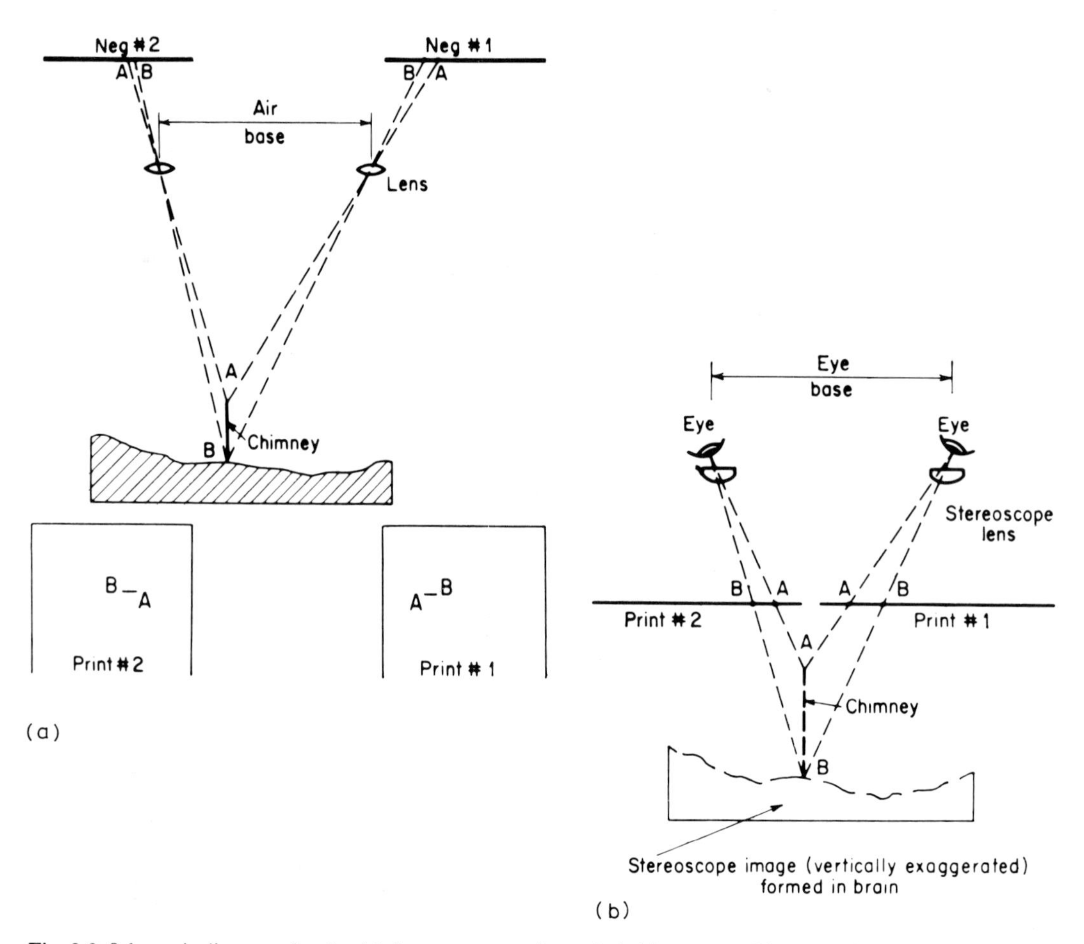

Fig. 8.2 Schematic diagrams showing (a) the appearance of a vertical object, e.g. a chimney on both negative and positive. Note the lean of the chimney top on the prints owing to height displacement. (b) The stereo image formed in the brain by viewing the prints through a stereoscope.

easily obtained from scanner images. Further discussion of stereoscopic images can be found on p. 138.

In practice these nine elements assume a variety of ranks of importance. Consequently, the order in which they may be examined varies from one type of imagery to another, and from one type of study to another. Sometimes they can lead to assessments of conditions not directly visible in the images, in addition to the identification of features or conditions that are explicitly revealed. The process by which related invisible conditions are established by inference is termed 'convergence of evidence'. It is useful, for example, in assessing social class and/or income group occupying a particular neighbourhood to note features such as detached or tenement types of building forms, proportion of open space, trees, size of gardens, road patterns, traffic and proximity to industrial plants or railroad facilities (see also Chapter 17). Likewise in estimating soil moisture conditions in agricultural areas the tonal or colour registration of vegetation, species identification, existence of field drains, riverine features, saline encrustations, and land use provide contributory evidence for interpretation of the soil moisture regime (Chapter 13).

Photo-interpretation may be *regional* or *spatial* in its approach and objectives, as in the case of terrain evaluation or land classification. Figure 8.3 is a sample photo-interpretation key for land use and vegetation mapping. On the other hand, photo-interpretation may be highly *site-specific*, and related to very precise goals. In the field of transport

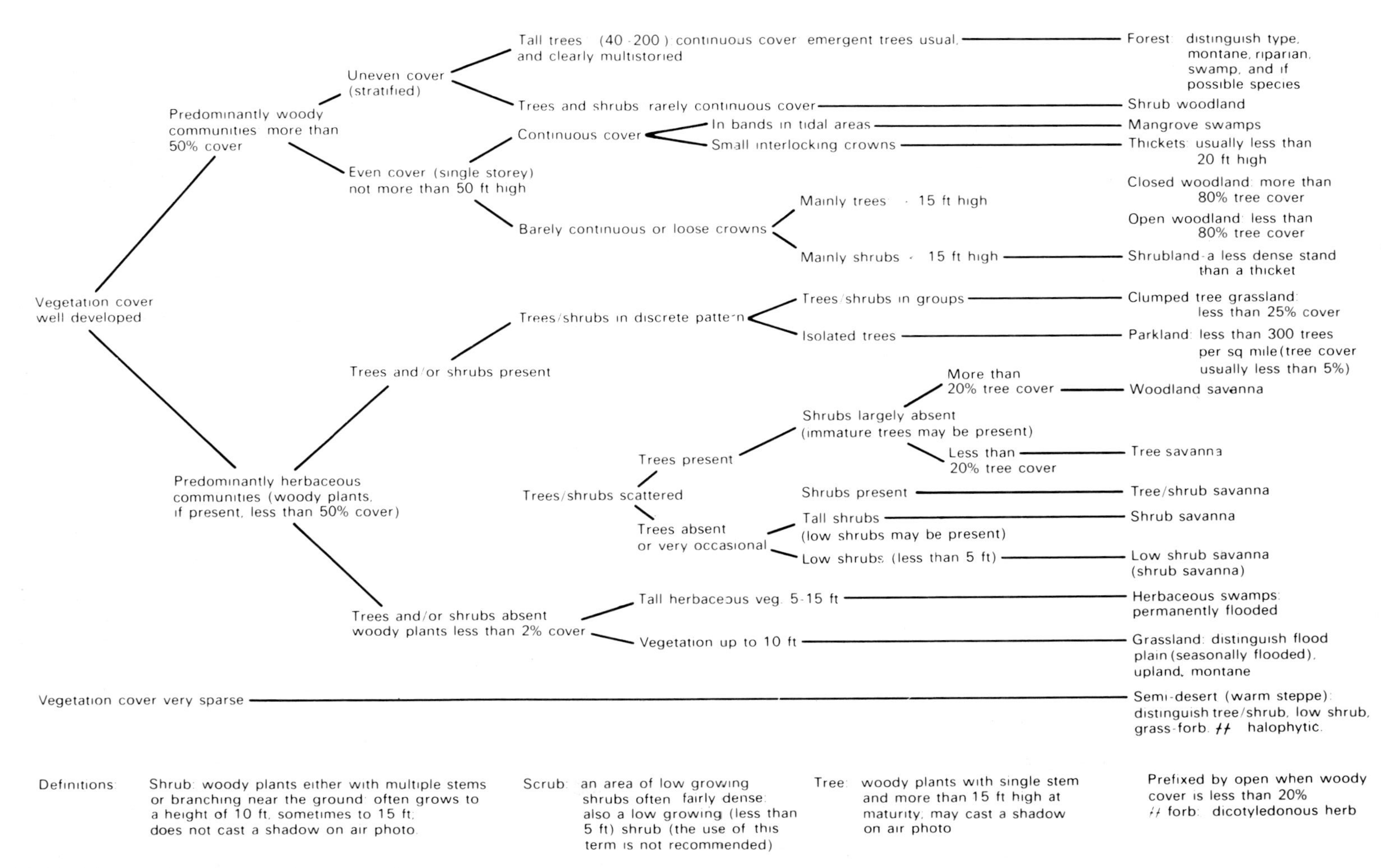

Fig. 8.3 A land use/vegetation key for mapping from aerial photographs in north-east Nigeria. (Source: Alford *et al.*, 1974.)

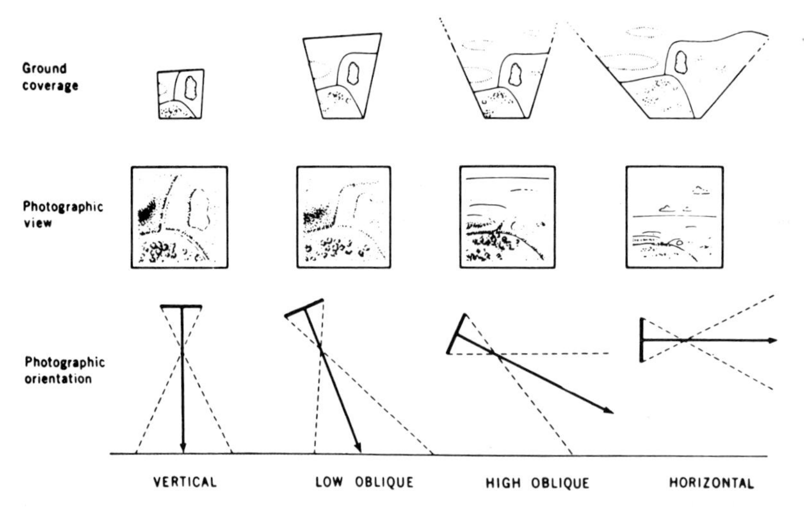

Fig. 8.4 The characteristics of vertical and oblique photographs taken from the air. The high oblique view is similar to that obtained from high ground or a tall building. The horizontal view is like that obtained from a slight eminence.

studies, for example, features such as car parks, railway yards or shipping berths are the subject of special study.

It is important to bear in mind that the degree of accuracy achieved in photo-interpretation may vary considerably depending on the nature of the subject, the type of photography and the skill of the interpreter. Consideration must be given, therefore, to the question of how much ground checking will be required to produce a result comparable in objective accuracy to a ground survey.

Whereas a map offers classification and symbolization of the features it records the photograph does not. This lack of symbolization makes a photograph more difficult to interpret but it also makes it potentially more flexible and informative for those prepared to set up their own classifications on the basis of the image characteristics.

8.3 Measurement and plotting techniques: selected aspects of photogrammetry

Various aspects of plotting from scanner data and scanner images are discussed elsewhere in relation to particular types of image data (e.g. radar, infrared linescan). This section will, therefore, concentrate on measurement and plotting from aerial photographs. The science of photogrammetry deals with the techniques involved and a detailed consideration of photogrammetric techniques cannot be given in the space of this volume. The reader is, therefore, directed towards further reading (see end of book); only selected points of background knowledge are presented here.

Aerial photographs may be either *vertical* or *oblique* according to whether the camera axis is vertical or not. A 'high oblique' includes the visible horizon whereas a 'low oblique' does not (Fig. 8.4). Oblique photography is most often obtained by single cameras but it is sometimes obtained by multiple camera arrangements of the trimetrogon type. In this a single, vertically orientated, camera is flanked on either side by oblique viewing cameras. These systems are cheaper than one using a single vertical camera, since this requires more flight lines to cover a given area. However the lateral distortions of scale are greater in obliques and less easy to correct.

It is important to recognize that all air photographs are perspective pictures and various types of planimetric inaccuracy are contained within them. These inaccuracies chiefly arise from scale distortions owing to height differences in the objects viewed and to tilt distortions (Fig. 8.4). In addition the planimetric accuracy of objects is affected by the displacement of the image owing to height variation (Fig. 8.2).

In order to overcome these problems of scale distortion and height displacement it is necessary to use overlapping photography capable of providing a stereoscopic image. Before considering the techniques used, attention must be focused on the

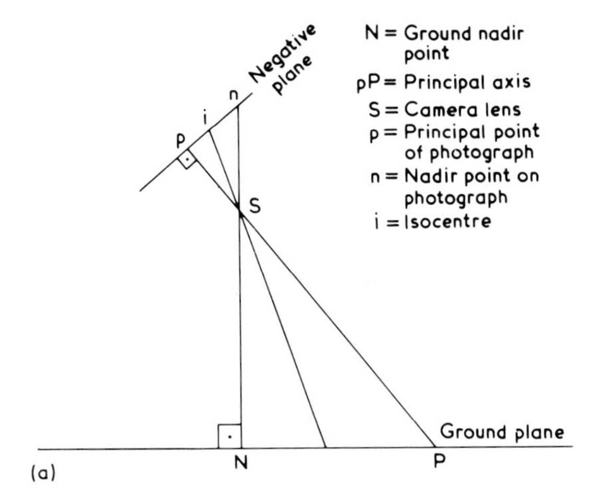

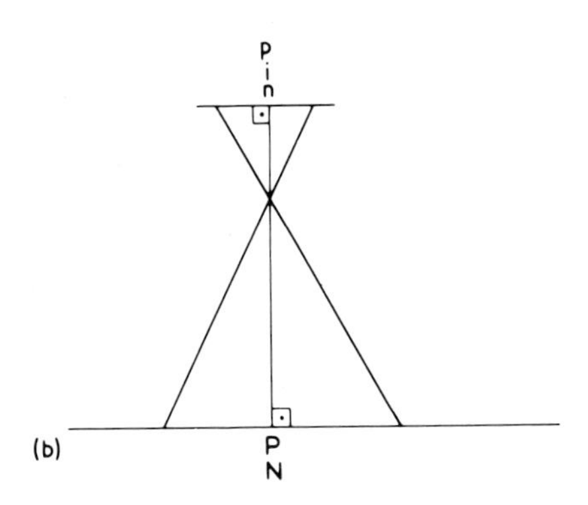

Fig. 8.5 The geometric characteristics of photographs. (a) Tilted photograph to show principal point, isocentre and nadir point. (b) Vertical case when principal point, isocentre and nadir point coincide.

principal geometric characteristics of air photographs which are fundamental to the photogrammetric processes. The geometrical properties of the principal point, isocentre and nadir point are particularly important (Fig. 8.5). Briefly stated they are as follows:

1. A photograph is angle true with respect to its isocentre for all points in the plane of the isocentre.
2. Straight lines drawn through the photographic nadir point pass through features which would lie on a surveyed straight line on the ground.
3. When a photograph is vertical the isocentre and nadir point are coincident with the principal point which then assumes their properties.

In these circumstances angles measured from the principal point are as truly representative of angles between features as if they had been measured with a theodolite on the ground. Straight lines measured from the principal point also cross detail which would lie on a surveyed straight line on the ground.

These properties are only completely true when the camera axis is perfectly vertical. However, providing the tilt of the camera does not exceed 2° and the height variation is small in relation to the flying height (less than 10%) the properties can be held to apply.

Thus it is possible to use the principal points of a number of photos as though they were plane table stations. Rays can be drawn from the principal point on each photograph through points of detail. Intersections can then be made by tracing off the rays from each photograph as they are successively placed under a transparent overlay along a base line. This method is termed the radial line plot and it provides a manual method of plotting the true positions of objects from air-photos.

Certain mechanical aids are used in this work by commercial firms. The rays are scribed on transparent film templets and then slots are cut to correspond to the rays. The technique is termed the 'Slotted Templet' method. The rays also can be scribed mechanically by the Radial Line Plotter apparatus.

Any method requiring measurement of the height of objects from air photographs is also dependent on stereoscopic cover between successive images. For this reason most vertical photographs are taken overlapping along the flight line (vertical line overlap). In this system features are photographed from successive camera positions (S_1, S_2, Fig. 8.6). Surface features (e.g. A, B, Fig. 8.6) will occupy different positions in relation to the principal points P_1, P_2 according to the heights of the features above a given datum. Thus the features A, B are imaged at the same photo point a, b on S_1 but at different positions on S_2 (a′, b′). If we add the distances P_1a and P_2a′ on the two positives this gives us a measure of the parallax of A. Similarly if we add the distances P_1b and P_2b′ we can obtain

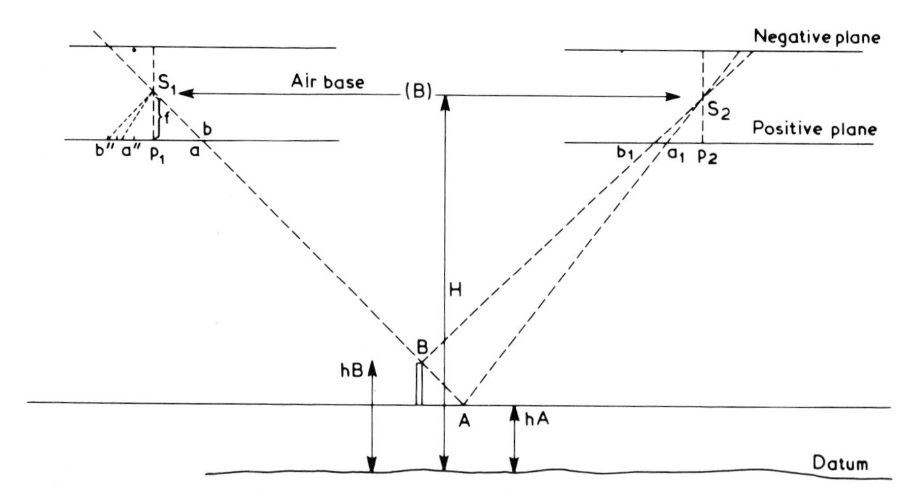

Fig. 8.6 Diagram to show the parallax of points A and B on two photographs taken at two positions S_1 and S_2.

a measure of the parallax of B. The difference between these two parallax measurements arises from the height difference between A and B, therefore differences in parallax can be used to calculate height differences from air-photos.

It is perfectly feasible to measure parallax on photographic prints with a ruler but generally speaking the differences are small so that a more accurate method is demanded. Therefore the Parallax Bar (or stereometer) has been developed which allows small differences in spacing to be measured with a micrometer gauge.

For contouring work it is necessary to use optical–mechanical devices that will allow an operator to view the stereomodel (Fig. 8.2) together with movable markers which can be adjusted in turn to positions A or B. When viewed stereoscopically such markers appear as a single image placed directly on the feature in the stereomodel. If the markers are moved closer together (i.e. parallax increased) the 'fused' mark will appear to float above the feature in the stereomodel. If they are separated (i.e. parallax reduced) the marks normally fail to be held in fusion by the brain and they separate on the model.

A skilled operator can manipulate a 'floating mark' of this kind so that it rises and falls on the model, just resting on the surface all the time. By doing so an automatic trace can be made which shows the true plan position of the features traversed. Alternatively the operator can place the floating mark at a known height and then move it across the stereomodel following the contours of the surface. By this means a contour map of the stereo image can be made.

Various types of 'floating' mark equipment are used in photogrammetry. Some instruments depend on the principle of anaglyph projection (i.e. one image red, the other blue, viewed through spectacles with one red and one blue lens). An example of such equipment is the multiplex projector. In this system the floating mark normally consists of a spot of light in the centre of a white disc on a movable tracing table.

Larger, and more expensive, manually operated instruments are used for precision work. Many of these have 'space rods' incorporated into the design which accurately reconstruct the angular relationships within the stereo model by mechanical rather than optical means (Fig. 8.7). In this equipment the operator manipulates handwheels and foot controls in order to move and adjust the floating mark on the stereomodel. As he does so the scribing pencil draws on the table beside the stereoplotter (Fig. 8.8).

Very high accuracies can be achieved by high-order plotting machines. The limitations in plotting accuracy are set by the nature of the stereomodel produced by the photography and by the ground control available for setting the model into its correct spatial orientation. Thus it should be borne in mind that accurate photogrammetric work also requires accurate ground control. This is normally achieved by establishing the heights of selected

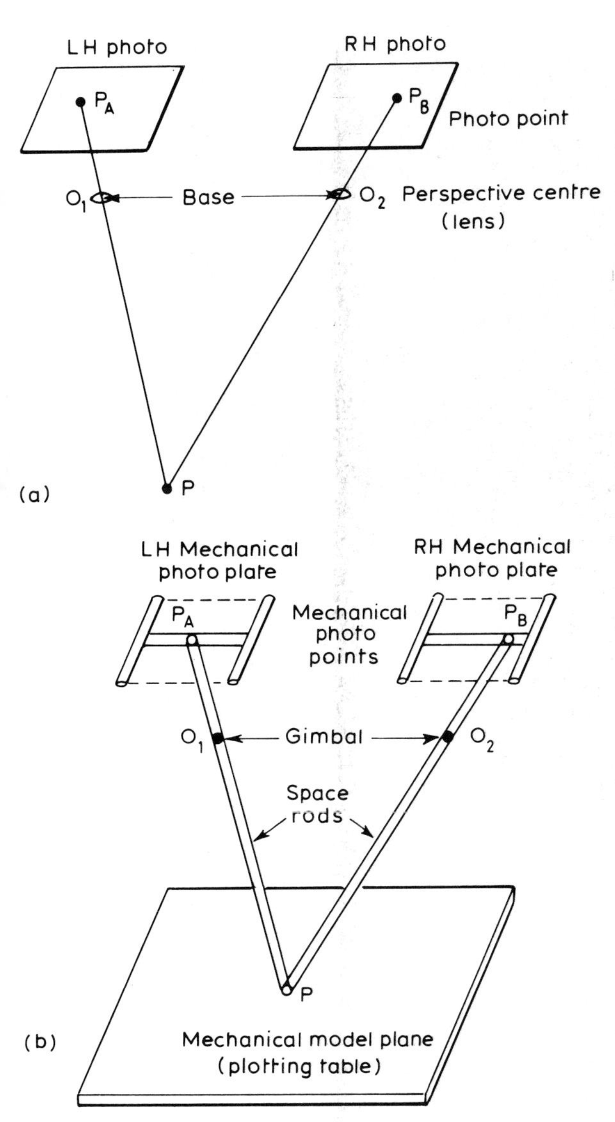

Fig. 8.7 Geometric and mechanical relationships in a stereo model and stereo plotter. (a) Geometric relationship between a point on the photograph and the corresponding point on the ground. (b) Mechanical representation of the image–ground relationship. (Source: Lo, 1976.)

well-defined features on the photographs. The operator can then orientate the stereomodel in relation to a true datum, and heights or positions are then maintained in true relationships within the model.

These manual techniques of photo-interpretation and photogrammetry are widely used by commercial air survey companies and many of our existing maps have been constructed by these means (Fig. 8.9).

Manual operation of a stereoplotter requires the employment of a highly skilled operator to control the work. As work loads increase there is a tendency to look for increasing automation in the photogrammetric process. Such automation is essentially a technical improvement on machine systems (Fig. 8.9). The goal has been to achieve automation of the three phases of machine operation, i.e. measurement, orientation and recording. There have been four main lines of development, consisting of:

1. photogrammetric digitizing of graphical outputs;
2. provision of analytical or computer-controlled stereoplotters;
3. orthophotoprojection systems;
4. automatic image correlation.

Although many semi-automatic systems are available in large commercial and national mapping agencies the individual user may find that the scale of his or her investigations does not warrant investment in such expensive equipment. The way forward may be that existing national and commercial facilities will be made available in a routine manner providing standard products for both large and small contracts.

It is important that the user should have a clear idea of the requirements for accuracy. In many cases the application of remote sensing data to environmental problems does not demand photogrammetric accuracy. In such cases the use of a device such as a Sketchmaster or a Transfer Scope (Fig. 8.10) will be quite adequate, providing that suitable base maps are already available on to which the photo or image detail can be transferred.

However, much more sophisticated systems are now becoming available for map-making from air photographs. One is the new breed of digital video plotters designed to make photogrammetry available as cheaply as possible through low-cost image display and graphics facilities supported by a small digitizing tablet, and run on a personal computer equipped with basic industry-standard software (Fig. 8.11). The other is the new type of photogrammetric workstation exemplified by the Leica SD2000 (Fig. 8.12), in which all the photogrammetric functions (user interface, real time program, calibration, orientation, triangulation, DEM collection, etc.) are built into the control computer (a

Fig. 8.8 A Wild A.8 Stereo Plotter in which diapositives are viewed stereoscopically. The diapositives are mounted beneath the illumination lamps on either side of the apparatus. Note the 'space rods' projecting above the centre of the plotter.

Fig. 8.9 Wild A.5 Autograph plotters in operation with plotting tables seen in the foreground. A scribing pencil is automatically controlled as the operator manipulates the stereoplotter, placing the floating mark over the points of detail to be mapped. (Courtesy, Clyde Surveys Ltd, Maidenhead, Berks.)

Fig. 8.10 Bausch and Lomb Transfer Scope. A graphical data transfer instrument that combines optical views of stereoscopic image and the base map. Used widely to transfer detail from photographs to maps. (Courtesy, Survey and General Instrument Co., Edenbridge, Kent.)

standard PC). Such a *workstation* is, in effect, a 'three-dimensional digitizer', capable of running any applications software for mapping, Computer Aided Design (CAD) or Geographic Information System (GIS: Chapters 9 and 19).

When the basic procedures of photo-interpretation and photogrammetry were developing some 60 years ago, the amount and range of image data were limited. The explosive development of remote sensing techniques has necessitated the evolution of a wide range of preprocessing methods. These modern methods are designed to improve the precision of the data to be analysed and to allow surplus data to be discarded.

The new generations of satellites offer greater capabilities than aircraft could ever do, both for areas covered and for information provided, with corresponding increases in data flow. Increasingly, satellites rather than aircraft will constitute the primary sources of data for photogrammetric analysis and map production.

To support such activities, digital workstations, in which analog or optical data are replaced by digital images, have begun to evolve since 1988. These are extensions of existing digital image processing systems (Chapter 10). In them, stereo-viewing of satellite imagery is made possible either by polarization techniques or attached mirror stereoscopes. They are designed to act as universal facilites for purposes of monoplotting, stereo-plotting and rectification, but using digital satellite images, not air photographs, as the inputs.

One such workstation is the KERN DSP1 (Fig. 8.13). This is equipped with both image processing and analytical photogrammetric capabilities, having hardware for image movement and data recording (trackball and/or handwheels, footdisk and twin footswitches), and for stereoscopic observation (a retractable optical system to view images on a high resolution, split-screen monitor). All common image processing functions are available through either hardware or software, plus an optional transputer array to enable the acceleration of time-critical functions. Photogrammetric tests of SPOT-1 system confirmed that at least 1 : 100 000 scale planimetric requirements can be met thereby. Tests in France have led to the generation of 1 : 100 000 *Digital Elevation Models* (three-dimensional digital arrays of surface morphology from SPOT data of the Marseille region (Fig. 8.14) with no discernible differences between the satellite maps and existing ground survey maps. Tests of the accuracy of terrain height determinations from the SPOT products compared with those from

Fig. 8.11 The Digital Video Plotter (DVP) System. (Courtesy, Leica.)

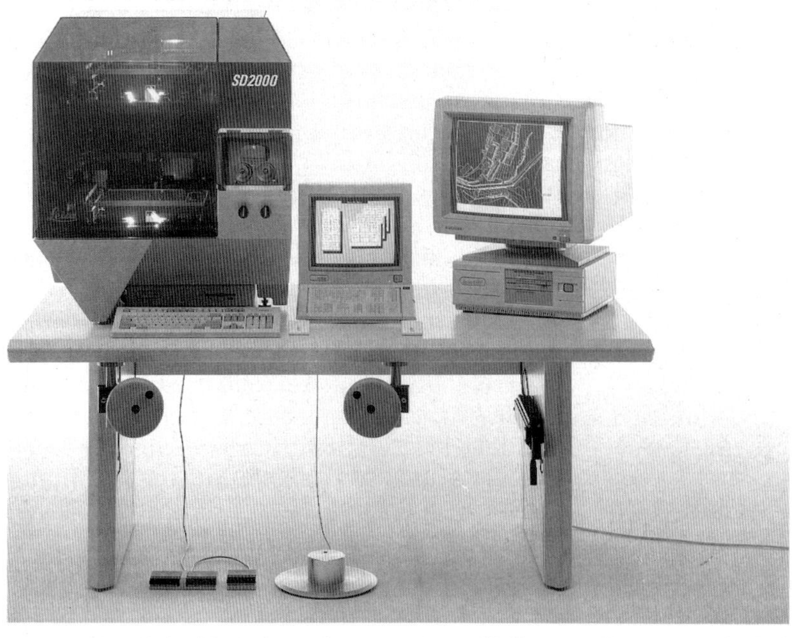

Fig. 8.12 The Leica SD2000 analytical stereoplotter. Unlike the DSP 1 and the DVP this uses film images as imput. (Courtesy, Leica.)

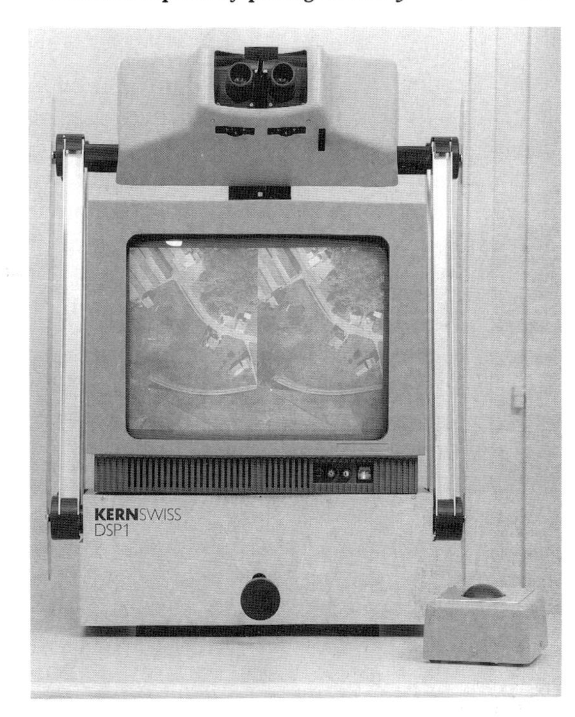

Fig. 8.13 The display and analysis of the KERN DSP 1 Digital Stereo Photogrammetric System. (Courtesy, Leica.)

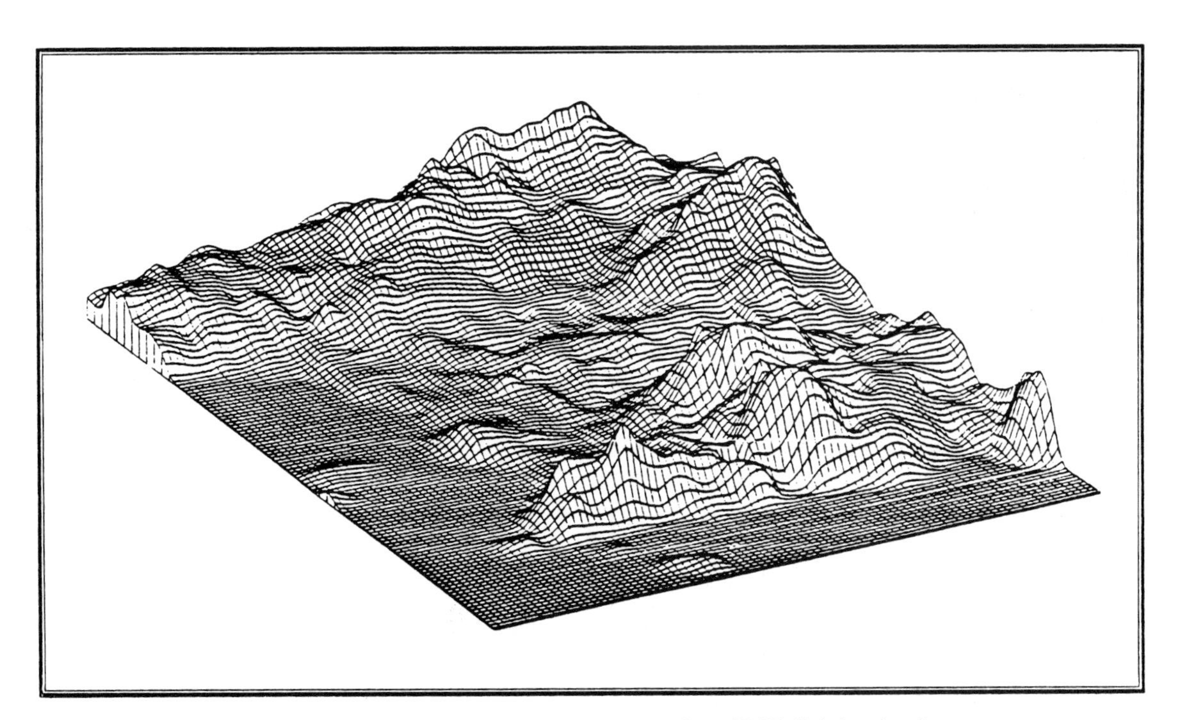

Fig. 8.14 A DEM of the Marseille area of southern France, generalized from SPOT digital analog data.

ground survey/air photography for over 33 000 point locations yielded SPOT-derived standard errors of only ±5 m. Such results give great confidence to the evergrowing number of companies and national government survey bureaux who, between them, are using SPOT data to improve – or even to generate – morphological and topographic maps of substantial areas of the world.

Thus great progress is being made with the development of analysis and interpretation systems for air photographs and Earth resources satellite imagery: clearly, SPOT is providing the spur for great advancement in these fields, for digital systems offer unprecedented flexibility to the operator, not only in analysis and interpretation *per se*, but also in respect of the generation of products for use with other types of data, whether *in situ* or remotely sensed.

9 *Digital data handling*

9.1 The need for numerical data manipulation

The analysis and interpretation of remote sensing data by manual means has much to recommend it. For example, the trained human eye can identify a very wide range of imaged features with both ease and accuracy. Consequently the interpretation of aerial photographs, space photographs, and other forms of remote sensing imagery is, perhaps, still best carried out by hand if the data sets are small and budgets are low. But the volumes of data generated by new remote sensing systems are growing rapidly and already far exceed the capabilities of trained interpreters in some environmental fields. For example, it has been estimated that a single low-altitude Earth-orbiting satellite, such as any of the NOAA (National Oceanic and Atmospheric Administration) weather satellites, have yielded something of the order of $10^{10}-10^{12}$ points of new data every day. Earth resources satellites planned for the future may be capable of even greater outputs, yielding in excess of 500 million data bits per second, i.e. between 10^{13} and 10^{14} new units of information every day. Still further increases may be expected as transmission bandwidth restrictions are eased. Computer-based interpretation techniques afford the only possible solutions to the problem of routine extraction of useful facts from such intense data streams, especially on a near real-time ('operational') basis. It is therefore fortunate that as we saw in Chapter 7, the three basic forms of remotely sensed data (image, analog and digital) are all interchangeable, for one consequence is that, if necessary, any remote sensing data set can be rendered into a digital form and thereafter processed automatically.

Another stimulus to the increasing use of computer analysis is the fact that numbers are required for environmental hypotheses to be tested rigorously. Although manual, '*eyeball*' or *subjective* analyses can provide much useful information, and are still the better suited to some types of tasks, numerical data arrays of suitable magnitude for statistical processing can now be compiled very quickly, and results from different areas, or different algorithms intercompared objectively. In particular, the eye is good at *feature recognition* and *image interpretation*, but relatively poor at assessing basic image characteristics such as *brightness* and *texture*. So, the choice often lies between range of features and accuracy of identification on the one hand, these benefiting from the skill and experience of the analyst, and numbers of cases, replicability and speed of operation on the other, these benefiting from computer-based automation.

To a great extent the success of '*automated*' or *objective* techniques of image analysis is dependent upon the type of question that is posed: increasingly, as a result of growing experience, acceptable answers are now being obtained in response to the posing of reasonable questions. For example, numerical methods are especially useful for the development of probabilistic statements about points, lines or areas, and for the simultaneous use of data from more than one waveband, or basic source. Meanwhile, in feature recognition and/or classification, complete accuracy is not to be expected whether manual or objective methods of image analysis are deployed. However, even in these cases quantitative techniques may be preferred to qualitative methods if only because they can be used to generate accuracy levels for the determinations that have been made, and related allowances can be made in the environmental models into which the results of remote sensing data analyses are fed.

Mainly in the last two decades a plethora of

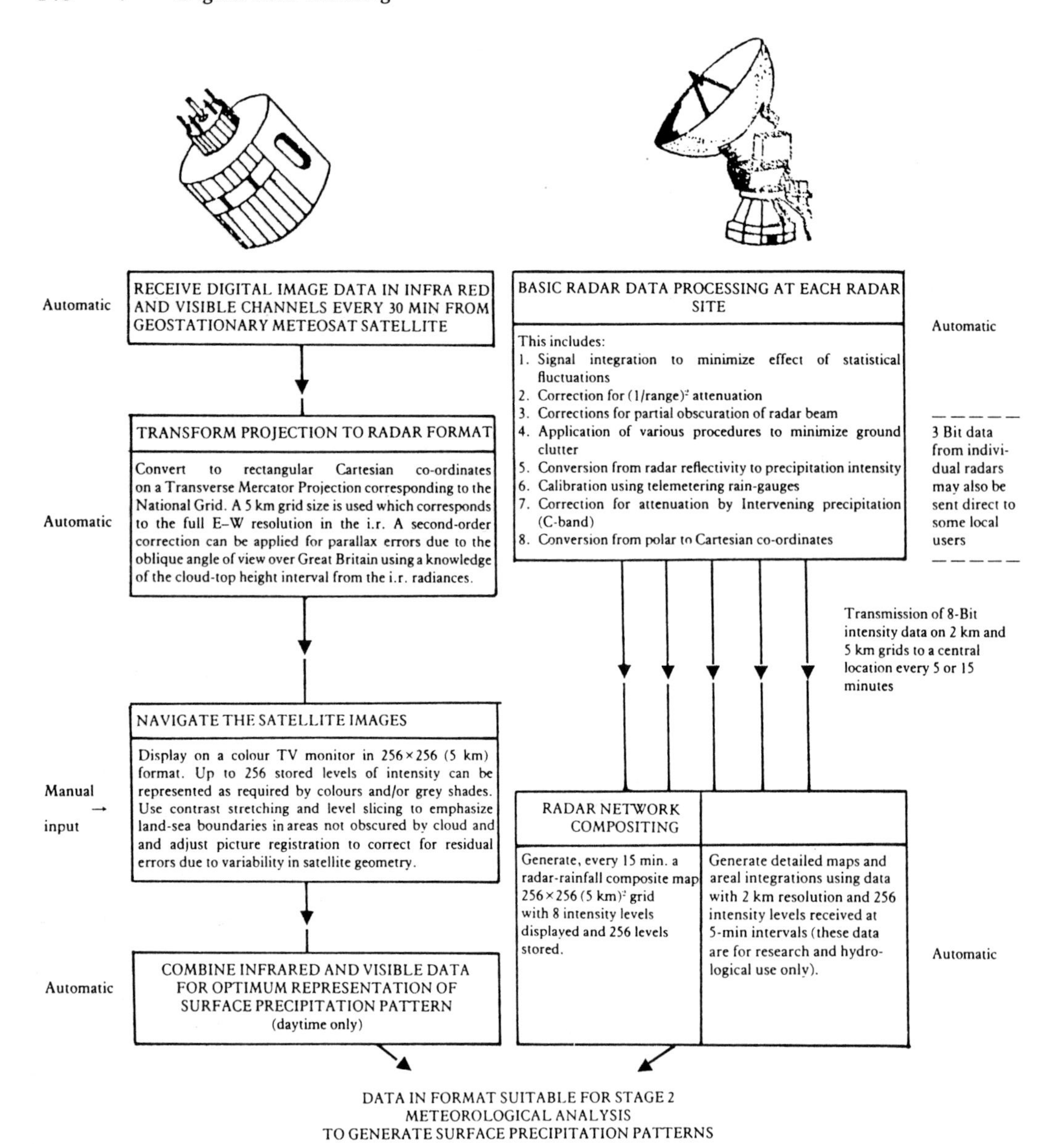

Fig. 9.1 Preprocessing of satellite and radar data for the 'FRONTIERS' plan to use radar and satellite imagery for very short-range precipitation forecasting (see also Chapter 12). (Source: Browning, 1979.)

numerical methods has grown up for use in remote sensing, stimulated by the satellite-led information explosion but applicable to most if not all forms of remotely sensed data and facilitated by the rapid rise of electronic computing. Complex sequences of digital techniques are commonly applied, not only in data preprocessing contexts, as exemplified in Chapter 7 (Fig. 7.5), but also for dependent data processing, analysis and product generation, as shown in Fig. 9.1. Given the already vast and often highly technical literature that has grown up in this field, our concern here is not to provide a digital

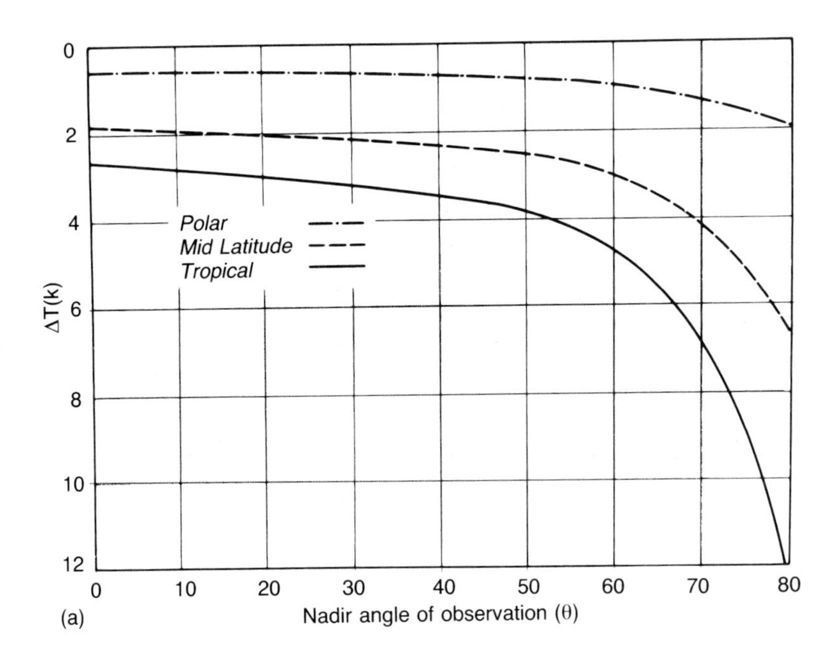

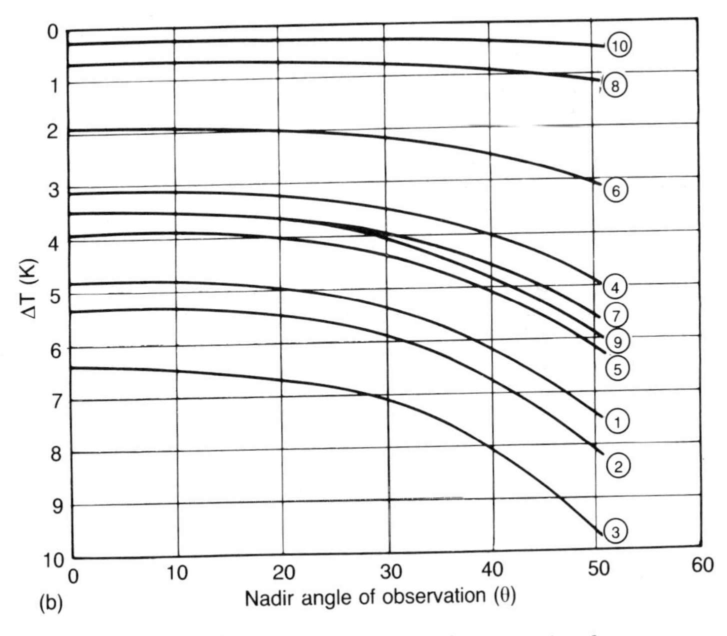

Fig. 9.2 Calculated departures of the THIR 11.5 μm channel equivalent black body temperatures from ground surface temperatures as functions of zenith angles for ten different atmospheric regions: (1) standard, (2) tropical, (3) subtropical summer, (4) subtropical winter, (5) mid-latitude summer, (6) mid-latitude winter, (7) subarctic summer, (8) subarctic winter (cold), (9) Arctic summer, (10) Arctic winter (mean). Unfortunately the structure of the atmosphere, being complex and rapidly changing, results in residual errors in such data even when adjusted by such detailed correction functions. (Source: Sabatini *et al.*, 1971.)

techniques 'how-to' manual, as the required length of this book alone strongly forbids this, but rather to give a structural overview of the nature and use of much more detailed numerical methods in remote sensing. This will serve as a springboard both into the journal literature, but more immediately, into Part Two of this book, where applications of environmental remote sensing –

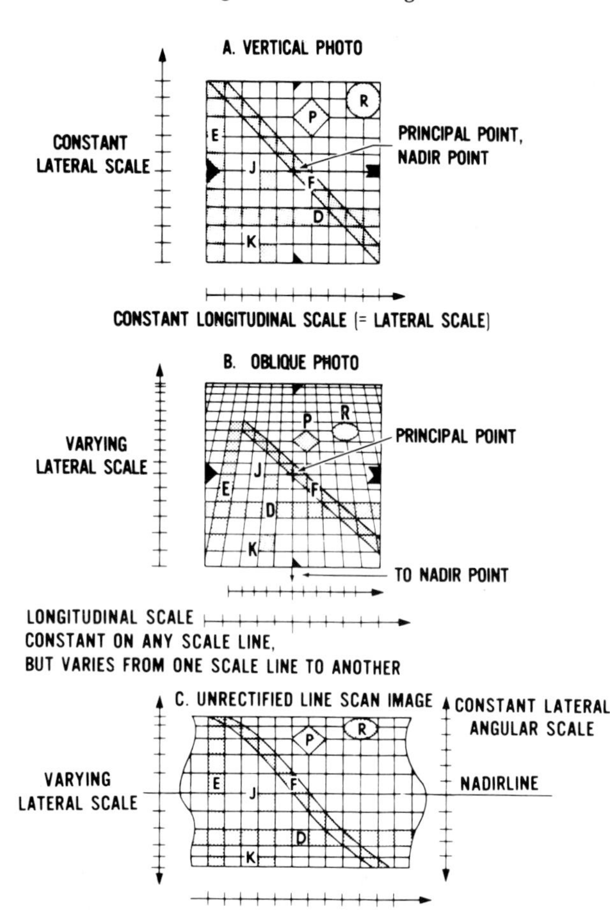

Fig. 9.3 To exemplify the differences in the scales and geometries of different types of remote sensing data from aircraft. The letters reference points or areas to show effects of different geometric distortions.

many of them heavily dependent on numerical methods – are reviewed, exemplified and discussed.

It should be remembered that four steps characterize much use of remotely sensed data: *data preprocessing*, *data processing*, *data analysis* and *data interpretation* (Chapter 7). Not least because the boundaries between these different steps differ from project to project, but more because the strongest affinities are between the first two and last two pairs, we will proceed in the present context to treat digital preprocessing and processing steps together, followed by analysis and interpretation steps towards the end of this chapter.

Finally, we will consider some increasingly important questions related to the operational implementation of remote sensing programmes before proceeding (in Part Two) to see how remote sensing is serving the community of environmental scientists today.

9.2 The quality and quantity of remote sensing data

Data from remote sensing missions are not always optimal in quality. Each remote sensing system has its own characteristic capability depending on such things as the instrument design, altitude of the platform, and performance of the recording equipment. Other less predictable, often extraneous, influences may adversely affect the quality of the retrieved data, as exemplified by Fig. 9.2. Preprocessing functions are necessary to reduce or eliminate such inadequacies. Some such methods involve photographic film or deploy optical or electronic analog systems (Chapter 7). Here we must concentrate our attention upon the digital approach, which is the more dependable.

9.2.1 Geometric corrections

Remote sensing data, whether obtained in snapshot, line-scan or digital array forms, come in a variety of scales and geometries, as exemplified for aircraft imagery by Fig. 9.3, and satellites by Fig. 9.4. Often these must be rectified before they can be compared in detail with existing data. The general problem may be exemplified by studies designed to investigate the relative merits of imagery from aircraft or satellite platforms for the identification and mapping of features at a selected level of detail in a particular region. The problem is most intransigent where:

1. imaged areas are topographically rough;
2. the images are obtained from systems with broad fields of view;
3. the images are obtained from low-altitude platforms, where image displacements owing to even quite modest topographic features may be considerable;
4. systems other than framing cameras are used.

Rectification can be performed by optical and electronic means (of which the former is usually cheaper), or by digital techniques. Basically the problem involves *bringing separate data into congruence*, that is to say ensuring that point-to-point

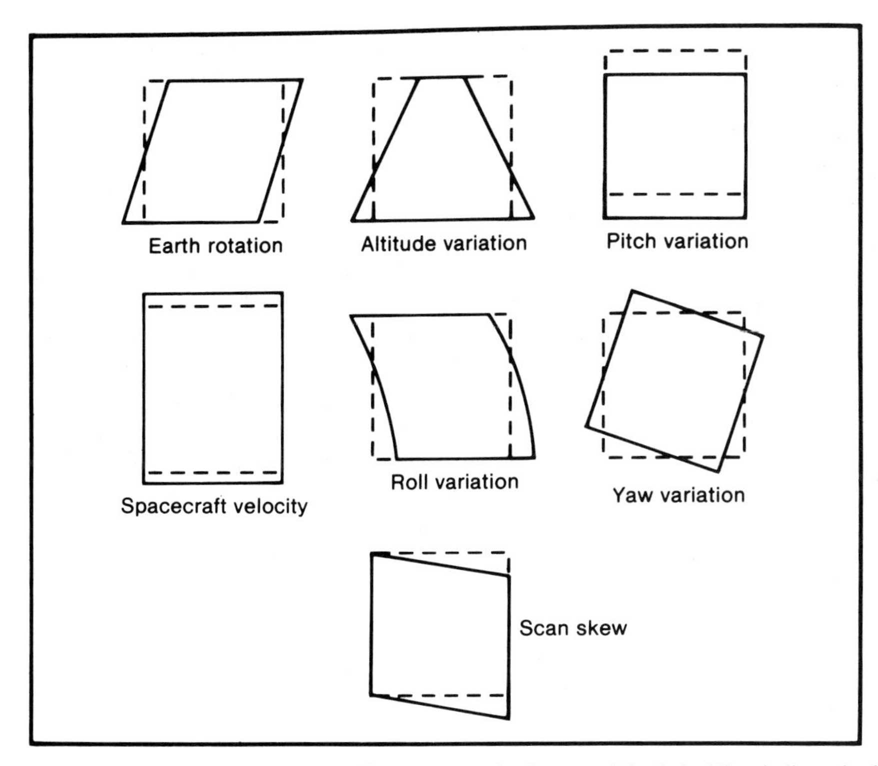

Fig. 9.4 Geometric distortions in satellite remote sensing imagery. The dashed lines indicate the desired image shape, the solid lines the actual shape. (Source: Harris, 1987.)

correspondences are achieved. These are necessary, within certain limits, in many remote sensing studies if point decisions or point comparisons are intended using statistical pattern recognition techniques, e.g. using the fast Fourier transform algorithm to provide the two-dimensional Fourier transform of individual images, and then to facilitate rapid cross-correlations between pairs of images or other data arrays.

Since the variations in the performance of an orbiting sensor system are proportionately less than those associated with lower-altitude systems, satellite images may be brought into acceptable congruence more easily and more cheaply per unit area than those from aircraft, using navigation and geo-registration (mapping) routines as mentioned above. The relative ease with which digital satellite data can be brought into congruence with a selected map projection is one of the principal advantages of many modern satellite remote sensing systems. Landsat is of sufficient significance today to merit special mention.

Although Landsat images are relatively free from panoramic distortions and relief displacements, they suffer from image distortions of both *systematic* (or predictable) and *random* (or unpredictable) types for which compensations must be made before really accurate analyses can be made of these data. On the ground, each Landsat MSS or TM pixel is basically trapezoidal (not square or rectangular) in form, as a result of the general viewing angle of its sensor system, although these effects are much more strongly marked with MSS than TM data (Fig. 9.4). Minor deviations from the norm stem from factors such as variations in the altitude, attitude and velocity of the satellite. Corrections for systematic distortions, such as the skewed-parallelogram effect of Earth rotation beneath the satellite during imaging, can be made relatively easily, in this case by offsetting each scan line by an appropriate amount.

Corrections for truly random distortions are more difficult, and are usually accomplished through reference to carefully selected *ground control points*, such as highway intersections, small water bodies, etc. For such points both image (column, row) and ground (latitude, longitude) coordinates are established, and the values then submitted to a least-squares regression analysis to determine coefficients for two 'transformation equations' which interrelate the satellite image and ground sets of coordinates. The full process by which Landsat geometric transformations are applied to the original data is termed 'resampling'. It involves:

1. The definition of a geometrically uniform 'output' matrix in terms of ground coordinates.
2. The transformation by computer of the coordinates of each output cell into corresponding coordinates in the image data set.
3. The transferral of each pixel value from the image data set to its appropriate place in the output matrix. The result should be a matrix of digital image data that is geometrically correct in respect of a set of ground coordinates.

9.2.2 *Radiometric corrections*

These are necessary to lessen the effects of a number of variations in radiation or image density. The aim is to achieve a product with a high level of invariance with respect to the radiance of the scene, whether that radiance stems from reflected or emitted energy. Common radiometric defects are associated with the following:

1. A decline in intensity away from the centre of an airborne light photographic lens.
2. Changing relations between the angle of view and the angle of solar illumination.
3. Variations in illumination along a sensor flight path. These are especially significant in connection with polar-orbiting satellites.
4. Variations in viewing angles of sensors scanning across the flight path of the sensor platform.
5. Variations arising from the performance of the sensor system, e.g. as they age.
6. Variations in photoprocessing and developing procedures (hand-copy imagery), and in preprocessing routines (digital data).
7. Atmospheric effects, e.g. those caused by haze, water vapour and subpixel clouds, which modify the radiation received by the sensor.

Correction of such effects is more difficult than in the case of geometric effects, for their distributions and intensities are often very inadequately known. Numerical methods of combating them include several 'normalization' techniques. Most involve the assessment of the nature and degree of the effect caused by one or more of the sources of variation in the recorded radiation density, followed by the development of appropriate physical remedies (e.g. filters or films), or numerical weighting functions (e.g. calibration curves and algorithms). In all these ways the original data may be improved – though care must be exercised because sometimes 'corrected' data are of a lower quality than the original.

9.2.3 *The reduction of noise*

Noise may be defined as any unwanted (either random or periodic) fluctuation of a signal which may obscure its basic form and make its analysis and interpretation more difficult. *Random noise* may be caused by the performance of remote sensing systems during recording, storing, transmission, and especially in ground reception of the data. *Periodic noise* may be caused by radio interference, and by certain components of the sensor/platform complexes themselves. In some cases it is possible to dampen random noise through the application of statistical smoothing functions to the data if these are presented in numerical form, but such noise can rarely be eliminated. Rather, efforts are necessary to keep it within acceptable bounds.

Periodic noise is easier to eliminate because its pattern in both one-dimensional and two-dimensional products is, by definition, more systematic. Figure 9.5 illustrates the noise outline for the tape recorder on NOAA 2. The rotating mirror in the NOAA 2 Scanning Radiometer, and the satellite's tape recorder drive motor, were controlled by a common power source oscillator. Consequently the noise from this source associated with the tape recorder – inherent in the data transmitted to the ground – tended to be periodic. Cloud pictures contained regular 'streaky' patterns, which, if untreated, would have been passed on in secondary (analytical) products. When any pattern of noise is known, appropriate modifications can be made to ground station hardware to lessen its effects.

In the case of Landsat 3 an unwanted image characteristic resulted from the improper functioning of one of the six identical detectors in the Multispectral Scanner: a noticeable banding appeared with a six-line interval. Similar problems

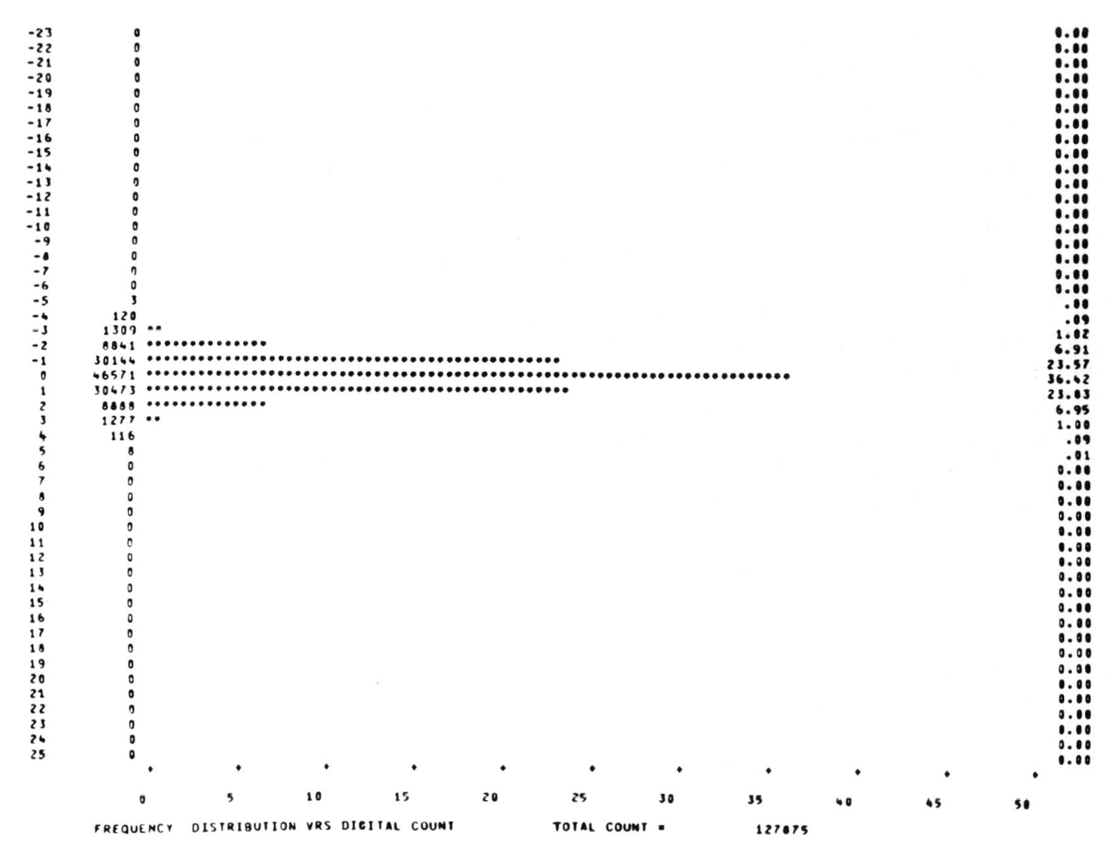

Fig. 9.5 Noise pattern induced by the tape recorder (absolute response) on NOAA 2 shown for a supposedly inert recording of night-time data from the visual channel scan sensor. Such effects may create systematic streaking in display products, and produce systematic contamination in the generation of more quantitative applications. (Source: Conlan, 1973.)

have arisen with Landsat TM data, but producing a 16-line banding effect because the TM scans 16 lines simultaneously and directs the energy on to different detectors. Several techniques are available to reduce the impact of these 'striping' effects (Fig. 9.6(a)). The simplest method is to multiply the pixel values in the deficient lines by normalizing factors to give adjusted values with mean and standard deviations equal to those in the rest (Fig. 9.6(b)). Unfortunately, all such methods yield improvements that are more apparent than real: there is no way in which fully perfect adjustments can be made for fundamentally defective data.

9.2.4 Data calibration

Substantial sections of this volume are concerned with the analysis and interpretation of remote sensing data in terms of familiar characteristics of human environments. Since our funds of environmental data from conventional sources involving *in situ* sensors, having been built up over a much longer period, are much greater than those of remote sensing data, it is common practice to calibrate the remote sensing data in terms of conventional parameters, although, as noted in earlier chapters, this may not always be simple or even sensible to undertake because of basic differences between remotely sensed and *in situ* data sets. However, there is now increasing interest in the analysis of the remotely sensed radiances themselves, without attempting to combine them with some other – probably quite different – type of data: environmental patterns mapped from satellite altitudes can be expressed simply in terms of the measured radiances (e.g. stratospheric radiances in terms of voltages), with no attempt to transform

Fig. 9.6 Landsat Band 7 images of the Woolacombe area of North Devon, England, 27 May 1977. (a) Original image showing Landsat 3 striping effects. (b) Digitally destriped image. (Courtesy, Space Department, RAE, Farnborough; Crown Copyright.)

the results into more traditionally recognized environmental units. At present this approach is restricted mainly to situations in which conventional ('ground truth') data are few or unavailable, e.g. in respect of middle and upper atmospheric features (Chapter 11), but may be expected to spread in the future as remotely sensed data rather than *in situ* data become more widely accepted as the norm.

Often *prelaunch calibration* procedures are carried out to provide the numerical voltage-to-temperature or voltage-to-brightness response relations for the sensor systems. For example, the calibration programme for operational weather satellites involves the preparation of families of calibration tables for each sensor for converting normalized raw counts into effective black body radiative temperature responses in the case of infrared sensors, or into calibrated brightness responses for the visual channel. These values are then corrected for any bias as a function of sensor temperature, a measurement regularly available in the telemetry data.

The final step involves those corrections necessary for the interpretation of measurements of the natural Earth scene. For example, allowances must be made for atmospheric attenuation in the case of infrared sensors. This 'limb darkening' correction is generally taken as a function of local zenith angle, with the assumption that water vapour is the primary absorbing constituent (Fig. 9.2). For the visual channel on current NOAA satellites the algorithm in use at present is a simple cosine function of the solar zenith angle.

9.2.5 *Data compression*

The reduction of the raw data obtained from a remote sensing system into a volume no larger than that required for a particular programme of analysis and interpretation may be effected by one or more of a wide range of standard statistical techniques. These include *smoothing and averaging* processes, *sampling techniques*, and *feature extraction* methods. Smoothing or averaging processes are commonly applied by numerical means to data in either one-dimensional (e.g. line scan) or two-dimensional (data array) forms. Supplementary processing may be carried out by the computer to ensure that the assumptions of subsequent numerical processes might be met; for example, weighting factors may be applied to normalize the frequency distribution of the new, reduced population.

If the original data are to be reduced by selection instead, the principles and practices of *statistical sampling theory* are invoked. The problems of feature extraction are not difficult to resolve providing that the features to be removed for further study are related to simple characteristics of the raw data themselves, for example, regions of radiation temperatures above a given threshold in infrared studies, or areas with selected ranges of brightness in densitometric analyses of conventional photographs. Further data-reduction processes may then be applied to the information selected for careful scrutiny. However, the problems of selective feature extraction are very difficult when combinations of data characteristics are invoked simultaneously. Some of the attendant difficulties of pattern analysis and pattern recognition are discussed in section 9.3.

9.2.6 *Image enhancement*

Specific users of remote sensing data often require that the features of special interest to themselves be emphasized or enhanced at the expense of other, e.g. background, features. Such enhancements can be carried out photographically (Chapter 7) or digitally. The commonest digital procedures include *density slicing*, *edge enhancement* (image sharpening), *contrast stretching* (increasing the range of image tones), and *change detection* by image addition, subtraction or averaging. The digital approach has the advantage of flexibility, whilst the photographic approach has the advantage of cheapness (Chapter 8). The digital approaches may be further detailed as follows:

1. Density slicing. Given image data presented in a digital form it is possible to simplify the information by reducing the number of classes through the selection of appropriate threshold levels. Sometimes a very simple two-class classification is adequate, for example, where a categoric yes/no type of decision has to be made. Numerous digital systems have been developed for rapid density slicing. In such systems any greyscale level or levels may be selected on a photographic or electrical imprint, with output

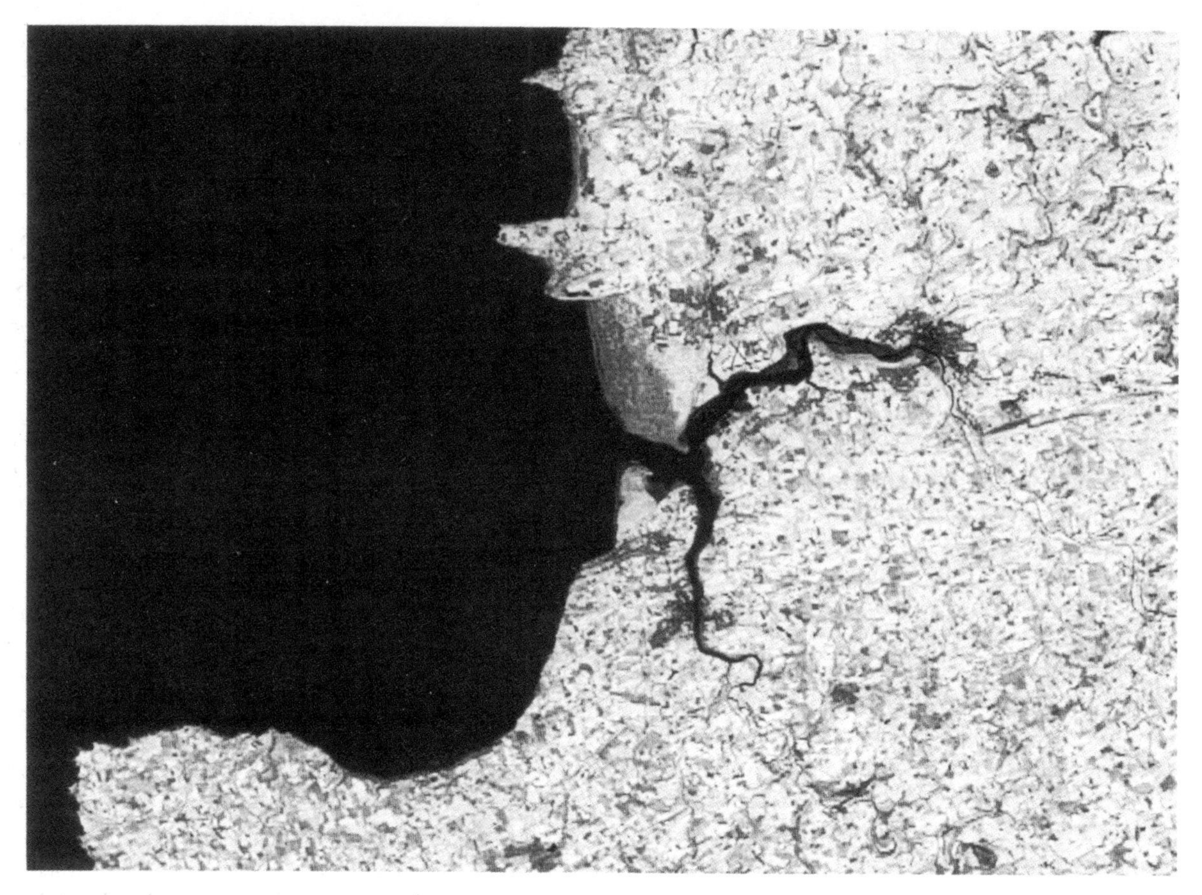

Fig. 9.7 The Landsat 3 image shown in Fig. 9.6, after destriping and edge enhancement. (Courtesy, Space Department, RAE, Farnborough; Crown Copyright.)

via a line printer on to a flatbed or drum plotter, or directly on to a photographic film writer. The procedure involves the development of suitable algorithms and computer programs. 'Quantization noise', or spurious contouring, may result if the slices are not chosen carefully.

2. Colour enhancement. Here preselected colours are accorded to chosen categories of contents in remotely sensed images. In this way, more obvious distinctions can be made between different features of each scene (Plate 4, colour section).
3. Edge enhancement. This can be used to sharpen an image by restoring high-frequency components through the removal of scan-line noise, or to emphasize certain edges to aid interpretation by taking the first derivative in a given direction. Unfortunately the computer cannot easily distinguish desired lineaments from others of similar greyscale change, and these will be enhanced as well. A common method of edge enhancement of Landsat images involves the doubling of the deviation of each pixel, or selected pixels, from the local averages. These averages can be computed from 'pixel neighbourhoods' whose shapes and sizes can be varied to suit the requirements of the data user. Figure 9.7 illustrates one edge-enhancement process and its results.
4. Contrast stretching. Any image of low contrast or any portion of the greyscale of an image may be enhanced in digital processing through suitable adjustments to the array of picture points. Many satellite sensor systems (e.g. Landsat MSS and TM) are designed to accommodate a wide range of scene illumination con-

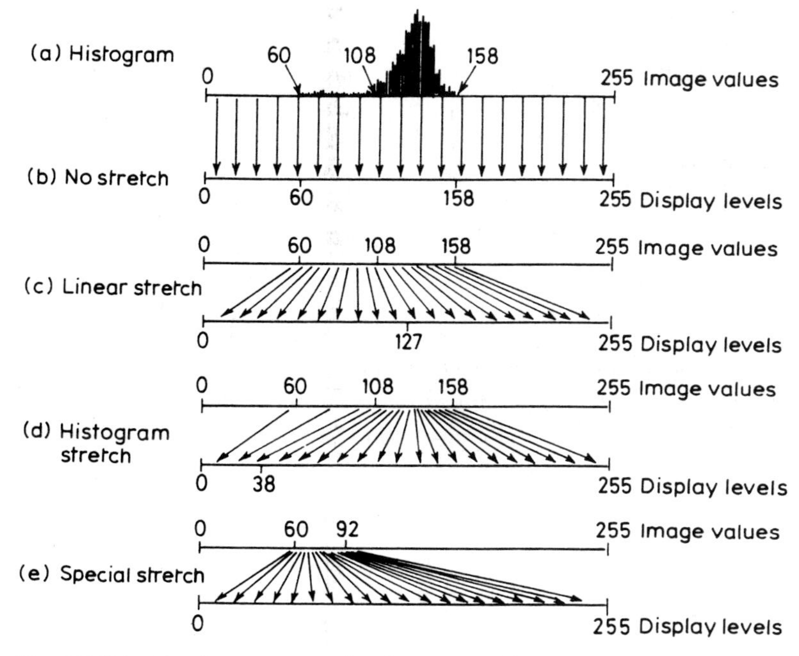

Fig. 9.8 Principle of contrast stretch enhancement. (Source: Lillesand and Kiefer, 1987.)

ditions. Consequently the pixel values in many scenes spread over relatively narrow portions of the full ranges that are possible. Contrast stretch enhancements expand the ranges of the original pixel values to increase the contrast between features with similar spectral responses, and thereby facilitate feature recognition and image interpretation. Figure 9.8 illustrates some of the available contrast stretching processes. The *linear stretch* (Fig. 9.8(c)) expands uniformly the observed histogram of brightness levels in a selected scene to cover the full range of available values. The *histogram stretch* (Fig. 9.8(d)) is better still, for image values are assigned here on the basis of their frequency of occurrence: more display values (and hence more radiometric details) are assigned to the more frequently observed levels in the histogram. For more specialized analysis and interpretation, use may be made of a 'special stretch' (Fig. 9.8(e)) in which features represented by a narrow range of brightness levels are greatly enhanced through the expansion of this narrow range to occupy the full range of display values, to the exclusion of other, perhaps previously dominant, features of the scene.

5. Special contrast stretching. In the previous category we considered procedures designed to accentuate intensity contrasts in individual waveband data sets. Sometimes it is valuable to accentuate the contrasts between different bands of image data. One common example is the *ratio image*, which is generated by computing the ratio of digital values for two selected spectral bands for each pixel in an image. A useful characteristic of the products of spectral contrast stretching of this type is that they are relatively free of extraneous effects, such as differential illumination (e.g. relief-induced light and shadow) across a scene. However, it is often difficult to determine which is the most appropriate choice of wavebands for a particular application. So, statistical techniques have been developed to transform pixel values on to alternative sets of measurement axes so that they may be described – and later reproduced in image form – in the most efficient manner possible for the task in hand. Such techniques are illustrated by Fig. 9.9, which relates to a simple, two-channel data set. In Fig. 9.9(a) a random sample of pixels from a selected image have been plotted according to their observed

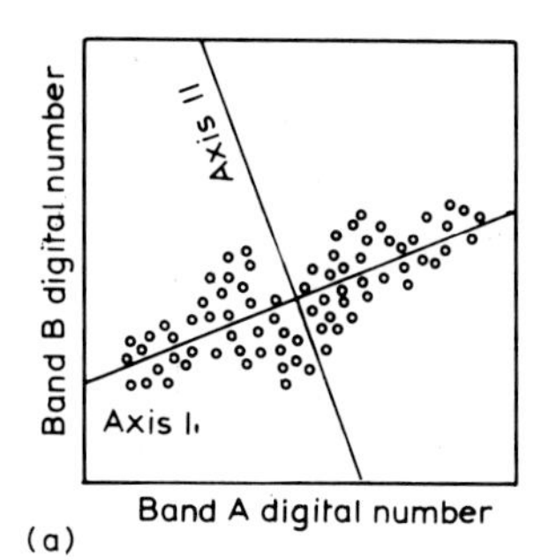

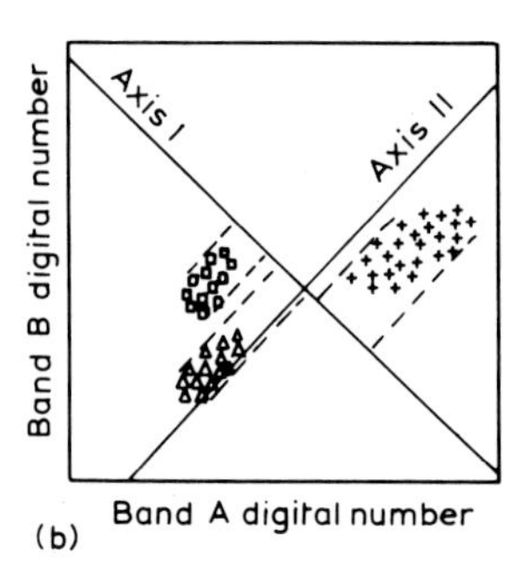

Fig. 9.9 Rotated coordinate axes used in multispectral transformations. (a) Principal components analysis. (b) Canonical analysis. (Source: Lillesand and Kiefer, 1987.)

values. New axes (I and II) are introduced to provide a more efficient description of the plotted data. Such axes are characteristic of *principal component analyses*, which can be undertaken for multidimensional space: each axis or component accounts for successively smaller portions of the observed variance in the data set. In Fig. 9.9(a), axis I accounts for much more of the variance than axis II. A principal components' enhancement can be generated by displaying the image in terms of the new pixel values, expressed in relation to axes I and II, not A and B as before. Where more prior information is available for a region, training set pixel values (from sites of known feature type) can be used as the basis for *canonical analysis*. In this case new axes are positioned so that the discrimination of given feature types can be maximized. In Fig. 9.9(b) three feature types can be discounted on the evidence of axis I alone.

6. Change detection. Change detection is generally carried out by hand, using interactive man–machine methods, for example the *flicker procedure* whereby the first and second images are presented alternatively to the observer many times a second. Apart from simple techniques, such as the detection of greyscale change from one photograph to another, digital procedures have not yet been developed with notable success in this interesting and important, but rather intransigent field.

9.3 The analysis and interpretation of selected features

9.3.1 Quantitative feature extraction

Quantitative feature extraction from remote sensing data involves four steps, namely:

1. the development of methods whereby significant features or variables may be isolated;
2. the preparation of appropriate quantitative data arrays;
3. the determination of the essence of the structure of the data;
4. the reduction of the information so that only the essential point in, or structure of, the data is carried forward into the final stage of decision classification.

The first step involves the logical or *syntactical* structuring of the design of the experiment. This is the planning stage, in which consideration is given to what will be measured and why; how the data will be sampled and aggregated; how many variables should be used; which methods of data retrieval, preprocessing, and data compaction will be employed; and how the final analytical stages will be organized. The second step involves some form of line-by-line data scanning to quantify the variables to be used in the analysis. These variables may be uncorrected greyscale (from a photograph, colour waveband or multispectral channel), a normalized variable, or some more sophisticated derivative from the initial data set. In most remote sensing applications the principal variables used in each channel will be tone, texture or height measurements. These relate to greyscale, greyscale organization and variability, and stereoscopic characteristics. Resolution in density, or *greyscale* (the number of detectable shades of grey, e.g. 16 in a 4-bit, or 256 in an 8-bit system), may limit the subsequent stages. On other occasions limitations are imposed by the analyst, for example with the number of final categories required for successful completion of his or her task.

Increasingly, data from different wavebands are being combined for subsequent analysis and interpretation, e.g. by *band ratioing* in which data from two wavebands are combined by addition, subtraction or division (Chapters 13 and 16), or in *multispectral feature space* (Chapters 12 and 17).

The third stage involves the search for key indications of the phenomena or relationships under scrutiny. Often it is sufficient to select a few aerial photographs from a large set, or the indications from some, rather than all, the multispectral channels available. Extraneous or highly cross-correlated data can be eliminated, reducing the measurements or variables to the minimum, which is substantially orthogonal in feature space. Thus the fourth step of the extraction process is reached.

9.3.2 *Data analysis and interpretation*

In many programmes of data analysis and interpretation attention is focused either upon point features, or upon non-point features having length (lines), or length and breadth (areas). In order that image contents may be interpreted and classified correctly, it is common practice for 'training sets' of data to be compiled (see also pp. 113–14). These are subsamples whose identification is completely known. Where data from two or more wavebands are available it is now commonplace to consider the relations between and amongst digital values (or 'vectors') in multidimensional space. Some approaches frequently adopted in Landsat image analysis are illustrated by Fig. 9.10. Here, for the sake of simplicity, pixel values from only two bands are considered; similar processes can be applied readily to all available data bands.

In these and other ways classifications of data from known regions can be used as stereotypes of points, lines, or areas, by comparison with which unknown areas can be assessed. Thus, as we have noted before, new data may be interpreted through their resemblance to sets of templates, the known 'fingerprints' in a 'supervised' approach. The statistical parameters of each training set are calculated by analog or digital computer and can involve many variables which may have a bearing on the characteristics of the key site, ground truth station, or ground survey area. These include such factors as topography, vegetation and soil content, farming practices and environmental management as well as time of day, season of the year, and prevalent weather conditions. In calculating the statistical parameters of the training set, the greater the number of characteristics considered at a point, or the greater the number of resolution elements within an area, the better the classification accuracy will be.

Training sets are commonly used where multispectral scanner data are available. The data-display techniques that are used to facilitate feature extraction are numerous, and include histograms for each category and by each channel (Fig. 9.11).

From such data manipulations matrices of correlation and covariance may be compiled. The training set data usually involve *parametric discriminants*, using *Bayesian probabilities* and *maximum likelihood functions*. An extra advantage of these methods is that they commonly use thresholding procedures in which items unlike the sample are assigned to a discard class, which then may be analysed in greater detail if necessary.

Many methods are being used for comparing new or unclassified data with established training sets. These include non-parametric procedures, such as *linear discriminant functions*, which are much faster in terms of computational time than the maximum likelihood decision rule, which in practice often becomes a quadratic rule requiring a large number of multiplications for each decision. Some workers have used *composite sequential algorithms* based on clustering techniques. Others have developed models based on boundary conditions. Still others have developed 'table look-up procedures', much faster than maximum likelihood functions in practice and nearly as accurate. Operations in the spatial frequency domain have been investigated for areal analysis and classification.

Analog computer techniques can be operated rapidly to provide image comparisons based on brightness levels, areas above threshold brightness levels, and the numbers and shapes of selected picture features. Unfortunately such systems lack the flexibility of digital computers, and each new capacity or operation demands new hard-wired components.

Conventional training-set techniques are, however, not without disadvantages. In particular, presupposing that detailed knowledge is not sufficiently detailed to ensure the representativeness of the training set, the ground truth sites or areas must be chosen carefully. Furthermore problems may be

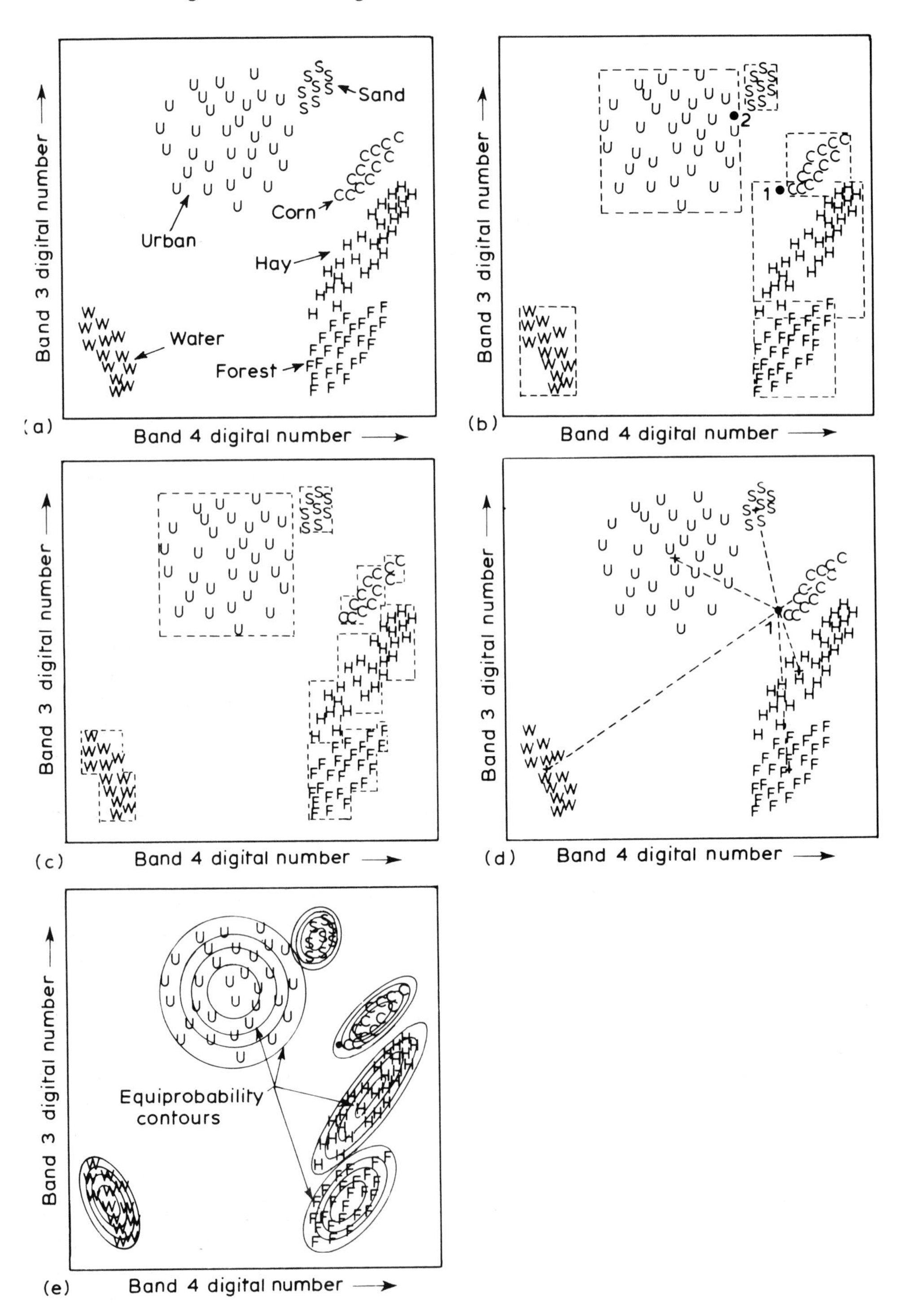

Fig. 9.10 Some common Landsat data classification strategies. (a) Pixel values from two wavebands are plotted on a bispectral scatter diagram. (b) A 'parallelepiped' classification where pixels are grouped on the basis of the highest and least class values in each band. (c) A 'parallelepiped' classification whose stepped boundaries represent each group of pixels more precisely. (d) A 'minimum distance to means' classifier where an unclassified pixel (1) is compared with established class means by computing distances between it and their centres. (e) An 'equiprobability contour' to accommodate new data, like (1), using maximum likelihood rules. (Source: Lillesand and Kiefer, 1987.)

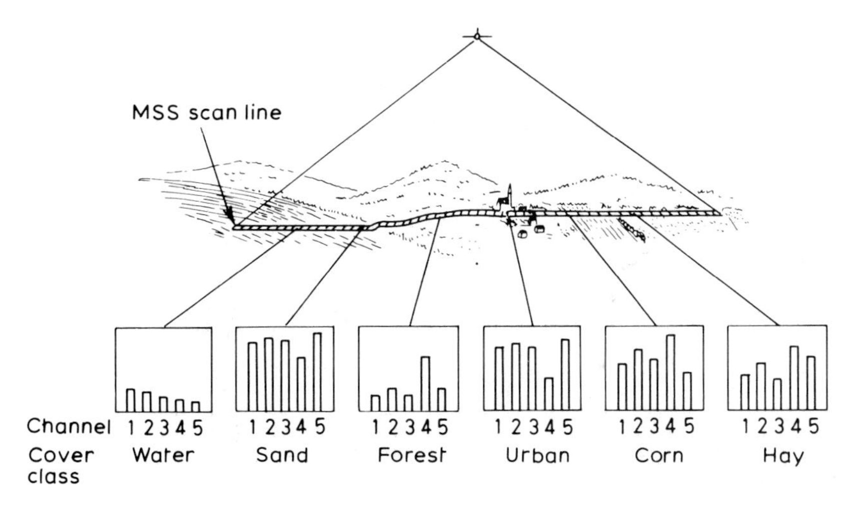

Fig. 9.11 Selected MSS measurements made along one scan line. Channels cover the following spectral bands: 1, blue; 2, green; 3, red; 4, reflected i.r.; 5, thermal i.r. (Source: Lillesand and Kiefer, 1987.)

caused by variability in time and space within the training area. Consequently efforts are being made to develop techniques that are described as *not supervised*, for application to areas for which training sets are unavailable. Such techniques involve *clustering algorithms*, perhaps based on interactive routines whereby selected features are identified by eye and mapped. The technique relies on the fact that the multichannel radiances of different features tend to cluster in different places in the corresponding multidimensional space. This allows the computer to calculate the classes present, and the resolution elements that belong to them. At a later stage the classes can be identified by reference to new ground truth, or through comparisons of sample data with a bank of spectral signatures. The particular advantages of such a method are that it is quick and cheap, no prior knowledge of the site is required, human bias is minimized, and future ground control stations can be located more objectively than in the supervised techniques.

9.4 Automatic decision/classification techniques

Often in the early stages of any remote sensing programme some retracing of one's steps is unavoidable in the march towards an *operational (i.e. routine) feature recognition system*. In the early stages of research, most attention is accorded to image analysis rather than feature recognition. Often the best methods of data processing and analysis can be established only by trial and error, by the comparison of results obtained by different means. The expenditure of time, money and effort, and the complexity of the computer hardware and software to complete each method of approach must be assessed in terms of the resources that would be available operationally, and in terms of the requirements of the consumer, all to be judged finally by the additional criteria of accuracy, frequency and resolution.

Once the appropriate method of data analysis has been established for any application the chief aim of most operational programmes is the correct recognition, identification, or classification of the contents of fresh imagery (Plates 2 and 3, colour section). We have seen that numerical techniques are superior to manual or photographic techniques in that their decision and classification rules are, by definition, quantitative or statistical, whereas those for the human interpreter are judgmental and substantially qualitative. Against this there is the chief disadvantage of automatic techniques: that allowances may be difficult to make for phenomena or conditions which are unevenly distributed or are present only from time to time. Such 'environmental noise' can be edited out by manual or man–machine mix (interactive) interpretation techniques, but may yield spurious patterns or results in automatic procedures.

Two cases in point are *background brightness*,

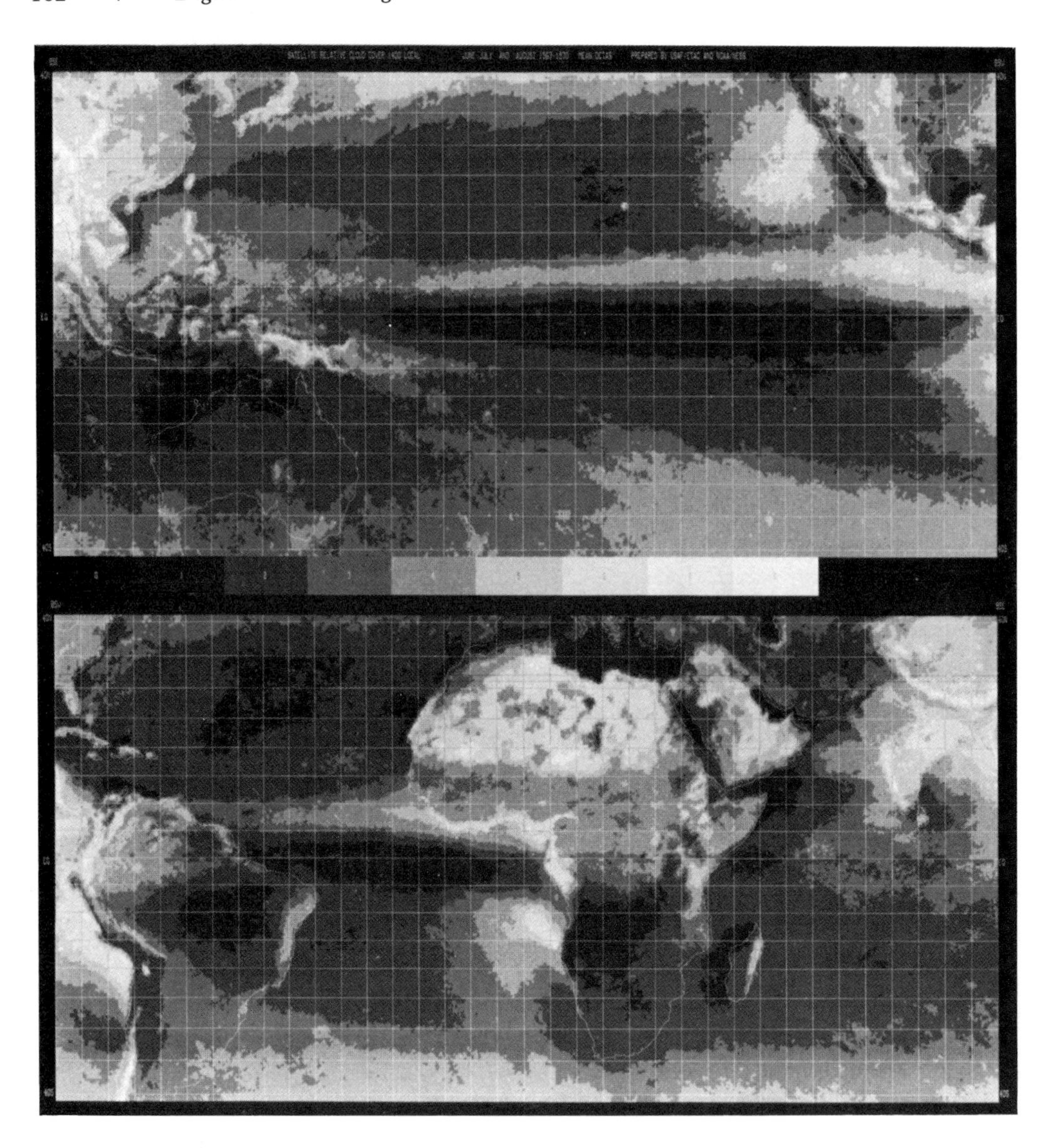

Fig. 9.12 Photographic computer maps from ESSA Satellites, averaged for the season of June, July and August for a four-year period (1967–1970). The use of the Mercator projection facilitates studies of tropical brightness patterns in general and trans-equatorial links in particular. Background brightness has been retained, hence, for example, the brightly reflective desert areas of the Old World. The South Asian summer monsoon cloud over the Indian subcontinent contrasts strongly with the more typical east–west oriented ITCB clouds across the Atlantic and Pacific Oceans. (Courtesy, NOAA.)

which appears as an unwanted feature in climatological cloud imagery from weather satellites (Fig. 9.12) or, conversely, *foreground brightness* related to clouds in Earth surface investigations. In satellite meteorology cloud cover maps (nephanalyses, section 10.3.1) can be prepared by hand to show, among other things, the percentage cloud cover categorized by broad classes. These can be summed and averaged to yield mean cloud cover maps exclusive of most background brightness effects. However, it was recognized even at an early stage in the evolution of satellite meteorology that digital

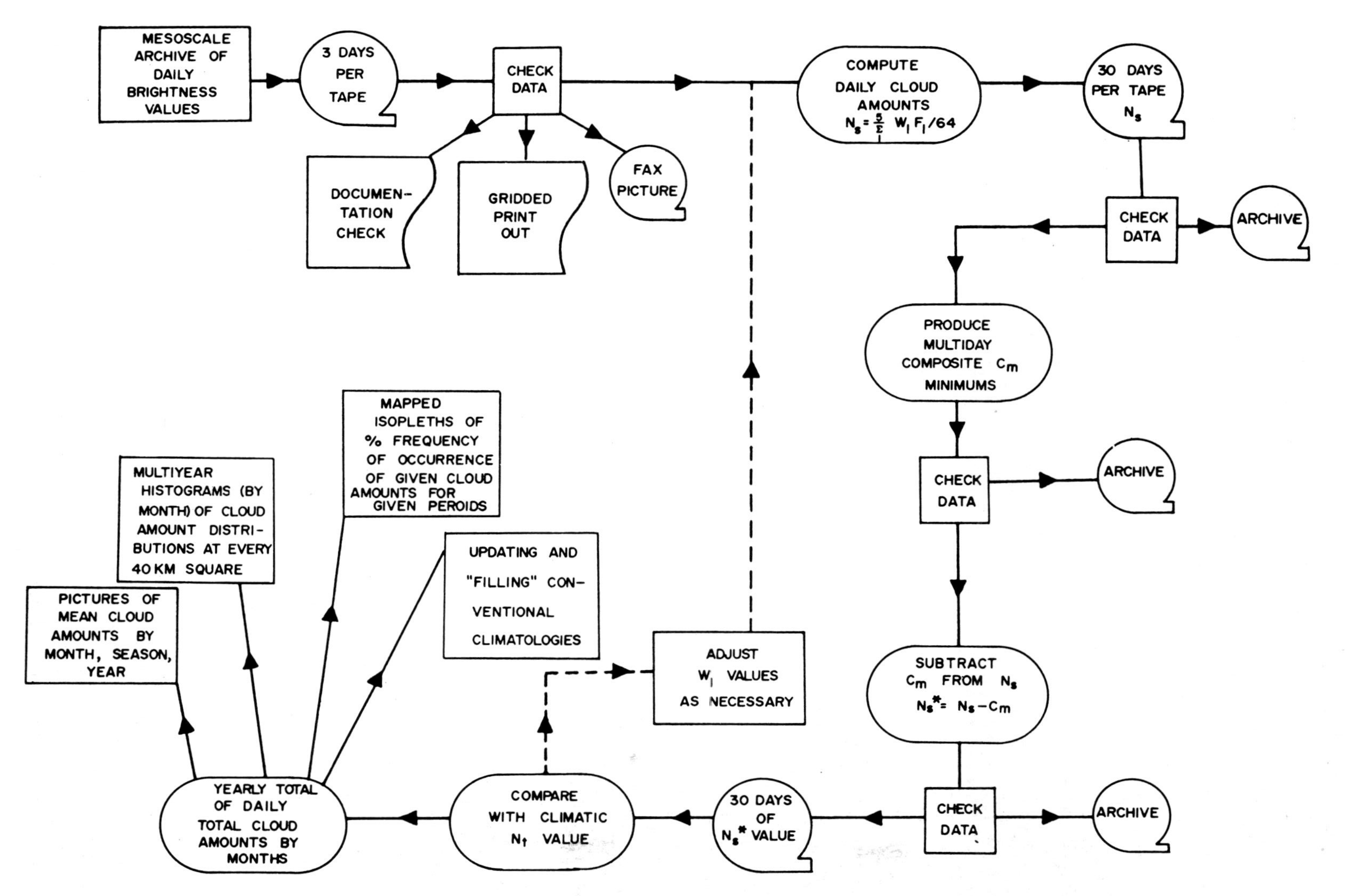

Fig. 9.13 Flow diagram for the production of a climatology of total cloud amount from the mesoscale archive of the weather satellite data. (Source: Miller, 1971.)

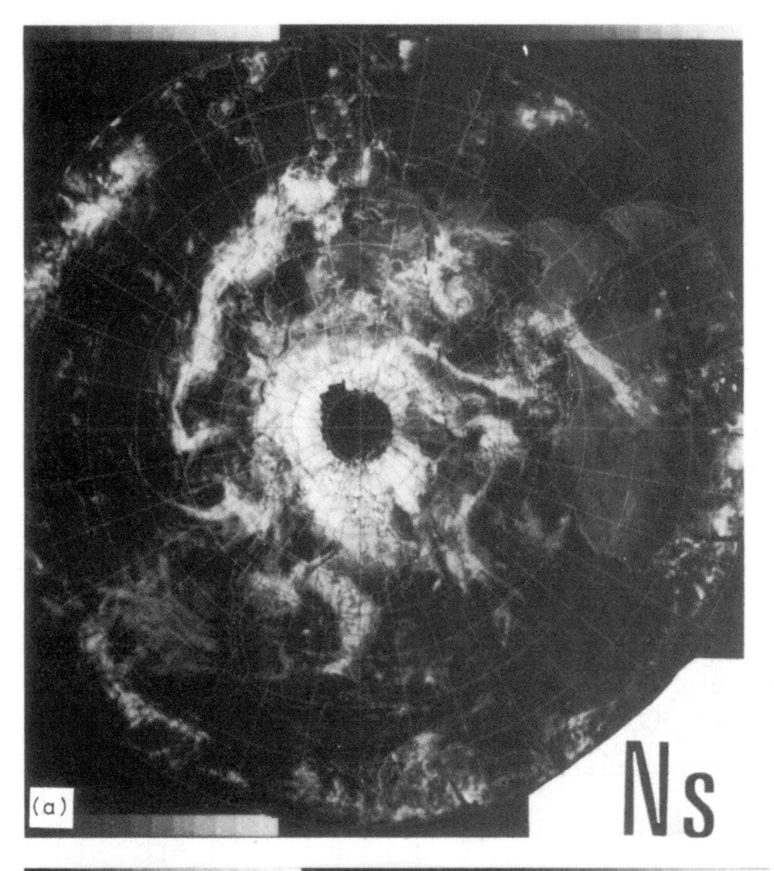

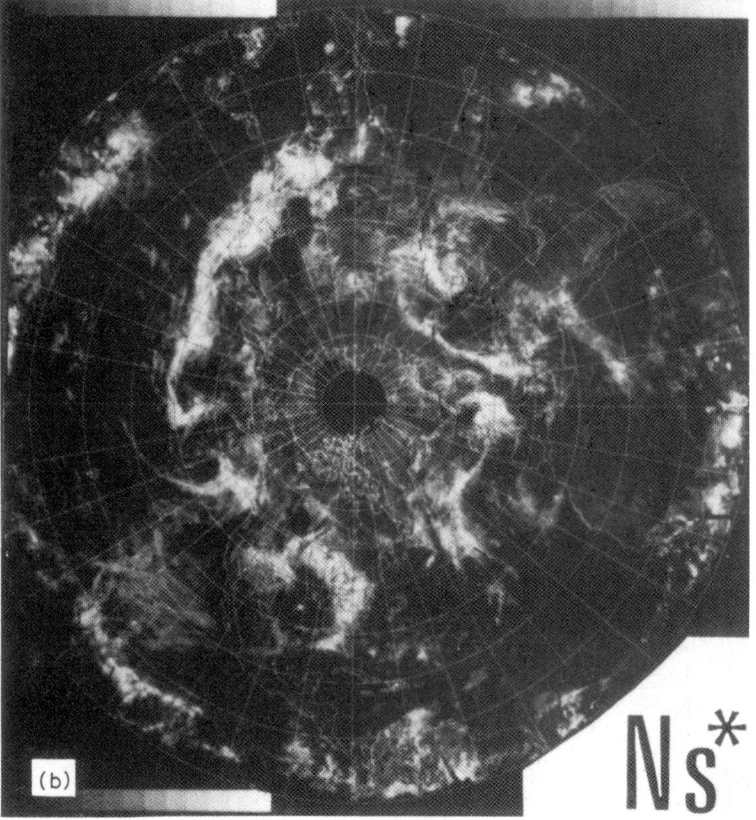

Fig. 9.14 (a) The total Earth-cloud scene represented in the mesoscale resolution of computer-rectified, mapped imagery from operational weather satellites. Note the high brightness of the Arctic ice and snow and the Saharan sands as well as the clouds. (b) The cloud scene (a) remaining when the 30-day composite minimum cloud amounts were subtracted from the total Earth-cloud scene. (Courtesy, US Air Weather Service, Scott Air Force Base, Illinois.)

techniques (Fig. 9.13) could provide acceptable results through the inclusion of a stage at which the minimum brightness level for each picture through the period in question is considered to be due to the albedo of the surface of the Earth. This level is especially high in frozen regions, over sandy deserts, and where sun glint from water surfaces is strong. By subtraction of 'brightness minima' from the daily mean brightness levels for every picture point a value more representative of mean cloudiness itself is identified (Fig. 9.14 (a and b)). Unfortunately it is obviously more problematic to apply a technique like this to mapping of cloud cover on a short-term (e.g. daily) basis, or to cloud inventorying over ice, snow, or deserts. Other (e.g. multispectral) approaches are necessary for such purposes, through which information from several wavelengths is considered.

As remote sensing becomes recognized ever more widely as a useful or potentially useful tool for many types of Earth resources investigations and environmental monitoring programmes so the number and range of related methodologies is expanding rapidly. Numerous examples of automatic decision/classification techniques will be introduced in Part Two of this book.

9.5 End products of numerical processing

At the end of any decision-making operation involving the classification of image contents, an output is produced. This may be in one of a variety of forms, including graphical, cartographic and tabular modes:

1. Graphical outputs. These usually consist of histograms of the frequency of occurrence of identified categories of features, or related categories.
2. Map outputs. Maps are popular and common, manually, interactively or automatically generated products of environmental remote sensing programmes. They may be qualitative (showing types of features, e.g. crops), quantitative (showing values of features or conditions, e.g. surface brightnesses) or selected combinations of the two. The mapped patterns may be isoplethed or choroplethed, they may be single value or multiple category presentations, they may be based on absolute values, ratios, percentages, and changes through time; they may depict only point, line or area features, or all three simultaneously. In short, maps are highly versatile forms of digital programme outputs, and are justly significant in remote sensing as a whole.
3. Tabular outputs. Once again a variety of forms and contents are possible. For example, the accuracy of identification may be summarized in matrix form, giving correct identifications on the diagonals, and errors of omission and commission on the *x*- and *y*-axes. Such tabular outputs can be prepared easily by computer. This may form part of the checking ('verification') procedure by which the programme results are assessed against those obtained by other, especially conventional, methods. Any remote sensing programme for the analysis and interpretation of the target area stands or falls finally on its ability to provide worthwhile, especially cost effective, results. Verification techniques often lead to the improvement of the procedures of analysis and interpretation by highlighting the errors in existing techniques or routines.

Last but not least in this section emphasis must be placed on the need in operational contexts not only to effect suitable *calibrations* of the interpretational algorithms, but also to effect verifications or *validations* of the end products, i.e. to attest their performance of the programme as a whole. Much effort is now being put into the improvement of algorithms whose basis were lain in the earlier, pioneering years of remote sensing.

9.6 Systems for digital data analysis and interpretation

Over the last 30 years image processing generally advanced from being a subject area dominated by electronic engineers and computer scientists with access to powerful – and very expensive – mainframe resources to one that involves school children playing with education software (e.g. 'games') on small but competent microcomputers. This progress has had one of its most profound influences on environmental remote sensing. In the 1960s a few national and international remote sensing laboratories were equipped with large,

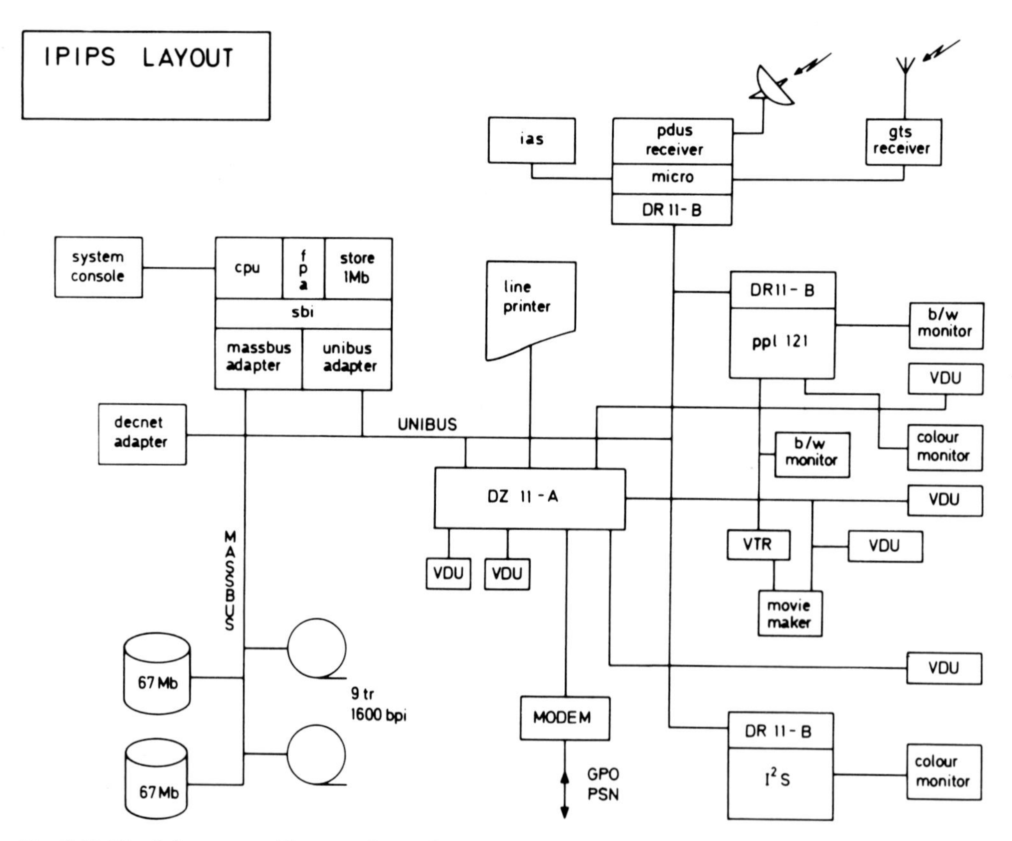

Fig. 9.15 The Infrastructure Planetary Image Processing System (IPIPS) schematic layout. (Source: Muller, 1988.)

costly computer systems providing relatively little computing power (it has been aptly said that 'rooms were filled with 32 K of RAM'!), whilst most other remote sensing practitioners had to try to make progress using optical projection systems and densitometric equipment of various degrees of sophistication – ranging from single spot microdensitometers to scanning (one-dimensional) microdensitometers or even, perhaps rotating drum (two-dimensional) systems if they were lucky.

By the end of the 1970s semiconductor memory had become cheap and fast enough to permit the construction of *'framestore systems'* (capable of retaining the very large data sets that typify modern satellite remote sensing), by which existing computers could be enhanced. About the same time it was realized that to meet the demands of cheaper RAM a higher speed input/output channel was needed to take data from slow storage devices, such as Winchester magnetic disks, via the central processing unit (CPU) to the image displays. The fusion of these two technological advances together with *'virtual storage techniques'* meant that the huge volume of data reaching Earth from space could be processed efficiently up to about 4.2 gigabytes for 32-bit processors. An example of such a combination of 'virtual readily available' memory management plus actively buffered framestores is the IPIPS systems of University College London, developed in the early 1980s (Fig. 9.15).

During the 1980s, increasingly more capable, yet more compact and more readily affordable systems began to be mass produced, the best-known including those from the USA company International Imaging Systems (I^2S), the Canadian firm DIPIX, and the UK GEMS systems from the Computer Aided Design Centre, Cambridge. Most of today's teachers of remote sensing will have learnt their trade on one or other of these systems. Each is a cunning blend of computing hardware and software designed to perform many computing operations in parallel under the management of a host computer (Fig. 9.16). A typical procedure is to read remote sensing image data in digital form from a computer

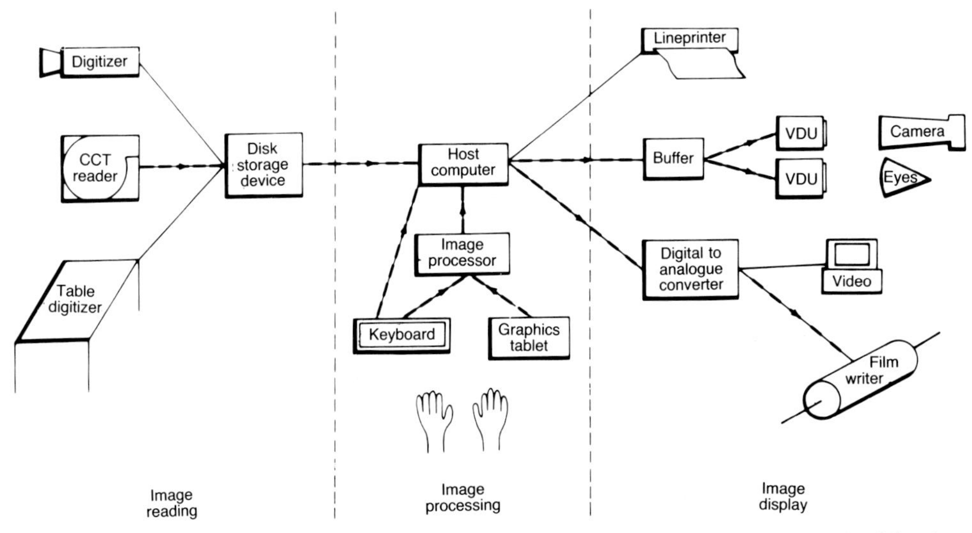

Fig. 9.16 The components of a large digital image processing system, comprising three methods of data input and five methods of data output. The dashed lines with arrows indicate a probable processing route for digital satellite sensor data. (Source: Curran, 1985.)

compatible tape (CCT) into a disk storage device before part or all of these data are called into the host computer by the operator-controlled keyboard. The image then can be processed from either the keyboard and/or the graphics tablet as the data moves to and from the host computer and image processor. The results of each processing stage are displayed on a visual display unit (VDU) screen, from which 'hardcopy' can be produced either by photographing it, or via a film writer via a D–A converter. The satellite imagery displayed on the VDU may be overlain with other types of data (e.g. point or line data) input from a digitizer, which may be either automatic or manual. The system as a whole is *interactive*, i.e. the operator may not only command and guide the processes undertaken, but also may interact with the information displayed on the VDU through joy-sticks, trackball and/or light-pens or tables.

Most recently, as small desk-top computers (often called 'personal computers' or 'PCs') have dramatically increased in processing ability as a result of advances in microchip technology and even more dramatically decreased in terms of their relative prices, so a new revolution in image processing has dawned. Whereas the earlier systems were based on specialized hardware, and therefore intrinsically hardware-limited, the new systems are based on ordinary PC or (slightly larger) 'work-station' computers, and are limited much more by the purchaser's ability to afford the required image processing and Geographic Information System (GIS) packages – or to write suitable software himself or herself. Many such packages of software are available, from commercial companies (e.g. ERDAS), or through government-released 'public domain' channels (e.g. LUCID), to cover all the common image processing needs. (Table 9.1). However, for specialized applications, or for advanced research, the best option is still to prepare software locally.

The rapid rise in the availability of very competent PC-based systems has helped remote sensing to become more and more a 'distributed' rather than a 'focused' or 'centralized' pursuit. This in turn has greatly aided the spread of operational remote sensing practices and programmes, e.g. in local agency or authority offices, and also the penetration of remote sensing into secondary and even lower levels of the educational hierarchy. Coupled with a simultaneous reduction in both the relative and, even the absolute costs of direct data gathering from satellites through the new generation of cheap ground reception facilities available for some satellite systems (e.g. Meteosat and NOAA), these digital data processing analysis and interpretation advances will certainly lead to a further spectacular growth of remote sensing in the foreseeable future.

9.7 Remote sensing and GIS

If maps, graphs, tables and images are the only forms in which data are available to the environmental scientist, combination or recombination exercises can be extremely difficult. As one recent author has exclaimed 'The eye becomes confused, and the mind tends to "boggle" when more than two or three maps are visually compared.' Fortunately one of the biggest and potentially most important developments in the realm of numerical data handling involving remote sensing during the last decade has been the rise of 'Geographical Information Systems' (GIS). As more and more information concerning any area on Earth has become available, so it has become increasingly necessary for environmental scientists, engineers, planners, etc. to attempt to combine subsets of the total available information in a variety of different ways (Fig. 9.17).

Fortunately for us all the primary problem of spatial data combination has been addressed with ever-increasing success since the 1970s as a result of rapid increases in computer memories and affordable computing power. Today it is possible to unite any spatial data that can be referenced by

Table 9.1 Application software for a PC-based image processing system: an abbreviated listing for ERDAS: (a) first and second-order categories; (b) third-order categories for one set of first and second-order categories, namely the 'image processing module' (Courtesy ERDAS)
(a)

First-order categories	Second-order categories
Image processing module	Display programs
	Enhancement programs
	Classification programs
	Geometric correction programs
Geographic information system module	Display programs
	Data entry programs
	Analysis programs
Utilities module	General programs
	Colour modification programs
	Cursor control programs
	Diagnostic programs
	Menu files
Colour hardcopy module	Scale mapping programs
Polygon digitizing module	Digitizing programs
	Display programs
	Gridding programs
	Measurement
Video digitizing module	Video digitizing programs
Tapes module	Input programs
	Output programs
	Utility programs
Topographic module	Input programs
	Analysis programs
	Surfacing programs
Three-dimensional module	
Software tool kit module	
Software subscription service (SSS) module	

Table 9.1
(b)

First-order category	Second-order categories	Third-order category
Image processing module	Display programs	Interactive display, ROAM, ZOOM
	Enhancement programs	Convolution filtering
		Screen and disk
		Destriping
		Linear Combinations
		Screen and disk
		Multiply images
		Band Ratioing
		Screen and disk
		Histogram equalization
		Principal components analysis
		Contrast stretch
		Level slicing
		Linear
		Non-linear
		Piecewise
		Negate
		Bias
		Shift
	Classification programs	Maximum likelihood, supervised
		Minimum distance
		Clustering, supervised
		Thresholding
		Classification overlay
		Interactive training field selection
		Rubber band
		Parallelepiped classification
		Boxcar
		Two-dimensional ellipse plots
		Scatter diagrams
		Add signature statistics
		Append signature statistics
		Contingency matrix analysis
		Save and retrieve signature library
		Edit signatures
	Geometric correction programs	Compute coordinate transformation
		Ground control points, RMS
		Accuracy
		Coordinate conversion
		Lat-long. utm. stage plane
		Screen image rotation
		Geometric correction
		User-specified pixel size
		Cubic convolution
		Bilinear interpolation
		Nearest neighbour
		Error estimation for reserved GCPs
		Error plotting on display

(a)

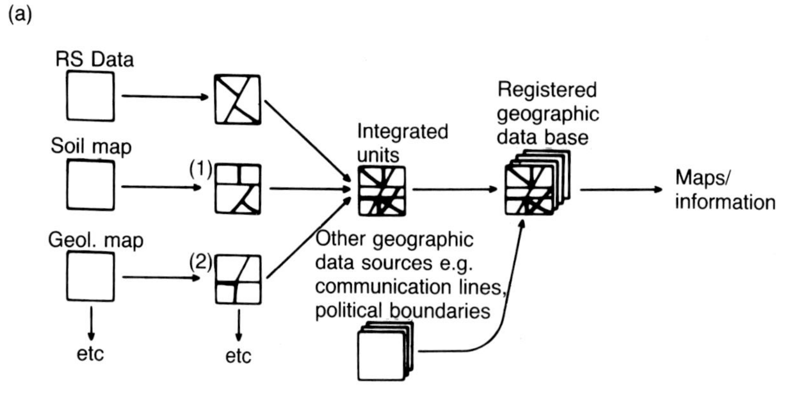

(b)

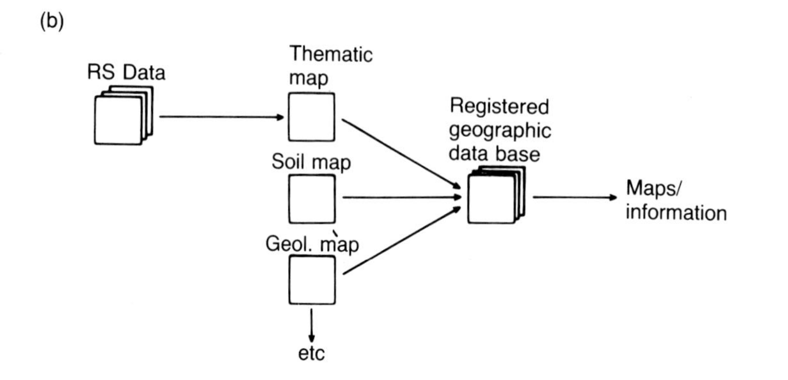

(c)

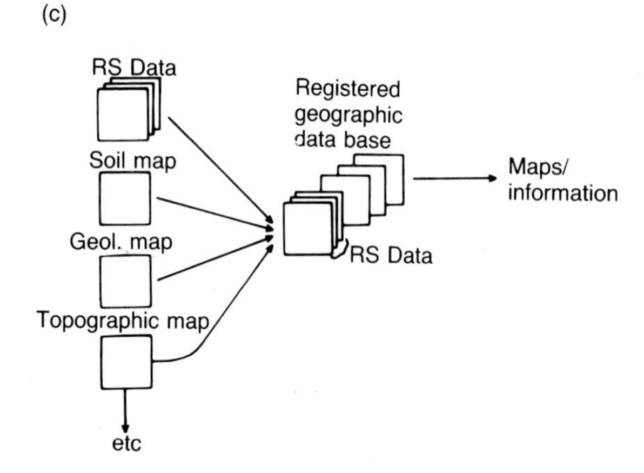

Fig. 9.17 Methods of integrating remotely sensed data (RS data) into a geographic information system. (Source: Curran, 1985.)

geographic coordinates. The three basic processes for this are as follows:

1. Data encoding, i.e. the subdivision of spatial data into small polygonal or grid units. The latter are regular, and therefore cheap to store, manipulate and handle; the former are necessary where shapes are irregular, and must be tailored to fit Earth surface features, whether natural (e.g. drainage basins) or artificial (e.g. administrative units).
2. Data management, i.e. the filing of data ready for subsequent recall and manipulation. In practice this involves detailed planimetric bases

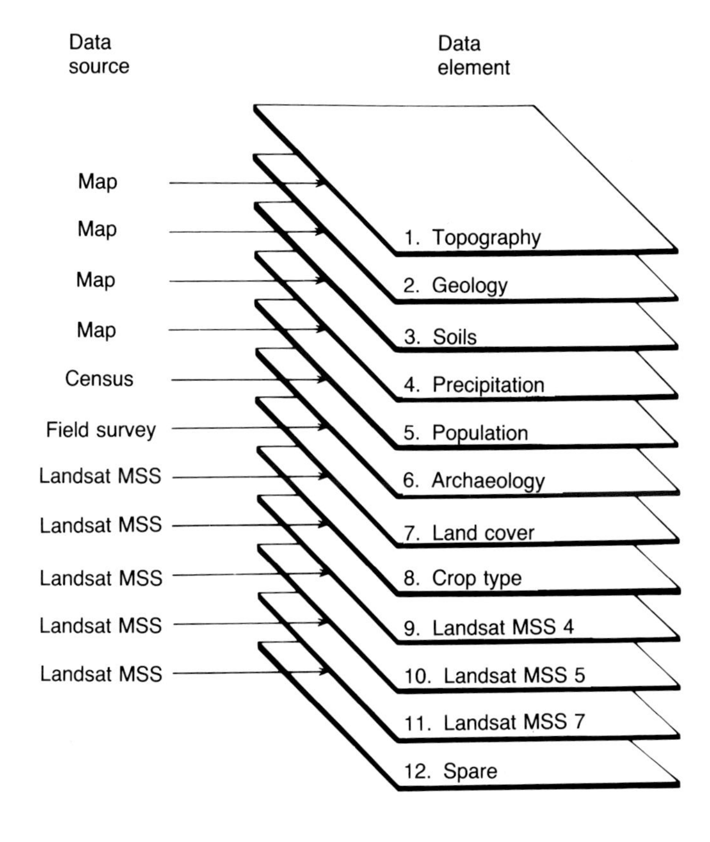

Fig. 9.18 Integration of Landsat MSS data with other spatial data sources to derive data elements for input to a 12-level geographic information system. (Source: Curran, 1985.)

(e.g. maps), plus a number of planes or overlays, which may be selected for use either singly or in any required combination, as exemplified by Fig. 9.18. Both the concept and use of GIS are infinitely variable in terms of the types of information involved. Increasingly today remotely sensed data or derived products are available as potential GIS planes.

3. Data manipulation. The commonest elements of software packages designed for such processes are retrieval, transformation, storage, searching, analysis, measurement, compositing and modelling.

At first GIS were developed for greater flexibility and effectiveness in the storage, management and use of relatively historic data sets, e.g. topographic maps, administrative boundaries, communication networks, etc. However, its potential with respect to aspects of surface change was quickly realized, e.g. land use/land cover changes as evidenced by comparative analyses of Landsat or SPOT data from different years. Most recently, it has become recognized that the basic structures of GIS are well suited to the incorporation of data not only from Earth resources satellites, which evidence relatively slow changes, but also from environmental satellites, whose measured parameters change more rapidly, e.g. snow cover, soil moisture status, etc., and even very rapidly, e.g. cloud cover and precipitation.

Therefore we may expect to see speedy advances in the development of combined GIS/Remote Sensing Systems, permitting remotely sensed data to be analysed in many different ways, viewed very flexibly from different angles (Plate 4, colour section), and readily integrating data from conventional sources with those from remote sensing sources. In many cases the remote sensing data will comprise relatively dynamic upper layers in each GIS, strongly complementing the more static and historic lower layers.

Although the integration of remote sensing data into GIS systems is not easy, several methodologies to accomplish this have been developed already. Perhaps the most successful presently available is based on grid squares, and the conversion of remotely sensed data into geometrically corrected *thematic maps* prior to integration with other data types. Alternatively, remotely sensed data may be geometrically corrected and registered for direct integration with other data elements. Some argue that this supports a greater flexibility of analysis, and improves the objectivity with which the remote sensing data are classified.

For the future, rapidly increasing computing power, especially if applied in a parallel rather than a sequential form will be necessary to deal with the expected great increases in remote sensing data from space, and to permit the improvements in numerical methods required to improve the range and detail of end products, and to increase the mutual interdependence of remote sensing and other environmental data within the GIS context.

Part Two Remote Sensing Applications

10 Weather analysis and forecasting

10.1 Remote sensing of the atmosphere

10.1.1 Advantages of remote sensing systems for weather studies

We may be asked why remote sensing should be welcome in support, or even in place, of conventional weather observation techniques, for these are particularly well-developed and long-established environmental monitoring activities in many areas of the world. This question is vital, for the costs of remote sensing systems can seem high in comparison with *in situ* sensor systems. The advantages of remote sensing systems for weather studies include the following, although of course, not all may necessarily apply to a particular investigation:

1. The remote sensor does not need to be carried into the medium that is to be measured.
2. This measurement system, unlike many *in situ* sensor systems, does not modify the parameter being measured, since it is, by definition, remote from it.
3. A high level of automation can be easily achieved in the data-gathering process.
4. It is possible to scan the atmosphere by remote sensing directly in two or even three dimensions. This is better than the single point, and/or line measurement capability of most *in situ* sensor systems.
5. Integration of given parameters along lines, over areas, or through volumes, is often obtained readily in the outputs of remote sensing systems.
6. Relatively sophisticated and elusive parameters, such as the spectrum of atmospheric turbulence and momentum flux, may be available as direct outputs from remote sensing systems.
7. Single instrument packages may be able to provide data for the whole globe, whereas each *in situ* sensor package is limited to very local application.

The three most common types of platforms for atmospheric remote sensors are ground-based structures, aircraft and satellites. Ground-based remote sensing systems are generally not very well-developed, although they have a bright future in meso- and microscale investigations, plus local and regional short-term forecasting. In view of the relatively broad and varied development of radar to operational or near-operational levels, we will dwell mostly on radar as an example of a ground-based system. Aircraft-based systems are important in commercial and military aviation, e.g. for special investigations of dangerous weather phenomena (Fig. 10.1), and in-flight planning for passenger jets, but are much less cost-effective for both general atmospheric monitoring and research than either *in situ* or satellite remote sensing systems. Consequently their treatment in this chapter will be very brief. Fortunately satellites are now fulfilling much of their great potential as vehicles for both operational and research meteorology. Therefore much of the discussion that follows deals with the involvement of satellite systems in modern atmospheric science. We shall see that by far the greatest volume of atmospheric remote sensing data obtained today is gathered from a relatively small number of weather satellites, and that their prominence in remote sensing of the atmosphere is sure to grow as larger and better equipped satellites take to orbit.

10.1.2 Ground-based remote sensing systems

Remote sensing of the atmosphere from the ground has been, and still is, a fragmentary affair. This is

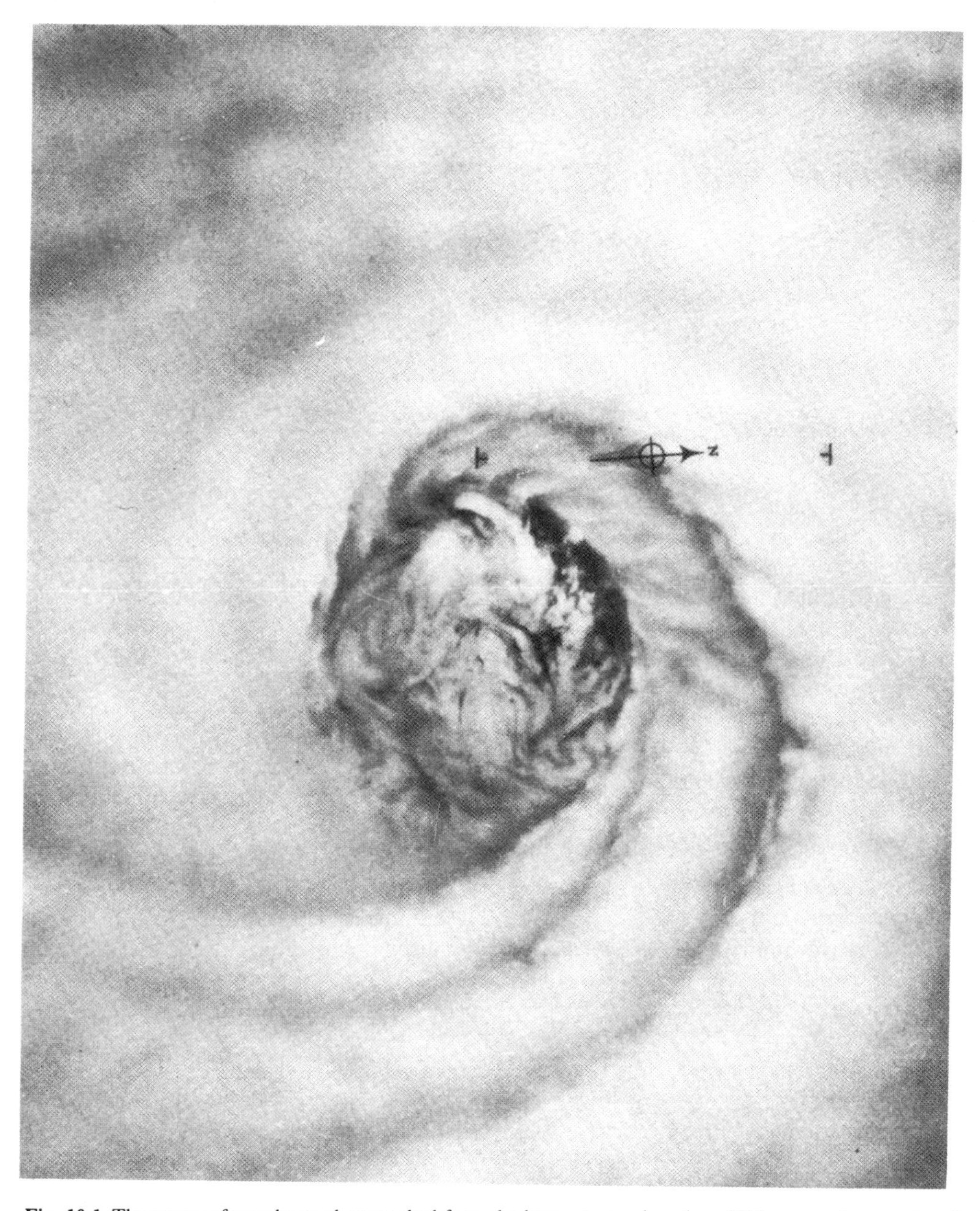

Fig. 10.1 The centre of a typhoon photographed from the lower stratosphere by a USA reconnaissance aircraft. The wall clouds around the eye have a marked helical arrangement. (Courtesy, USWB.)

partly because conventional weather observatories provide data of the types commonly required by standard forecasting procedures, and partly because of the complicating factor of the geometry of the Earth's surface. However, data from satellites have added new dimensions to our view of the atmosphere, and therefore to weather forecasting, even in areas generously supplied with conventional stations. Indeed, satellites are so efficient at global observation that related economies have proved possible in respect of *in situ* observations, e.g. in the eastern North Atlantic from which most weather ships have now been removed.

The earliest form of artificially assisted remote

sensing of the atmosphere was cloud photography from the ground. Later, balloons and aircraft were used to provide new views of these very significant atmospheric aggregates of water droplets. The invention of films of different speeds and sensitivities, camera and radiometer filters to eliminate specific wavelengths of radiation, and time-lapse techniques in photography have all made important contributions to the meteorology of clouds. However, there are limits to the value of the essentially local information such studies of the clouds can provide. One result is that today cloud photography from the ground is dominated by the amateur enthusiast, not the professional weather observer. At principal meteorological stations three aspects of the clouds are routinely observed, namely:

1. the proportion of the sky that is cloud covered;
2. the type or types of clouds present at different levels;
3. the height(s) of the cloud base (or bases if the cloud is multilayered).

Of these the last is the only one which is sometimes evaluated by instrumental remote sensing from the ground: the height of the cloud base may be assessed either by searchlights ('ceilingometers') or lasers.

Without doubt the most important ground-based system of atmospheric remote sensing – and that in most widespread use – is radar (an acronym for 'radio direction and ranging'). This active microwave system was quickly developed for weather analysis and short-term forecasting after the end of World War II. Today it is used to identify and examine a wide range of atmospheric phenomena, including raindrops, cloud droplets, ice particles, snowflakes, atmospheric nuclei, and regions of large index-of-refraction gradients. Basically, the most widely used types of weather radar consist of a *transmitter*, which produces very short bursts of power at the selected frequency, an *antenna*, which radiates the energy and intercepts each reflected signal before the next pulse of energy is transmitted, a *receiver*, which amplifies and transforms the received signals into video form, and a *screen*, on which the returned signals can be displayed (Chapter 4).

strength of the echo from a target, and its distribution: no account need be taken of the wavelength or frequency of the returning radar wave by comparison with the transmitted wave. But for some studies it also may be important to know the direction and rate of movement of a phenomenon with respect to the radar location. In such cases the *phase* of the received signal must be compared with the transmitted signal: here a 'coherent' radar system is required, designed to exploit the 'Doppler shift' effect. This is the observed change in frequency of radiant energy reaching a receiver when the receiver and the target are in motion relative to each other. Coherent, or Doppler, radar is now used in meteorology for a wide range of purposes. These include the measurement and prediction of turbulence and updraft velocities, e.g. at major airports, the evaluation of particle-size distributions for the local atmosphere through tracking natural scatterers, and measurements of wind speed through tracking artificial targets, such as rawinsonde balloons.

Returning to non-coherent radars some of their principal applications are listed below, along with the more common types of radar display units. These include:

1. The examination of echo intensities in chosen directions. For such purposes the simple A-scope may be used (Fig. 10.2(a)). This presents the back-scattered energy in profile form, and permits the operator to compare the echo with the transmitted pulse, which also appears upon the screen.
2. The identification and distribution of rain areas, and the recognition of the types of rainfall included in them. Plan Position Indicator (PPI) display units (Fig. 10.2(b)) are in quite widespread global use for such purposes. By photographing the radar screen at intervals the development and movement of rain areas can be followed and predicted. Plan Position Indicator displays are widely used in operational meteorology for short-term rainfall forecasting and the monitoring of severe storms, and in research for time-integrated studies of radar echo patterns, satellite cloud images, and/or *in situ* data.
3. The study of vertical profiles through the atmosphere. These may be obtained by vertically

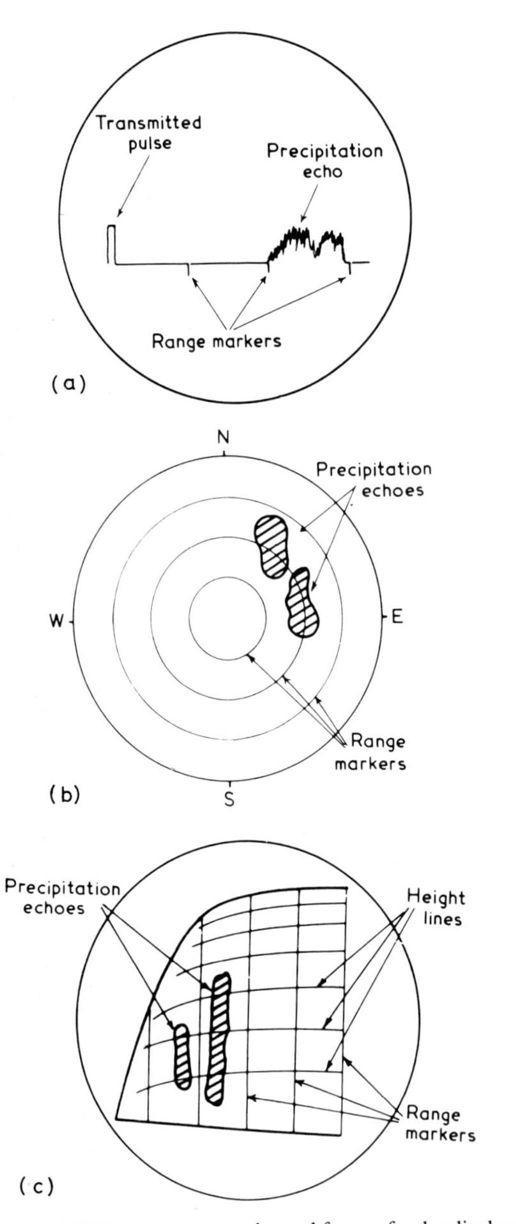

Fig. 10.2 The more commonly used forms of radar displays: (a) the A-scope; (b) the plan position indicator (PPI); (c) the range-height indicator (RHI).

scanning radar, whose results may be displayed conveniently on a Range Height Indicator (RHI) (Fig. 10.2(c)). This, like the PPI, involves 'intensity modulation', i.e. the intensity of the bright spots on the visual display unit corresponds to the strength of the reflected radar beams. The RHI system is used to study such phenomena as convective cloud growth and development, or, in cloud-free conditions, atmospheric wave motions and turbulence, as well that most intriguing yet elusive phenomenon known as 'clear air turbulence' (CAT).

However, the widest areas of application for ground-based radar – increasingly in conjunction with other types of meteorological data gathering systems – are those which involve precipitation. Examples of recent research programmes in these fields are given in Table 10.1.

10.1.3 Aircraft-borne systems

For many years larger aircraft of both military and civilian varieties have been equipped with weather radar for detailed in-flight safety corrections to their flight paths. Such corrections may be made, for example, to avoid unnecessary encounters with powerful convective clouds of the cumulonimbus and cumulocongestus types, where strong thermal uplifts and pronounced atmospheric turbulence may occur.

In atmospheric research, airborne radars of the non-coherent type fitted with vertically scanning antennae have been used to study the development and spread of precipitation in active mid-latitude fronts. The use of airborne Doppler radars for such meteorological purposes is relatively new, an added analytical complication accompanying the movement of the sensor platform as well as the target. We may expect useful results from such systems by the end of the present century.

10.1.4 Weather satellite systems

Remote sensing of the atmosphere received a tremendous fillip with the launching of the first specialized weather satellite, TIROS 1, in April 1960. It was not until the satellite era had opened that investigations of weather by remote sensing means became a practical possibility on anything other than a very piecemeal, localized scale. Prior to 1960 a number of spacecraft and satellites had made some preliminary observations of the Earth's atmosphere from orbital altitudes, e.g. high-altitude rockets of the Viking series, and members of the American Vanguard and Explorer satellite families, but the first TIROS (Television and Infrared Observation Satellite) was the true harbinger of satellite meteorology. By the time of writing well over 100 satellites have been launched for the primary purpose of monitoring the Earth's atmosphere. About half of these have been 'operational' satellites, intended to back up and extend conventional weather observing facilities, and so improve weather forecasts. The rest have been 'research and development' (R and D) satellites, test-beds for new concepts, equipment and instruments. The principal atmospheric satellites that have been concerned with observing weather patterns directly affecting us are listed in Table 10.2. Other atmospheric satellites have investigated solar phenomena and patterns in the upper atmosphere. In this account we will concern ourselves only with those viewing our immediate atmospheric environment – the weather in which we are normally immersed.

It was during the 1970s that weather satellites from countries other than the original 'big two' in space (the USA and the USSR) began to make significant contributions to the World Meteorological Organization's 'World Weather Watch' (WWW). Of special importance in this respect have been geostationary satellites operated by the European and Japanese Space Agencies, namely Meteosat and GMS (or 'Himawari', appropriately meaning 'Sunflower') respectively. Since data from the Western weather satellites have been much more readily available than from the others, we will focus the remainder of our discussion in this chapter and the next mainly on the Western weather satellite programmes and their outcomes, except in certain special instances.

Most of the weather satellites have occupied low-altitude, near-polar, sun-synchronous orbits. These are vital for coverage of polar and high–middle latitudes, which the growing group of geostationary satellites have been unable to monitor from their high-altitude equatorial orbits. However, both types of satellites are vital for adequate global weather coverage in the WWW. This has called for four or five geostationary satellites to provide complete round-the-clock coverage of low and middle latitudes (effectively between about 50–55°N and S), imaging the Earth and its cloud cover at frequent intervals (generally every 30 min; Fig. 5.19 and Fig. 10.3). Unfortunately, such a coverage has not been achieved fully yet, except for a part of the FGGE (First GARP Global Experiment) year of 1979, mainly because of political and economic problems, though premature failure of GOES satellites has been the cause of recently restricted coverage affecting the American continents. However, ATS, SMS and GOES satellites have given good, and mostly continuous, coverage of the Americas, the western Atlantic and eastern Pacific Oceans for many years, nowadays well supported by Japanese GMS satellites over eastern Asia, and Meteosat satellites over Africa. However, the Russian satellite intended to cover the Indian Ocean region by late 1978 did not materialize, and its slot has been incompletely filled by satellites of the Insat series and data from these are not easy to obtain outside India itself.

In the areas not well covered by geostationary satellites, i.e. especially in middle and high latitudes, many satellite data users continue to rely upon the American NOAA near-polar orbiters to meet their imagery requirements. With only commendably brief exceptions, there have been two such satellites in continuous operation since 1966. Tiros-N type NOAA satellites currently provide 6-hourly imagery of the entire globe through two spacecraft orbiting at right angles to each other.

Through the long period of meteorological satellite operations there has been continuous improvement of the observing systems. Although it would be interesting to trace these developments, and of considerable value to relate them to the application and exploration of remote sensing principles we reviewed in Part One, strictures of space preclude such a study in the present context. The US National Space Science Data Center issues an annual *Report on Active and Planned Spacecraft and Experiments*. Payloads of any past, present and planned weather (and other Earth observation) satellites are all detailed in volumes in this series of NASA publications. For present purposes, summaries of representative observing systems on NOAA, Meteosat, Nimbus and American military DMSP (Defense Meteorological Satellite Program) satellites must suffice to exemplify and clarify the value of meteorological satellites to weather and climate studies past and present. For the future, complex and far-reaching plans are being drawn up for the next generation of operational weather satellites and supporting research systems in the US Climate Program. These will be summarized at the end of this chapter.

Table 10.1 Examples of recent and current atmospheric observational programmes involving weather radar (After Collier, 1989)

Title of experiment	Operational period	Location	Precipitation measurement		Observational systems deployed	Aims
			Rain type	Measurement type		
COPT-81 Convection Profonde Tropicale 1981	May–June 1981	Northern Ivory Coast, Africa	Tropical convection	Three-dimensional structure	Mobile ground-based dual Doppler radars; surface observations; radiosondes; acoustic sounder	Development of a better understanding of the dynamic and theoretical features of precipitating convection in continental tropical regions
COST-73 European Weather Radar networking Project	1986–1991	North-western Europe including northern Sweden and Finland	All types	Mainly surface, but some three-dimensional information	Conventional and Doppler ground-based radars	Establishment of international radar data exchange
European Mesoscale Frontal Dynamics Project	Autumn–Winter 1987	South-western England and north-western France	Mid-latitude frontal systems	Three-dimensional structure	Radiosondes; dropsondes; conventional, dual polarization and Doppler ground-based radar; surface observations	Improve understanding of frontal structure and the prediction of the associated mesoscale rainfall distributions

The German Front Experiment	1987	Germany, Switzerland and Austria	Mid-latitude frontal systems	Three-dimensional structure	Polarimetric and conventional radar, surface and upper air observations	As for European Mesoscale Frontal Dynamics Project
GALE – Genesis of Atlantic Lows Experiment	15 January–15 March 1986	East Coast, USA	Extra-tropical cyclones	Three-dimensional structure through storm life	Ground-based Doppler and conventional radars; buoys; research vessels; radiosondes; rawinsondes; surface observations including lidar; aircraft; balloon and satellite measurements	Study of mesoscale and air – sea interaction processes in winter storms with particular emphasis on their contribution to cyclogenesis
COHMEX – Co-operative Huntsville Meteorological Experiment	Summer 1986	Northern Alabama and central Tennessee, USA	Subtropical convection	Three-dimensional structure throughout storm	Airborne advanced microwave moisture sounder (AMMS) (92 and 183 GHz); mobile ground-based dual-wavelength-and-polarization radar	Investigation of summertime subtropical convection
AMEX – The Bureau of Meteorology Research Centre (BMRC) Australian Experiment	October 1986; January–February 1987	Northern Australia	Tropical convection – monsoon	Three-dimensional structure	Radiosondes; ground-based radar and a ship radar; surface observations	Improving understanding of the physics and dynamics of tropical weather systems

Table 10.2 A summary of important meteorological satellite famillies, 1959–1990

Family name	Country of origin	Number launched	Approximate period covered	Special remarks
Vanguard	USA	1[a]	Feb–Mar 1959	Early experimental satellites with primitive visible and infrared imaging systems
Explorer	USA	2[a]	Aug 1959–Aug 1961	
Television and infrared observation satellite (TIROS)	USA	10	Apr 1960–July 1966	First purpose-built weather satellites
Cosmos	USSR	23[a]	Apr 1963–Oct 1983	Some weather satellites in this large, cosmopolitan Russian satellite series, including experimental operational weather satellites
Nimbus	USA	7	Aug 1964–present	Principal American R and D weather satellite
Environmental survey satellite (ESSA)	USA	8	Feb 1966–Jan 1972	First American operational weather satellite
Molniya	USSR	8	Apr 1966–May 1971	Dual purpose communication/weather observation satellite
Applications technology satellite (ATS)	USA	4[a]	Dec 1966–June 1981	First meteorological geostationary satellite to test SMS concepts
Meteor	USSR	>40	Mar 1969–present	Current Russian operational weather satellite series
Improved Tiros observational satellite (ITOS)	USA	1	Jan 1970–June 1971	NOAA prototype
National Oceanic and Atmospheric Administration satellite (NOAA)	USA	5	Dec 1970–Dec 1979	Second generation American operational weather satellite series

10.2 Representative weather satellites

10.2.1 Operational polar orbiters: The NOAA payload

Current NOAA satellites are 'third-generation' operational polar-orbiting environmental spacecraft designed to improve on and extend the functions of their predecessors, ESSA 1–8 and NOAA 1–5. They are co-operative efforts of the USA, UK and France which capitalize on experience gained by these three nations in earlier satellite and sensor operations and experiments. The sun-synchronous NOAA satellites have been designed to operate in orbits inclined at 99° to the Equator, at altitudes of about 850 km, giving southbound Equator crossings between 0600 and 1000 Local Solar Time (LST), and northbound crossings between 1400 and 1800 LST. The most versatile instrument of the NOAA satellites is the AVHRR (Advanced Very High Resolution Radiometer), designed to provide swaths of imagery for central processing (after onboard storage) and analog-type Automatic Picture Transmission (APT) imagery at 4 km resolution, and digital-type High Resolution Picture Transmission (HRPT) imagery at 1.1 km resolution. The APT and HRPT imagery (Fig. 10.4) is intended for local users, and stored data for global archiving. As mentioned earlier the AVHRR instrument is being upgraded to a six-channel instrument in 1994 or 1995, and to a seven-channel ('VIRSR') instrument before AD

Table 10.2 *Continued*

Family name	Country of origin	Number launched	Approximate period covered	Special remarks
Defense Meterological Satellite Program (DMSP)	USA	12	Feb 1973–present	Military weather satellites, often equipped with newer sensors than contemporary civilian satellites
Synchronous Meteorological Satellite (SMS)	USA	2	May 1974–Dec 1975	Testbeds for operational geostationary weather satellite systems
Global Operational Environmental Satellite (GOES)	USA	8	Oct 1975–present	First operational geostationary weather satellites
Geostationary Meteorological Satellite (GMS)	Japan	4	July 1977–present	Geostationary weather satellites covering East Asia/Pacific
Meteosat	Western Europe (ESA)	5	Nov 1977–Nov 1979	Geostationary weather satellites covering Africa/ Europe
TIROS-N (prototype; series numbered NOAA-6 *et seq*)	USA	5	May 1979–present	Third generation American operational polar-orbiting weather satellites
Insat	India	3	August 1983–present	Geostationary weather satellites covering India/ Indian Ocean
Fengyun (FY)	China	2	Sept 1988–1991	First Chinese polar-orbiting weather satellites

[a] Families that have included satellites designed for non-meteorological purposes also; numbers of weather satellites only listed here.

2000. Many countries are now able to obtain (digital) HRPT data direct from NOAA satellites; cheaper reception facilities are required to access (analog) APT data, and such facilities are now commonplace in schools, colleges and other institutions not requiring the absolute calibration which HRPT alone provides.

Meanwhile, the equally important NOAA TOVS (Tiros Operational Vertical Sounder) system is comprised of three complementary sensors, each providing integrated radiances from narrow columns of the atmosphere for vertical profiling purposes (Fig. 10.5). These are:

1. The Basic Sounding Unit (BSU), or High Resolution Infrared Sounder designed to profile temperatures from the surface to 10 mbar, water vapour in three layers of the troposphere (both under cloud-free conditions), and the atmosphere's total ozone content.
2. The Stratospheric Sounding Unit (SSU) designed to extend temperature profiles higher up the atmosphere.
3. The Microwave Sounding Unit (MSU) for broad temperature profile retrieval even in the presence of clouds.

The Tiros-N satellites also carry ozone and Earth radiation budget monitoring equipment, plus a Data Collection System (DCS) to acquire information from fixed and free-floating terrestrial and atmospheric platforms. Platform location is also possible by ground processing of the Doppler measurements of radiowave carrier frequencies. Thus, data can be collected from physically remote sites, and movement of bodies (e.g. trans-Atlantic

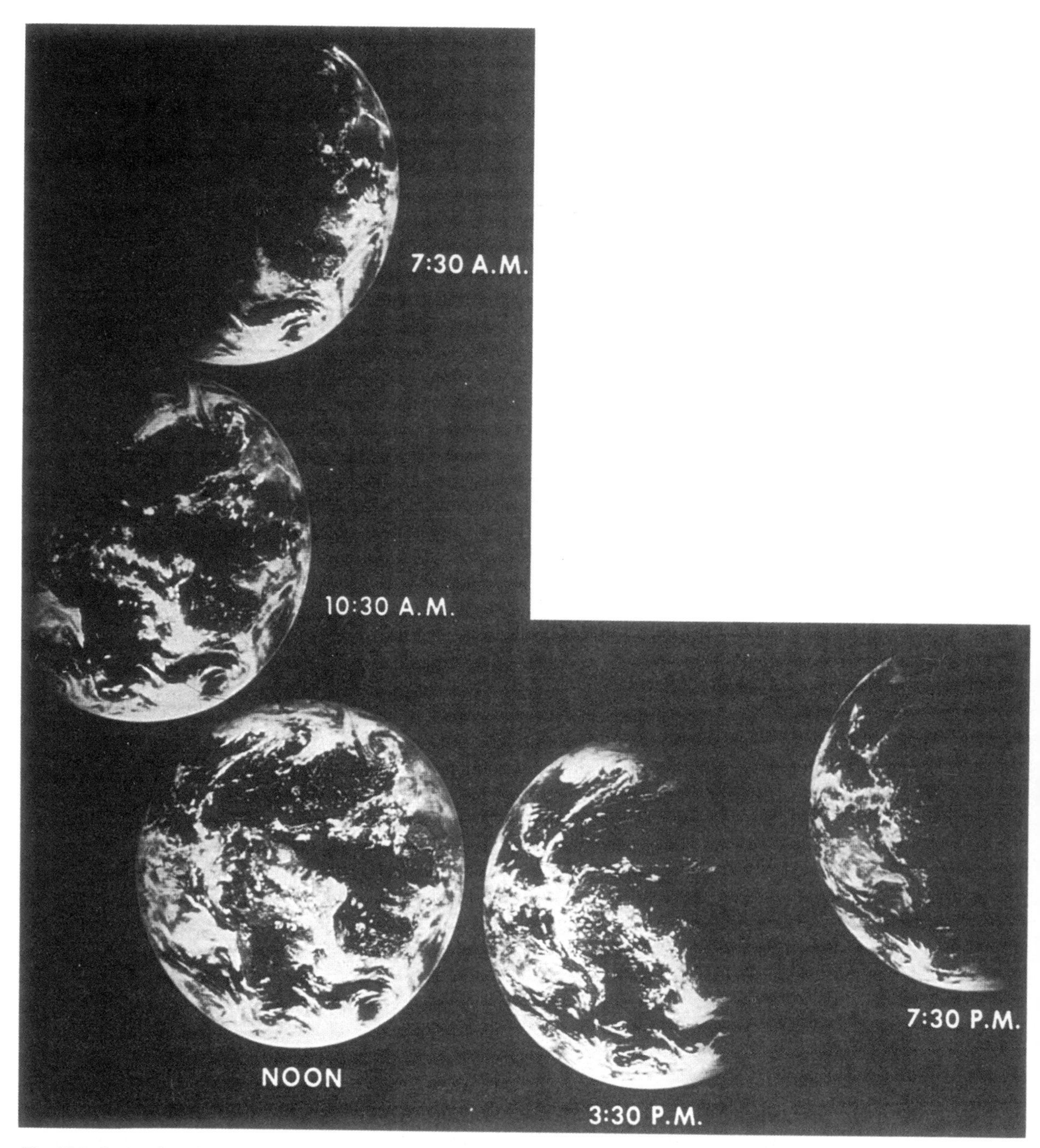

Fig. 10.3 Geosynchronous satellite views of the illuminated portions of Earth, in the visible portion of the spectrum, in orbit above the mouth of the Amazon. These pictures show the passage of daylight from dawn to dusk around the globe. Times are local. (Courtesy, NASA.)

yachts) can be tracked using these facilities (Fig. 6.12), affording very welcome life-saving search and rescue opportunities.

10.2.2 Geostationary satellites: the Meteosat 4 payload

An imaging radiometer, designed to give frequent (usually half-hourly) images of the visible disc of the Earth (about one-fifth of its surface area) constitutes the principal payload on this satellite, as on its predecessors, Meteosat 1, 2 and 3. This radiometer consists of a large (400 mm aperture) telescope with a step-scan motor synchronized with the spin of the satellite to facilitate the construction of each separate image from 2400 adjacent scan lines from one extreme of the visible disc of the Earth to the other, collected over 24 min at a satellite rotation rate of 100 rev min^{-1}. At the end of each individual imaging sequence the telescope is returned to its starting position in readiness for repetition of the process. The optically collected visible and infrared signals are converted into analog electrical signals in three wavebands, one each in the visible (0.4–1.1 μm), infrared window (10.5–12.5 μm) and water vapour absorption (5.7–7.1 μm) regions giving resolutions of 2.25, 4.5 and 4.5 km, respectively, on the ground below at the subpoint. A Data Collection System is also available for interrogation of Data Collection Platforms (DCPs). These are conventionally equipped surface weather stations able to transmit their observations by radio to the satellite overhead. This in turn is able to collect such data from a large number of DCPs in very short periods of time, returning them to a central facility on Earth for routine analysis.

10.2.3 Research and development satellites: the Nimbus 7 payload

Research and development satellites, of which there have been many, have differed much more widely from one another than have the operational polar orbiters or geostationary satellites. The most effective family of R and D satellites has been the Nimbus family, which functioned for almost a quarter of a century from the launching of Nimbus 1 in 1964. Nimbus still merits special mention today because of the large literature of research results based on their splendid data. The following summary of sensors carried by Nimbus 7, the last of the family, elucidates the type and range of problems thought both scientifically significant and technically possible for satellite research and development in the 1980s. As we will see later, the range to be investigated in the foreseeable future on the planned polar platforms is even broader.

1. The Scanning Multichannel Microwave Radiometer (SMMR). This measured radiances at five wavelengths in ten channels to provide data on sea-surface temperature, cloud liquid-water content, precipitation (mean droplet size), soil moisture, snow cover, and sea ice.
2. The Stratospheric and Mesospheric Sounder (SMS). This measured vertical concentrations of H_2O, N_2O, CH_4, CO, and NO, and the temperature of the stratosphere, to approximately 90 km.
3. The Solar Back-scattered Ultraviolet/Total Ozone Mapper System (SBUV/TOMS). This measured direct and back-scattered solar ultraviolet radiation, to provide data on global variations of solar irradiance, vertical distribution of ozone, and total ozone.
4. The Earth Radiation Budget (ERB) sensor. This measured short and long wavelength upwelling radiances and direct solar irradiance to provide data on the solar constant, Earth reflectance, emitted thermal radiation, and anisotropy of the outgoing radiation.
5. The Coastal Zone Colour Scanner (CZCS). This measured chlorophyll concentration, sediment distribution, gelbstoffe, ('yellow stuff') concentration as a salinity indicator, and the temperature of coastal waters and open ocean.
6. The Stratospheric Aerosol Measurements II (SAM II) experiment. This measured the concentration and optical properties of stratospheric aerosols as a function of altitude, latitude and longitude. Tropospheric aerosols were observable if no clouds were present in the field of view of the sensor.
7. The Temperature–Humidity Infrared Radiometer (THIR). This measured the infrared radiation from Earth in two spectral bands (6.7 μm, and 11 μm), both day and night, to provide three-dimensional maps of cloud cover, temperature maps of clouds and of land and

Fig. 10.4 Examples of NOAA AVHRR–HRPT images: (a) infrared.

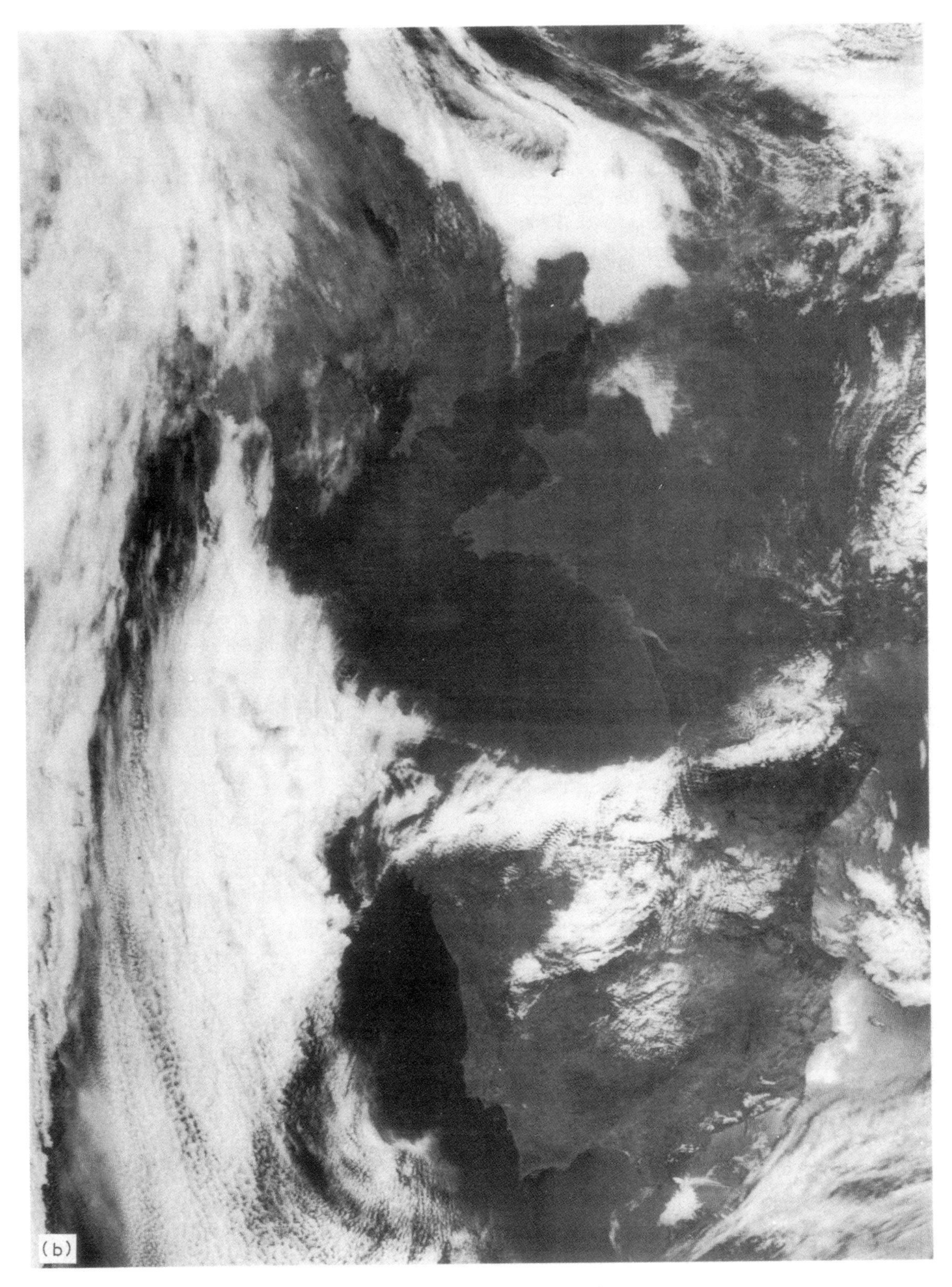

Fig. 10.4 cont. (b) Visible, 10 May 1978. Note especially the greater sense of 'depth' in the infrared cloud field owing to the physical relationship between cloud-top temperature and cloud-top height, which is not found in visible cloud images. (Courtesy, Department of Electrical Engineering and Electronics, University of Dundee.)

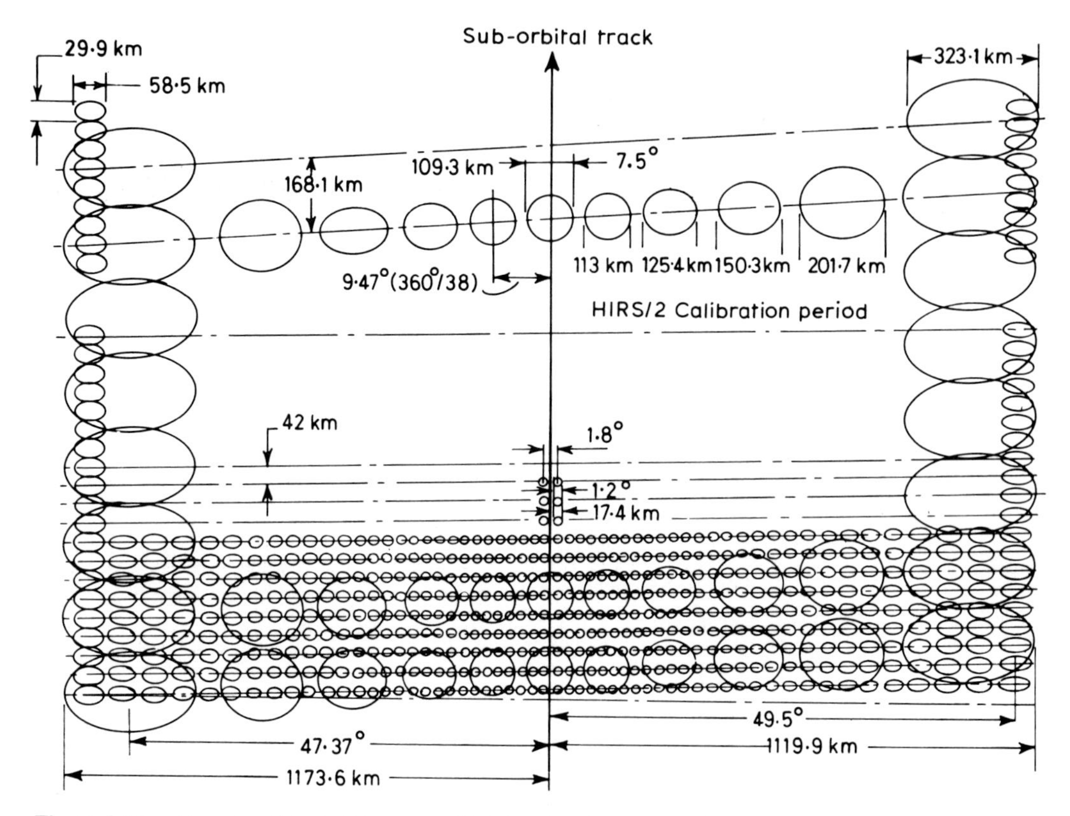

Fig. 10.5 Tiros Operational Vertical Sounder HIRS/2 (smaller element) and MSU (larger element) scan patterns projected on Earth: the microwave sensor integrates radiation over larger areas because naturally emitted target radiation is weaker in the microwave than in the infrared. (Source: Schwalb, 1978.)

ocean surfaces, and maps of atmospheric moisture.

8. The Limb Infrared Monitoring of the Stratosphere (LIMS). This surveyed globally selected gases from the upper troposphere to the lower mesosphere. Inversion techniques were used to retrieve gas concentrations and temperature profiles from its data.

10.3 Satellite data applications in applied meteorology

10.3.1 Routine data products and their uses

We have seen that, for each satellite system, there is a central ground facility which receives most or all of its observations. Here standard data preprocessing, processing and analysis is carried out. The result is a wide range of products for subsequent use by the meteorological community. The breadth of such operations, designed to meet the needs of the major (hemispheric) weather forecasting centres of the WWW, are admirably exemplified by Tables 10.3 and 10.4 and illustrated by Fig. 10.6. Of special significance in the first are the soundings or *vertical temperature profile retrievals*, which routinely augment radiosonde data in mapping the three-dimensional state of the atmosphere through reference to the relief on a recognized series of constant pressure surfaces ('significant levels') upwards through the atmosphere. Of special note in the second are *satellite-derived winds* (Fig. 11.11); these are of particular value in initializing numerical forecasting models in areas sparse in radiosonde data, and are obtained by cross-correlating individual clouds in successive pairs of geostationary images.

More locally, in national and regional meteorological bureaux with APT or HRPT (direct read-out) facilities, reliance is still placed on manual analysis and interpretation of cloud images in both the visible and the infrared. It is generally accepted

Table 10.3 NOAA products and/or services based on operational NOAA satellite data (Source: Oliver and Fairbridge, 1987)

Sounding products

1. Layer-mean temperatures (K)
   a. Layer precipitable water (mm)
   b. Surface–700 mbar; 700–500 mbar; above 500 mbar
   c. Tropopause pressure (mbar) and temperature (K)
   d. Total ozone (Dobson units)
   e. Equivalent black body temperatures (K) for 20 HIRS/2 stratospheric channels
   f. Cloud cover
2. Thickness (m) and layer-mean temperatures (K) between selected standard pressure levels
   a. Precipitable water (mm)
   b. Tropopause pressure (mb) and temperature (K)
   c. Cloud cover
3. Clear radiances
   a. (mW/m^2 sr cm^{-1})
4. Earth-located, calibrated radiances from SSU plus selected HIRS/2 and MSU channels
5. Earth-located sensor output, with calibration parameters appended, for HIRS/2 and MSU (DPSS Level I-b data base).

Oceanographic and hydrologic products

	Accuracy goals
1. Sea-surface temperature observations	±1.5°C absolute ±0.5°C relative
2. Sea-surface temperature regional scale analysis	±1.5°C relative
3. Sea-surface temperature global scale analysis	±1.5°C relative
4. Sea-surface temperature climatic scale analysis	±1.5°C relative
5. Sea-surface temperature monthly observation mean	±1.5°C relative
6. Weekly composite surface-water temperature analysis	1.5°C absolute 0.5°C relative
7. Ice concentration and coverage analysis[a]	±5 km
8. Surface-water temperature analysis[a]	±5 km
9. Snow and ice melting conditions[a]	Contour location within ±5 km

Earth heat budget product list

1. Heat budget parameters
   a. Daytime longwave flux
   b. Night-time longwave flux
   c. Reflected energy or equivalent (albedo, absorbed)
   d. Available solar energy (calculated field to be included in output form)
   e. Angular data

Mapped or gridded initial imagery products (non-quantitative)[b]

1. Hemispheric mapped polar mosaics IR and VIS
2. Mercator-mapped mosaics IR/VIS
3. Polar-mapped 'chips' IR/VIS
4. Mercator-mapped 'chips' IR/VIS
5. Polar-mapped composites IR/VIS (minimum brightness and/or maximum temperature)
6. Pass-by-pass gridded imagery VIS/IR (one satellite)
7. Imagery from limited area coverage (LAC) data: both recorded and direct readout (ungridded)

[a] From hand analysis of full-resolution imagery.
[b] From one satellite.

Table 10.4 Products and/or services based on operational GOES satellite data (Source: Oliver and Fairbridge, 1987)

Meteorological products and/or services
GOES imagery (IR and visible)
Movie loops with annotation and/or narrative
United Press and Associated Press print and captions
Local television support
Hurricane intensity classification
Satellite interpretation messages (SIM)
SELS and SIGMET charts
Special message to the Agency for International Development (AID)
Local thunderstorm, high wind, icing warning service
Meteorological consultation
Cloud motion winds (full disc)
Cloud motion winds, United States Pacific Coast (200-mile limit)
Cloud motion winds, Gulf of Mexico
Cloud motion winds, Gulf of Alaska
Computer-derived cloud motion vectors (picture/pair winds)
Cloud-top height data
Rainfall estimates (Scofield–Oliver techniques)
Surface frontal and pressure analysis
Moisture analysis
NMC support
Satellite cloud PROGS for NMC Aviation Weather Branch
NMC support (satellite cloud PROGS for quantitative precipitation)
Support to NMC (surface and upper air analyses)
Oceanographic and hydrologic
Ocean current messages
Oceanographic consultation
Flash-flood precipitation, amounts, and estimates
Rainfall estimates
Snow melt enhanced IR images
Great Lakes ice analysis
Regional snow maps
River basin snow maps
Agricultural
Experimental: freezeline analyses and tracking solar insolation estimates
Other products and/or services
Numerical grid corrections
WEFAX program
EDIS archive
Data collection system
VISSR data base

that the analysis of images for conventionally classified cloud types is easier in the visible than the infrared; however, infrared images are advantageous in that they are available for night as well as day, and because their contents, relating to target emission rather than reflection, are physically more meaningful, revealing the temperatures of radiating surfaces. Where cloud is absent, land and sea-surface temperatures can be evaluated; where cloud is present, evaluations can be made of the heights of the cloud tops above the ground, as in infrared images that have been objectively enhanced to make their subjective interpretation and subsequent use easier (Fig. 18.2).

Table 10.5 Characteristics of clouds portrayed by satellite visible images

Cloud type	Size	Shape (organization)	Shadow	Tone (brightness)	Texture
Cirriform	Large sheets, or bands, hundreds of kilometres long, tens of kilometres wide	Banded, streaky or amorphous with indistinct edges	May cast linear shadows, especially on underlying cloud	Light grey to white, sometimes translucent	Uniform or fibrous
Stratiform	Variable, from small to very large (thousands of square kilometres)	Variable, may be vertical, banded amorphous, or conforms to topography	Rarely discernible except along fronts	White or grey depending on sun angle and cloud thickness	Uniform or very uniform
Strato-cumuliform	Bands up to thousands of kilometres long; bands or sheets with cells 3–15 km across	Streets, bands, or patches with well-defined margins	May show striations along the wind	Often grey over land, white over oceans, due to contrast in reflectivity	Often irregular, with open or cellular variations
Cumuliform	From lower limit of photoresolution to cloud groups, 5–15 km across	Linear sheets, regular cells, or chaotic appearance	Towering clouds may cast shadows down sunside	Variable from broken dark grey to white depending mainly on degrees of development	Non-uniform alternating patterns of white, grey and dark grey
Cumulo-nimbus	Individual clouds tens of kilometres across. Patches up to hundreds of kilometres in diameter through merging of anvils	Nearly circular and well-defined, or distorted, with one clear edge and one diffuse	Usually present where clouds are well-developed	Characteristically very white	Uniform, though cirrus anvil extensions are often quite diffuse beyond main cells

Table 10.5 summarizes the characteristics of clouds imaged by satellites whereby the principal cloud types may be identified visually. For many years it has been routine for forecasting offices to analyse the cloud contents of weather satellite images for the preparation of simplified clouds charts or *nephanalyses*. Figure 10.7(a) summarizes a standard nephanalysis key, whilst Fig. 10.7(b) is an example of one such chart. It is important that each nephanalysis should be prepared in the same way, even when different analysts are involved. The information content of a nephanalysis includes cloud type, the degree of cloud cover, an indication of the structure of the cloud fields, interpreted features, such as cyclonic vortex centres and jet streams, and boundary lines for ice and snow as well as clouds.

It has been estimated that the information content of a nephanalysis is some two orders of magnitude less than that of the image on which it is based. However, it is usually sufficient for the purposes of short-term forecasters, who may compare the nephanalyses with conventional weather charts for the day, and correct these where necessary in the light of the satellite information. Especially where the forecasting area includes sub-

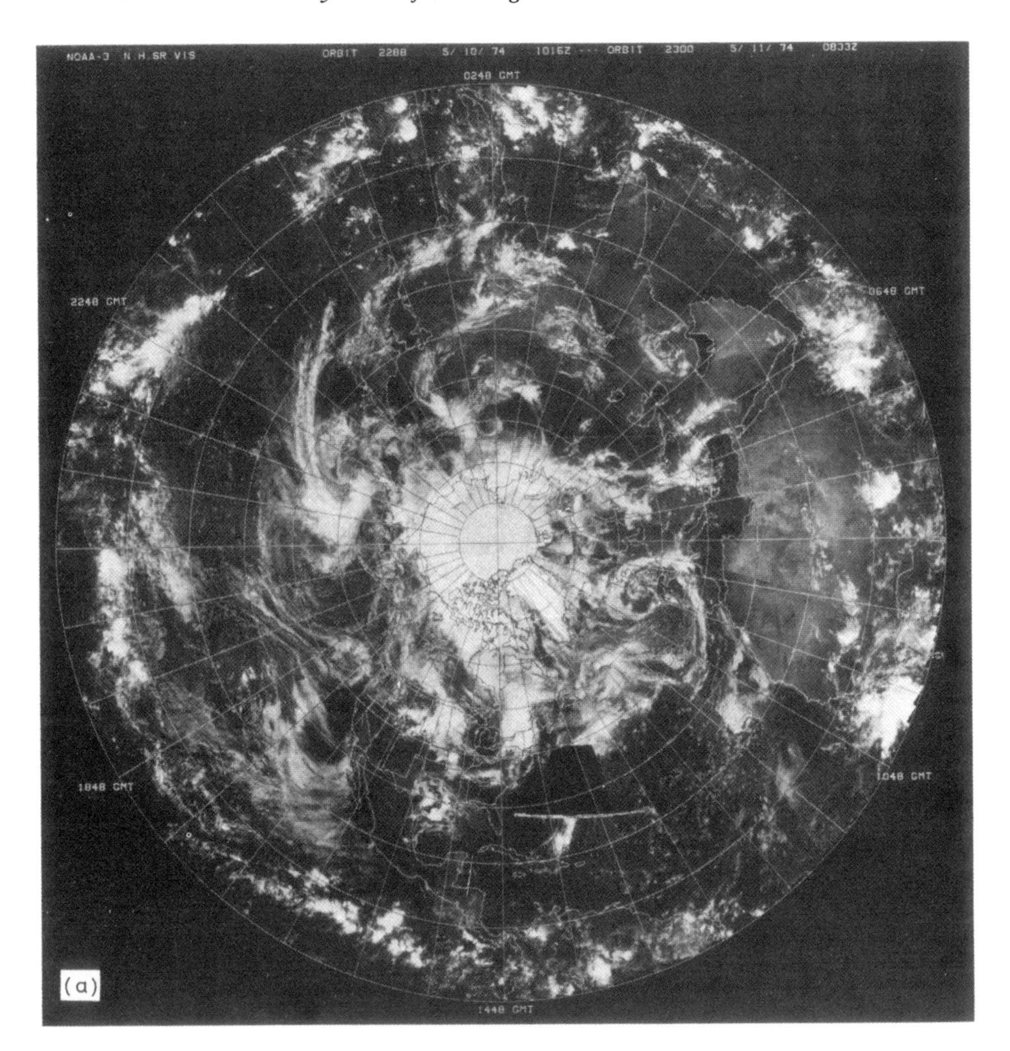

Fig. 10.6 Computer-mapped polar stereographic NOAA images: (a) visible.

stantial sea areas and/or coastal zones the satellite image or nephanalysis may be the salvation of a forecast through the correct or improved identification and positioning of, say, approaching fronts, or ridges of clear, anticyclonic weather. Current computer-based forecasting may provide good predictions of expected contour patterns for selected pressure surfaces, but much detail, e.g. frontal analysis, is still undertaken mostly by hand. In many smaller or poorer countries most of the forecasting procedure is still manual. In such respects and in such areas weather satellite imagery is of great value to the weather forecaster.

On the other hand, especially so far as the major numerical weather forecasting centres are concerned (e.g. Washington, DC and Moscow for the Northern Hemisphere, and Melbourne, Australia for the Southern Hemisphere), satellite data are now considered most valuable as digital inputs in the pools of weather observations by which the numerical models of the atmosphere are initialized. Thus the impact of satellites on weather forecasting is now difficult both to isolate and evaluate: satellites have become accepted as such vital and valuable sources of atmospheric data that advanced forecasting systems and procedures could not do without them.

10.3.2 Special forecasting applications

In many forecasting situations some weather structures are of particular significance, because extreme

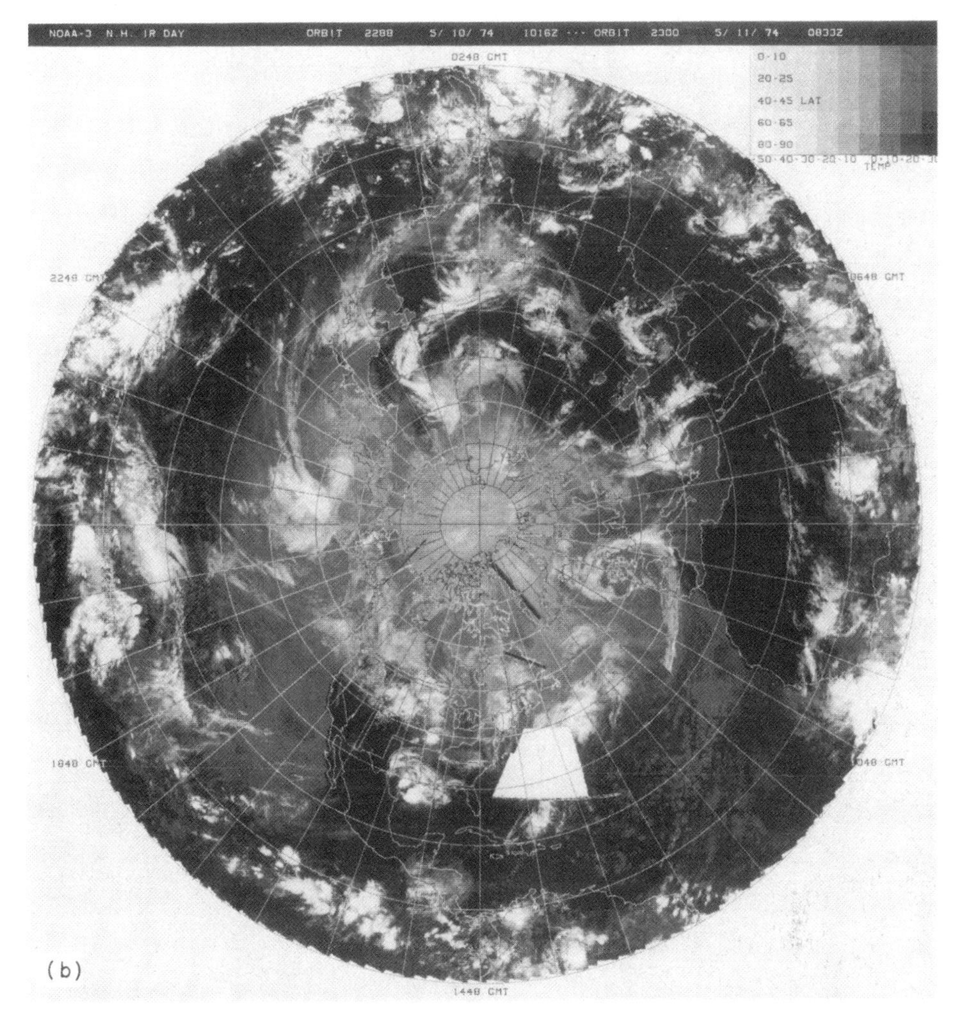

Fig. 10.6 cont. (b) Infrared. Each day one visible mosaic is prepared, for archival purposes, for either hemisphere, and two (one night and one day) infrared mosaics.

events are associated with them. Over the years, research in many centres, but most especially in the Satellite Applications laboratory of NOAA/NESDIS (the US National Environmental Satellite Data and Information Service), has established that a very wide range of synoptic weather systems and structures can be identified in satellite imagery, whether visible, infrared or microwave. Prominent amongst them are vorticity patterns. A wide range of tropical and extratropical cyclones of different types and at different stages of development have been recognized and modelled. *Frontal forms*, and the results of *frontogenesis* and *frontolysis*, often can be identified. Various cellular cloud patterns are known to be related to different instability conditions. In the upper troposphere, upper level troughs and jet streams can be identified and classified. Life cycles of convective cloud cells have been evaluated, and their manifold, and often locally very intense, organization into 'mesoconvective clusters' (MCCs) have been recognized and schematized only since geostationary satellite data became regularly available in the 1980s. Anticyclones have always been more difficult to analyse, for they are characteristically less cloudy, but their fringing clouds have been used to differentiate between many different types of ridges and anticyclones. Last, but by no means least, many cloud features have been recognized as arising as a result of particular interactions between the atmosphere and the underlying topography, significantly improving much local weather forecasting.

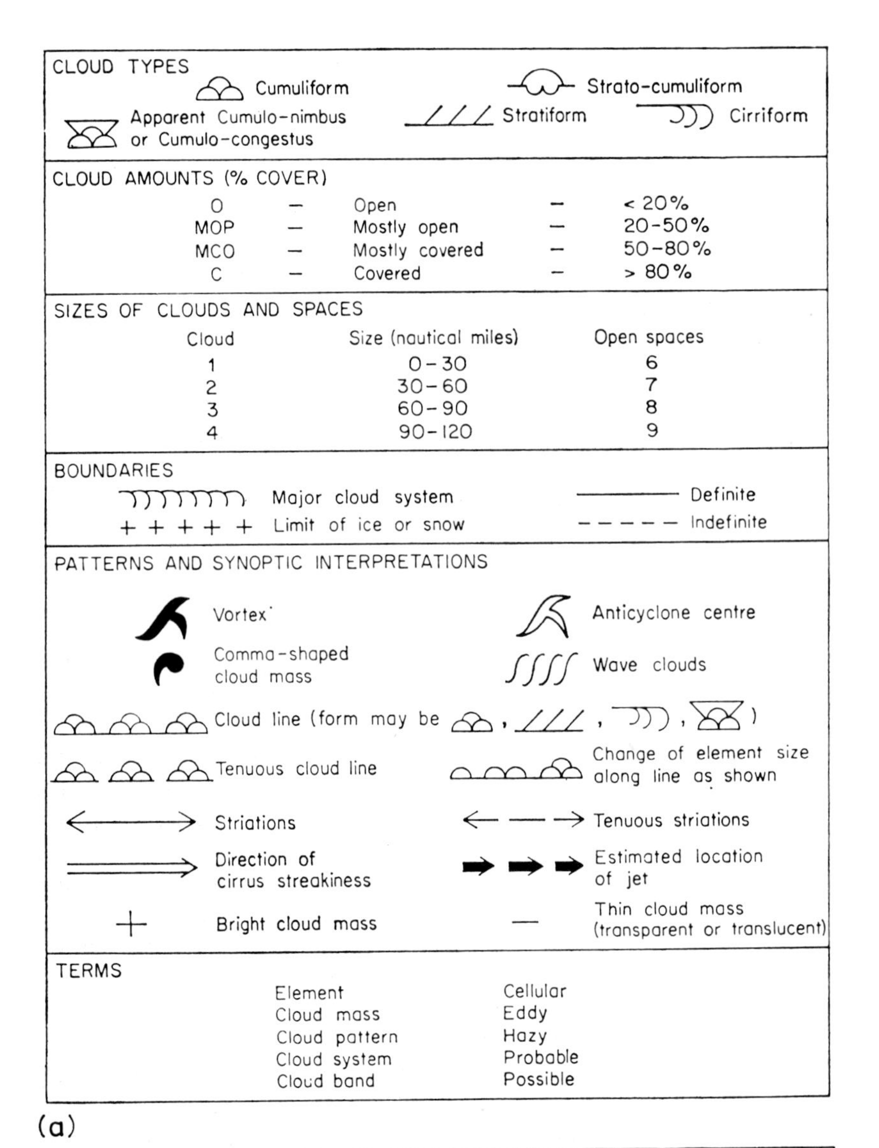

(a)

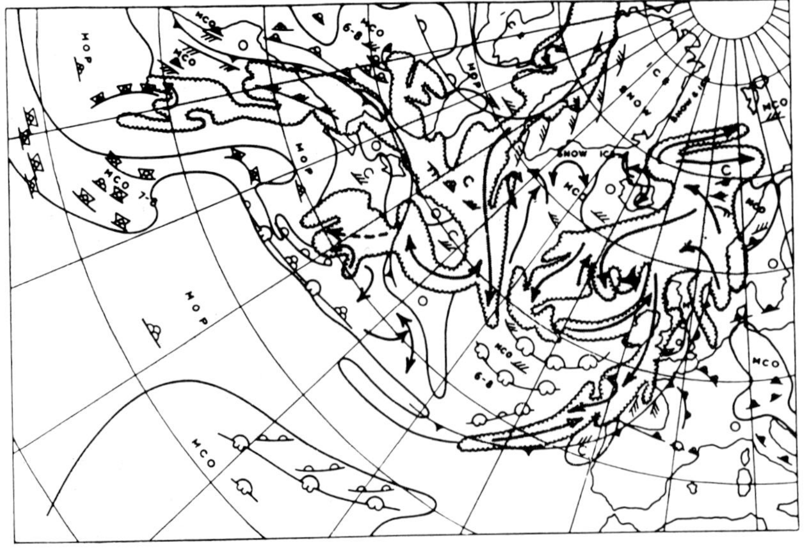

(b)

Fig. 10.7 (a) The standard nephanalysis key. (b) A sample standard nephanalysis summarizing cloudiness over the North Atlantic, on 17 July 1967. (Source: Barrett, 1970.) A more detailed scheme was suggested by Harris and Barrett in 1975.

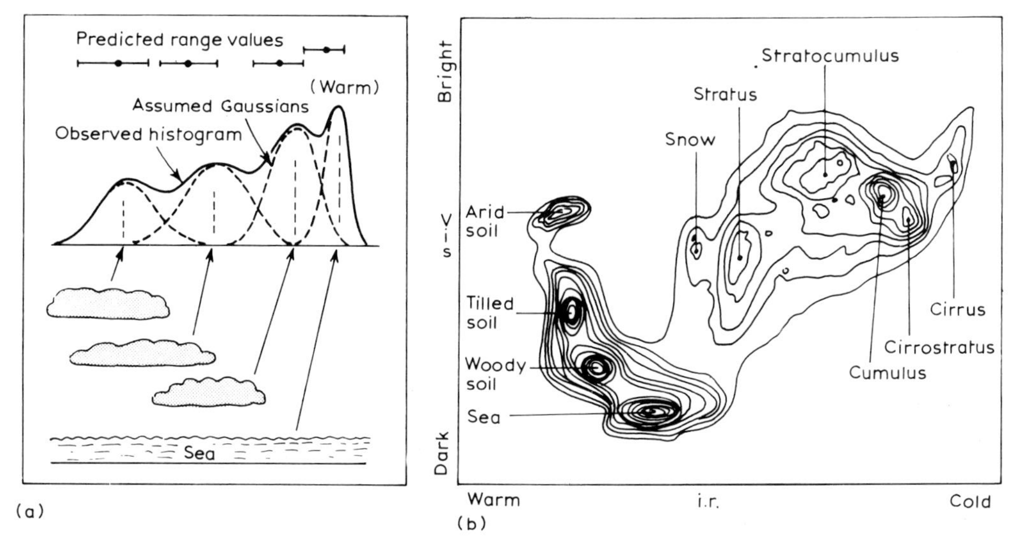

Fig. 10.8 (a) An idealized schematic of the Meteosat imagery one-dimensional histogram analysis concept. (b) A two-dimensional histogram for visible and infrared radiances observed by Meteosat. (Source: Fusco *et al.*, 1980; copyright, ESA.)

Weather satellites have almost certainly justified their costs since the mid-1960s through improvements in hurricane monitoring and forecasting alone. Probably no hurricane or tropical storm anywhere in the world has gone unnoticed since the first operational polar-orbiting satellite system was inaugurated by two ESSA satellites in February 1966. In Chapter 18 (Hazards and Disasters) we will return to this important topic in much greater detail.

10.3.3 Research and development

Three aspects of research and development particularly deserve our attention here. One involves new modes of data analyses, another the search for an improved understanding of the structure and behaviour of the atmosphere, and the third the exploration of new types of data. Analytically, one interesting and potentially useful line of research has been directed towards the development of an automatic ('objective') nephanalysis procedure. We have seen already how so-called 'three-dimensional nephanalyses' are prepared from infrared data, but these classify cloud areas in terms of cloud-top temperatures alone. Automatic identification and mapping of cloud type using single waveband data is probably impracticable, although some success has been achieved (where spatial resolution in not critical) by methods based on both *image brightness* (which is a point or pixel characteristic) and *texture* (a characteristic that can be evaluated only for groups of pixels). An approach using *bispectral* (visible and infrared) data has been applied to Meteosat data at the European Space Operations Centre (ESOC) in Germany. This involves the production of both one and two-dimensional histograms of target radiances. In the first (Fig. 10.8(a)), a segment description provides predicted radiances, whilst a real segment yields an observed histogram which is processed to reveal the underlying Gaussians. From these, the contributions from high, middle and low cloud and the sea surface can be estimated. In the second case (Fig. 10.8(b)) a combination of the results of one-dimensional analyses of visible and infrared data yields a two-dimensional histogram in which the characteristics of several types of clouds and surfaces are typified. In this way new Meteosat data can be classified automatically in terms of the types of surfaces which are visible in the imagery. Unfortunately this approach cannot be used at night.

A second approach to objective cloud classification is applied to multispectral data, e.g. from the NOAA AVHRR–HRPT, by exploiting the known differences in radiant emissions from different types of clouds in several wavebands simultaneously (Fig. 10.9). The way forward in the operational use of weather satellite data in meteorology will certainly involve increased use

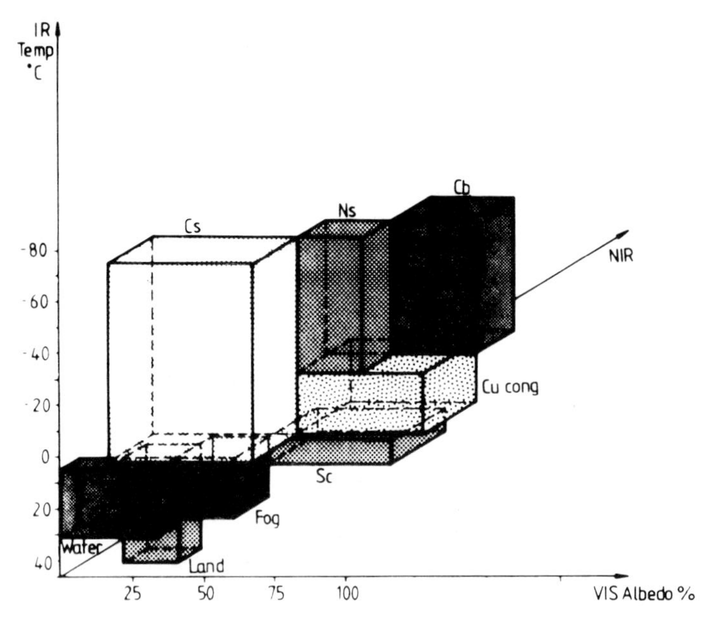

Fig. 10.9 The classification in a three-dimensional intensity space: water, land and six cloud-type classes. (Source: Liljas, 1987.)

of multispectral data through the classification of even four-dimensional or multidimensional space, leading to increased automation of multispectral image data analysis and interpretation. The trend in instrument design and development is towards systems with higher data rates and more spatial and spectral information: to make best use of the data, more use must be made of computers to handle and process them. The 'Autosat' system of the UK Meteorological Office is one example of an objective multiple-output system for effective and efficient use of AVHRR–HRPT data. Others, too, are being developed for multiple environmental applications.

In *synoptic meteorology* the biggest advances in our knowledge and understanding of the behaviour of the atmosphere through the first 25 years of satellite operations were undoubtedly made in the tropics. Exciting progress was made with the identification and interpretation of major weather features, such as the Inter-Tropical Cloud Band (ITCB) (Fig. 10.10), and the South Asian summer monsoon. One result has been a new 'scale-interaction' approach to tropical meteorology. This arose from the realization that the basic building blocks of tropical disturbances (whether 'linear' or 'revolving') are relatively small, short-lived 'hot-towers' of tropical convective clouds. The influence of these, the principal dynamic links between the lower and upper troposphere, ranges from the meso- to the subhemispheric scale. Geostationary satellites have played a special part in the assessment of tropical and subtropical cumuliform convection, for the related clouds are extremely dynamic: they may develop and dissipate within a matter of hours (Fig. 10.11). Much has been learnt about mesoscale convective activity also, by classifying clouds according to the time evolution of their volumes. Techniques have been developed to infer the nature and intensity of vertical mass circulations within and around cumuliform cloud clusters. Related studies of the effects of larger scale circulations on cumulus convection off West Africa revealed remarkable periodicities not only of the expected diurnal mode, but also over 4–5 days. So it appears that larger scale processes have an important effect on the space and time distributions of deep convection more locally, with implications for both tropical and extratropical weather and climate; however, both the nature and relative significance of each process of this kind remains to be established.

Other areas in which weather satellites are now promising much to the meteorologist include middle

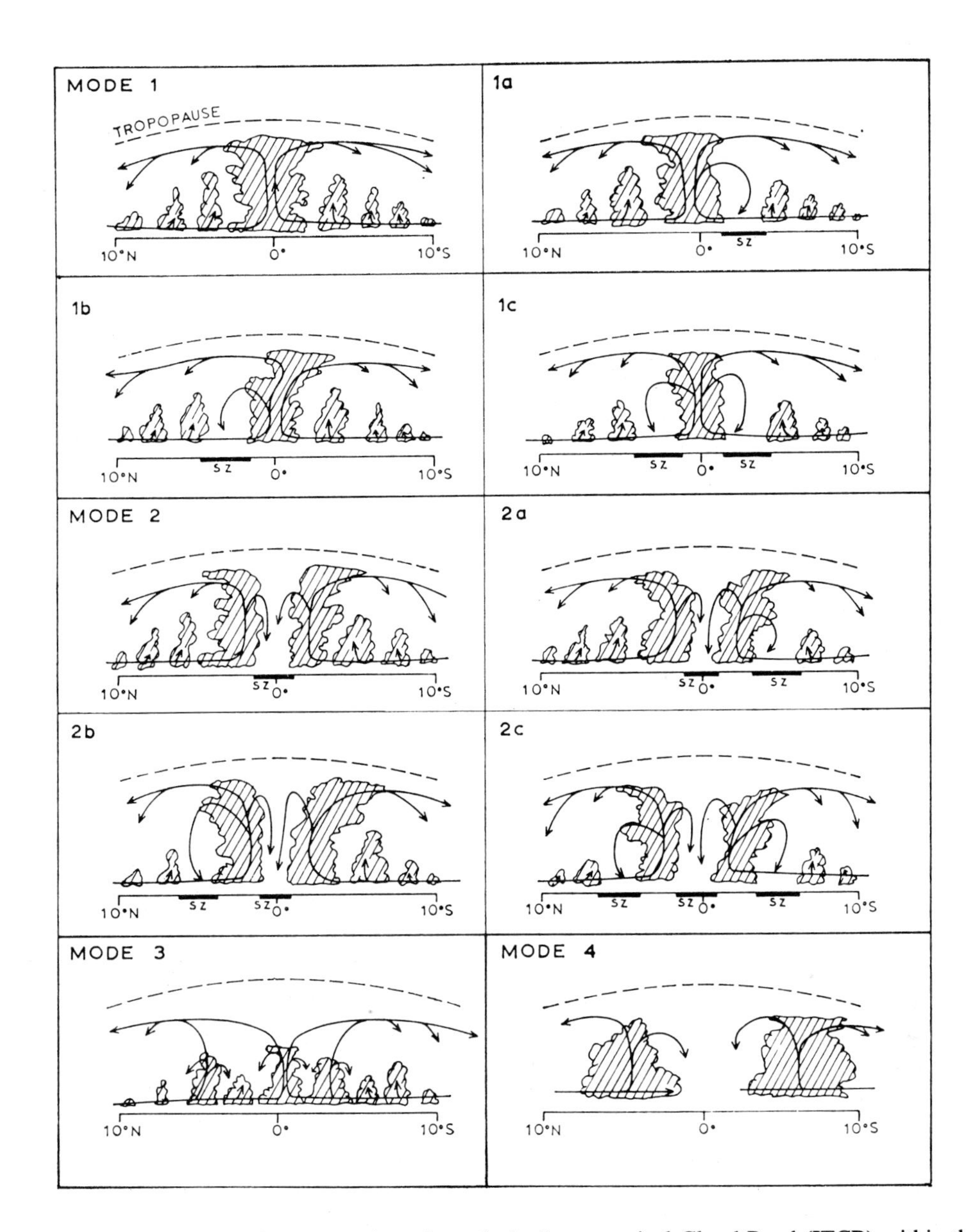

Fig. 10.10 Idealized meridional cross-sections through the Inter-tropical Cloud Band (ITCB) within the Inter-tropical Convergence Zone (ITCZ) from the evidence of meteorological satellite cloud imagery, illustrating the possible range of forms that the equatorial instability axis may adopt. The classification is based jointly on the cloud organizations and subsidence zones (SZ) that develop roughly parallel to the main instability belts. The main distinctions are, therefore, between arrangements involving single instability axes, double instability axes, shear-line axes with subdued instability cloudiness, and patchy instability cloudiness lacking linear organizations.

latitudes of the Southern Hemisphere, and the north and south polar regions. After satellites had helped elucidate the life-cycles of mobile frontal depressions in mid-latitudes of the Northern Hemisphere in the 1960s, it became apparent in the 1970s that mobile Southern Hemispheric depressions (Fig. 10.12) differ in many respects of their structures and life-cycles from their Northern Hemispheric counterparts (Fig. 10.13). Maps of cyclogenesis, vorticity distributions, and cyclolysis over such remote areas can be compiled much more satisfactorily from satellite than conventional data.

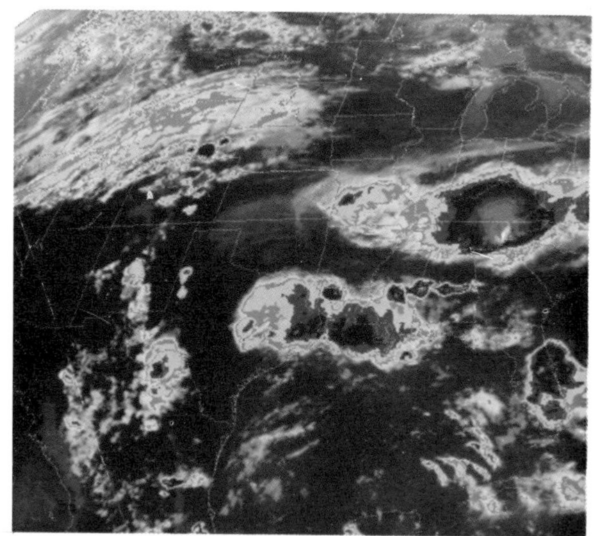

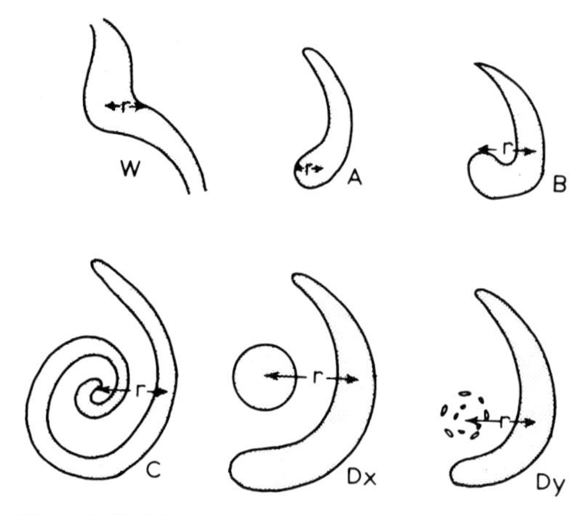

Fig. 10.12 Schematic cloud patterns associated with defined vortex types in mid-latitudes of the Southern Hemisphere. Cloudy areas are stippled. r indicates distance taken as radius from vortex centre. For non-frontal vortices, r is distance to edge of cloudmass. (Source: Barrett, 1974; after Streten and Troup, 1972.)

Similarly, associated frontal troughs, and their intrusions into polar latitudes (revealed well by the infrared imagery, which can distinguish cloud from ice and snow even during the polar night), can be studied best from weather satellite evidence. The 1980s and early 1990s are adding much to our knowledge of the relatively small, almost only mesoscale, cloud and weather vortices of high middle latitudes, around the fringes of the polar regions. Efforts are now being made to elucidate atmospheric structures and associated variations in weather within the polar regions themselves.

Certainly yet more discoveries remain to be made. For researchers in synoptic and subsynoptic meteorology, archives such as that of computer-rectified, brightness-normalized polar-orbiting satellite imagery at NOAA/NESDIS, at Camp Springs, Maryland, USA is a scientific pot of gold. So too are the hard-copy browse files of geostationary satellite data held in various centres around the world. However, perhaps the most promising and useful development of recent years with satellite meteorological research implications has been the continuing fall in the relative costs of direct data reception and allied computing facilities: more and more local centres are now able to access satellite data directly from the satellites themselves, making the task of research case selection both easier and cheaper, and the analysis of digital data the exception rather than the rule. This is important, for, as we have seen before, numerically based remote sensing is both more precise and more readily reproducible. For such reasons alone satellites will continue to expand, and illumine meteorology for long into the future.

But one other type of development is destined to play a huge part in this process, hopefully even before the year AD 2000: the third and last aspect of 'R and D' that deserves special attention in this chapter involves the technological advances which are now beginning to expand our meteorological horizons into regions of the spectrum hitherto largely or completely neglected. This progress may be exemplified by today's high level of interest in passive microwave data, from both *imaging* systems, such as the SSM/I (the Special Sensor Microwave Imager on the US military DMSP-Block 5D-2 satellites commencing with F8 launched in the summer of 1987), and *profiling* systems, such as AMSU-A and AMSU-B, the Advanced Microwave Sounding Units of USA civilian operational weather satellites. These systems, functioning at relatively long wavelengths (low frequencies), are providing much new information on features within, and under, clouds. Global distributions of water vapour, rain areas and rain rates, land snow cover, sea ice, surface wind speeds and other useful atmospheric, geophysical and even botanical parameters can be mapped from passive microwave images (Plate 5, colour section), usually whether cloud is present or not; similarly some vertical structures of the lower troposphere are amenable to passive microwave evaluation with a spatial resolution not possible in infrared.

Through microwave channels a fuller picture is being built up of the atmosphere and the surface of the Earth, both as separate and interdependent entities: this is significant to us all, for weather is not only of direct importance to human life and activity, but also of indirect importance through its impact on other facets of our environment. In turn, the weather is affected by the surface of the Earth, and the versatile data from passive microwave sensors are certain to elucidate such influences on the atmosphere (see also section 12.2.2), all the more so as the spatial resolutions of passive microwave systems (at present mainly between about 12 and 60 km) become improved, and their temporal resolutions (at present only about one to four times per day) increase as more satellites fly with these sensors on board.

In Chapter 19 atmospheric sensors planned for EOS satellite systems now under development are listed. With the expectation of such excellent facilities just round the corner, much interest, even excitement, clearly remains in store for tomorrow's satellite meteorologists.

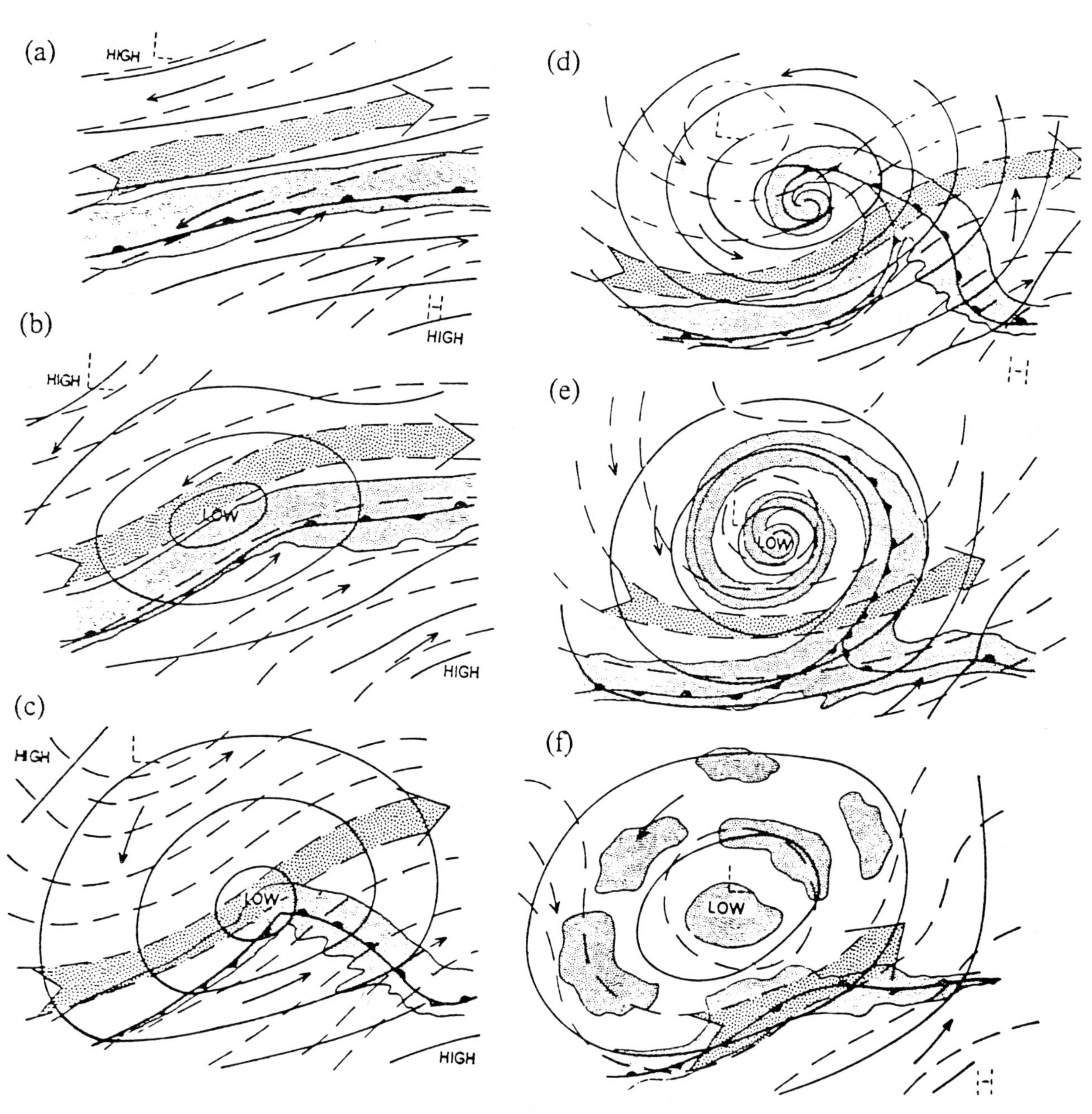

Fig. 10.13 A life-cycle model for mobile, frontal, extra-tropical cyclones in the northern hemisphere. Finely-stippled areas are major bands of clouds; broken lines and letters represent mid-tropospheric contours and winds; continuous lines and letters represent sea-level isobars and winds; coarsely-stippled areas are likely jetstream cores. The model combines conventional and satellite evidence for the evolution of these storms. (From Barrett, 1974).

11 Global climatology

11.1 Climatology emerges from the shadows

The recent rise of interest in climatology has been so startling that it is no cliché to suggest that this is truly the present Cinderella of the planetary and environmental sciences. From a position of lowly obscurity, climatology has risen to be recognized and fêted by many in the course of a single decade: from a situation of poverty and obscurity it has found itself being transported to the environmental ball, with surprise but also pleasure, and there to be acknowledged and respected by many. Funding for its activities may not yet permit it to advance in a gilded coach, but it no longer has to jog along on its traditional pumpkin.

What has prompted this amazing change? Primarily two things: the first a growing realization that mankind may now be exercising powerful, probably detrimental, and possibly irreversible, influences on the lower atmosphere, e.g. through practices that seem to be promoting *global warming*; the second was the relatively sudden discovery of a long unsuspected yet apparently already major impact of our activity upon the middle atmosphere – the '*ozone holes*' over the polar regions of the world (Plate 6, colour section).

By 1989, the UK's Natural Environment Research Council were prompted to state in their policy document 'Air of Change' that 'The study of climate . . . is now of unprecedented importance . . .' and NASA's Earth System Sciences Committee, in their widely acclaimed 'Blue Book' entitled *Earth System Science: a Closer View* (1988) stated that '. . . the examination of global climate change is crucial to an understanding of the Earth system today' having become 'a challenge of the greatest urgency.'

Fortunately, as we saw at the end of the previous chapter, many of the tools that will be needed to meet this challenge more fully than before will soon become available on satellites and space stations; fortunately, the young science of satellite climatology is already sufficiently well developed to provide a sure foundation for the greatly increased efforts that are now beginning to be expected of it.

11.2 Global climate monitoring: the nature of the problem

Towards the end of the previous chapter, reference was made to studies of atmospheric variations which extend not over hours or days, but over months, years or even longer. If our objects of interest vary over these time-scales, automatic extraction of key features of satellite imagery, smoothing, and reduction of the instantaneous data, are usually essential if the more persistent features, patterns, or trends they contain are not to be obscured by the statistical noise of short-term weather variations and their effects.

Studies of the atmosphere over periods upwards from about five days in length may be considered climatological rather than meteorological. As a consequence of the different time-scales involved, the physical factors thought significant by climatologists may be different from those which are basic to meteorology. Although the advantages of satellite remote sensing systems for weather studies (Chapter 10) are advantages for climate studies also, a number of additional advantages that are specifically climatological can be listed for emphasis. The chief of these advantages are as follows:

1. Weather satellite data are much more nearly complete on a global scale than conventional data.
2. Satellite data for broad, even global-scale areas, are more homogeneous than those

collected from the much larger number of surface observatories.

3. Satellite data are often spatially continuous, in sharp contradistinction to data from the open network of surface (point-recording) stations.
4. Satellite observations are complementary to conventional observations, especially elucidating some parameters and phenomena which are otherwise very difficult to observe or measure.
5. Satellites can provide more frequent observations of some parameters in certain regions than is usual by conventional means.
6. Data from satellites are all objective (unlike some conventional observations, e.g. visibility and cloud cover), and immediately amenable to computer processing.

However, it should not be thought that satellite data are an immediate, universal panacea for the problems of the climatologist. Indeed, several new problems are posed to the would-be climatological user of satellite data on account of their own intrinsic characteristics. The chief of these problems include:

1. The vast quantities of new data points involved.
2. The physical meanings of satellite data, which are often different from those of conventional observations.
3. How best to develop techniques to analyse the new types of information.
4. The resolutions of the data. These are not always optimal for climatological uses.
5. Differences and/or inconsistencies in the performances of satellite and ground systems, which complicate data evaluation and analysis.
6. Difficulties in deciding how best to present results of satellite observation and monitoring.

Notwithstanding such problems, many climatological products of interest and value have begun to appear since the runs of satellite data have lengthened sufficiently to make them possible. These products include inventories of parameters as significant as the *net radiation balance* at the top of the Earth's atmosphere (the primary forcing function of the Earth's atmospheric circulation patterns) and the *distribution of cloud cover* (a big influence on the albedo of the Earth–atmosphere system and its component parts, as well as being an indicator of horizontal transport patterns of latent heat). In pre-satellite days certain components of the radiation balance (shortwave (reflected) and longwave (absorbed and re-radiated) energy losses to space) (Fig. 11.1) could be established only by estimation, not measurement. Therefore the only comprehensive maps of global cloudiness compiled in pre-satellite days depended heavily on indirect evidence, and could not be time-specific. Now, satellites are affording us more comprehensive, and much more dynamic, views of global climatology than were possible before their day.

Satellites are also assisting the rise of applied macroclimatology, for example, the development and testing of computer models of the Earth's atmosphere for the improvement of both extended and long-range weather forecasting. In such models the net radiation balance at the top of the atmosphere is one of the outputs. By comparing this with satellite-observed net radiation patterns new insight is gained into the efficiency of the computer model, and aspects of it which might be improved.

In the past, the development of climatology as a worthwhile and distinctive field of study was singularly hampered by an inadequacy of data. Satellites are helping enormously to correct this traditional deficiency. Let us review in more detail the areas of climatology to benefit most, beginning with energy and radiation budget modelling, which has benefited much from analyses of data from measurements from radiation budget sensors on TIROS, Nimbus and NOAA satellites, and is benefiting greatly from data provided by the dedicated Earth Radiation Budget Satellite (ERBS) launched in 1984, and from NOAA-9 and 10, which carry an Earth Radiation Budget Experiment (ERBE) package. Data from these sources are now being analysed.

11.3 The Earth/atmosphere energy and radiation budgets

11.3.1 Introduction

A distinction is sometimes drawn between:

1. A *radiation budget* (or radiation balance). This is the relationship between radiation received by the Earth and its atmosphere from the Sun, and that emitted and reflected by the Earth and its atmosphere in return (Fig. 11.1(a)).

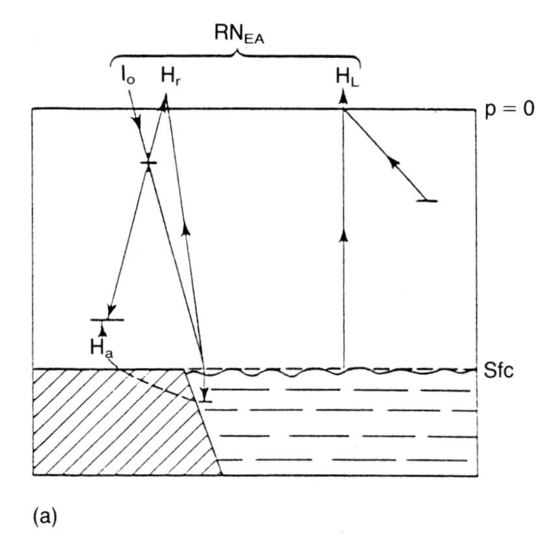

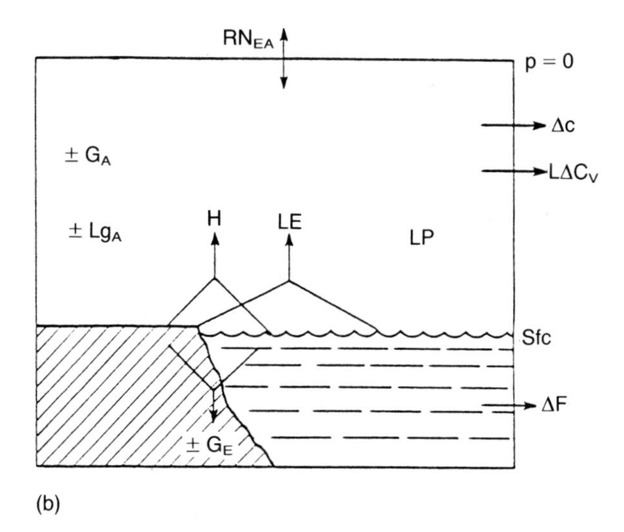

Fig. 11.1 Major components of (a) the radiation budget and (b) the energy budget of the Earth-atmosphere system. (Source: Vonder-Haar, 1968.)

2. An *energy budget* (or energy balance), which is the equilibrium known to exist when all sources of heat gain and loss for a given region or column of the atmosphere are taken into account. This balance includes advective and evaporative terms as well as a radiation term; that is to say both horizontal movements of energy and the part played by absorptive gases in the atmosphere are considered as well as the upwards/downwards radiation interchanges (Fig. 11.1(b)).

Much progress has been made in reassessing the Earth/atmosphere radiation balance from satellite data sources. However, one substantial problem in energy budget modelling which has not yet proved amenable to solution by remote sensing since the advective processes involved in the heat balance include movements within both the atmosphere and the oceans, is that these are hard to evaluate from satellite data alone. Therefore *estimates* of these quantities are usually used instead of measurements, the estimates being based on rates of atmospheric and ocean water flux, established by *in situ* sensors.

11.3.2 The global picture

Consider that the net radiation balance depends on three quantities, namely:

1. The *solar constant* (a measure of the flux of solar energy into the Earth's atmosphere).
2. The *planetary albedo* (the percentage ratio of that solar energy scattered and reflected by the atmosphere and Earth surface, to the total solar radiation incident upon it). Albedo depends largely upon the solar content, the inclination of the Earth to the Sun's rays, and the reflecting capabilities of the Earth surface and its cloud cover.
3. *Longwave (re-radiated) energy losses to space.*

Outward-looking, solar observatory satellites evaluate (1) more accurately than before through their measurements in the ultraviolet and visible wavebands. Infrared radiometers are providing vital measurements of (2) and (3) through appropriate visible and infrared channels. Figure 11.2(a–c), illustrates some of the first (satellite-derived) mean annual geographical distributions of

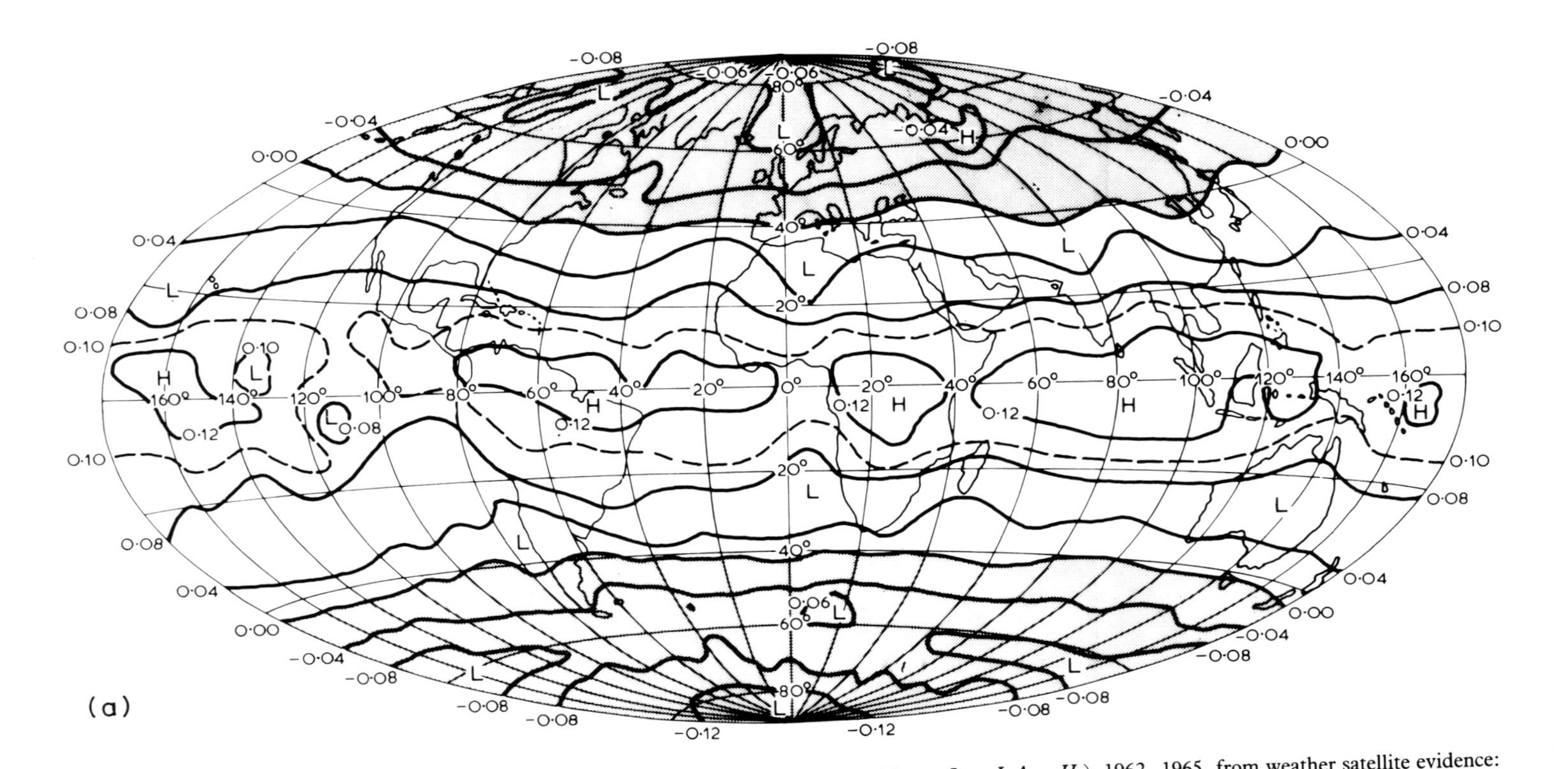

Fig. 11.2 Mean annual patterns of components of Earth atmosphere radiation budget ($RN_{EA} = I_0 - I_0A - H_1$), 1962–1965, from weather satellite evidence: (a) planetary net radiation balance (RN_{EA}) at the top of Earth's atmosphere.

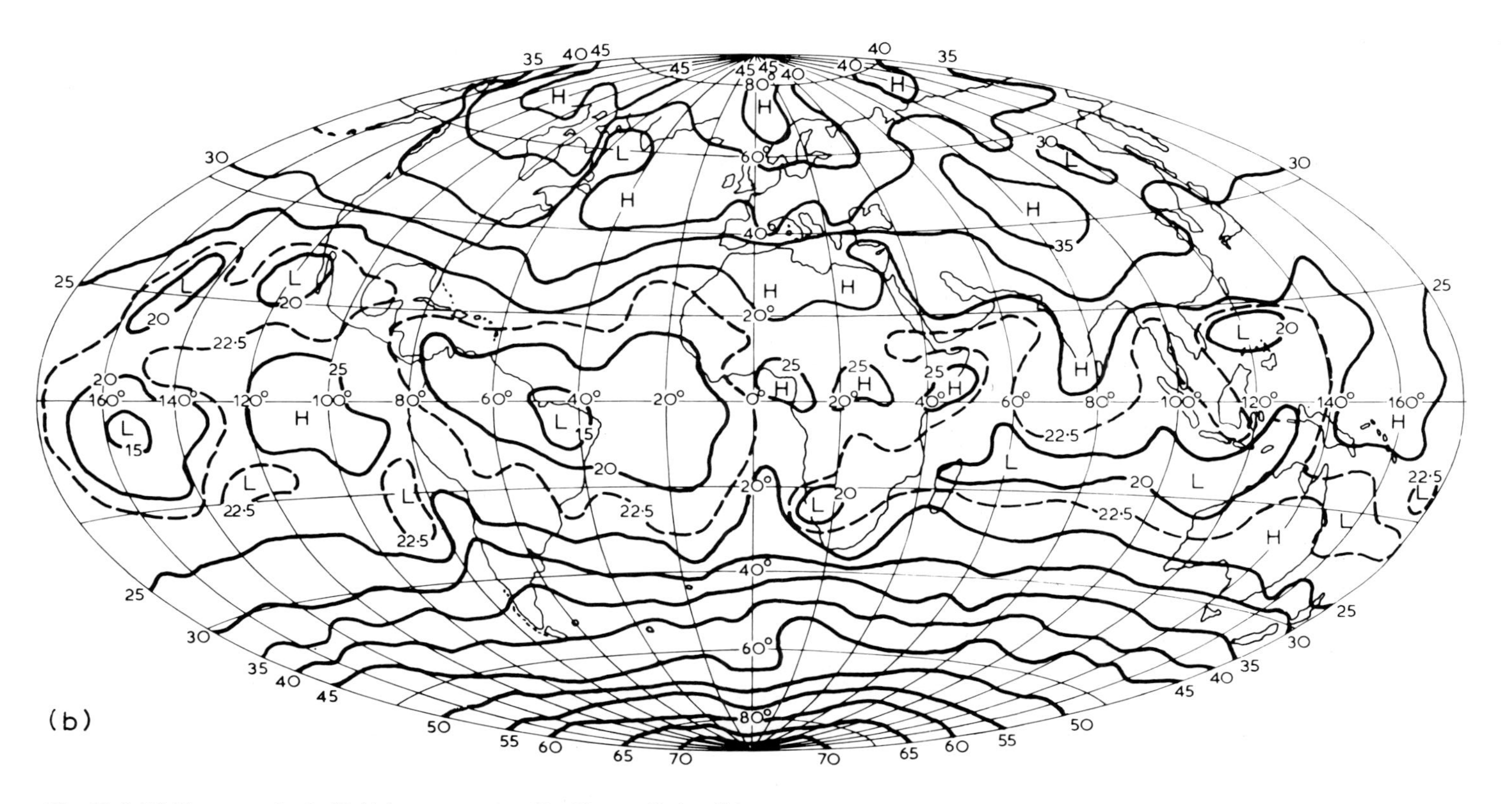

Fig. 11.2 (b) Planetary albedo (I_0A) in percentages of incident radiation (I_0).

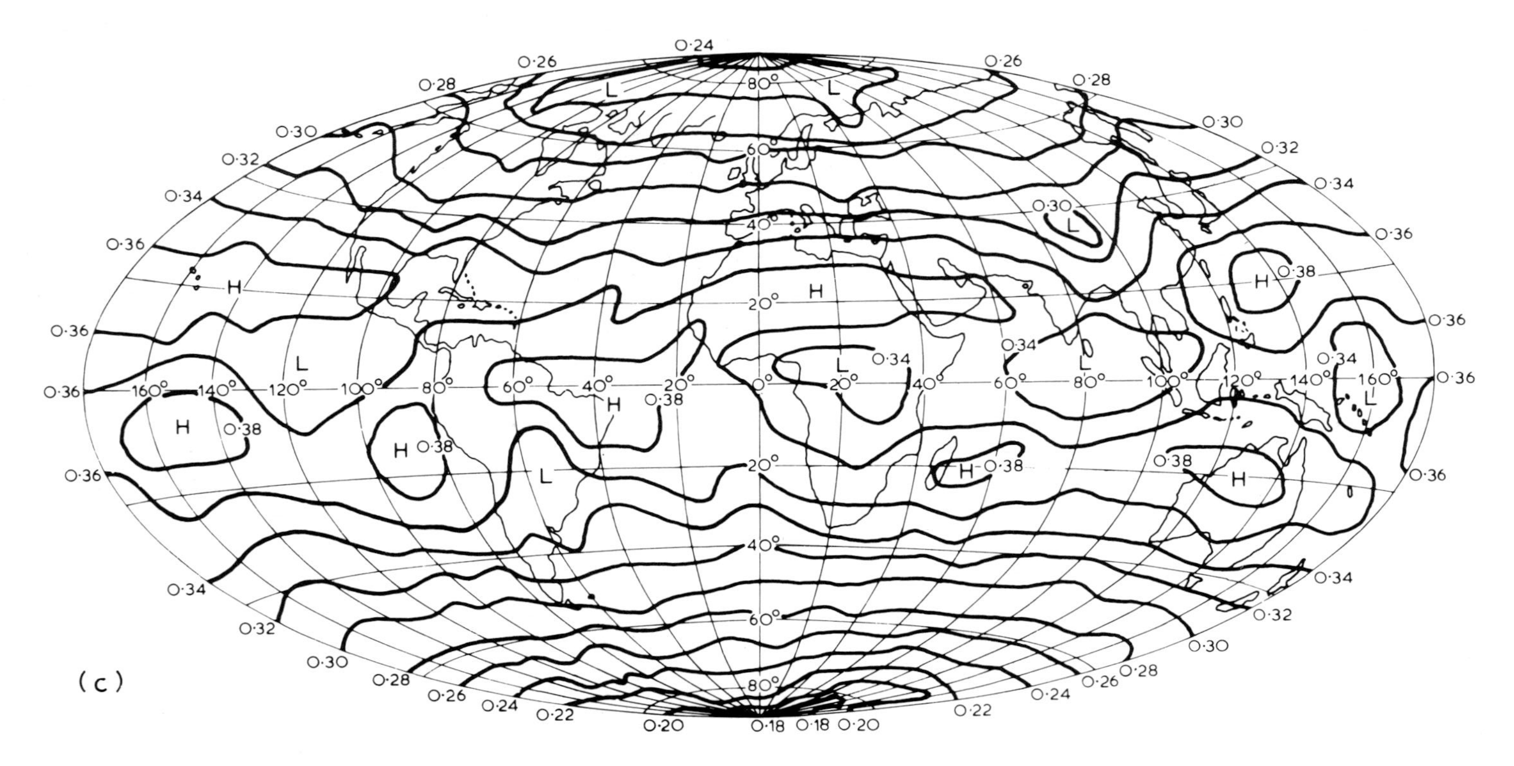

Fig. 11.2 (c) Longwave radiation (H_I). (After Vonder Haar and Suomi, 1971.)

Table 11.1 Mean annual and seasonal radiation budget data for the Earth–atmosphere system, from first generation weather satellites[a] (Source: Vonder Haar and Suomi, 1971)

	DJF	MAM	JJA	SON	Annual
Northern Hemisphere					
I_0	0.34	0.56	0.65	0.42	0.50
H_a	0.24	0.39	0.48	0.31	0.36
H_r	0.10	0.18	0.17	0.12	0.14
A	0.29	0.31	0.26	0.27	0.28
H_L	0.32	0.33	0.34	0.34	0.33
RN_{EA}[b]	−0.07	0.06	0.13	−0.03	0.02
Southern Hemisphere					
I_0	0.69	0.43	0.32	0.58	0.50
H_a	0.46	0.30	0.25	0.41	0.35
H_r	0.22	0.13	0.07	0.17	0.15
A	0.32	0.30	0.22	0.29	0.29
H_L	0.33	0.32	0.32	0.34	0.33
RN_{EA}[b]	0.13	−0.02	−0.07	0.06	0.02
Global average					
I_0	0.51	0.50	0.49	0.50	0.50
H_a	0.34	0.35	0.37	0.36	0.35
H_r	0.16	0.15	0.12	0.14	0.15
A	0.31	0.31	0.25	0.28	0.29
H_L	0.32	0.33	0.33	0.34	0.33
RN_{EA}[b]	0.03	0.02	0.03	0.02	0.02

[a] I_0 = incident solar radiation (cal cm^{-2} min^{-1}),
H_a = absorbed solar radiation (cal cm^{-2} min^{-1}),
H_r = reflected solar radiation (cal cm^{-2} min^{-1}),
A = planetary albedo (%),
H_L = emitted infrared radiation (cal cm^{-2} min^{-1}), and
RN_{EA} = net radiation budget of the Earth – atmosphere system (cal cm^{-2} min^{-1}).
[b] Probable absolute error of ±0.01 cal cm^{-2} min^{-1}.

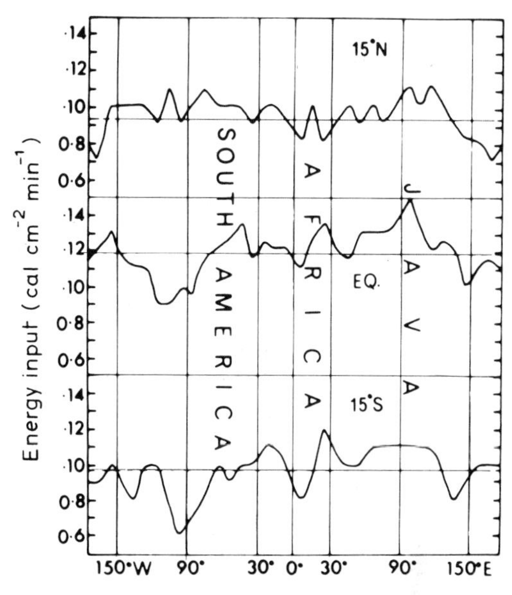

Fig. 11.3 Longitudinal variations in net inputs of energy (RN_{EA}) (mean annual case), evaluated from five years of satellite data. (Source: Vonder Haar and Suomi, 1971.)

the parameters involved in the annual radiation budget ever to be compiled. Table 11.1 integrates these parameters for global and hemispheric areas, using non-SI units. The most significant conclusions from studies of these kinds have included the following:

1. The net radiation balance of the Earth (and that of either hemisphere separately) is in radiative balance. Thus there seems no requirement for an annual net energy exchange across the Equator as assumed by many earlier workers.
2. Both hemispheres are darker and warmer than the widely accepted estimates by London (1957) in pre-satellite days, by some 5% and 15 cal cm^{-2} min^{-1} respectively. If the Earth–atmosphere system is both warmer and darker than was previously thought, then the system must accommodate – and probably transport – about 15% more energy in each hemisphere.
3. Each hemisphere has nearly the same planetary albedo and infrared loss to space on a mean annual basis. Since the surface features of the two hemispheres are quite different, clouds would seem to be the dominant influence on the energy exchange with space. Fortunately, these can be quite conveniently and accurately evaluated from satellites.
4. Significant variations in global averages may occur from year to year, e.g. the average annual albedo may vary by as much as 2 or 3%.

11.3.3 Regional details

Although some regional differences are apparent in Fig. 11.2, we need to view the world at a larger scale if more detailed pictures are required. Important findings in the tropics have been that the Earth is especially darker and warmer in low latitudes than previously believed. The lower tropical albedos (24% instead of pre-satellite values of 34% or more) that are now deployed in radiation

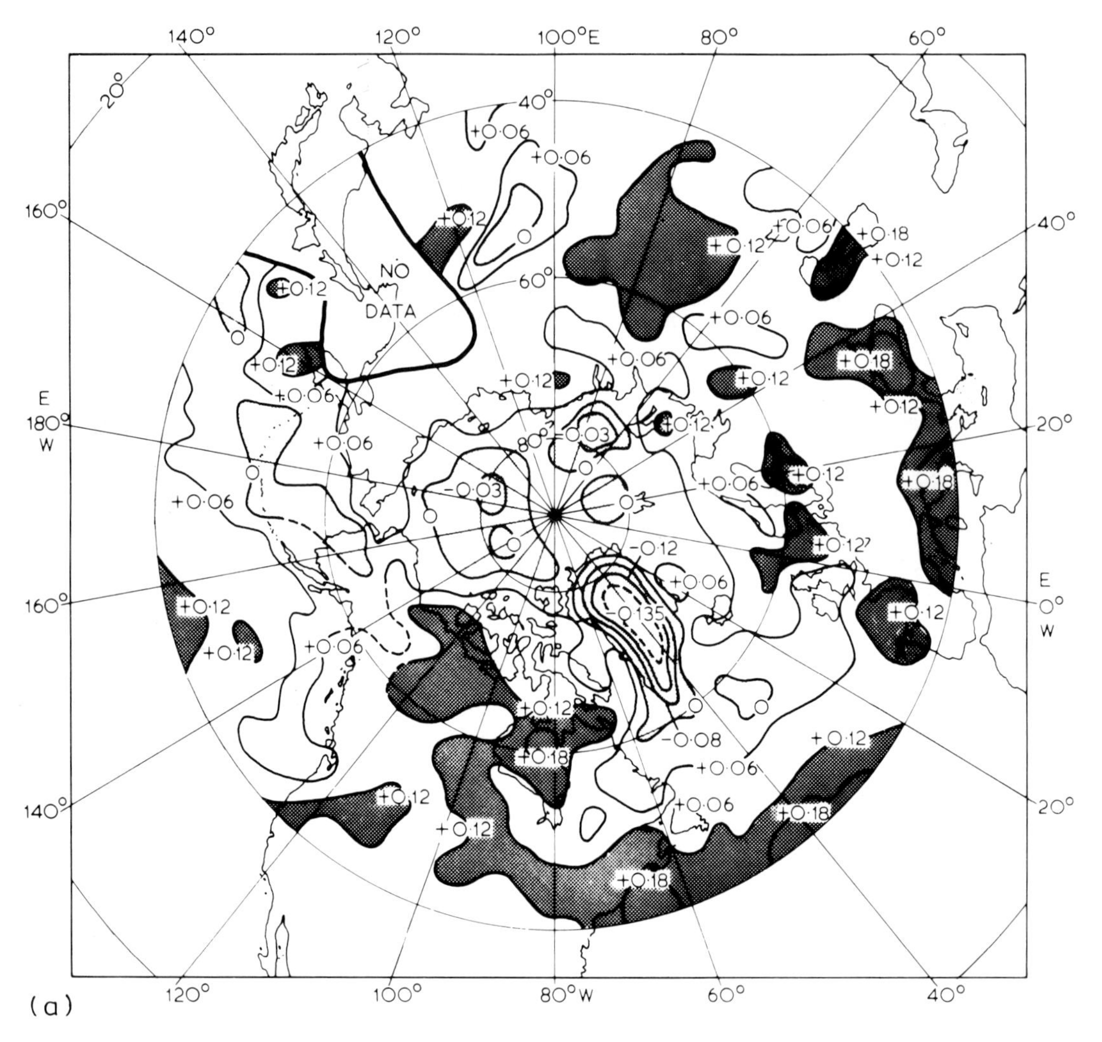

Fig. 11.4 (a) Net radiation balance (RN_{EA}) of the Earth–atmosphere system over the Arctic based on Nimbus 2 data, 1–15 July 1966 (cal cm^{-2} min^{-1}). Areas in surplus are stippled.

budget evaluations are attended by higher absorption and longwave radiation levels. These, in turn, indicate higher rates of energy exchange from the ocean to the atmosphere, and greater energy exports from low to higher latitudes. Thus quite simple satellite findings can have profound implications for both regional and global climate modelling. This is true in both north–south and east–west directions. Fig. 11.3 portrays the variation of net radiation with longitude for three latitude circles. It is clear that some of the variations are related to the contrast between land and sea, whilst others seem to be related to air mass and circulation differences.

The influence of even more localized complexities of geography on the net radiation balance is exemplified by Fig. 11.4(a and b). The earliest observations from Nimbus 2 quickly revealed that the radiation balance in middle and high latitudes of the Northern Hemisphere in a summer month assumes a highly fragmented pattern. Some areas – even within the Arctic Circle – experience a net radiation gain, while the chief area of net radiation loss is over the Greenland ice-cap, neither over, nor very near, the pole. Generally the areas of net gain ('sources' of radiation energy for the general circulation) indicate little cloud cover, and the areas of net loss (the radiation 'sinks') indicate a high percentage of cloud, or, in the Arctic Ocean, permanently frozen surfaces.

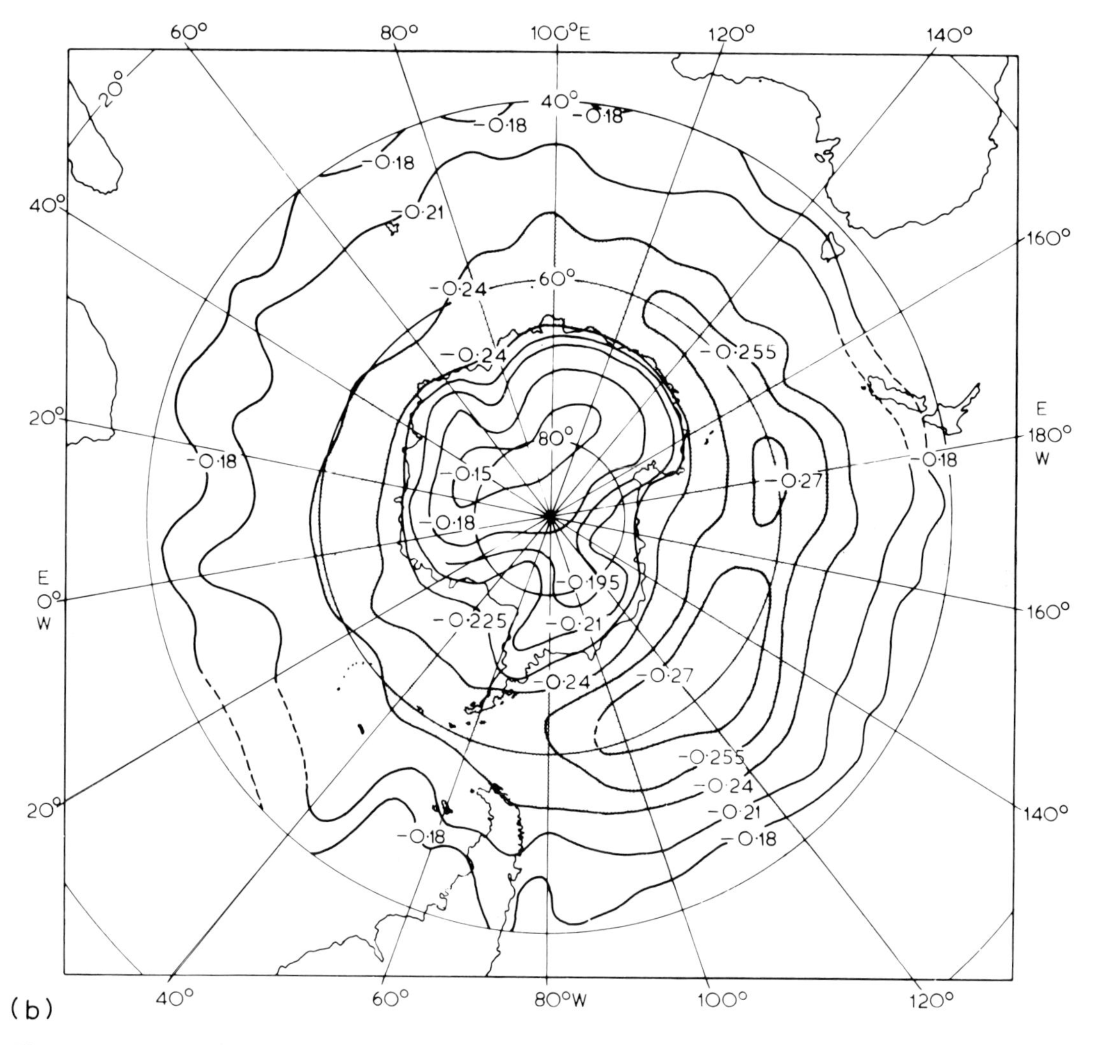

Fig. 11.4 cont. (b) As (a) but over the Antarctic. (Source: Barrett, 1974.)

In the same period, winter held the south polar region strongly in its grip. Figure 11.4(b) shows the whole area south of 40°S to be in net radiation deficit, though, curiously at first sight, the deepest radiation sink is not over Antarctica itself, but over the surrounding southern oceans. This is the zone in which most of the drastic environmental changes take place from one high season to the next. Most of Antarctica is always frozen, and radiation losses in winter are relatively small since there is little stored or advected energy available to be emitted as longwave radiation (heat energy) to space. Meanwhile, the surrounding oceans thaw in summer but freeze in winter, and much energy stored through the summer months is lost as longwave radiation, especially just before the phase change of water from liquid to solid. From the meteorological point of view, when frozen, the oceans act more like continents: as they freeze, their albedos suddenly and dramatically increase, and the difference in received radiation and radiation lost by reflection combines with that lost by emission from remaining patches of open sea to tilt the net radiation budget to its early winter level of considerable deficit.

11.3.4 Local and temporal details

Analyses of satellite radiation budget data have also been made for selected localities around the world, revealing great differences, as exemplified by Fig. 11.5. Further, data from geostationary satellites have been utilized recently for the first time in

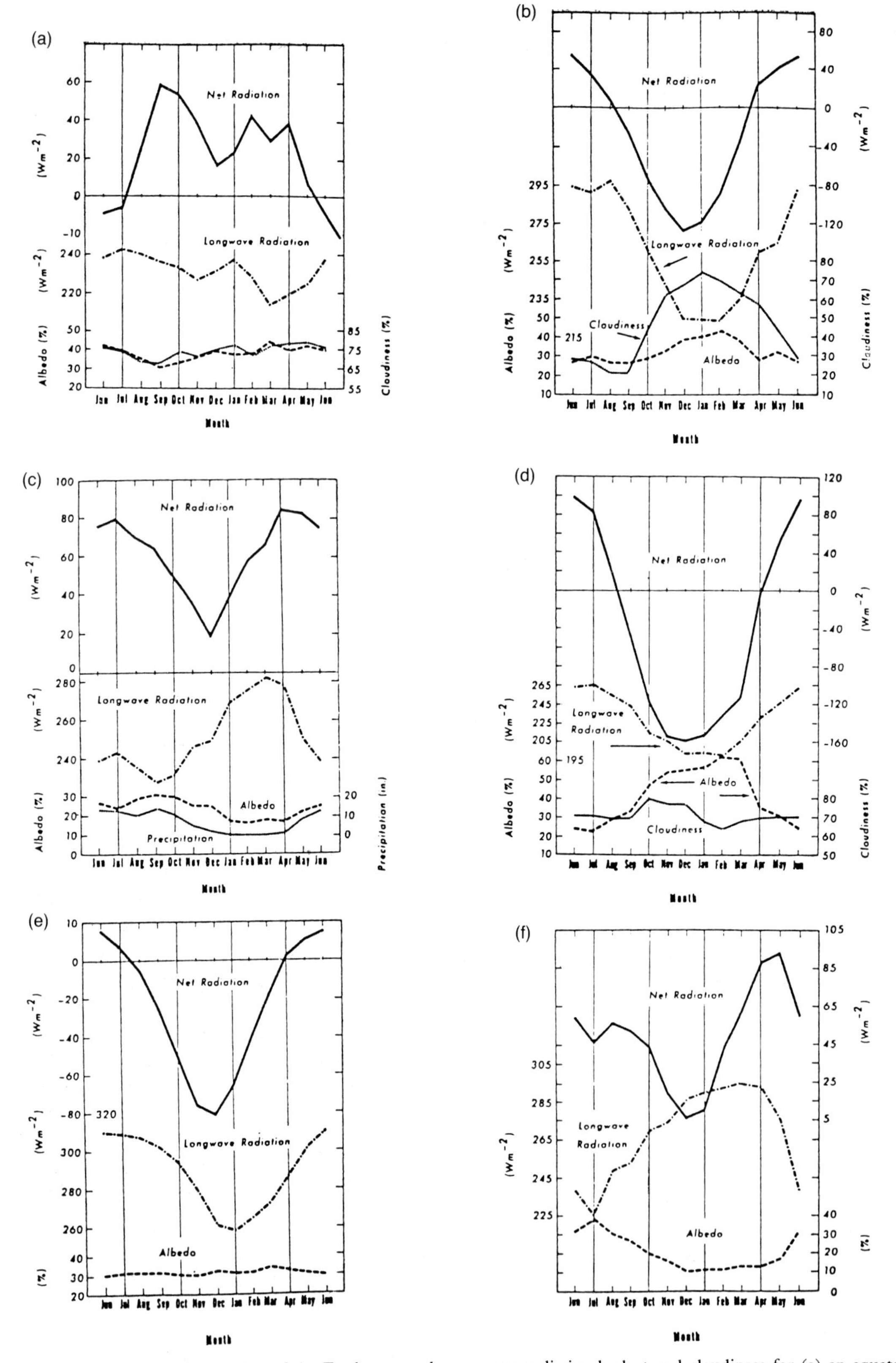

Fig. 11.5 The annual variation of the Earth–atmosphere system radiation budget and cloudiness for (a) an equatorial location (Equator, 67.5°W – Sao Gabriel, Brazil); (b) a subtropical continental location (40°N, 55°E – Aidin, USSR); (c) an equatorial monsoon location (10°N, 107.5°E – Saigon, Vietnam); (d) a mid-latitude continental location (52.5°N, 82.5°E – Barnaul, USSR); (e) a tropical continental location (25°N, 32.5°E – Aswan, Egypt); (f) an equatorial monsoon location (15°N, 70°E – Indian Ocean). (——) Net radiation; (–·–·–) longwave radiation; (–––) albedo. (Source: Ohring and Gruber, 1983.)

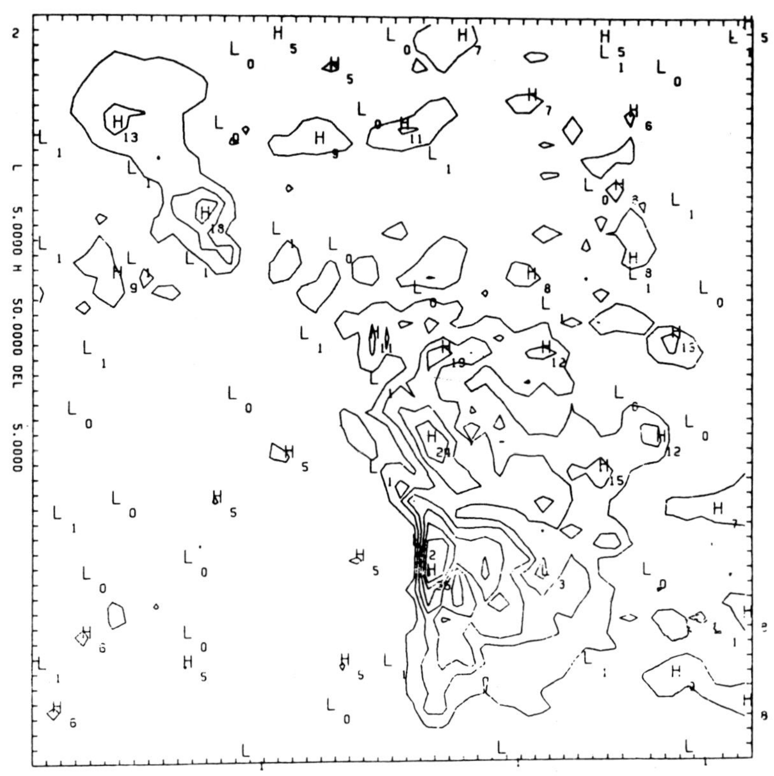

Fig. 11.6 GOES-derived diurnal Fourier amplitudes for RN_{EA} over South America and neighbouring regions, determined from 24 hourly observations on each of ten days in November 1978. (Source: Tanczer *et al.*, 1981.)

radiation budget studies so that short-term temporal detail also may be embroidered on our appreciation of radiation budgets and their component parts. It is becoming clear that some unsatisfactory assumptions have had to be made when polar-orbiting satellite data have been used to evaluate such terms. Research with GOES and Meteosat image data has suggested that:

1. outgoing infrared flux, albedo and net radiation budgets show considerable diurnal variation (Fig. 11.6);
2. the variations differ according to the underlying surfaces and/or types of cloud;
3. infrared flux tends to be fairly constant over sea surfaces and low clouds, but alters over land, especially over deserts, rising from a trough in early morning to a peak in early afternoon;
4. the albedos of deserts, high and low clouds change little, but show a marked diurnal cycle over sea surfaces.

Such variations are significant, not least because they imply that biasing errors exist in all radiation budget parameters evaluated from polar-orbiting satellite data. In time, it is hoped that correction factors might be determined from geostationary satellite data so that polar-orbiting satellite radiation budgets might be made more accurate and realistic. Research into comparisons between budget parameters assessed by the two types of meteorological satellites is continuing, including attempts to merge data series from different satellites and even different sensors. Such merged series will support much-improved time-series studies to reveal variations in radiation budget components through time.

The future is certain to reveal an increasingly intricate and kaleidoscopic view of atmospheric radiation and energy budget distributions and interrelationships. As atmospheric energy is the fuel for atmospheric motion, and this, in turn, drives many secular variations of climate, the place of such studies in the very forefront of satellite climatology is assured.

11.4 Global atmospheric moisture distributions

11.4.1 Water vapour

Water vapour is of key significance as it is the raw material from which clouds and precipitation are formed. It is important, too, in the absorption of both short- and longwave radiant energy in transit through the atmosphere, and in the transport of latent heat in both vertical and horizontal planes. Unfortunately, infrared radiation to space within the 5.7–6.9 μm water vapour absorption waveband is not easy to analyse. First, this energy emanates mostly from water vapour rather than clouds, but is emitted partly from within upper layers of clouds when clouds are present. Therefore both the cloud cover in each column of the atmosphere, and the temperatures of any cloud tops present, must be known if cloud contamination is to be removed from water vapour waveband radiances. Only then can these be interpreted accurately in terms of water vapour concentrations. Second, the water vapour concentrations themselves (especially in clear, i.e. cloud-free, columns of the atmosphere) are not easily assigned an altitude, although it is well known that most water vapour radiation to space is from the upper troposphere.

European Meteosat satellites pioneered water-vapour absorption imaging from geostationary altitudes, providing hourly information for approximately one-quarter of the global surface from 1978 to the present time. These data are now being appraised both meteorologically and climatologically. The TIROS Operational Vertical Sounder (TOVS) on the new Tiros-N third generation operational polar orbiters provides a three-level profile of atmospheric water vapour, and is already being used to improve routine global maps of precipitable water.

11.4.2 Clouds

Many satellite-based methods have been developed to provide mean cloud maps or other forms of time-averaged cloud displays. In the first decade of weather satellite operations the most popular techniques were based on manual nephanalyses (p. 194), deploying sets of weighting factors through which the nephanalysis categories of cloud cover (commonly C, MCO, MOP, O and Clear) might be translated into acceptable cloud percentages. Such techniques differed mostly in their sampling procedures. Some cloud climatologies were based on nephanalysis information at selected intersects of latitude and longitude, completing them by interpolating 'isonephs'. Others were based on estimates of the mean cloud cover in each grid square of given size, presenting their information either in *chloropleth* (area shaded) or *isopleth* forms. Such approaches yielded some very valuable and instructive results, and may still prove useful in low-cost working environments.

However, any maps prepared manually are subject to errors arising from the need for subjective judgements to be made by their compilers. It has become clear that nephanalysis-based mean cloud maps tend to overestimate cloud cover at the upper end of the range, and to underestimate it at the lower end. This is because the satellite imaging systems, and the dependent nephanalyses, may eliminate small breaks in cloud fields where the proportion of cloud-covered sky is high, and fail to resolve and represent small cloud units where the sky is mostly clear. Despite these *beam-filling* problems the patterns revealed by earlier hand analyses of mean cloud cover were of great interest and value as new climatological features (such as the split ITCB, for example) were first revealed by them. Indeed on the credit side, subjective cloud maps may be contaminated less by background effects caused by Earth surface phenomena than can be the case with automated ('objective') procedures, as we saw in Chapter 9.

Perhaps the earliest objective technique for mean cloud mapping achieved some success in the late 1960s. This involved the production of 'multiple exposure average' pictures for a selected major region of the world for a chosen period of time. A photographic plate was exposed to each of a number of daily computer-rectified, full-resolution picture products (4096^2 picture points) in turn. Multiple exposure averages, whilst being interesting climatological statements, suffered from two inherent characteristics: first, they did not differentiate between brightness due all or mainly to highly reflective Earth surfaces (e.g. ice, snow and deserts) as distinct from that due to clouds; and second, they were qualitative, not quantitative, products. More recently, efforts have been made to develop techniques that eliminate such problems.

We mentioned in Chapter 9 that local brightness values, derived from visible imagery from polar-orbiting meteorological satellites have been composited to give daily cloud pictures of the sunlit Earth (p. 162). By the early 1970s the resulting 'mesoscale archive' had become the basis for the most comprehensive and successful approach to the mapping of relative cloud cover on a climatological time-scale that had yet been achieved. Indeed, it is a great pity that this excellent scheme has been discontinued: as a basis for climate change studies it would have been very useful indeed. The mesoscale matrix was comprised of 512^2 unit areas for each hemisphere compared with the 4096^2 points in the full resolution matrix. A grid square on the mesoscale matrix was one sixty-fourth the area of an American Numerical Weather Prediction/Global Weather Center grid square. Using a set of empirically derived weights, the daily relative cloud cover was estimated in octas (eighths) for each mesoscale area. By saving the daily values by the month, for the entire period of the record, these values were grouped in a ten-class frequency distribution including 0–8 octas of cloud cover, plus one class (9 'octas') for missing data. A range of mean ('relative') cloud cover maps was prepared for 1967–1970, as illustrated by Fig. 11.7(a–d).

Comparisons between picture brightness values (converted to 'relative cloud cover') and conventional observations were subsequently made by staff at the USAF Environmental Technical Applications Center. These studies revealed some differences between point observations of total cloud amount made at the surface, and the satellite cloud cover estimates: the satellite and surface observations were similar in pattern, but their point-to-point values often differed. Such differences between satellite and surface observations can be ascribed primarily to differences between the field of view of the surface observer and the angle of view from the satellite. To some extent the lower response threshold and resolving capability of the satellite sensor also may be important. Further, 'background brightness' affected most notably by strongly reflective surfaces, such as ice, sand and snow, contaminates 'relative cloud cover' maps unless specifically removed, e.g. by filtering out static areas of bright pixels. In this way it is possible to remove persistent brightness associated with the surface of Earth, leaving the more mobile brightness caused by clouds. This is already achieved in more specialized studies, for example, ice margin mapping.

During the 1980s new objective cloud climatology procedures have been developed – the need for automatic approaches being even greater for climatological than meteorological applications because of the much greater volumes of data to be analysed in the former. One new approach, which has much to recommend it because of the additional information it provides over and beyond mean cloud cover and cloud-top temperatures (both of which can be derived from infrared images), is based on sounding (profiler) data. These provide information on mean cloud-top pressures also, as shown by Plate 7 (colour section). Features in these images may be compared usefully with those in Figs 9.12 and 9.14 although the effects of the different seasons involved should be borne in mind.

11.4.3 Rainfall

The possibility of mapping rainfall (especially areas and rates of rainfall) from satellite data has received ever-increasing attention during the 1980s, for rainfall is a key environmental parameter poorly monitored by *in situ* observation systems. Although it is not recorded very directly by satellite observations, its importance has led to the development of a range of techniques for the evaluation of rainfall from satellite data, especially satellite imagery. Several aspects of rainfall hydrometeorology are now amenable to improved analysis using data from satellites:

1. mapping the boundaries of areas likely to be affected by rain;
2. mapping totals of rainfall accumulated through unit periods of time;
3. assessing extreme (intense) rainfall events;
4. assessing the climatology of rainfall distributions;
5. forecasting rainfall, especially in areas open to systems from relatively poorly observed regions.

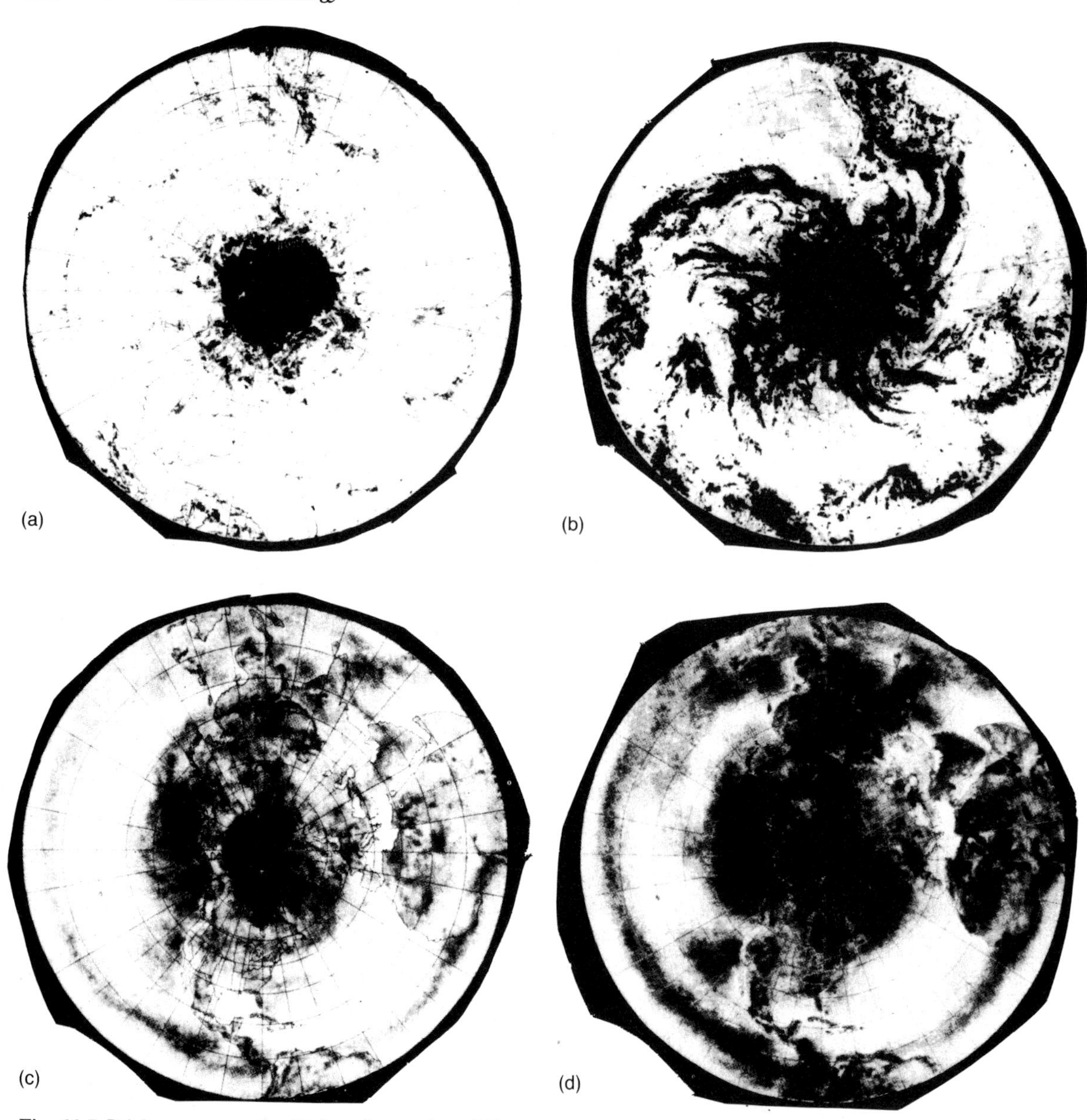

Fig. 11.7 Brightness composite displays from polar-orbiting weather satellite photography. (a) Five-day minimum brightness composite. (b) Five-day maximum brightness composite. (c) Thirty-day average brightness composite. (d) Ninety-day average brightness composite. (Courtesy, NOAA.)

During the first decade of satellite meteorology the satellite was viewed generally as an *alternative* to the ground observing station for local environmental information. However, during the second and third decades, the two have been viewed more as *complements* to each other. Today the general-purpose satellite rainfall monitoring methods best-suited to operational use are those which *integrate* evidence of rainfall from surface stations and satellite data, to give an analysis superior to those that could have been prepared from either type of information alone. The ground station (e.g. raingauge, streamgauge, etc.) has the advantage of being able to provide quantitative data *aggregated through time*, but each represents rainfall variations only at a single location; the satellite has the

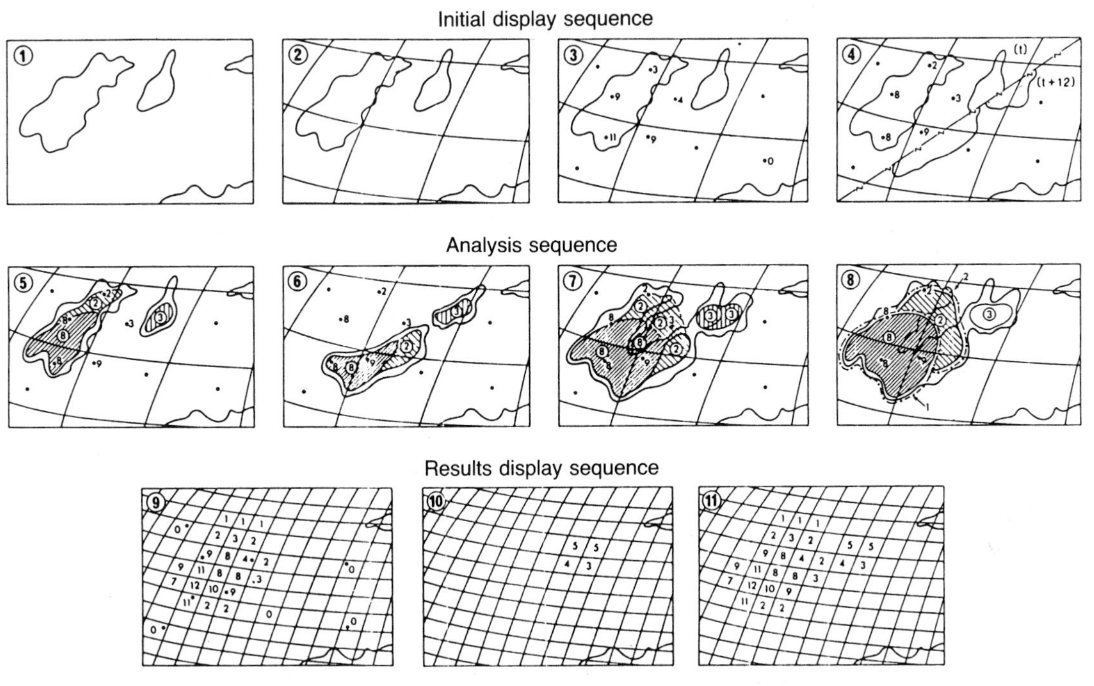

Fig. 11.8 Stages in BIAS, the Bristol/NOAA Interactive scheme. (Source: Barrett, 1986.)

advantage of being able to provide *areally complete information*, though only for separated points in time.

All satellite-based rainfall algorithms relate more or less indirectly to rainfall as measured on the ground. Therefore, the practice has grown to use satellite data to fill gaps in the conventional data network, calibrating the remote sensing data by reference to available *in situ* observations. Methods of improved rainfall monitoring by satellite are based on visible, and/or infrared and/or microwave image data, and include the following:

1. Cloud indexing methods. Satellite visible and/or infrared cloud images are ascribed indices relating to cloud cover, and the probability and intensity of associated rain. Different methods have been used to calibrate the indices to give final rainfall estimates. In its most developed form the cloud indexing approach has become a virtual extension of classical synoptic meteorology, involving expert interactive combinations of remotely sensed and *in situ* data. (Fig. 11.8).
2. Cloud climatology methods. The basis of such methods is some general relationship between the satellite imagery and rainfall measured on the ground, e.g. the frequency of occurrence of cold cloud tops regressed against regionally averaged raingauge data, or numbers of satellite-identified 'rain-cloud days' translated into rainfall estimates through pixel-resolution fields of climatic mean rainfall per rain-day statistics. Climatology thus dictates the amounts of rain deemed likely to be carried by the satellite-identified 'probably precipitating' cloud systems. Use of such methods is now widespread for periods of ten days and upwards and, although often quite simple, they perform well over longer time periods or large areas on the ground.
3. Life-history methods. These are based on the three premises that significant precipitation falls mainly from convective clouds, that these clouds can be both distinguished from others in satellite images, and their individual developments can be tracked through time. The estimation of rainfall then rests on suitable calibrations of cloud performances and their modelled changes as the life-cycles of the clouds unfold. These methods are for use in convectional regimes,

and may be implemented either objectively (by computer) or subjectively (manually), the latter being used especially for severe high-intensity rainfall situations (Chapter 18).

4. Bispectral and multispectral methods. Here images from different regions of the spectrum are analysed objectively together to (more confidently) identify probably precipitating clouds than might be possible using a single region of the spectrum. As with other methods, the results must then be scaled according to some conventional (gauge and/or radar) data. One of the commonest assumptions in such methods is that heaviest precipitation falls from clouds that are both *bright* and *cold*. An operational difficulty with techniques based on this assumption is that visible imagery is not available at night. Future methods may be expected to combine not visible and infrared data, but infrared and passive microwave.
5. Cloud model methods. These are more physically based, whether utilizing infrared and/or microwave image data as the basic inputs, and are essentially research techniques being developed in man–machine modes in the search for more elegant formulations of relationships between clouds and rainfall. At present these methods are typified by a high degree of rigour but also by very complex algorithms. Their performances do not yet exceed those of the more empirical approaches, although the researchers expect that they should do so eventually.
6. Passive microwave methods. Radiometers on recent Nimbus and present USA military satellites of the DMSP family have measured naturally emitted radiation from the Earth and its atmosphere, or related to the ice crystals and hydrometeors within the rain clouds themselves. Some data (e.g. from 19, 37 and 85 GHz) have been processed successfully to yield rain area and rain-rate products, especially over sea areas. One area in which passive microwave data are being used climatologically is over the North Sea (Fig. 11.9), where previously the distribution, variation, and even mean annual rainfall amounts were very poorly known. More problems are found over land where background radiation is stronger and more variable, but much progress has been made especially since the inauguration of NASA's DMSP-dependent 'WetNet Project' in 1989: research with multiband microwave data is continuing apace.
7. Active microwave methods. These are expected to become available in the future if tests of planned satellite-borne radar systems prove to be a success. The first of these may be the planned American/Japanese Tropical Rainfall Monitoring Mission (TRMM) satellite, expected to be in operation between 38°N and 38°S by the mid-1990s with the object of improving our knowledge of climatological rainfall patterns and processes in the tropics and subtropics.

It is the *cloud indexing* and *cloud climatology* types which have, in one form or another, been most widely used to date, shown most flexibility, and yielded most results in support of continuous operational rainfall monitoring programmes. For example, they have been used in support of irrigation design in Indonesia, water resource evaluation and management in Oman, desert locust control in north-west Africa, crop prediction in the Sahel, and general environmental assessment in tropical Africa and the Caribbean. Experience confirms that rainfall maps based on both conventional and satellite data are more realistic in spatial detail than maps based on gauge data alone (Chapter 18), and therefore reveal more accurately the areas of average, above average and below average precipitation. Given suitable arrangements for the acquisition of input data, e.g. from increasingly cost-effective satellite data reception facilities, satellite rainfall monitoring methods can be tailor-made to suit local needs and provide information in near real-time if required.

In the foreseeable future it is likely that passive microwave methods will become increasingly important, either by themselves at least for global oceanic rainfall climatologies, or combined with geostationary infrared data, and/or radar and/or raingauge data for meteorological uses. For example, the relatively physically direct but rather infrequent passive microwave data may be used to establish instantaneous rain areas and rain rates; subsequently these may be moved and/or developed using the less physically direct but much more frequent infrared data from geostationary

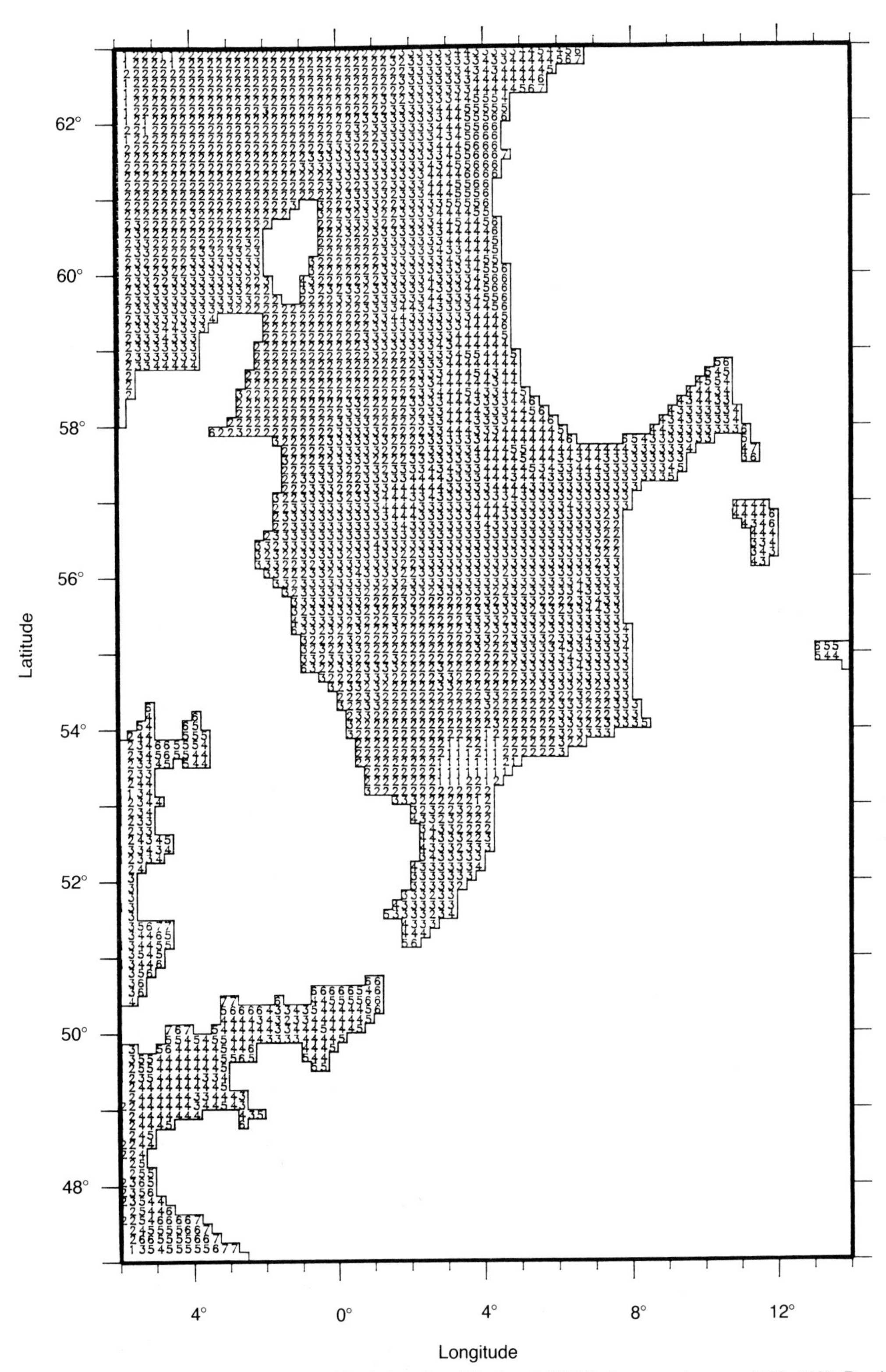

Fig. 11.9 Mean annual rainfall over the North Sea, from Nimbus-7 SMMR data over the years 1978–1987. Previously mean annual rainfall in this region could be estimated only from coastal station data, and spatial patterns were largely unknown.

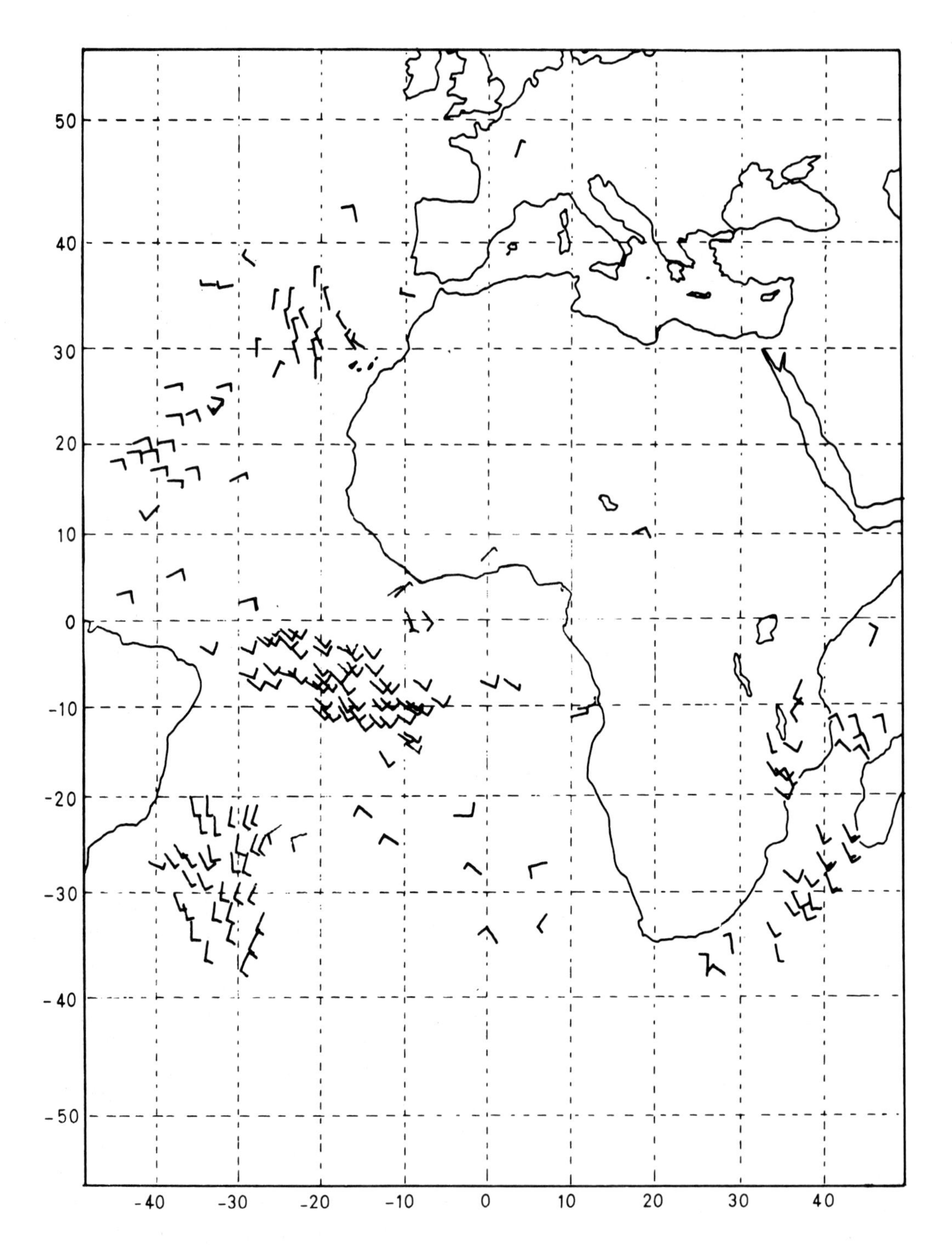

Fig. 11.10 'Meteosat Winds' for 2 July 1979 at 1200 GMT (700–1 000 mbar layer established by tracking cloud elements in successive images). Such maps are prepared routinely in the Meteorological Information Extraction Centre (MIEC) at ESOC from Meteosat data for the synoptic hours of 00 and 12 GMT, for high levels (700–400 mbar). (Copyright, ESA.)

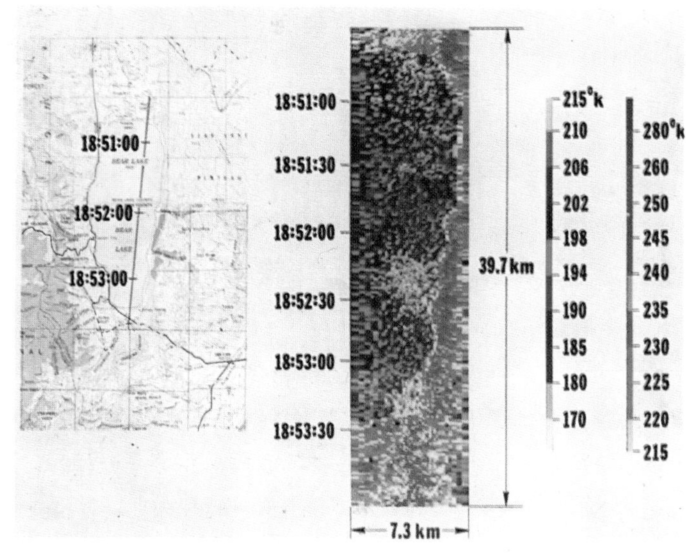

Plate 1 The flight path and microwave image (1.55 cm waveband) for 3 March 1971, Bear Lake, Utah/Idaho, from an altitude of 3 400 m. (Source: Schmugge *et al.*, 1973.)

Plate 2 Composite image made by computer from Bands 4, 5 and 7 of 43 Landsat scenes. (Courtesy, BNSC)

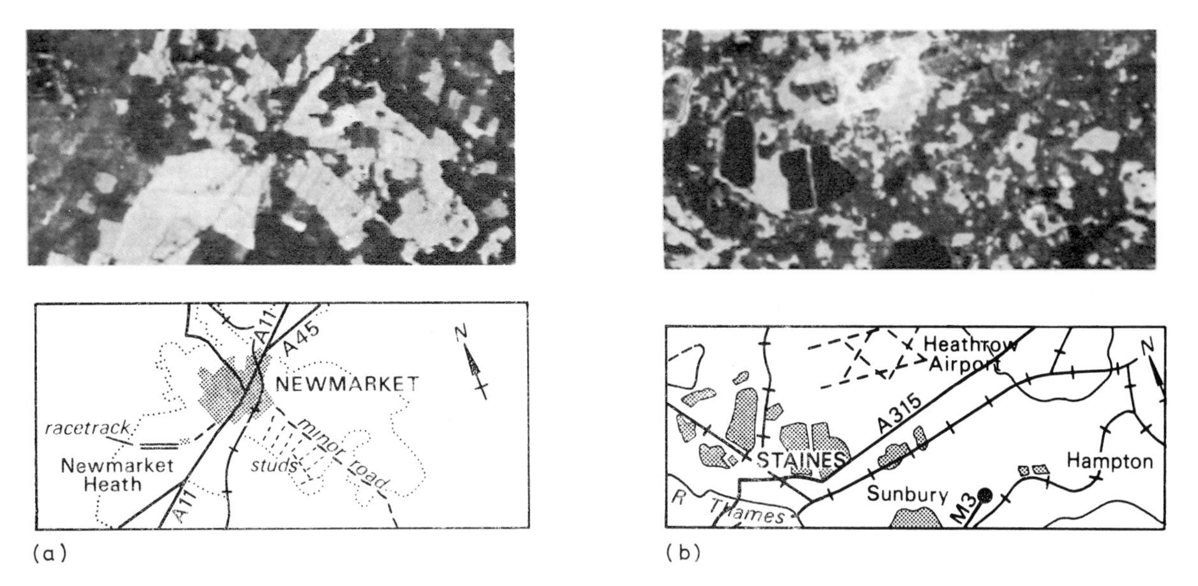

Plate 3 Examples of various land use categories obtained by computer processing of Landsat 1 data for the UK Department of Environment. The processing minimizes degradation of the picture detail. Scale of reproduction 1 : 250 000. See key maps for selected features. (a) Urban area of Newmarket. (b) Heathrow airport and adjoining areas. Rural areas are imaged in red and green, reservoirs in black, and construction works in white. (Source: Smith, 1975.)

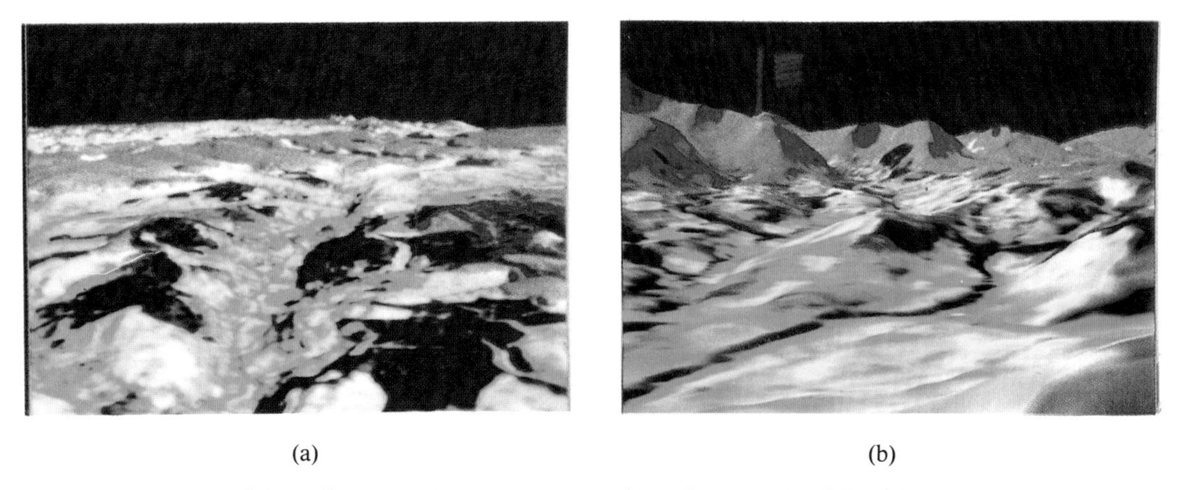

(a) (b)

Plate 4 Example of three-dimensional image sections of the South Wales DTM (North Brecon Beacons) overlaid by co-registered Landsat TM images: (a) view to the north-east along the Vale of Neath towards the Brecon Beacons; (b) view to the south-west from Brecon to the Brecon Beacons. (Source: Barrett *et al.*, 1991.)

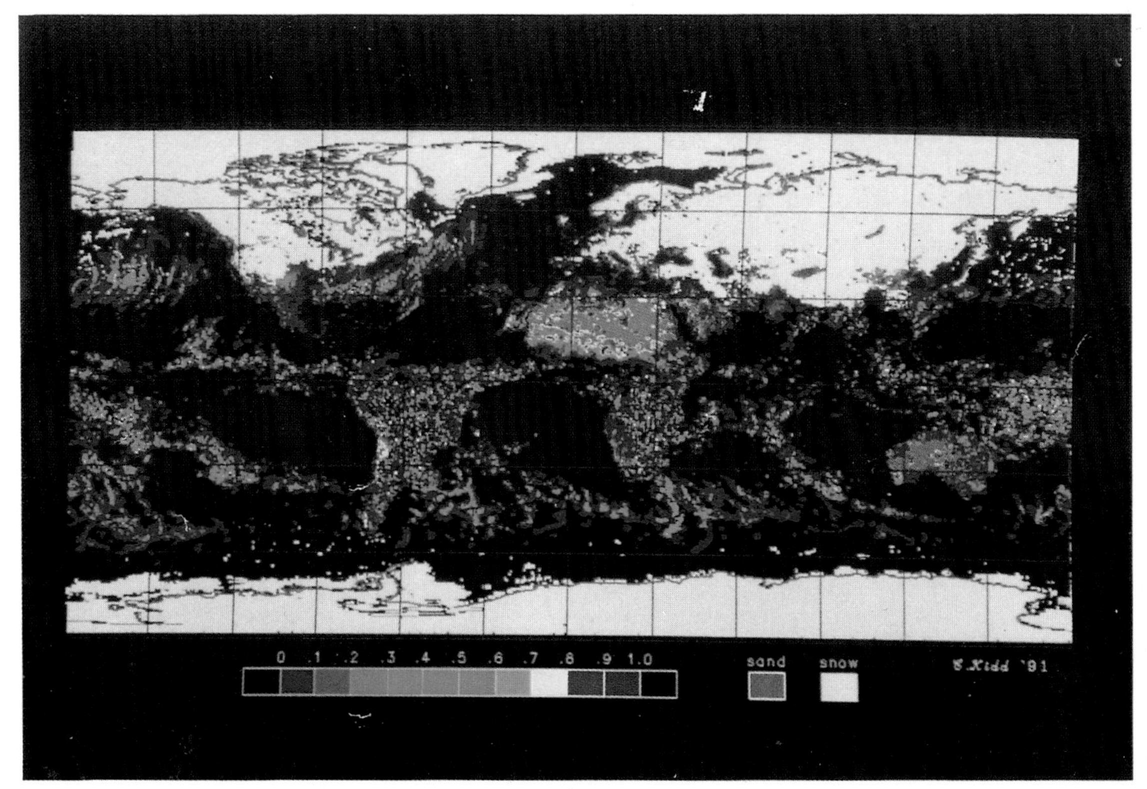

Plate 5 Global precipitation averages (mm h^{-1}), plus areas of snow and ice (white) and areas of desert (orange), for the ten-day period from 9 to 18 February 1988. Derived from DMSP-SSM/1 passive microwave data. See text for discussion. (Source: Barrett *et al.*, 1991.)

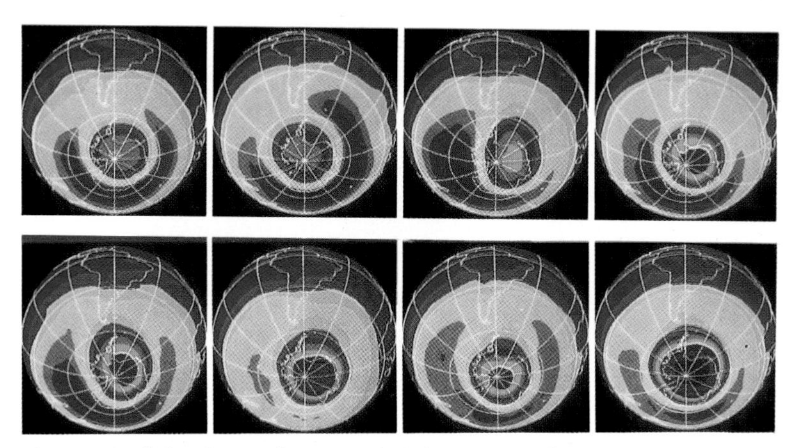

Plate 6 Development of the Antarctic 'ozone hole', annually from 1981–88, as seen by TOMS. (Courtesy, NASA).

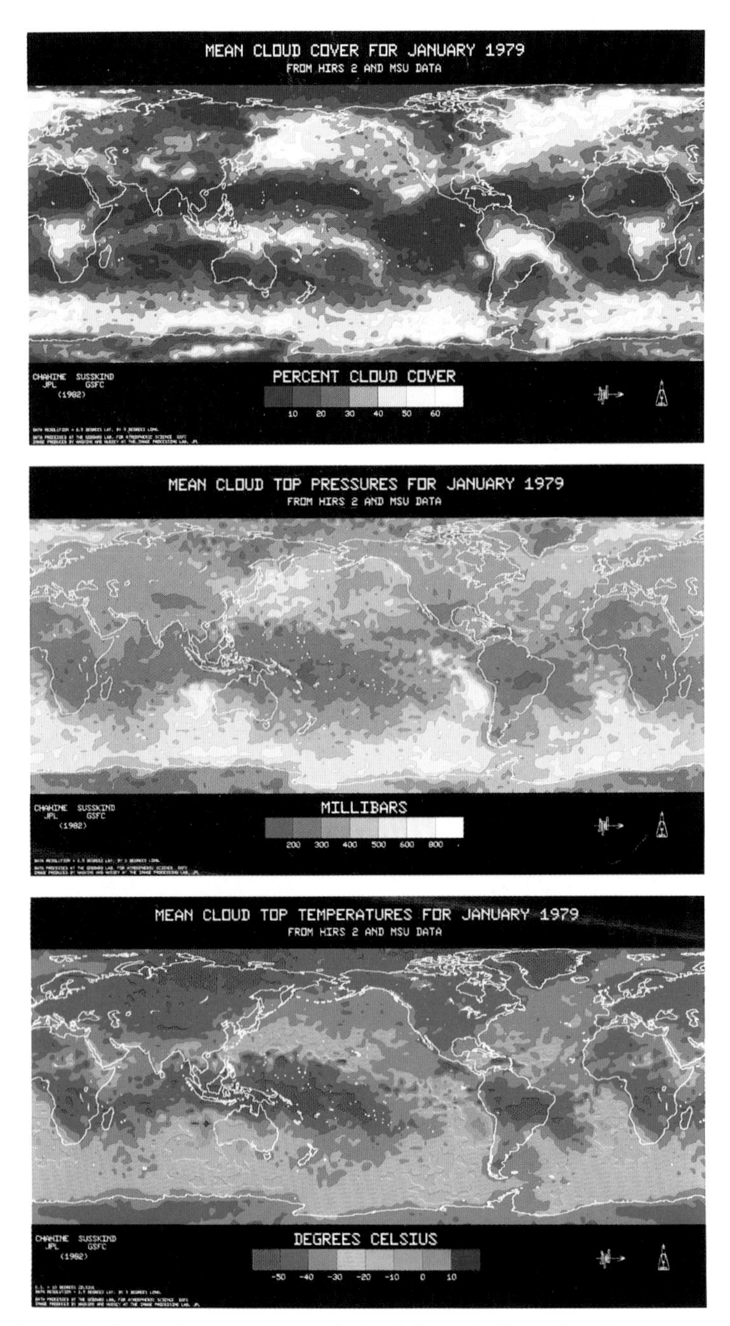

Plate 7 Global cloud climatology products derived from infra-red and microwave sounders on NOAA polar-orbiting satellites (see Section 11.3.2). (Courtesy J. Chahine, JPL Pasadena, CA).

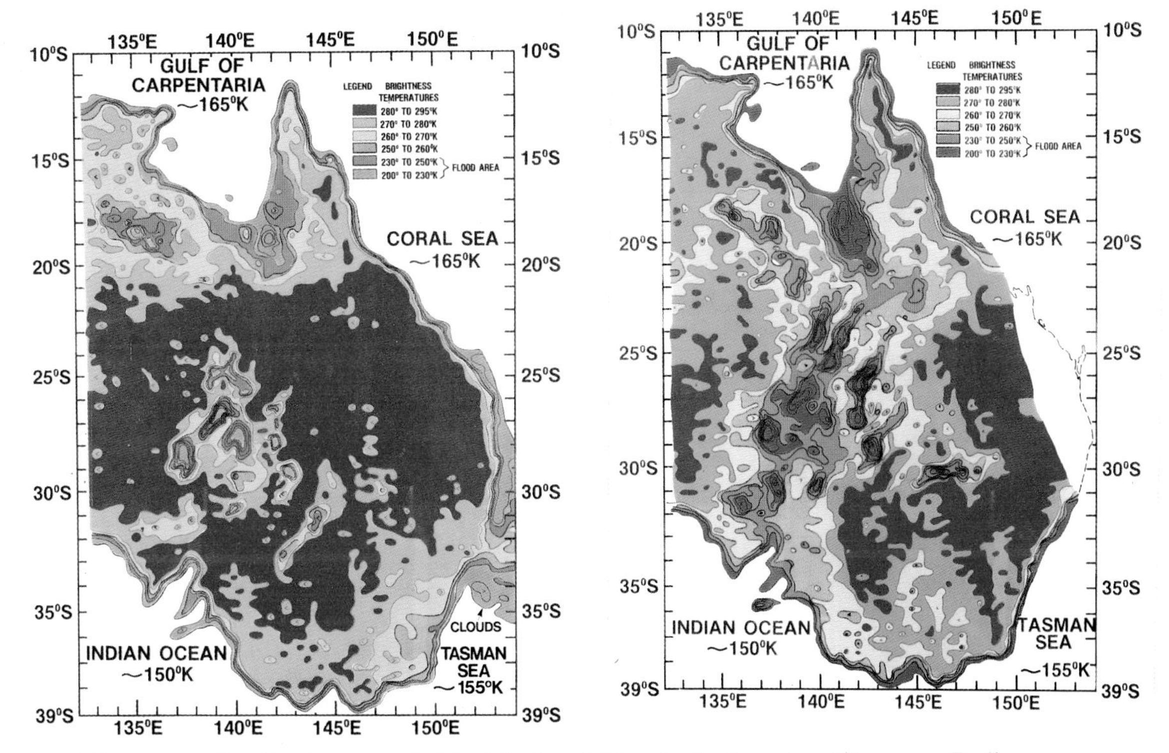

Plate 8 East Australian floods as revealed by Nimbus 5 Electrically Scanning Microwave Radiometer (19.35 GHz): (a) day, 0225–0240 GMT, 1 February 1974; (b) day, 0215–0230 GMT, 11 March 1974. Blue (250–230 K) and purple (230–200 K) surface brightness values indicate flooded surfaces. (Source: Allison and Schmugge, 1979.)

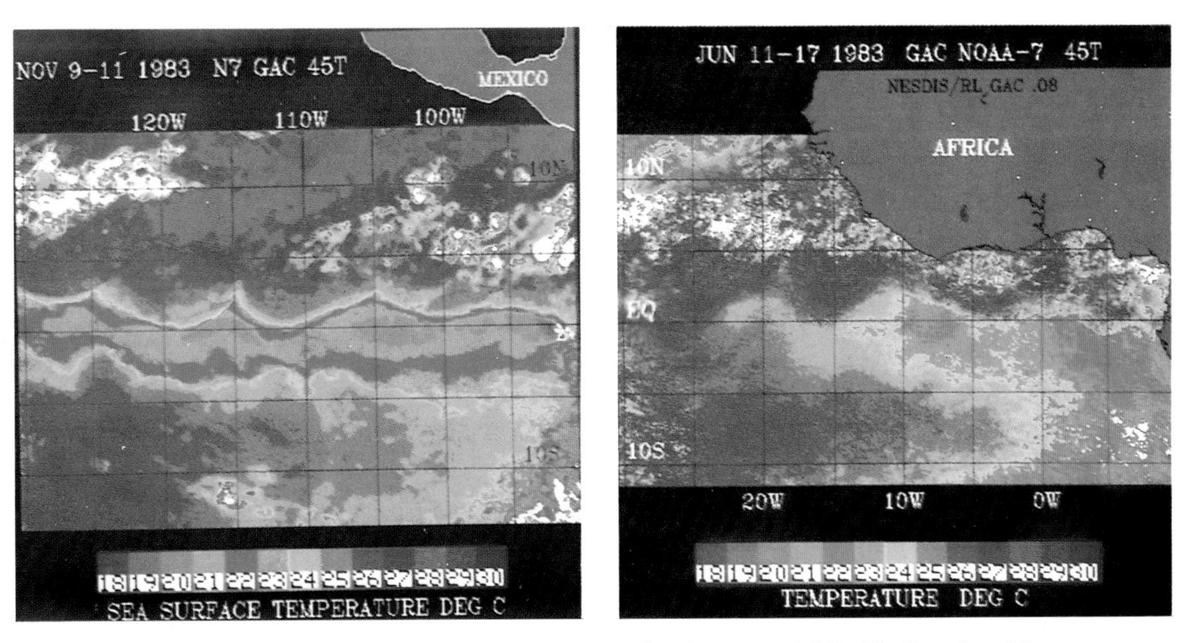

Plate 9 Colour-coded multiday maximum SST composites for the equatorial Pacific from 9 to 11 November 1983 (left), and the equatorial Atlantic from 11 to 17 June 1983 (right), from NOAA-7 AVHRR infrared data sets. These show westward moving equatorial oceanic longwaves discovered by satellite data analyses. (Source: Legeckis, 1986).

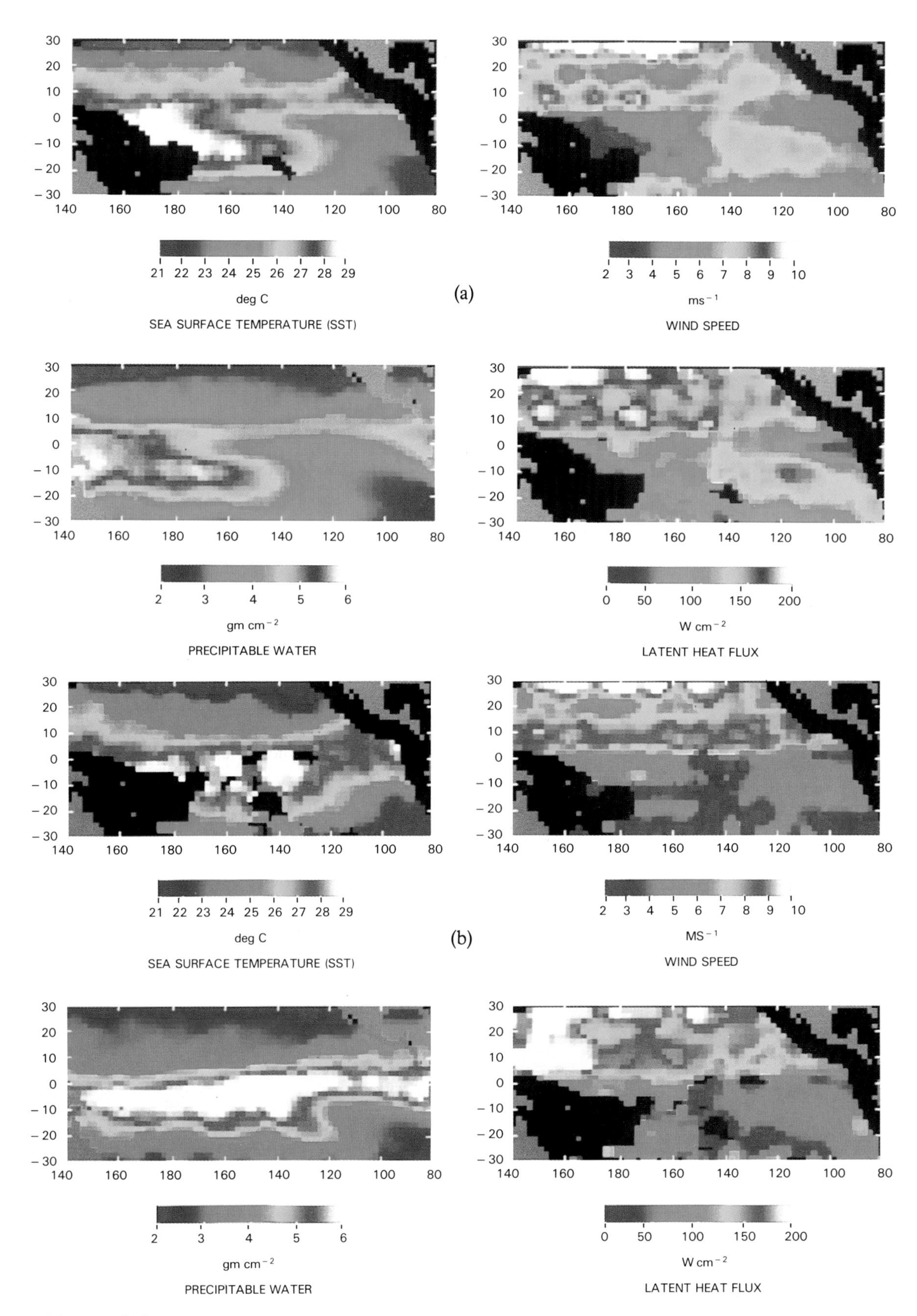

Plate 10 SMMR-derived products for (a) non-El Niño Januaries of 1980 and 1981 (averaged), and (b) the extreme El Niño January of 1983, showing four interrelated parameters. (Courtesy, NASA.)

Plate 11 Infrared colour image of part of Exmoor National Park, England. Dark brown and greenish grey colours show moorland of heather, bilberry, gorse and grasses. Red tones show reclaimed fields of pasture. Pale grey fields have been cut for hay. Deciduous trees appear pink in tone. (Source: Exmoor National Park.)

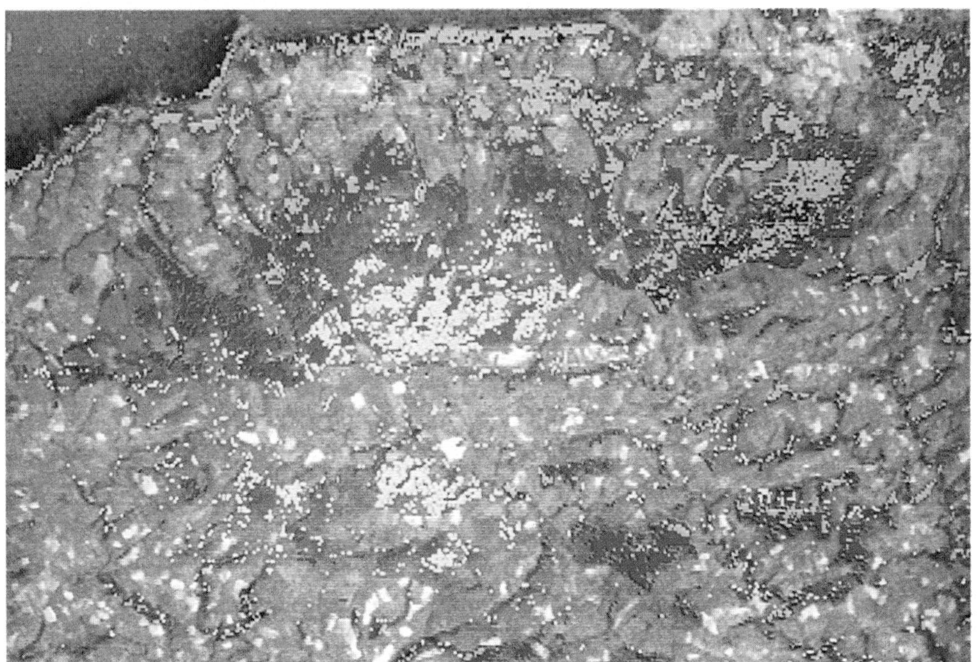

Plate 12 Colour image from display screen of IDP 3000 processor showing Exmoor National Park as analysed using supervised classificatory systems. Yellow, grass moorland; purple, heather moorland; green, woodland; red, farmland. (Source: Exmoor National Park.)

Plate 13 (a) This thermal picture shows the heat loss from a large industrial complex. (b) For comparison, this shows a conventional aerial black and white photograph of the same site. The colour image depicts the variation in grey tones on the black and white print. Each colour represents a discrete temperature banding, which varies between sites, from grey and light blue (the coolest) to purple and white (the warmest).

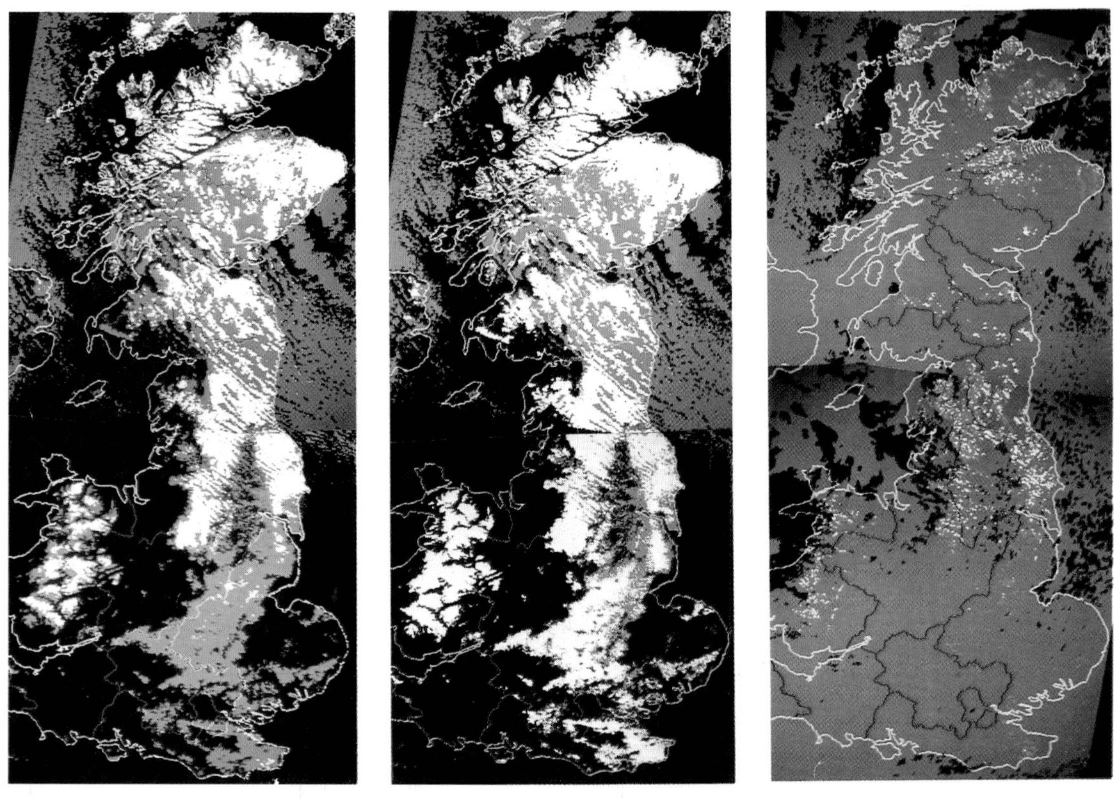

Plate 14 Snow area estimates for 27 February 1986 showing (a) partial snow cover (blue), complete snow cover (white) and cloud (grey); (b) surface temperature below 0 °C (white) and above or equal to 0 °C (blue); and (c) zones of melting snow (blue) and accumulating or stable snow cover (yellow). (Source: RSV, University of Bristol)

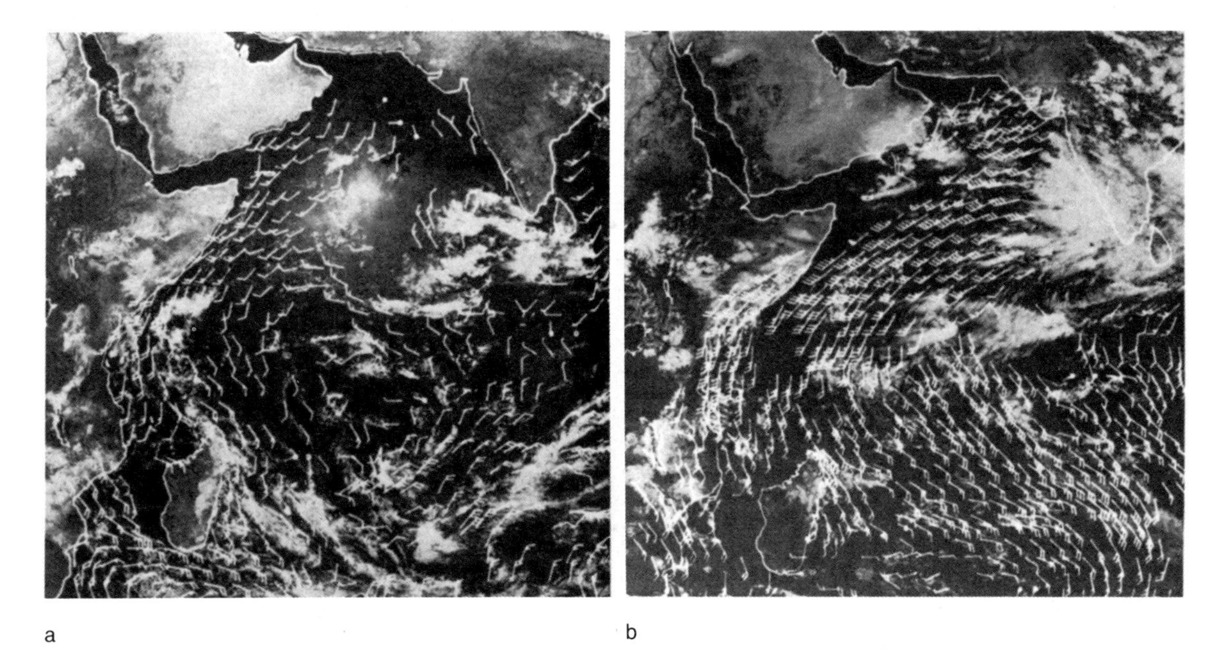

Fig. 11.11 Low-level wind fields derived from GOES-IO for (a) 17 May and (b) 21 June 1979, prepared in support of FGGE investigations. (Source: Desbois, in Tanczer *et al.*, 1981.)

satellites, until the next passive microwave images become available. Ideally, passive microwave imagery is needed from geostationary platforms. This, however, will not become available for many years because the weak passive microwave signals present difficult problems in sensor/platform design and engineering. More immediately, the chief aims must be to develop techniques which can be applied with equal confidence to all kinds of rain systems, and everywhere across the globe using presently available data: despite much effort, even these have so far proved to be elusive goals.

11.5 Wind flows and air circulations

Several techniques have been developed for assessing the speed and direction of wind flow from satellite evidence, particularly from geostationary satellite infrared or water vapour channel imagery, although to date these have been utilized more in meteorological than climatological contexts. For example, estimates of maximum wind speeds in hurricanes are made routinely by the US Weather Bureau (p. 377). Cross-correlation techniques are used by major meteorological satellite data processing centres (e.g. the European Space Operations Centre at Darmstadt, Germany) to identify geostationary satellite-imaged cloud elements at selected levels in the troposphere and evaluate their direction and rate of movement from successive geostationary cloud images (Fig. 11.10). However, interest is now being shown in these products by climatologists. Their value for air flow analyses is greatest in the tropics (Fig. 11.11), e.g. in relation to the major monsoonal systems, as such data can be processed to yield a variety of wind and circulation parameters, including wind speed and direction, relative and absolute vorticity, and horizontal divergence, especially when combined with other data (e.g. satellite cloud images themselves). Daily wind charts for tropical regions are now improved regularly through satellite inputs. Since mean monthly, seasonal and annual maps of isotachs and streamlines are prepared from such base data, satellite winds are already contributing implicitly to wind and circulation climatologies,

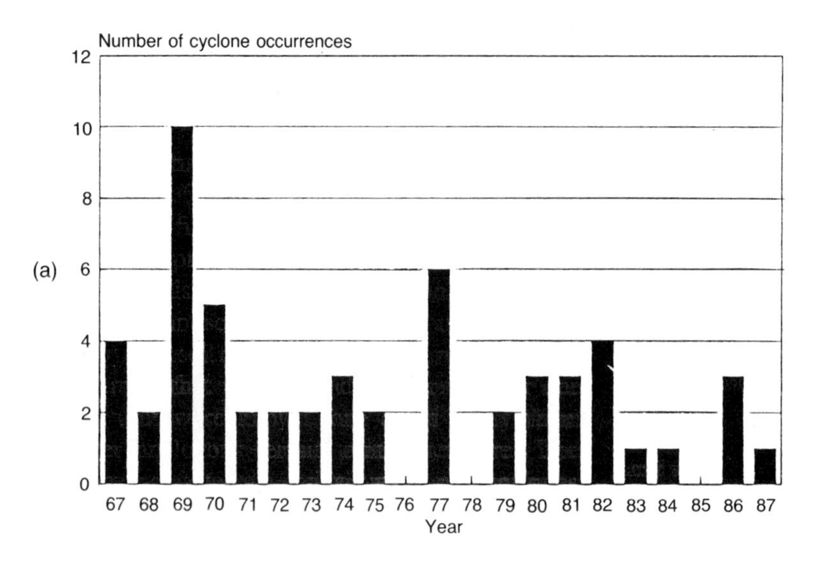

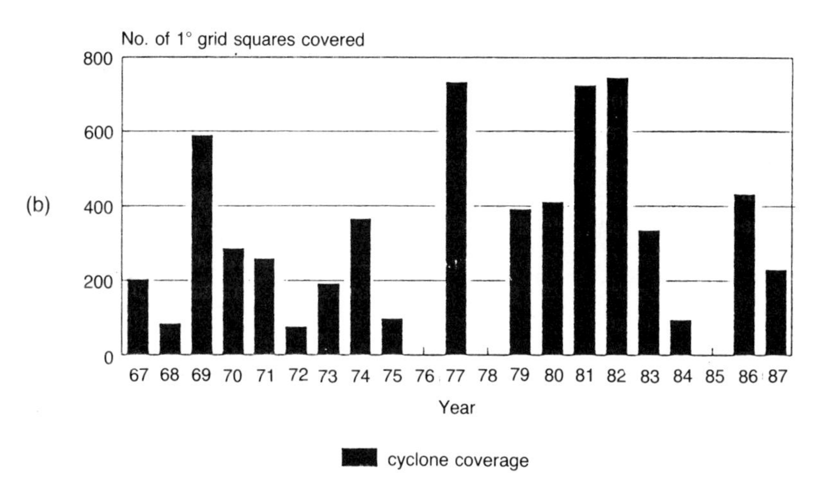

NB. coverage at least once in the year (1978 satellite data missing for whole year; 1976 & 1985 no cyclones occurred)

Fig. 11.12 NOAA-based climatological analyses of tropical cyclones in the Arabian Sea, 1967–1987: (a) number of storms; (b) canopy areas.

especially in some of the more conventional data-sparse regions of the world.

11.6 The climatology of synoptic weather systems

Remembering that it is possible to identify many *atmospheric weather systems* in weather satellite images, especially through their attendant cloud and radiation temperature fields, it is not surprising that many new climatological facts have emerged from studies of their time frequencies and spatial distributions through extended periods. Understandably the most interesting of such studies have been carried out for the more inaccessible regions of the world, although worthwhile results have

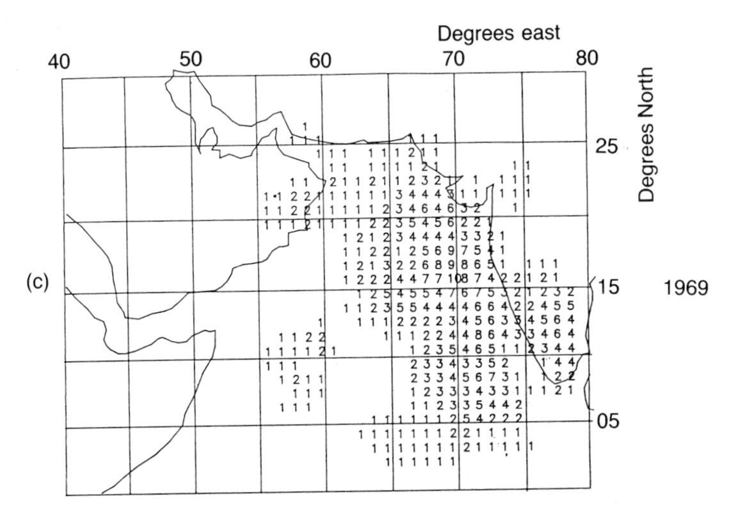

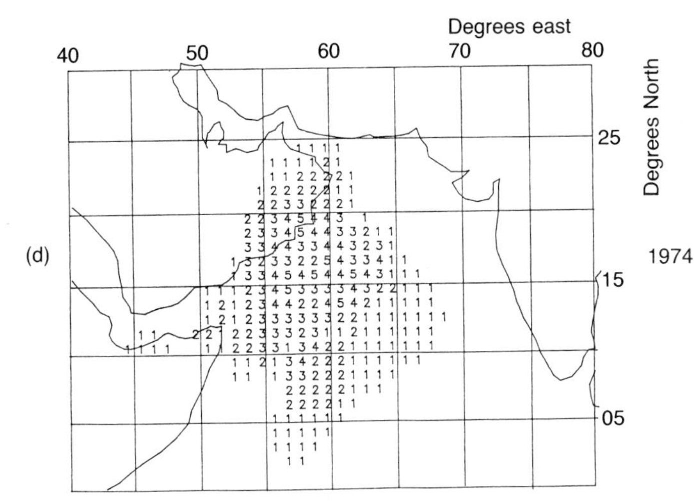

Fig. 11.12 cont. (c and d) Canopy frequency of occurrence maps for two strongly contrasting years, namely 1969 (c) and 1974 (d). (Source: Barrett and Power, 1990.)

emerged for some better known land areas, and in the realms of jet-stream climatology also.

As mentioned in Chapter 10, particular light has been thrown upon the tropical climatology, e.g. through the structure of the Inter-Tropical Cloud Band (ITCB), a prominent feature on visible and infrared images of the tropics. The position, range of forms, and secular shifts of the ITCB have been elucidated around the globe, and regional and seasonal contrasts described. Lateral to the ITCB, wave-like cloud features, associated with perturbations in pressure and wind fields, have been identified not only in the tropical North Atlantic (where these 'easterly waves' were first described) but in most if not all of the other tropical ocean regions too. The climatology of tropical revolving storms (especially hurricanes) has been much improved (Fig. 11.12 and Chapter 18), as well as that of mid-latitude vortices and frontal bands, especially in the Southern Hemisphere (Fig. 11.13). There is now growing interest in satellite evidence of global change through the longer runs of satellite data now becoming available virtually for all levels of the atmosphere, and the world as a whole. Interest is growing in respect of possible relations among them, e.g. in terms of strengths and frequencies of synoptic weather systems ('atmospheric teleconnections'), and possible applications of such knowledge and understanding in climate prediction. These types of studies have begun to assume much greater significance with the satellite climatological con-

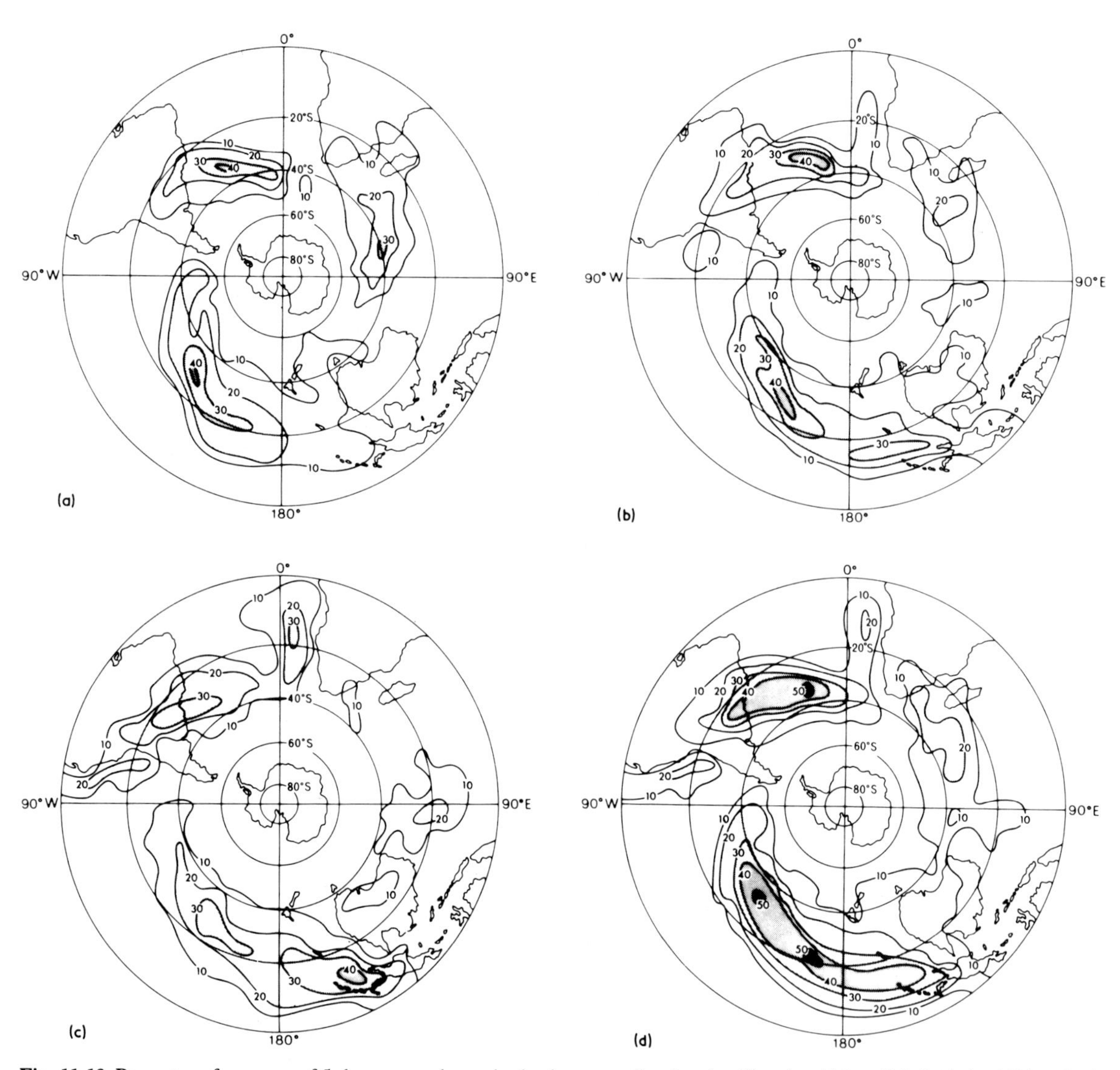

Fig. 11.13 Percentage frequency of 5-day averaged mosaics having axes of major cloud bands within a 5° latitude by 10° longitude square for (a) summer (December–March), (b) intermediate season (April, May, October, November), (c) winter (June–September), and (d) annual. Data based on November 1968 to October 1971. (Source: Streten and Troup, 1973.)

firmation of the polar 'ozone holes', and with the rise of interest in possible global warming.

11.7 The climatology of the middle and upper atmosphere

Many modern authorities believe that surface weather and climate is greatly influenced, or even largely controlled, by situations and events in the upper layers of the atmosphere. Thus *vertical interactions* are being sought with increasing vigour. Since few radiosonde balloons penetrate far into the stratosphere, pre-satellite studies of the atmosphere above the tropopause depended heavily on expensive research rocket campaigns. From these, valuable vertical profile data were obtained, but understandably their spatial coverage was very poor. Today we have a much clearer picture of the middle and upper atmosphere. Sensor systems, such as the CO_2-absorption waveband Selective Chopper Radiometers (SCRs, Chapter 3) of the Nimbus family, provided data which have formed

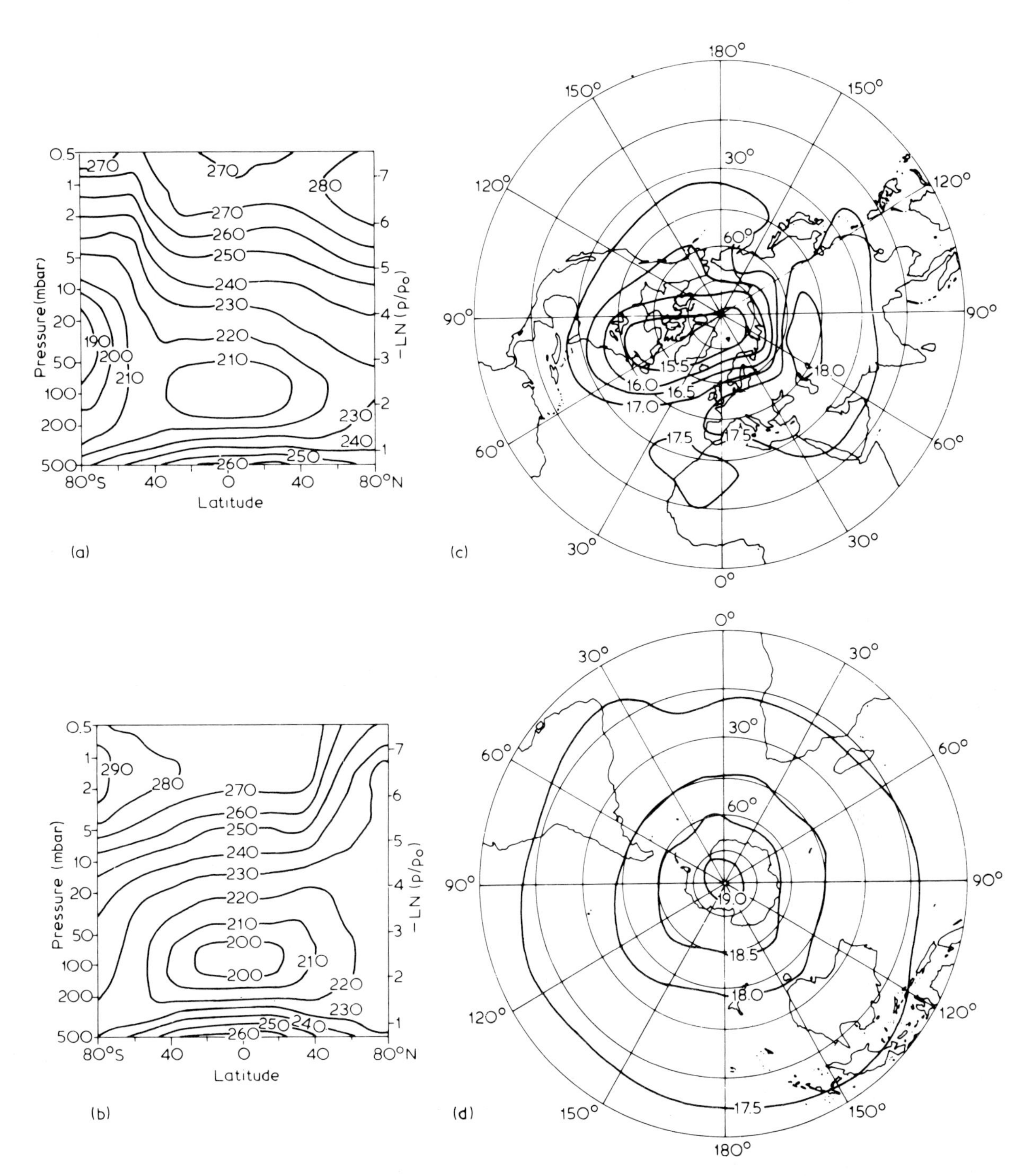

Fig. 11.14 Examples of meridional cross-sections of zonally averaged temperatures (isopleths give temperatures in K): (a) 16 July 1970; (b) 21 January 1971, obtained from Selective Chopper Radiometer (SCR) on Nimbus 4, and thicknesses of the 10–1 mbar layer at pressure surfaces from the same source; (c) Northern Hemisphere; (d) Southern Hemisphere, both on 4 July 1970. (Source: Barnett *et al.*, 1973.)

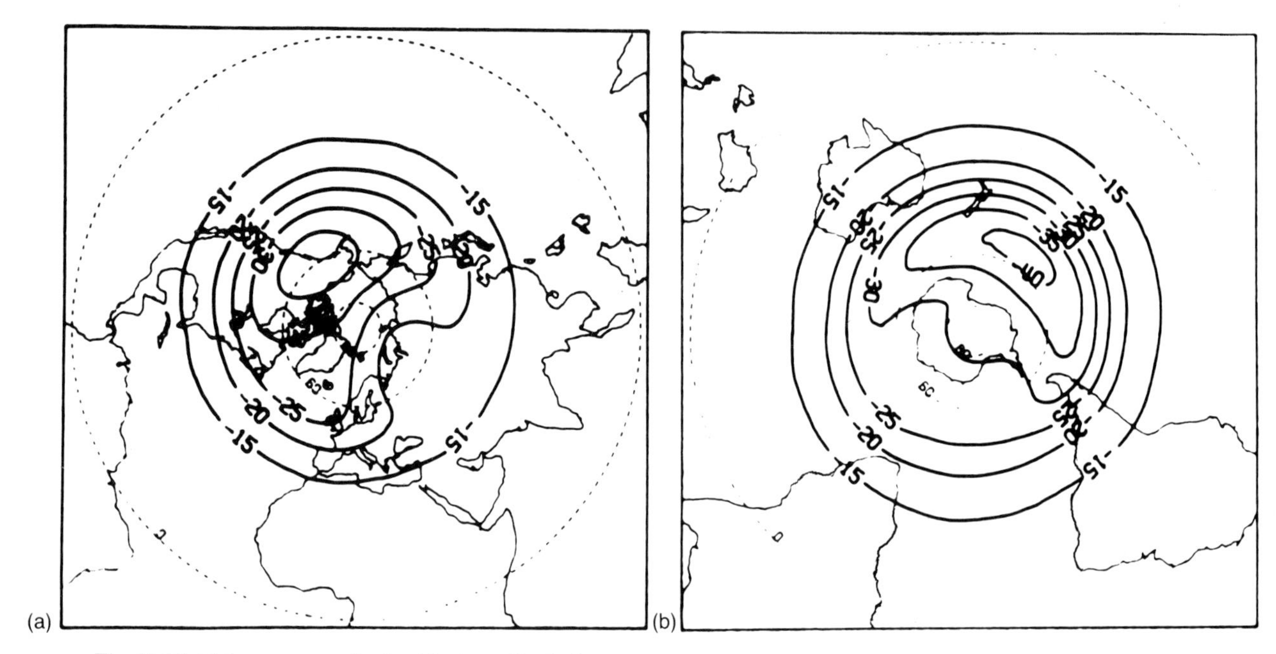

Fig. 11.15 (a) January mean 2-mbar Northern Hemisphere temperature, 1979–1986. (b) July mean 2-mbar Southern Hemisphere temperature, 1978–1985. Both are constructed from data gathered by NOAA operational sounders. (Source: Rao *et al.*, 1990.)

the bases for historic series of mean meridional profiles (Fig. 11.14(a and b)) and maps of selected layers high above Earth's surface (Fig. 11.14(c and d)). The availability now of multiyear data sets is helping to establish the mean circulation patterns at very high altitudes (Fig. 11.15) – and is revealing that the individual months exhibit a great deal of variability.

Present high-altitude sounders (e.g. the MSU and SSU of NOAA satellites, as described in Chapter 10) will help resolve some of the outstanding problems in upper atmosphere research. Prominent among these are key questions concerning the directions and degrees of vertical interactions: e.g. does weather and climate develop principally from the top down, or the bottom up? The answer to this may lie in the middle atmosphere (the upper stratosphere and lower mesosphere). Other key questions involve the aerosol (small particle) contents of the atmosphere, and how these may be changing – for it is known that, like clouds, these can cause substantial warming or cooling of the Earth surface and atmosphere. Indeed, aerosol effects in total are comparable in magnitude with those of atmospheric CO_2. Satellite estimates of both cloud and aerosol distributions in space and time will help improve the modelling of their effects on climate variations, and in predicting future changes of climate.

One thing is clear: through the eyes of a weather satellite system we are able to study the atmosphere much more comprehensively than before, either as a single, complex fluid continuum, or as a sum of many interrelated and more or less interdependent parts. The 1970s and 1980s were particularly exciting years for the meteorologist; perhaps the 1990s will be a golden age for climatology from satellites.

12 Water in the environment

12.1 The importance of water

Water is the most ubiquitous yet most variable of the mineral resources of the world. Alone among the constituents of our environment water may be found simultaneously in one area as a liquid, a gas, and a solid. Unlike most other Earth resources, water (on land and in the atmosphere) is continuously variable in its availability in one state or another. A further complication is that, geographically speaking, its patterns of concentration are always changing.

Apart from the purely scientific interest that water arouses, we have a keen practical interest in its behaviour and distribution on account of its significance as a basic requirement of living organisms, and the consequent dependence of we ourselves and our economies upon it.

For these reasons – both academic and intensely applied – the satisfactory monitoring of environmental water is one of the most difficult yet pressing problems confronting remote sensing as a distinctive discipline today.

Study of water in the environment is the preoccupation of hydrology, which has been defined by the WMO as 'The science which deals with the occurrence and distribution of waters of the Earth, including their physical and chemical properties, and their interaction with the environment'. This modern science, along with its natural bed-fellow water management, is growing rapidly in importance because of ever-increasing demands for adequate and suitable supplies of water for human consumption, domestic use, agriculture and livestock husbandry, even for commerce and transportation. For convenience, the areas of chief concern to hydrology and water management can be subdivided into:

1. hydrometeorology, which is concerned primarily with the exchange processes in hydrological cycles;
2. surface hydrology, which focuses on water at or near the surface of the land;
3. hydrogeology, which is concerned with water at or below the surface of rock and regolith;
4. oceanography, which deals with the nature and behaviour of the world's largest water bodies, although many would argue that this is a distinctive, even a separate, science in its own right.

We will review the use of remote sensing in each of these fields in turn.

12.2 Hydrometeorology

The hydrological cycle of the Earth–atmosphere system can be represented in simplified form as shown in Fig. 12.1. This cycle is composed of three types of component namely *storage*, *transport*, and *exchange processes*. By far the greatest storage of water is in the oceanic girdle of the globe, although significant amounts of water are retained temporarily within the rocks of the Earth's crust, in ice caps and snow fields on the surface of the Earth, in soils and regoliths, in fresh water reservoirs, such as lakes and inland seas, and in the atmosphere. Water is transported by circulation systems in the atmosphere and oceans, and by movements over and near the surfaces of land areas, chiefly under the influence of gravity. It is left to the exchange processes to effect transformation of water from one physical state (solid, liquid or gas) into another. These include precipitation, evaporation and evapotranspiration.

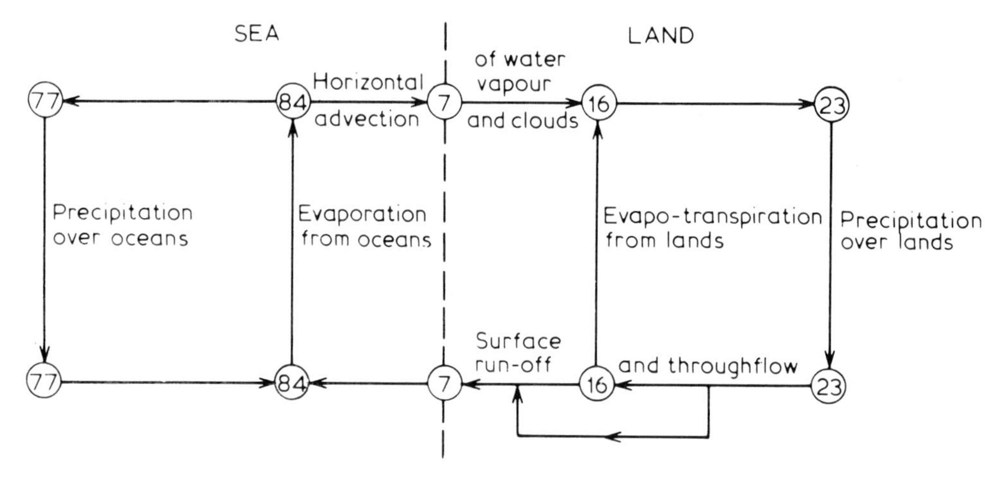

Fig. 12.1 A summary of the global hydrological cycle (assuming 100 water units).

12.2.1 Precipitation monitoring

In any review of the hydrological cycle a convenient starting point is the water supplied to land and sea surfaces by the exchange process of precipitation. Since we have already reviewed the use of satellites for improved rainfall monitoring (Chapter 11), we may now proceed to consider how surface-based remote sensing systems may be used for this purpose too, albeit in a more local and selective fashion. Microwave radar has been used for rainfall monitoring since the late 1940s. Today, operational use of rainfall radar is generally confined to parts of North America, Europe and the Far East, but it has a vital part to play in meteorological and hydrological research, particularly as follows:

1. Radar systems can provide frequent, repetitive and real-time estimates of rainfall intensities over selected areas. For many problems, such as river management and the control of soil erosion, the short-period rainfall intensity is an important factor.
2. Radar systems can give estimates of the total rain that has fallen over, say, a basin or a catchment through a specified period of time. Such information is often vital to water management programmes generally, and predictions of the likely flows of streams and rivers in particular.
3. Some types of radar systems can give information on the movement and development of rain cells within clouds or cloud systems ('Doppler' systems: Chapter 10). This is important in relation to our knowledge and understanding of the structure and behaviour of rain clouds and rain-cloud systems, and therefore, to short-term rainfall forecasting.

By established practice, rainfall data are mostly obtained from networks of raingauges, preferably of a continuous-recording type. However, it is common for rainfall stations to be quite widely spaced. Especially when rain falls from convective systems, e.g. thunderstorm clouds, raingauge data give very misleading pictures of the patterns of rainfall across wide areas. Dependent depth/area (volume) estimates of rainfall through periods of time may therefore be quite inaccurate. Shortly after the end of World War II it was recognized that radar was capable of observing the location and areal extent of rain storms, and that in conjunction with some raingauge data for calibration, more accurate maps of rainfall might be obtainable from a suitably calibrated radar network than from raingauges alone.

If we know the technical specification and performance of a radar set, the radar equation basic to most rainfall studies may be written in simplified form as follows:

$$\bar{P}_r = \frac{cZ}{r^2} \qquad (12.1)$$

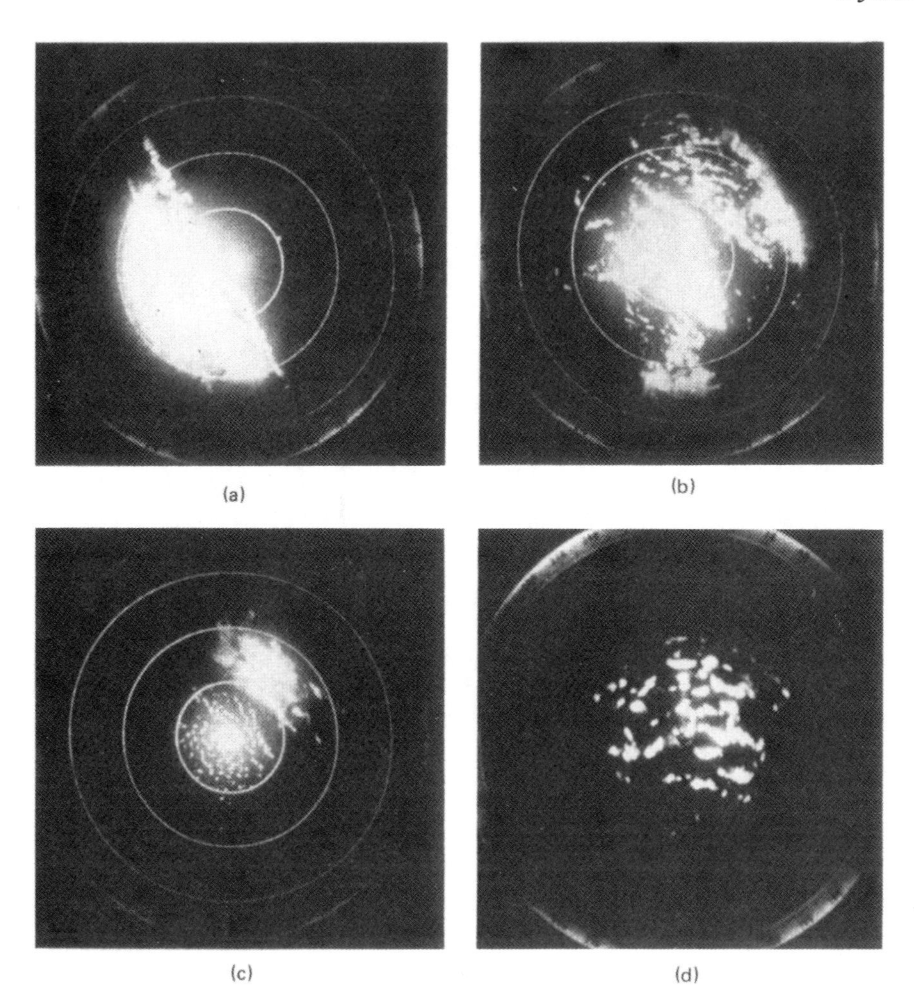

Fig. 12.2 Typical maritime precipitation radar patterns interpreted in terms of the cloud characteristics with which they are likely to be associated. (PPI displays, 150 mile range.) (a) Continuous stratiform cloud; (b) ragged stratiform patches; (c) small convective cells; (d) large convective cells. (Source: Nagle and Serebreny, 1962.)

where $\bar{P}_r$ is the mean received power (the 'signal') from the target, c is a constant (the speed of light), Z a reflectivity factor relating to the type of precipitation in the target area, and r the range (or distance) between the radar and the reflecting shower of rain. Whether the output required is rainfall intensity or amount, the reflectivity factor is of particular significance: it is where the raindrops are large and numerous that Z is high, and where they are small and few Z is low. Quantitative estimates of rainfall (R) may be derived from the target echoes by evaluating the relationship $Z = aR^b$, where a and b are local constants, empirically established. In middle latitudes, typical solutions for different types of rain are $200R^{1.71}$ for orographic rain, and $486R^{1.37}$ for thunderstorm rain. Such relations may be substituted for Z in Equation 12.1 so that, for operational usage, the radar equation (in simplified form) becomes:

$$P_r = \text{constant} \times \frac{R^b}{r^2} \qquad (12.2)$$

The implication is that for different latitudinal zones and for different types of rain the radar signal is proportional to the rate of rainfall divided by the square of the range.

The principles summarized above have been deployed by many workers on both sides of the North Atlantic. One interesting pioneering project involved the Chester Dee in North Wales, under-

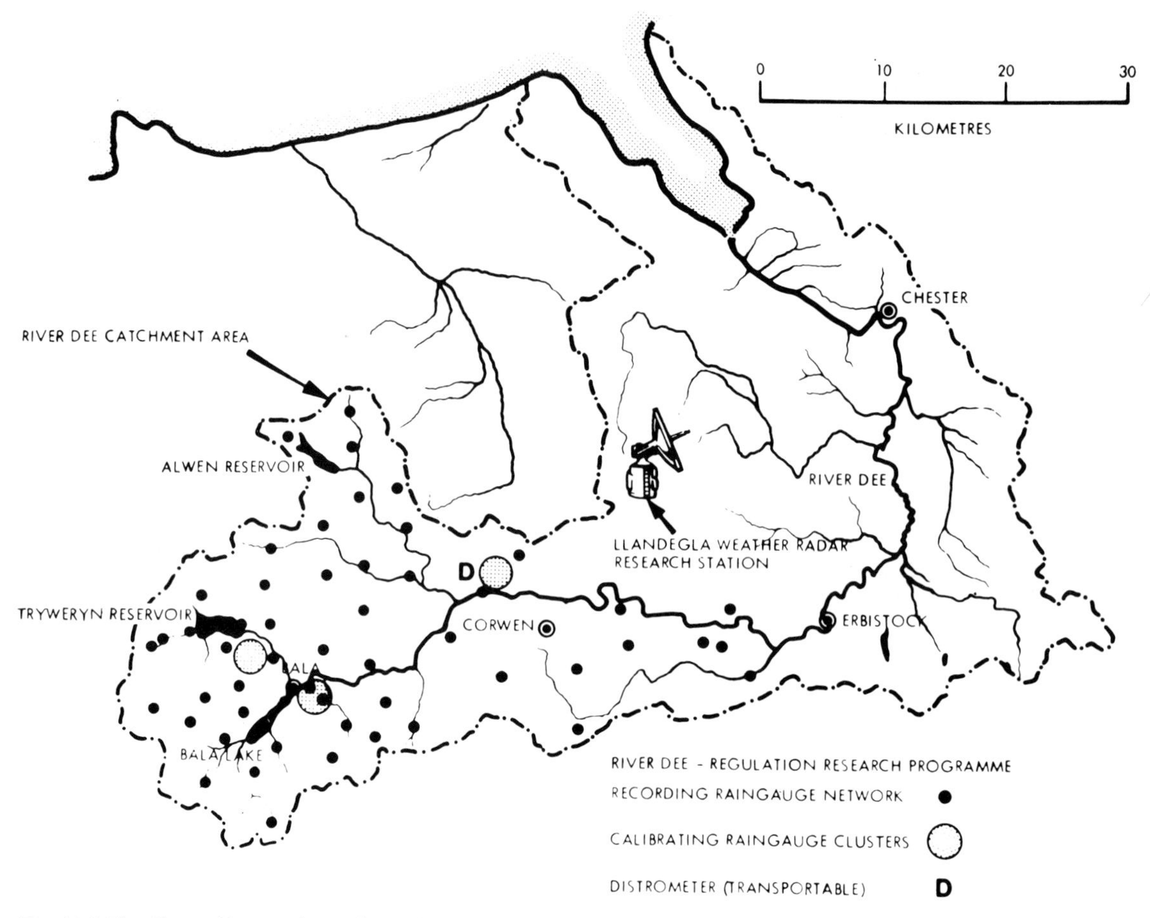

Fig. 12.3 The Chester Dee weather rader project.

taken cooperatively by the UK Meteorological Office, the Water Resources Board, the Dee and Clwyd River Authority and Plessey Radar. The area covered measured approximately 1 000 km^2. This was carefully instrumented to facilitate the necessary statistical comparisons between radar and conventional parameters (Fig. 12.3). Sample results for a particular rainstorm are illustrated by Fig. 12.4. A good measure of agreement was found between the two. Unfortunately in very mountainous, inaccessible terrain – the type of area from which radar estimates of rainfall might be most valuable – the narrow beam, low-elevation type of radar currently used in work of this kind would be largely ineffective. However, if a higher elevation beam were to be used instead, the distance over which the rainfall estimates would be valid would be relatively short, because ever higher regions of the atmosphere are sampled as range is increased. As a consequence, where there are large and important watersheds in mountainous regions, present systems and techniques allow only a small area to be covered by each radar set. Perhaps new techniques – possibly involving a form of 'Doppler' radar, which takes account of the frequency shift of the radar echoes as well as their direction and intensity (Chapter 10) – will be developed to deal with this problem.

Since the Chester Dee Project, very significant advances have been made in implementing operational radar rainfall systems in the more developed countries of the world, including the UK (Fig. 12.5). Important advances are now being made with the development of new technology radar networks, e.g. the digital 'NEXRAD' system in the USA (Table 12.1), and the curiously named

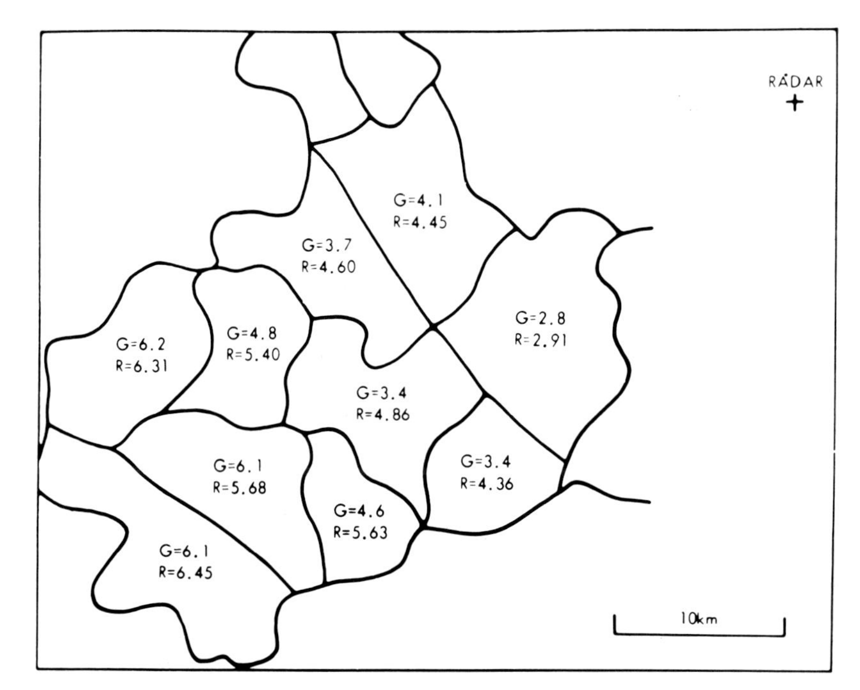

Fig. 12.4 The Chester Dee weather radar project representative results. Rainfall totals (mm) derived from radar (R) and raingauge (G) are shown for subcatchments for the period 1530–1630 GMT, 7 November 1971. (Courtesy, Plessey Radar.)

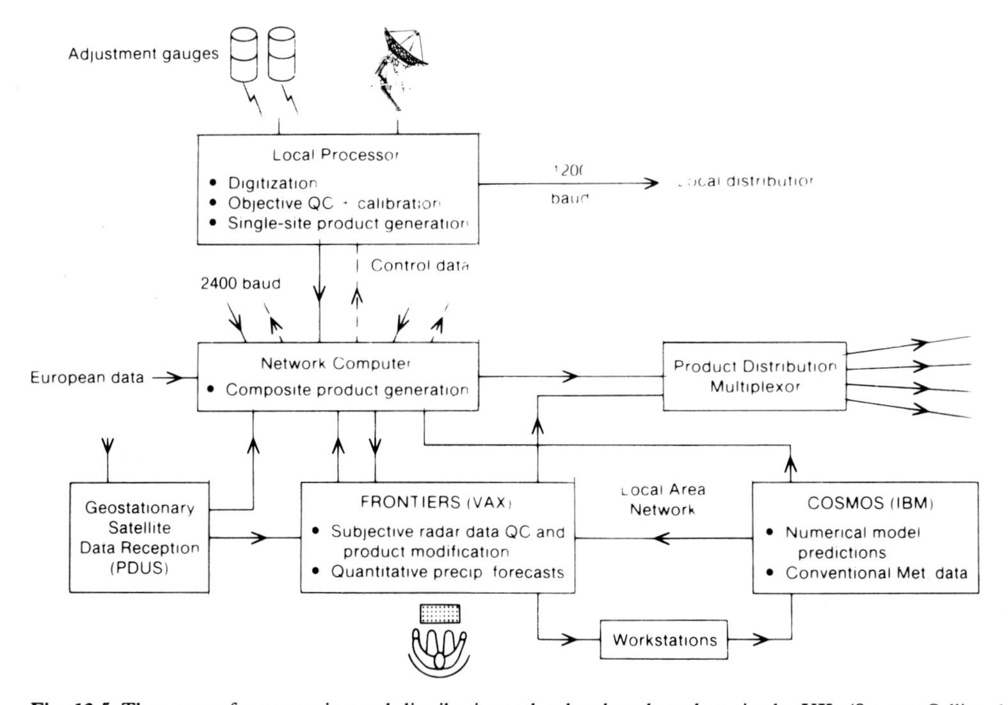

Fig. 12.5 The system for processing and distributing radar data-based products in the UK. (Source: Collier, 1989.)

Table 12.1 Key operational goals and objectives of the US NEXRAD system (Source: NEXRAD, 1984)

1. Increase the average tornado warning time from the present 1–2 min to at least 20 min
2. Improve the accuracy of descriptions of the location and severity of thunderstorms and the ability to distinguish between severe and less-than-severe storms
3. Improve the detection of damaging winds and hail
4. Improve the safety of aircraft operation by detecting and measuring the wind shear and turbulence associated with thunderstorms
5. Provide improved rainfall estimates for flash-flood warnings
6. Reduce the size of warning areas to minimize unnecessary warnings
7. Substantially reduce the number of false hazardous weather warnings
8. Minimize the failure to detect hazardous weather due to radar outages
9. Optimize the efficacy of information provided to forecasters and other personnel by improving distribution and display of radar information
10. Detect hazardous weather conditions throughout the 50 States and at overseas locations specified by users
11. Maintain annual operations and maintenance costs at the same level as that of the radar system to be replaced (excluding the cost for radars in areas not at present covered)

'COST-73' system now being extended across western Europe. However, radar costs are high and infrastructural requirements demanding; therefore most of the land areas of the world will not be covered by weather radar in the foreseeable future. Also, because of progressive down-beam changes in the relationships between the radar signal and the height of the precipitating layer in the troposphere, the areas over which rainfall can be estimated by radar quantitatively not qualitatively will still remain considerably less than 100%, even in North America and Western Europe (Fig. 12.6).

12.2.2 Ice and snow monitoring

Our second important consideration in this section is *frozen precipitation*, especially through its surface manifestations of ice and snow. Two aspects of ice and snow distribution transcend the rest, namely the significance of frozen water within the hydrological cycle, and the influence of sea ice on maritime activities. A substantial portion of the world's fresh water resource is stored temporarily in the form of snow, especially in high-mountain regions. Knowledge of the distribution of snow fields and their volumes in terms of *water equivalents* is required for stream flow and water storage forecasting. In turn this strongly affects power generation and irrigation programmes, water quality control, river management, manufacturing, recreation, and many other activities which contribute to a nation's economy. Sea ice imposes a seasonal control on shipping in many regions of the Northern Hemisphere. The opening and closing of ports is influenced thereby, and certain navigation routes can be used for only a part of each year, with or without the use of icebreakers.

It is now widely recognized that satellite remote sensing has useful potential for snow mapping and monitoring (Table 12.2), and Scandinavian and Alpine countries are prominent amongst those which have developed airborne techniques for estimating the volumes of snow accumulated on watersheds that serve hydro-electric power stations at lower altitudes. However, in recent years a growing number of lowland countries, including the UK, have also paid careful attention to the possibility of snow monitoring from satellites, as described below.

Several methods for snow mapping have been developed based on Landsat and local SPOT imagery to supplement or even replace surface or aircraft observations of local or regional ice and snow accumulations. Broad-scale ice and snow surveys have been based on Landsat imagery also, for example improved maps of poorly surveyed regions of Antarctica. It has been shown that such data can be processed by computer to give snow-cover area estimates with an accuracy of 93% or better down to the scale of individual pixels. Highly accurate positioning of the snowline is also possible – a most important consideration in run-off predictions and monitoring of the melting process in spring and summer. Until quite recently the biggest remaining problem involved the separation of broken cloud from ice and snow. However, today's additional information from 1.55 to 1.75 μm sensor channels had made clear dis-

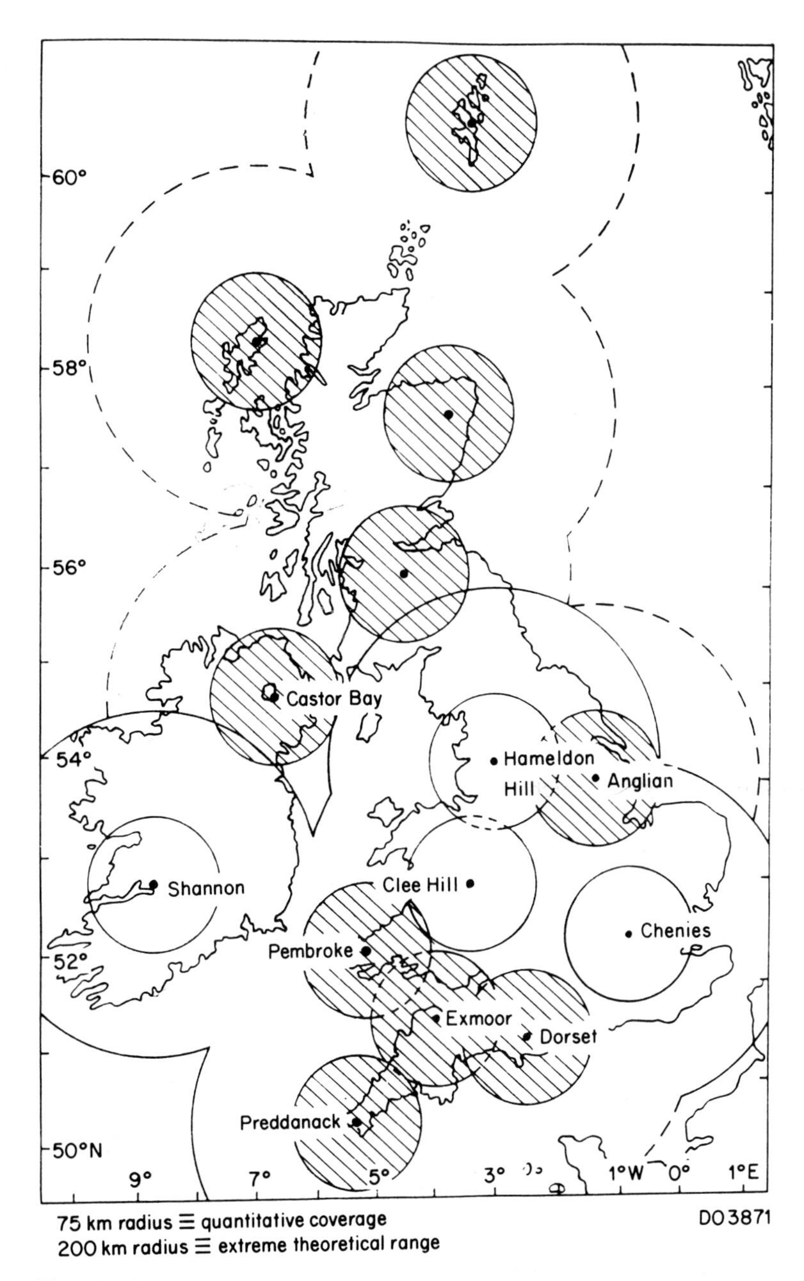

Fig. 12.6 Location of existing and proposed weather radars in the UK, showing areas of quantitative rainfall estimation (shaded) and qualitative rainfall estimation (blank). (Source: Collenge and Kirby, 1987.)

crimination of these features possible, e.g. through Channel 5 on the Landsat Thematic Mapper (Chapter 5).

Unfortunately, regular analyses of Landsat or SPOT data for snow monitoring is rarely feasible financially, and badly affected by cloud cover. Therefore, for repetitive ice and snow mapping over large areas, data from weather satellites, especially of the NOAA family, are more widely used. The North American nations, Canada and the USA, have long operated an expensive programme of ice reconnaissance in the St Lawrence region and northern coastal waters using aircraft to spot natural leads (navigational passageways) and

Table 12.2 Satellite sensor responses to various snowpack properties (After: Rango, 1980)

Property or performance	Waveband		
	Visible/near infrared	Thermal infrared	Microwave
Snow-covered area	Yes	Yes	Yes
Depth	If very shallow	Weak	Moderate
Snow water equivalent	If very shallow	Weak	Strong
Stratigraphy	No	Weak	Strong
Albedo	Strong	No	No
Liquid water content	Weak	Weak	Strong
Temperature	No	Strong	Weak
Snow–soil boundary	No	No	Weak (high frequency) to strong (low frequency)
All weather capability	No	No	Yes
Current best spatial resolution from space platform	10s of metres	100s of metres	Passive: 12.5 km (high frequency) to 150 km (low frequency)

passages through the ice. Satellite data were first used in support of this programme through the 'TIREC' (Tiros Ice Reconnaissance) Project in the early and mid-1960s. Today, NOAA imagery is analysed routinely – though still by hand – by the US National Weather Satellite Service to provide its customers with a range of medium to small-scale ice and snow reports. These include:

1. Northern hemispheric snow and ice charts. Polar-stereographic maps of snowfield and icefield boundaries and their relative reflectivities (three category classifications) on a scale of 1 : 50 m for use by the US Navy and various secondary users.
2. Basin snow-cover observations. Percentage snow-cover messages sent viz teletype, and maps sent via telecopier, for selected river basins in North America, primarily for use by the US Office of Hydrology, the Corps of Engineers, and the Soil Conservation Service.
3. Great Lakes and Alaskan ice charts. Detailed (1 km resolution) analyses of the boundaries and type or age of ice, revealing ice-fast and ice-free areas as well as ice concentrations and leads. The primary users include the US Navy and Coast Guard, the National Marine Fishery Service, and commerical marine transportation companies.

Elsewhere, fully objective (automatic) techniques based on NOAA–HRPT multispectral data are being developed for routine snow area and snow surface characteristics (surface temperature, melt/accumulation status, etc.) mapping and monitoring, e.g. for the National Rivers Authority in the UK (Plate 14, colour section). Evaluating snow depth is more difficult, being readily available for assessment from visible/infrared data within a very few depth categories only. Furthermore, cloud cover is still a nuisance, though much less so than in the case of the infrequently viewing Earth resources satellite systems. As with rainfall monitoring, new hope is emerging that passive microwave data will provide helpful supplementary information both spectrally and temporally (Fig. 12.7). For example, early assessments of *snow water equivalent* using SMMR and SSM/I data calibrated by ground data over Finland and the UK, and by aircraft over Canada, promise much for fully operational methods, as they can be operated under any cloud conditions, not only for dry (frozen) snow but also for wet snow, to quite shallow depths where vegetation is short or sparse. For best

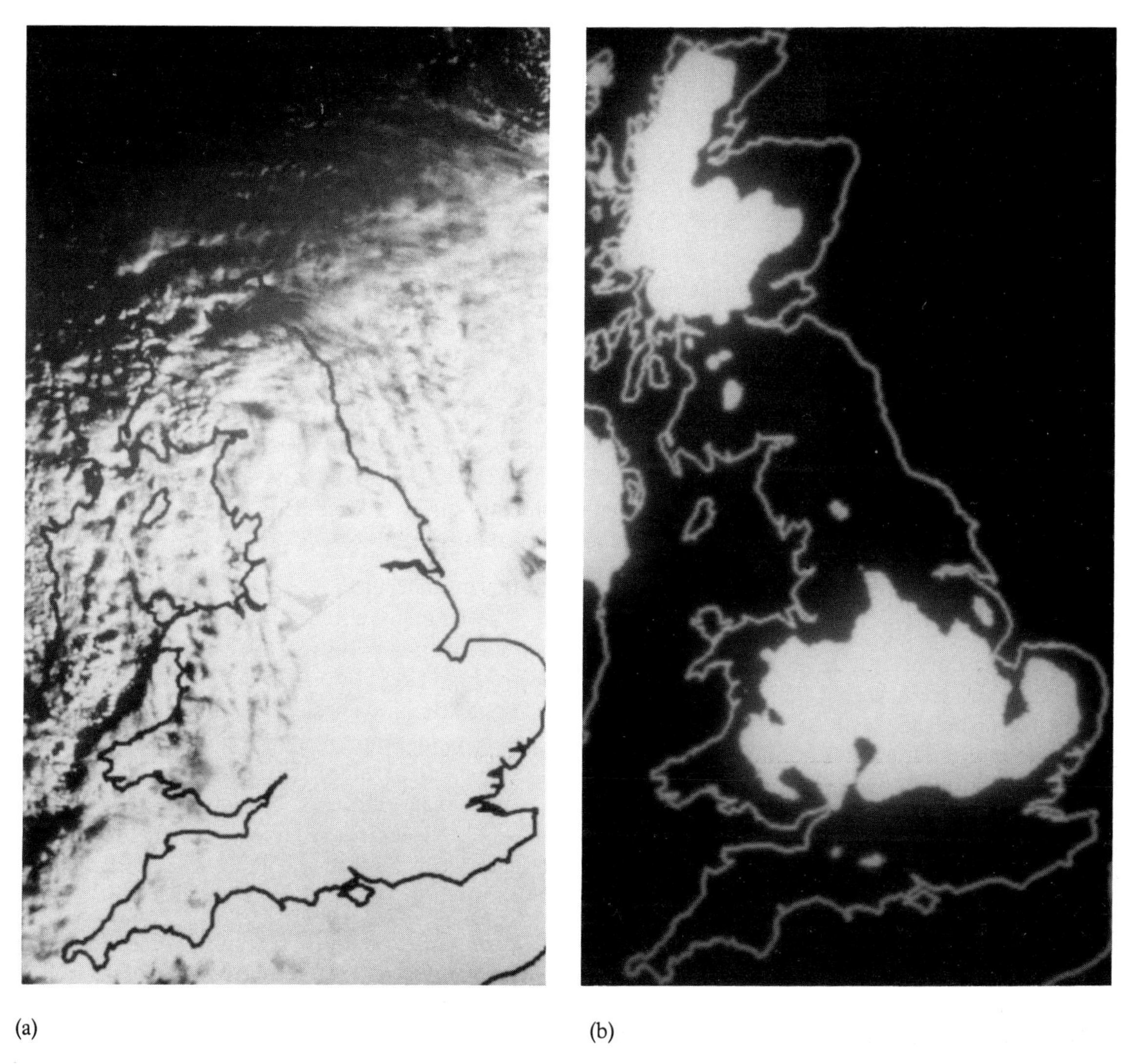

(a) (b)

Fig. 12.7 Computer maps of satellite imagery for the morning of 23 January 1988, exemplifying the value of (cloud penetrating) passive microwave data for snow monitoring: (a) a NOAA-AVHRR visible waveband image showing Scotland still in darkness and England and Wales covered by (frontal) cloud: conditions inimical to snow monitoring using visible/infrared techniques and (b) a DMSP SSM/1 passive microwave image for approximately the same time, but able to reveal the extent of snow lying beneath the clouds. (Courtesy R. Kelly and Hui Xu, RSU, University of Bristol).

possible monitoring of all snow conditions a combination of multispectral visible/infrared and passive microwave techniques seems most promising.

12.2.3 Evaporation and evapotranspiration estimation

These associated exchange processes play key roles in the hydrological cycle, and are of fundamental importance to agriculture through their effects on plant growth and performance. Evapotranspiration is the loss of water from the Earth's surface in vapour form. It occurs as evaporation from open water surfaces and as transpiration from living plants as part of their respiration and photosynthetic processes. As far as the local or regional water balance is concerned, evapotranspiration (ET) is a significant and often the dominant water flux leaving the Earth's surface. In arid regions, nearly all the input in the form of rain is lost through ET. Even in humid regions, half or more of the water balance can be attributed to ET. Monitoring evaporation and evapotranspiration by conventional (*in situ*) means is not easy, and by remote sensing means has been described as 'the most challenging of all the applications of remote sensing to hydrometeorology' (A. Rango, 1980). The most useful satellite observations for such applications are of measurements of solar radiation, surface albedo and surface soil moisture, temperature and type of ground cover, and potentially of microwave emission and reflection also, none of which can be obtained readily over wide areas from *in situ* sources. However, observations of other parameters that affect evapotranspiration have so far proved more intransigent from satellites, including water vapour gradients and winds. Thus approaches to evapotranspiration estimation using satellite remote sensing have had to work around these missing data.

One interesting approach to evaporation monitoring from satellites has exploited the *thermal inertia* properties of soils as they are expressed through diurnal changes of surface temperatures (see also section 13.5.2). These changes are governed by the radiation budget (related to the external environment of the soil) and thermal inertia (related to the internal characteristics of the soil): the first can be modelled and evaluated with or without remote sensing inputs; the second can be assessed from infrared imagery when this is processed appropriately. We know that the thermal inertia of unsaturated soils is influenced greatly by soil porosity, and that any soil experiences a diurnal thermal inertia cycle which is closely related to its porosity. The greatest separation between the thermal inertia cycles of different soils is found about the local solar maximum and minimum, i.e. about 0200 and 1400 hours local time. Much useful information relating to this problem was obtained from the Heat Capacity Mapping Mission (HCMM), whose sun-synchronous orbit was planned to image at these times. The TELL-US model was developed to yield the following parameters from twice-daily aircraft or satellite observations:

1. thermal inertia;
2. surface relative humidity;
3. daily estimated evaporation totals as illustrated by Fig. 12.8. Work to achieve the simplifications which everyday operations might require is now in progress. On the other hand exploitation of the more frequent radiation temperature measurements made by geostationary satellites, such as Meteosat, are likely to permit significant improvements in the accuracy of the results.

Turning next to evapotranspiration itself, within which the loss of water from vegetated as well as non-vegetated surfaces is involved, a comprehensive estimation model designed to exploit satellite data has been developed from a resistance form of the energy balance equation, in which

$$LE = R_n - G - QC_p(T_c - T_a)/r_H$$

LE is latent heat flux (a measure of moisture in the air), R_n is net radiation, G is soil heat flux, Q is air density, C_p is the specific heat of air, T_c is plant canopy temperature, T_a is air temperature, and r_H is thermal diffusion resistance. Key variables which can be monitored by satellites include R_n, T_c and T_a. This is very demanding, for it requires the radiation intensities observed by satellites to be translated not only into radiation temperatures, but also corrected for temperatures of a thin layer of the atmosphere at its very base, as with the temperature in the upper foliage of a crop or stand of natural vegetation. Since atmospheric trans-

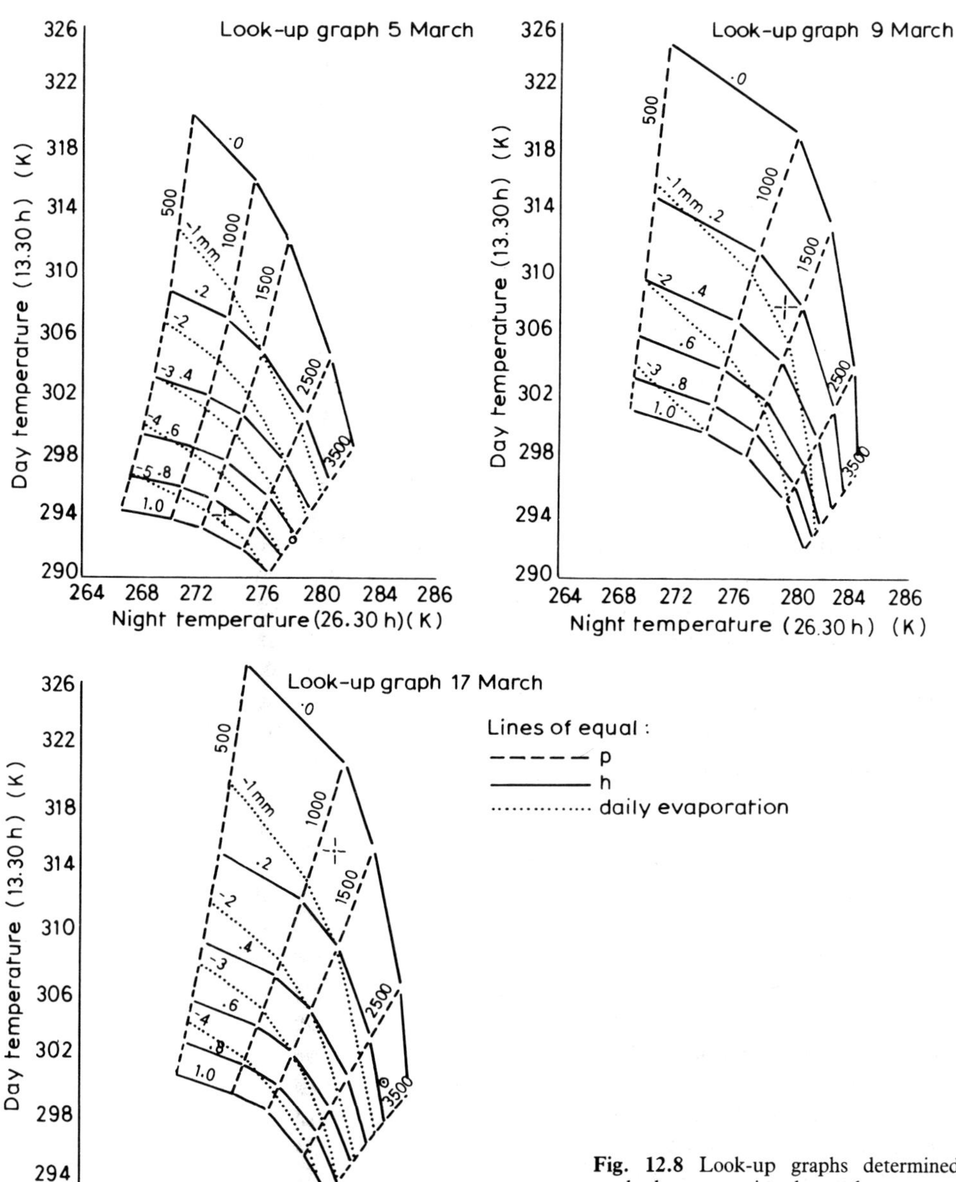

Fig. 12.8 Look-up graphs determined by the TELL-US method, representing three subsequent stages in drying of a test plot in the USA, March 1971. The crosslets mark observed day and night temperature pairs. From the graphs, values of thermal inertia, surface relative humidity and daily evaporation may be read off. (Copyright, EEC.)

mission and surface emissivities are quite variable, and canopy temperatures possess a high degree of spatial variation it is not surprising that results so obtained are thought to be too gross. It is recognized, too, that the evapotranspiration performance of a crop or vegetation area changes through the growing season. 'Leaf Area Indices' (relating the leaf area of a crop to the area of the ground on which it grows) evaluated from Landsat reflectance data are being utilized in attempts to tune such methods for these seasonal effects.

The major problems that all existing remote sensing methods of evapotranspiration have yet to overcome are:

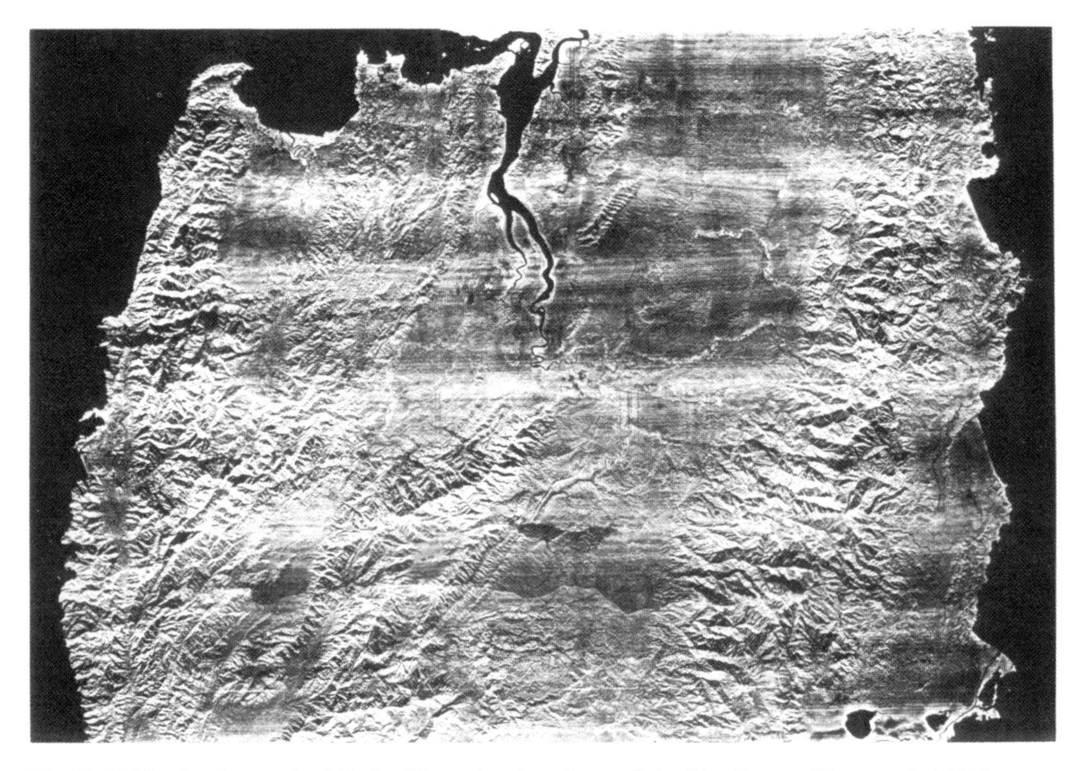

Fig. 12.9 K-band radar mosaic of Darien (Panama) and north-west Columbia. (Source: Viksne *et al.*, 1969.)

1. the process of transpiration is still not well understood and parameterized either for structured crops, such as cereals, or for complex vegetation, such as trees;
2. in the presence of vegetation, the surface temperature (T_S) estimated by a thermal infrared sensor is at an unknown level within the vegetation;
3. the most appropriate use of microwave observations of surface soil moisture in the presence of vegetation remains to be determined. For the future, we expect that the most practical method will probably use a multispectral approach including repetitive observations at the visible, near and thermal infrared, and microwave wavelengths. This will afford the possibility of estimating solar insolation, surface vegetative cover and/or albedo, surface temperature, and surface soil moisture from remotely sensed data and incorporating them into models of the type described here.

emerged': research continues, at present paying particular attention to the problems of how best to combine the point data available from the ground and the areal data from the satellites, and how to develop more physically-based, and therefore more realistic models.

The big prize of widespread application of the successful method awaits the end of all these researches.

12.3 Surface hydrology

Some overlap is inevitable between this section and the subsequent chapter on 'Soils and Landforms'. However, the emphases are different, for in the present context, the 'surface' is to be viewed only in respect of its significance for hydrology through run-off, infiltration, and/or standing surface water. We may review this large and rapidly expanding field under six subheadings, leading us from matters of observation, through modelling, to conclude with possibilities of hydrological prediction.

12.3.1 Watershed parameters

The chief object of terrain analysis is to recognize and map 'recurring landscape features', especially through geomorphic, soil and land use characteristics. Since terrain is, to a greater or lesser degree, dynamic, the advent of air survey and Landsat and SPOT image analysis methods has contributed dramatically to the ability of applied hydrologists to take account of the change factor in their assessments of significant watershed parameters. For example, deforestation and urbanization both encourage increases in the rate and volume of basin run-off, even in the absence of any change in precipitation.

In areas that are heavily clouded, recourse must be made to microwave imagery in the search for more detailed and/or more frequent information on remote and inaccessible watershed regions than could be obtained readily at ground level. A benchmark project codenamed RAMP (Radar Mapping of Panama) was initiated in 1965 by the US Army Engineer Topographic Laboratories to demonstrate the performance of high-resolution radar imagery in lieu of optical photography in heavily clouded regions. The sensing system was a K-band radar (see Fig. 2.3). This has better than a 99% cloud penetration capability, although it does not penetrate heavy rain or vegetative cover. Once-over coverage of the eastern end of the Republic of Panama was achieved in approximately 4 h of flying time. A number of interpretational overlaps were prepared from the final mosaic. These included surface drainage patterns, and types of surface drainage regions (Figs 12.9 and 12.10(a and b)) as well as general surface configuration (topographic provinces), vegetation and engineering geology.

It was possible to identify variations in the local and regional expression of *surface drainage*, for example differences in the density of streams and other drainage channels, and the depth of channel incisions. In turn such variations in the drainage patterns are diagnostic of specific terrain conditions, for example the nature and depth of soil cover, lithology, morphology, and prevailing structural and tectonic influences. The patterns present themselves in an infinite variation of density and habit (or form). The *density* is determined mainly by the lithologic character of the rock traversed – itself a matter of considerable interest to hydrologists – involving its hardness, porosity, solubility and consolidation. The *form* relates mainly to structure, involving faults, fractures and the attitude of bedding. Thus the drainage pattern overlay was the first to be developed from the imagery, being basic to most of the others. Similar radar mapping projects have since been undertaken in numerous regions, particularly in the tropics where surface hydrologic characteristics have been relatively little mapped, and where cloud is particularly persistent. Clearly side-looking radar (SLR) has a part to play in regional hydrological studies especially where cloud cover restricts the use of aircraft as platforms for conventional photography.

At the scales at which the more remote satellite systems operate most efficiently, the complexity of the total hydrological problem, and the detail required by many operational monitoring programmes, comprise formidable hurdles to be overcome if remote sensing is to replace more direct 'contact' methods of assessing key variables. Although the improved stereoscopic capabilities of SPOT satellites compared with Landsat have been of considerable help in such respects, this is one of the relatively few fields for which further technological development is urgently required (e.g. in the spatial resolution of multispectral image data, and in satellite-borne active microwave systems). Remote sensing for everyday hydrological use has been likened to the computer: it promises large increases in efficiency in specifying and tackling environmental problems, but much of this potential has yet to be fulfilled.

12.3.2 Drainage network features

Stream network analysis, involving the mapping, measurement, and classification of active drainage lines, especially in terms of stream lengths, complexities of stream patterns, and the location of associated features, including springs, ponds and lakes, is an important aspect of modern hydrology. It has been shown that these can be assessed very satisfactorily from Landsat MSS imagery at scales of 1 : 250 000 or even better, and from TM and SPOT imagery to scales of at least 1 : 50 000. Furthermore, changes in dynamic features, e.g. river channel characteristics, also can be evaluated therefrom. Many rivers exhibit frequent changes

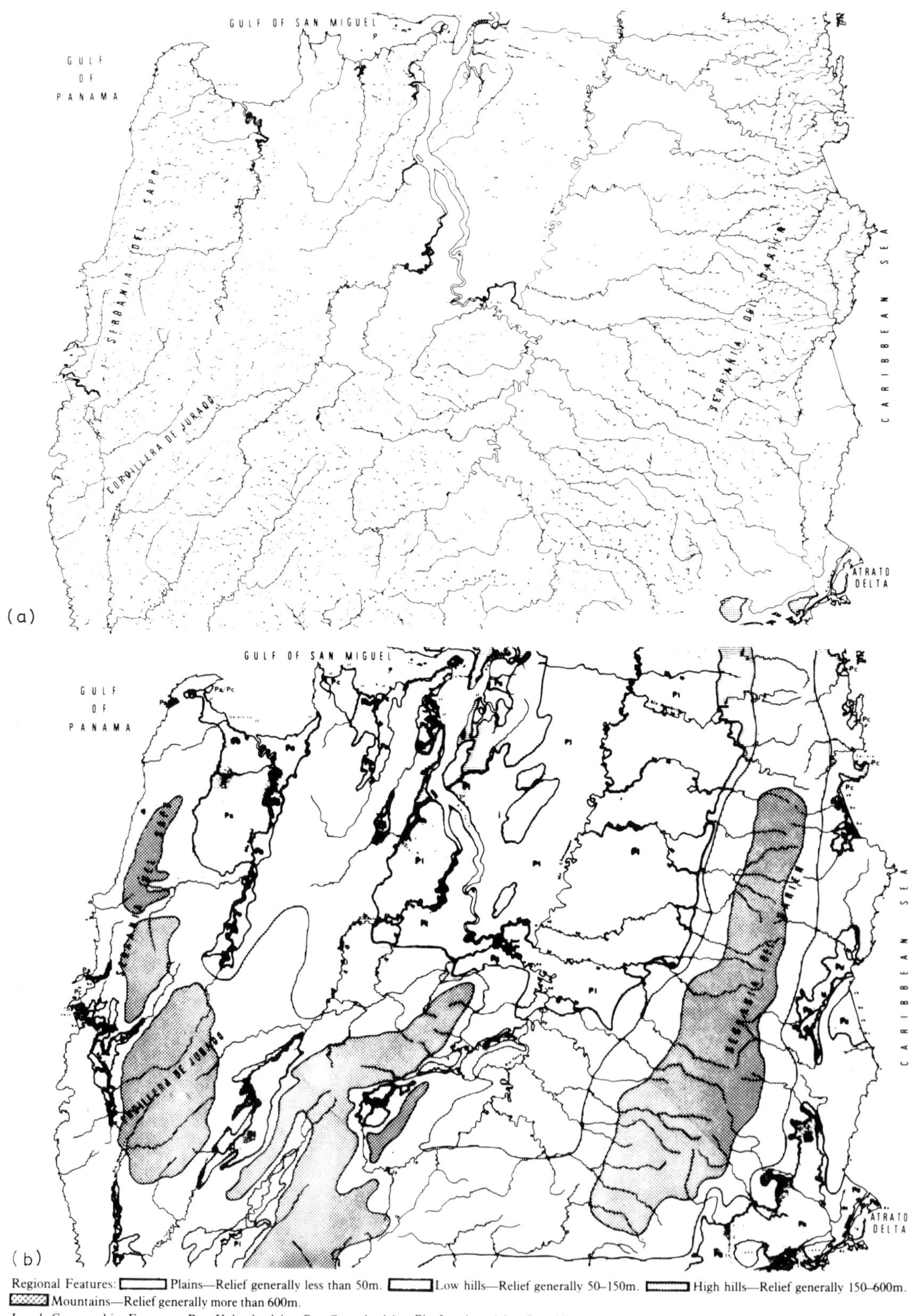

Fig. 12.10 Hydrologically significant distributions mapped from a radar mosaic of the Darien region of eastern Panama, and north-west Columbia (see Fig. 12.9): (a) surface drainage network; (b) the surface configuration. (Source: Viksne *et al.*, 1969.)

of course and interactions between interlacing ('braided') channels. Even manual analyses of Landsat scenes, displayed on simple colour additive viewers, have been undertaken profitably for many regions, for example the basin of the Kosi River in Bihar State, India, whose results revealed the following MSS band preferences:

1. Band 5: flood-plain features, e.g. abandoned channels, sandy islands, shoals, etc., and contrasts between them and nearby vegetated areas;
2. Band 7: active river courses (water is dark in the near infrared);
3. Band 4, 5, and 7 false colour composites: alluvial areas, and classes of vegetation.

Increasingly, interactive and objective analyses of satellite imagery are now being developed for such purposes.

12.3.3 *Flood-plain mapping and monitoring of floods*

Much of the world has still to be mapped on the ground at scales of 1 : 250 000 or better, affording much scope for Earth resources satellite imagery to be used to map important *flood plain features*, either for the first time or over even wider areas after each new major flood event. Unfortunately because of their low imaging frequencies, systems such as Landsat and SPOT are not well-suited to the mapping of floods, although under optimum arrival time and sky conditions such satellites can and have been extremely valuable sources of vital flood statistics. Lesser restrictions circumscribe the use of environmental (e.g. meteorological) satellite data for such purposes, but these data carry the additional disadvantage of a relatively low spatial resolution. However, Meteosat visible and infrared imagery have been used for the monitoring of change in large lakes and swamps, e.g. Lake Chad, and the Okavango Swamps in south-western Africa. Passive microwave data from ESMR on Nimbus 5 were analysed successfully in studies of the widespread floods in eastern and central Australia in 1974 (Plate 8 (a and b), colour section) and similar data from the DMSP–SSM/I sensors now regularly provide information on major flood events throughout the world. Data requirements for flood monitoring in basins larger than 1000 km^2, as proposed at a specially convened international planning meeting, are summarized in Table 12.3. This also provides in the right-hand column an updated indication of those elements for which there is, even at present, either a significant or a reasonable satellite monitoring capability. Big improvements should become possible with the advent of EOS satellites in the later 1990s.

12.3.4 *Wetlands monitoring*

Here the object of attention is the hydrology of coastal and freshwater lake regions. *Surface water area* assessment is of key significance, involving such things as seasonal susceptibility to flooding and the variation of water reserves in dams and reservoirs. As remarked earlier, surface water is prominent in near-infrared images, and from Landsat and SPOT water bodies as small as 1 ha can be located, and rivers only 20 m across have been directly traced. In coastal zones a wide range of important features may be identified and mapped reliably at 1 : 100 000 or better, including the marsh–water interface, the upper wetland boundary, and different plant communities within the marshy area. Successful estimates of water quality have been made through a variety of means, e.g. regressions of a 'quantitative brightness' parameter (the cube root of the product of standardized Landsat data in Bands 4, 5 and 6) against observed sediment levels in North American lakes. Impressive analyses have been made of ecologically and scientifically important coastal zones, such as coastal Queensland, Australia, including the Great Barrier Reef. Interest in remote sensing monitoring procedures has risen sharply as concern for the coastal environment generally has grown in recent years, and remote sensing is playing an increasingly important role in water quality evaluation and management strategies.

Most progress has been made with satellite evaluations of turbidity, suspended sediment, chlorophyll, eutrophication and water temperatures, and least with wetland-specific pollution problems. However, sources of pollution are often easy to identify, especially when pipes or open channels discharge into a lake or river. Non-point-source pollution can, perhaps, be evaluated better already by remote sensing than conventional means because of the broader areal coverage that aircraft and satellite data sets provide. Although absolute

Table 12.3 Data requirements for floods in basins larger than 1000 km^2 (After Ferguson *et al.*, in Salomonson and Bhavsar, 1980)

Hydrological element	Accuracy	Resolution (m)	Frequency	Present satellite capability
Flood plain area/ boundary	±5% area	10	5 years[a]	Good
Flood extent	±5%	100	≤4 days[b]	Partial
Precipitation	±2 mm if <40 mm ±5% if >40 mm	5000	6 h	Good
Saturated soil area	±5%	100	≤4 days[b]	Partial
Soil moisture profile	±10% field capacity	1000	1 day	Little
Snow covered area/snowline	±5% area	1000	1 day	Good
Snowpack water equivalent	±2 mm if <2 cm ±10% if >2 cm	1000	1 day	Partial
Snowpack free water content	±2 mm if <2 cm ±10% if >2 cm	1000	1 day	Little
Snow surface temperature	±1°C	1000	6 h	Good

[a] And after each major flood event.
[b] Depends on degree of flooding: with major floods, daily monitoring would be desirable.

accuracy is more elusive using remote sensing approaches, only large areas can be monitored economically by these means. As in so many other contexts, the most effective use of remote sensing data in made when used in conjunction with other data or information.

12.3.5 Mathematical modelling for flood hydrograph prediction

Run-off is the hydrologic variable most often used by hydrologists and water resource planners, and the ability to accurately forecast peaks on the *stream hydrograph* is, perhaps, the most conclusive test of applied hydrological knowledge and understanding. This ability is much sought after because of its significance for hazard advice and warnings, as well as in river management and control. The most important objections are the timely and accurate prediction of run-off at any given point in a drainage basin. The tools traditionally available for this include a wide range of equations and models, as well as streamgauges that measure streamflow or reservoir volume directly in real time. The principal limitation under which hydrologists work with respect to these tools is that none work well under all conditions. To improve their performance recourse to satellites may help, although it is important to note that, at present, run-off cannot be measured directly by remote sensing techniques. Therefore the role of remote sensing is generally to help in the estimation of equation coefficients and model parameters, including:

1. infiltration into the soil surface, and through the soil profile;
2. surface run-off;
3. channel flow.

Five steps have been proposed for the development of basin models to interrelate these and thus predict flood peaks accurately. For each of these steps, preferred remote sensing approaches can be identified, as follows:

1. Determine the physical extent of the basin (boundaries and area). Air or Earth resources satellite visible imagery should be used unless

a high incidence of cloud necessitates radar imagery. A low angle of solar illumination is best, for the evidence of shadows is especially useful in watershed determination.

2. Idealize the basin topography. Stereoscopic air photographs (or multifrequency radar images where slopes are shallow) are analysed for rectangular slope elements (areas and mean angles of slope).
3. Idealize stream channel geometry. Near-infrared imagery (e.g. air-photo and Landsat MSS Band 7) may be used both for high water-level determination and the analysis of stream patterns into rectangular line segments. Photogrammetric procedures must be followed for the analysis of stream cross-section at low water.
4. Quantify the hydraulic roughness of slope elements and stream channels. Radar is the most direct source of information on slope roughness, because the wavelengths used are of the same scale as the surface variations. Visible image analysis (e.g. air-photo, Landsat or SPOT) may be used to identify larger scale landscape or terrain units. Channel roughness may be assessed under clear water conditions by imagery in the blue/green region of the visible spectrum (c. 0.5 μm); where water is turbid, much greater difficulties are encountered.
5. Differentiate and delineate areas of impermeable surfaces and saturated soils. Both these conditions result in surface flow of additional water. Impermeable surfaces may be identified and assessed in infrared images on account of their low thermal inertias. Many of these surfaces are of human origin, and have sharp, rectangular, outlines. Wet soils are dark in visible images, and radar at wavelengths in excess of 20 cm has some ability to integrate subsurface moisture values.

In addition to the above, satellite remote sensing is beginning to contribute useful information with respect to *soil moisture* itself, and, even more importantly, to likely run-off from snow melt and rainfall. Improved satellite information on land use or land cover is also helping. For the future, additional information on such basic watershed-state variables as basic energy balance parameters and frozen soils may become available. All will become increasingly important as the scale of this type of inquiry and modelling continues to increase from that of the single basin to those of countries, continents and even the globe itself.

Our conclusion must be that remote sensing has a clear and varied applicability to surface hydrology, especially in poorly instrumented regions, but no single approach can provide all the additional information required from remote sensing sources. Perhaps more than in most other fields of environmental inquiry a *multisystems* approach is strongly indicated.

12.3.6 Hydrologic basin models

There are many types of hydrologic basin models, including:

1. physical (laboratory) models, based on hydrologic processes;
2. 'representative basin' models, based on empirical findings in a small number of well-instrumented basins considered to be characteristic of the broader areas in which they are set;
3. mathematical models, based on response, rather than causation. These include deterministic, parametric and probabilistic varieties.

In every case the aim is the same, namely to transform the *inputs* to a basin into *outputs* with the best possible representation of reality. Figure 12.11 depicts some of the key variables that are involved, with the interlinkages between them.

As we have seen earlier in this book, satellite remote sensing can already help to monitor precipitation inputs and, less well, estimate outputs via streamflow and evapotranspiration. Levels of interception, soil moisture and groundwater storage are strongly influenced by topography and subsurface geology; aircraft and satellites can be used to map and assess the significance of these important influences on local basin cycles. However, big strides forward are expected with the advent of EOS instruments in the late 1990s, especially the HIRIS (High Resolution Infrared Spectrometer) and improved (microwave) SARs, as suggested by Fig. 12.12.

12.4 Hydrogeology

Since hydrogeology is concerned predominantly with features at depth rather than at or very near

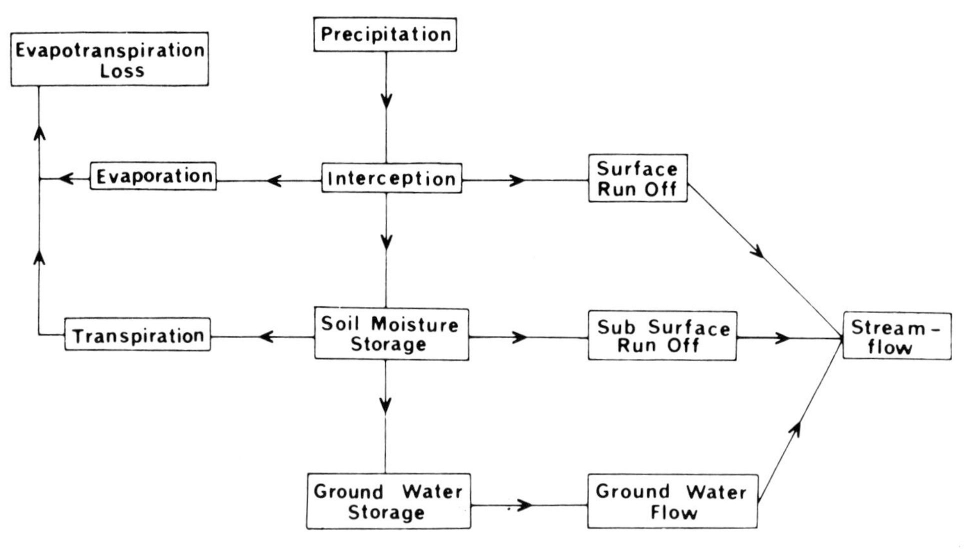

Fig. 12.11 A water-balance schematic for individual catchments. Certain of the components lend themselves more readily to assessment by remote sensing than others. (Source: Painter, 1974.)

the surface of the Earth this is the province of the geologist rather than the environmental scientist. Therefore our review of this economically important field will be very brief. Remote sensing texts relating more specifically to geology may be consulted for greater detail on this, and related, topics.

Groundwater is important to the hydrologist (as distinct from the hydrogeologist) because it is a store of water at least theoretically available for controlled surface exploitation. Now since groundwater is, by definition, subsurface, neither air nor satellite observations can be used to delineate aquifers directly. Rather, remote sensing may be of value in large-scale preliminary investigations of groundwater reserves. Important indicators of groundwater include anomalous soil moisture and morphologic (especially fracture) zones, vegetation types and densities, geological structure, and streamflow characteristics (especially the density of the drainage network).

Under reasonable conditions all such phenomena can be elucidated by air-photo interpretation procedures and photogrammetry. The use of Landsat and SPOT has been shown to be increasingly profitable, especially but not exclusively in large, poorly surveyed countries. Extensive groundwater surveys have been undertaken, for example, in India, based on Landsat MSS data as the primary data source. Ground surveys are planned for the areas of greatest promise, and test wells drilled in the most promising locations. Then subsurface information from the wells is correlated with the remote sensing indicators so that extrapolations can be made sideways into areas of apparently similar hydrogeology. Indeed, it may be remarked that this type of approach, which integrates remote sensing data with ground truth and then improves assessments of conditions between the ground control points using aircraft or satellite data alone, is fast becoming the classic way of capitalizing on remotely sensed data for environmental monitoring operations.

12.5 Oceanography

Oceanographic remote sensing can be subdivided into two broad categories related to the use, for such purposes, of either general environmental

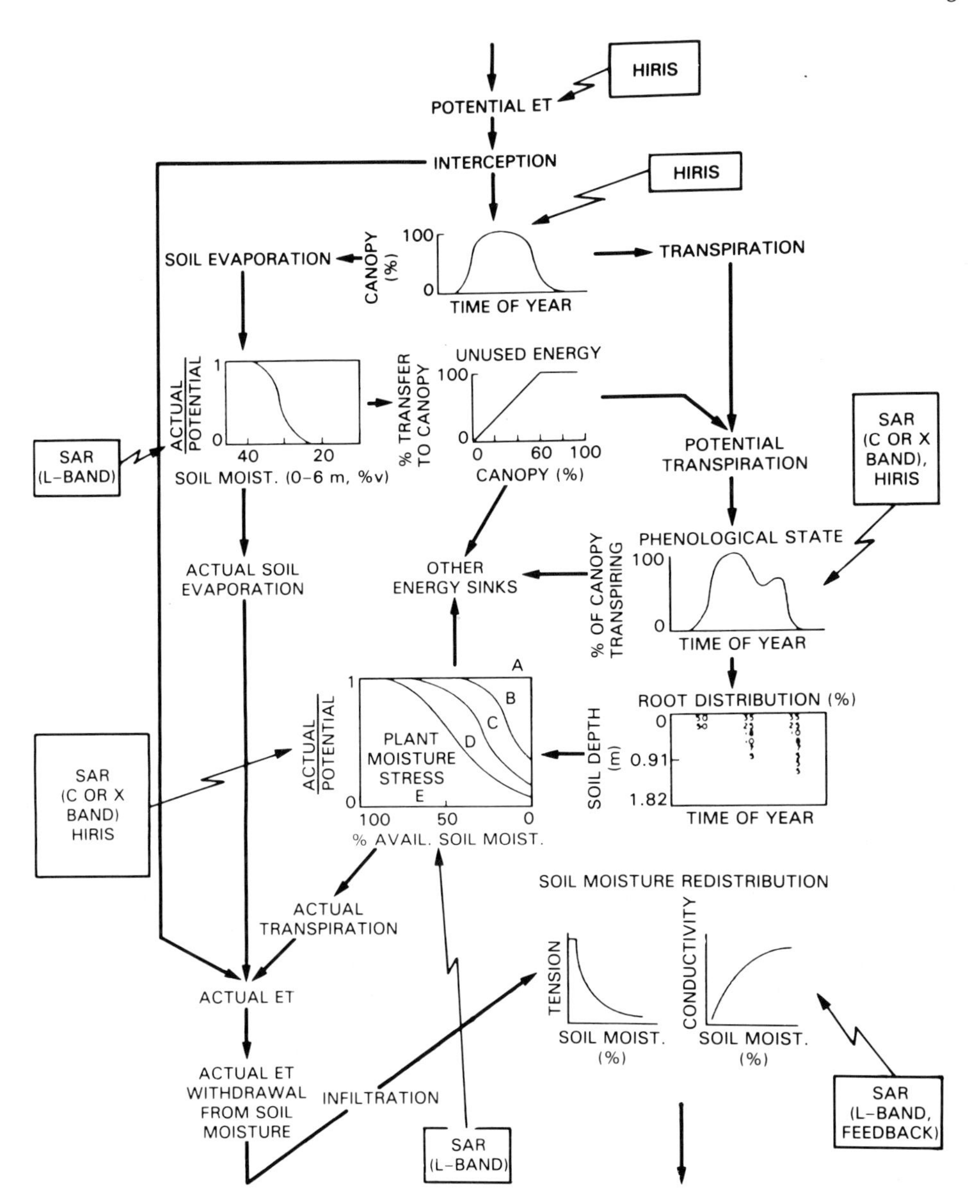

Fig. 12.12 Schematic of water-balance modelling to show synergistic roles of EOS instruments. (Source: Schultz and Barrett, 1989.)

satellites (e.g. NOAA, Meteosat) or more specialized satellites. Those designed with oceanographic applications especially in mind have included Seasat, which suffered such a disappointingly short life of 106 days, but which attracted great interest within the oceanographic community, the Japanese Marine Observation Satellites (MOS), and the European Space Agency's ERS family. For present purposes, therefore, it will be most convenient to review remote sensing applications in oceanography generally at first, before we summarize finally the contribution of these dedicated satellites to ocean science.

12.5.1 *Water temperature and circulation patterns*

Knowledge of *sea-surface temperature* (SST) distributions and changes over large areas is important

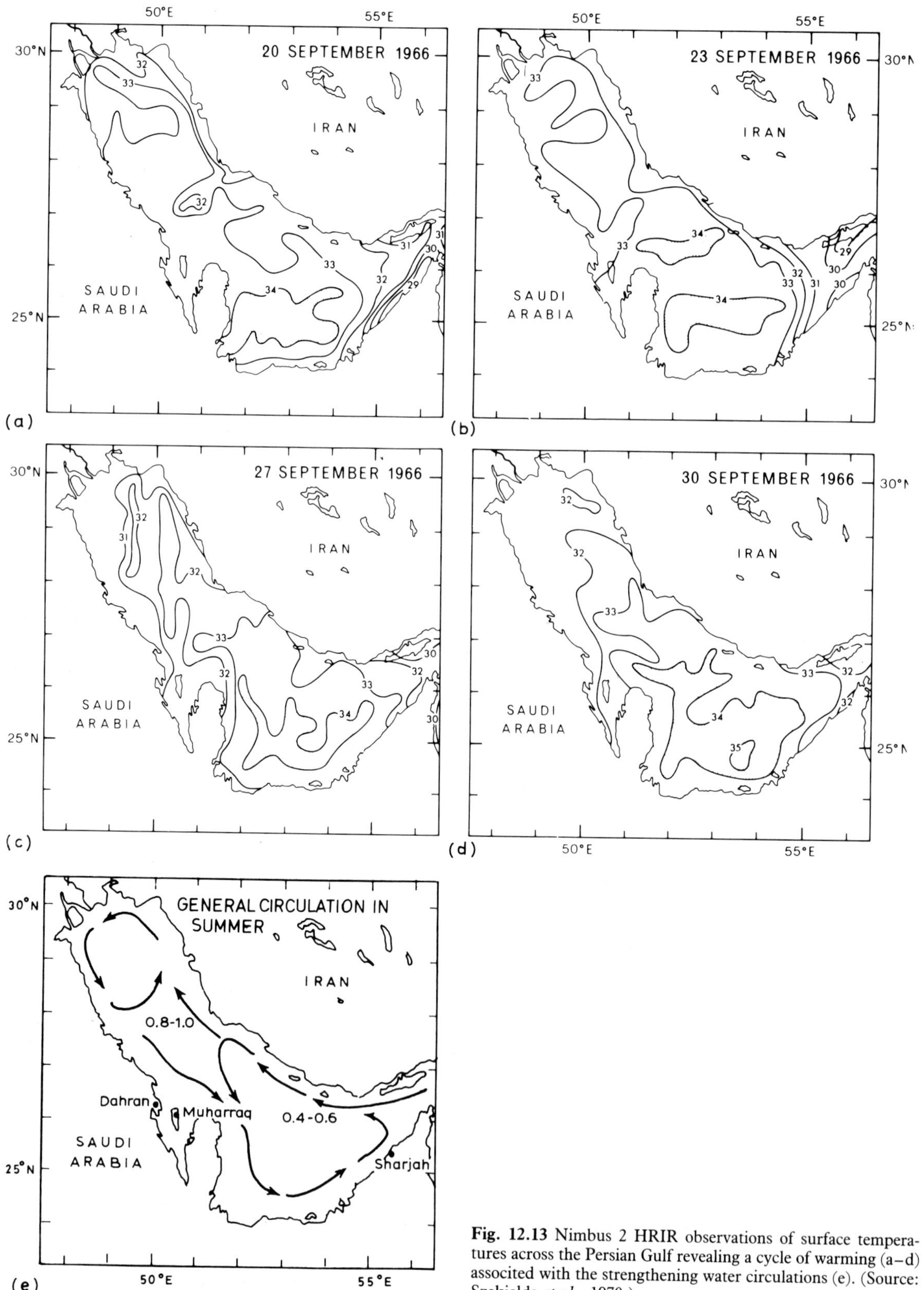

Fig. 12.13 Nimbus 2 HRIR observations of surface temperatures across the Persian Gulf revealing a cycle of warming (a–d) associated with the strengthening water circulations (e). (Source: Szekielda *et al.*, 1970.)

for proper understanding and accurate evaluation of many oceanic and atmospheric processes and phenomena, and for related economic activities, e.g. ship routeing and fisheries. These include the detection and monitoring of ocean currents, water fronts, upwelling zones and other thermal or motion systems, levels of air–sea energy exchanges and the initiation, development and dissipation of many atmospheric organizations at scales ranging from the local (1–10 km) through the meso-, to the synoptic and even global scales.

Any unique interpretation of infrared radiance measurements from a satellite in terms of sea-surface temperature requires a homogeneous target of known emissivity filling the radiometer field of view. Early studies of sea-surface temperatures using Nimbus satellite infrared data employed synchronized cloud photographs to reveal the extent to which radiation temperatures of target areas were contaminated by clouds. These, of course, reduce the temperature levels. Aircraft-borne support systems were used to assess the confidence with which prominent features (e.g. the boundary of the Gulf Stream) could be identified.

Later, temperature patterns were mapped, not only on individual days in various parts of the world, but also on a repetitive basis in selected areas in the search for secular patterns of temperature change. Figure 12.13 illustrates an application of satellite thermal infrared data to the study of short-period temperature variations. Such studies reveal the dominant circulation patterns of water at different temperatures, and assist us in assessing the parts they play in related systems, for example the Earth/atmosphere energy budget. In this case (the Persian Gulf) the incoming water from the Gulf of Oman was found to compensate for the strong local excess of evaporation over precipitation. The inflow current spreads along the Iranian coast until it turns cyclonically back along the Trucial coast. The water is warmed in this shallow region, and flows back eastward as warmer water to mingle with the currents along the Iranian coast. These circulations were previously unknown, but knowledge of this kind was invaluable in early 1991 in efforts to control and disperse oil spilled into the Persian Gulf during the 42-day Gulf War, which for many reasons was hailed as the first 'remote sensing war'.

For global operational mapping of sea-surface temperatures NOAA/NESDIS have developed methods for computer processing of data from operational polar-orbiting weather satellites. Their model is known as GOSSTCOMP (Global Operational Sea Surface Temperature Computation). Surface temperatures are derived by a histogram technique applied to 1024 instrument measurements with partially overlapping fields of view in areas of roughly 100 km^2 around the retrieval point. Corrections for atmospheric attenuation are computed from VTPR data. The model develops Earth-located values of sea-surface temperature between 8000 and 10 000 times daily. Computer products include photographic displays (Fig. 12.14) and gridded fields (map showing each individual temperature estimate). The accuracy of the results is checked twice daily by comparison with ship-board observations. It is generally within ±1.5°C of 'sea-truth'. More detailed products from other NOAA/NESDIS programmes include:

1. high resolution (1 km) maps of surface temperatures of the Great Lakes;
2. bulletins on the location of the Gulf Stream Wall, where the current achieves its fastest flow;
3. experimental Gulf Stream analyses, to show current speeds and areas of cold and warm water eddies;
4. analyses of the USA West Coast Thermal Front for the location of upwellings of cold water, where fish nutrients abound.

Routine methods for sea-surface temperature generation have been developed also based on geostationary satellite data. Of special importance is the interpretation of the initial histograms for clear-column radiances, which for sea-surface temperature monitoring should be uncontaminated by the effects of clouds. Figure 12.15 illustrates some common histogram forms emerging from this work. Where cloud cover is complete, or totally absent, the histograms are characteristically *unimodal*. In partly cloudy regions they tend to be *bimodal* or *multimodal*. The highest temperature mode corresponds to surface temperature, the lowest to extensive cloud contamination. Analysis of remote sensing histograms is a very necessary weapon in the armoury of the remote sensing expert in this and numerous other areas of this science.

Rather like satellite cloud image analyses that

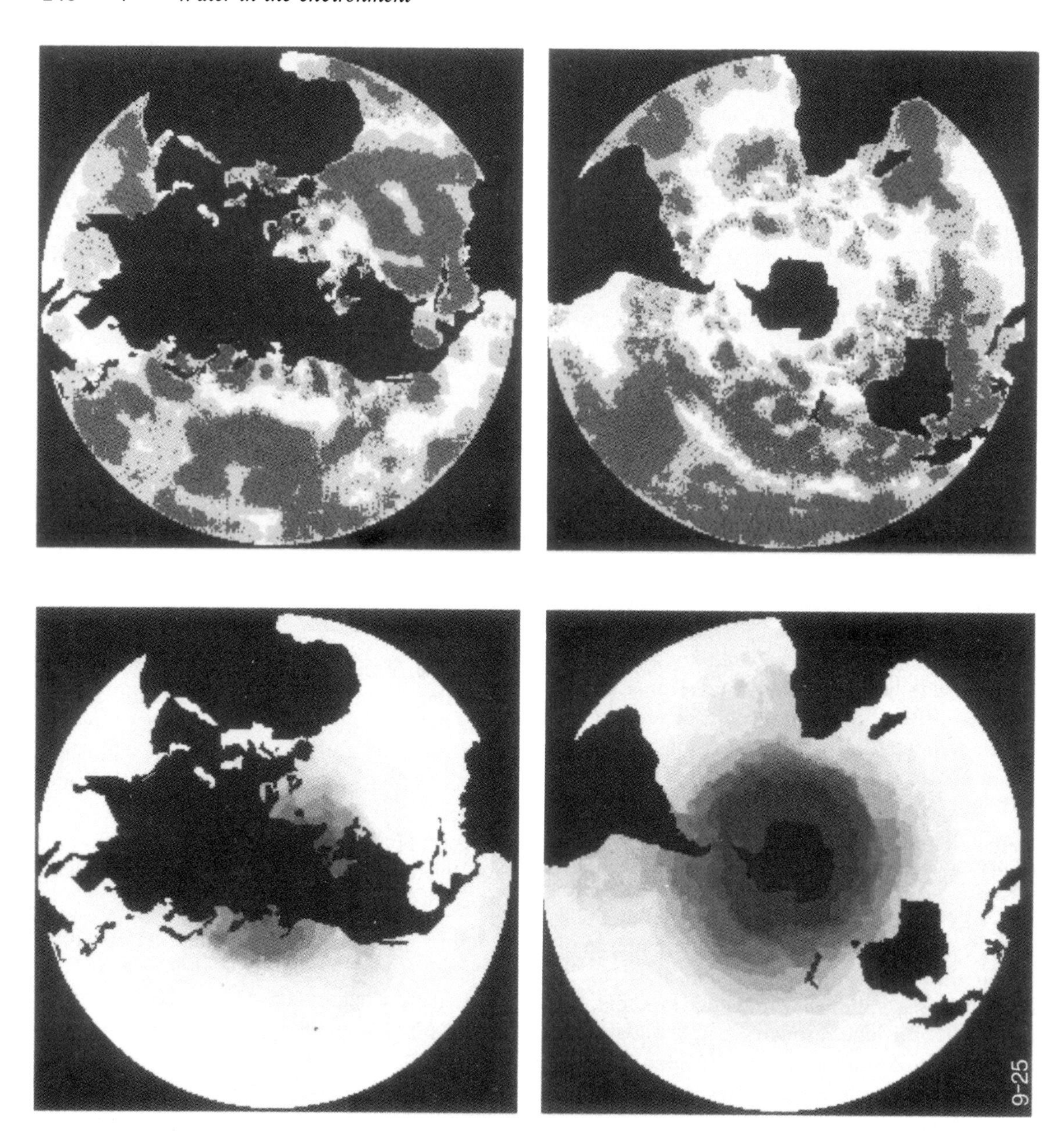

Fig. 12.14 Example of GOSSTCOMP 'quick look' photographic display of sea-surface temperatures (top) and spatial distribution of observations (bottom). (Courtesy, NOAA.)

have revealed features of the atmosphere which were previously unknown, so satellite sea-surface temperature studies have also revealed new features of the global oceans and their circulation patterns. Examples of these are westward-moving equatorial longwave patterns observed in both the Atlantic and Pacific Oceans, which have been identified through computer analyses of AVHRR data in the 3.55–3.93 μm region, using 10.3–11.3 and 11.5–12.5 μm data from the same sensor to effect atmospheric corrections. The resulting SSTs are composited in space and time by retaining the warmest observation for each 50 km pixel in the product, so eliminating most cloud effects. The resulting wave-like features (Plate 9, colour section) have a spatial scale of about 1000 km, and

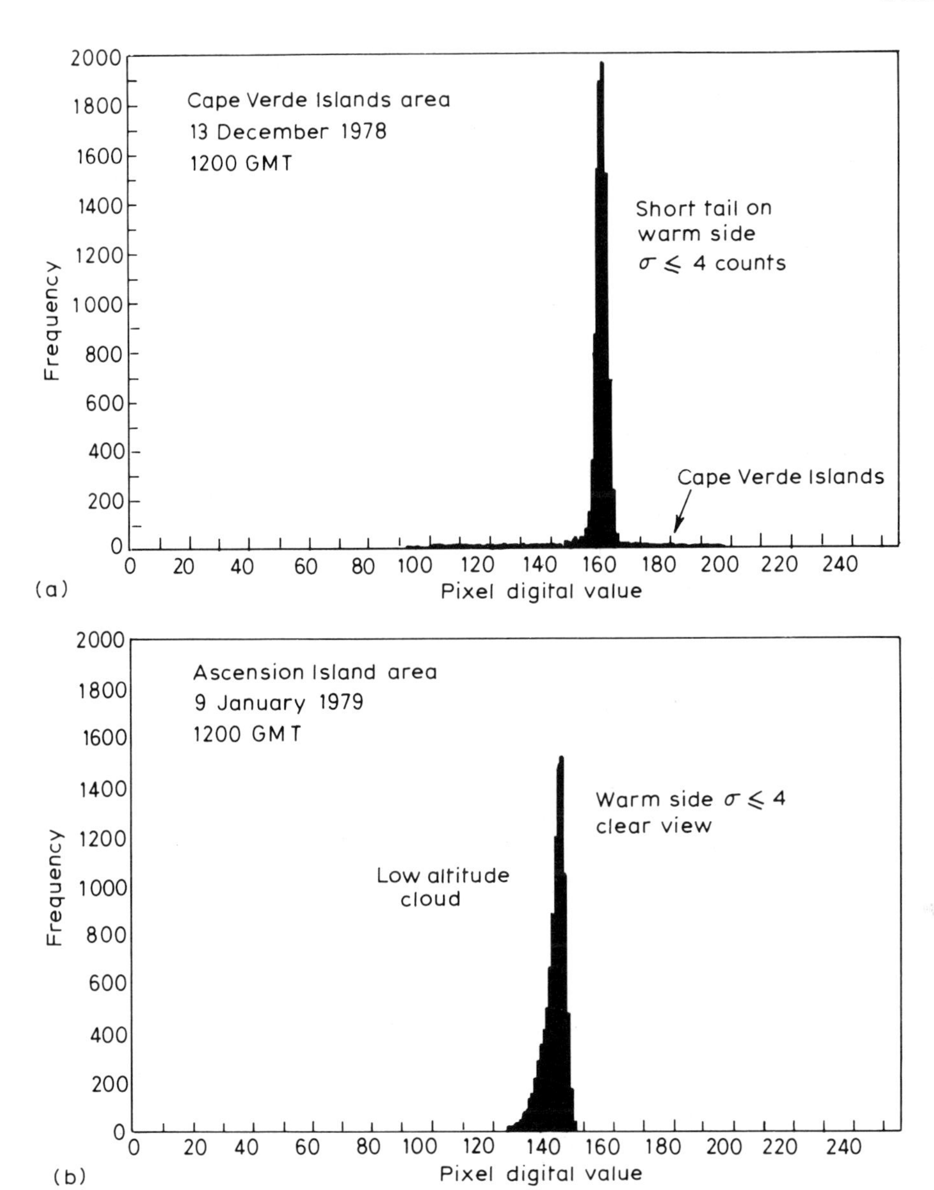

Fig. 12.15 Histograms for 100 × 100 pixel areas of Meteosat infrared window waveband data: (a) clear sea view; (b) clear sky plus low cloud; (c) clear sky plus partial cumuliform cover; (d) complete cloud cover. (Source: Williamson and Wilkinson, 1980.)

move westward with an average phase speed of 40 km day^{-1}. Since they are weak or absent in the Pacific in El Niño years, it seems likely that the waves derive their energy from the shear between the westward-flowing south equatorial current and the eastward-flowing north equatorial counter current and the trade winds that drive them. However, differences between the waves are apparent in the Atlantic and Pacific Oceans, and studies of these differences are now proceeding.

Increasing attention is also being paid to the use of such data for air–sea interaction studies, which may be expected to grow in significance in the immediate future. Here the El Niño is of special interest, not only because of the crucial role it is thought to play in the global weather and climate

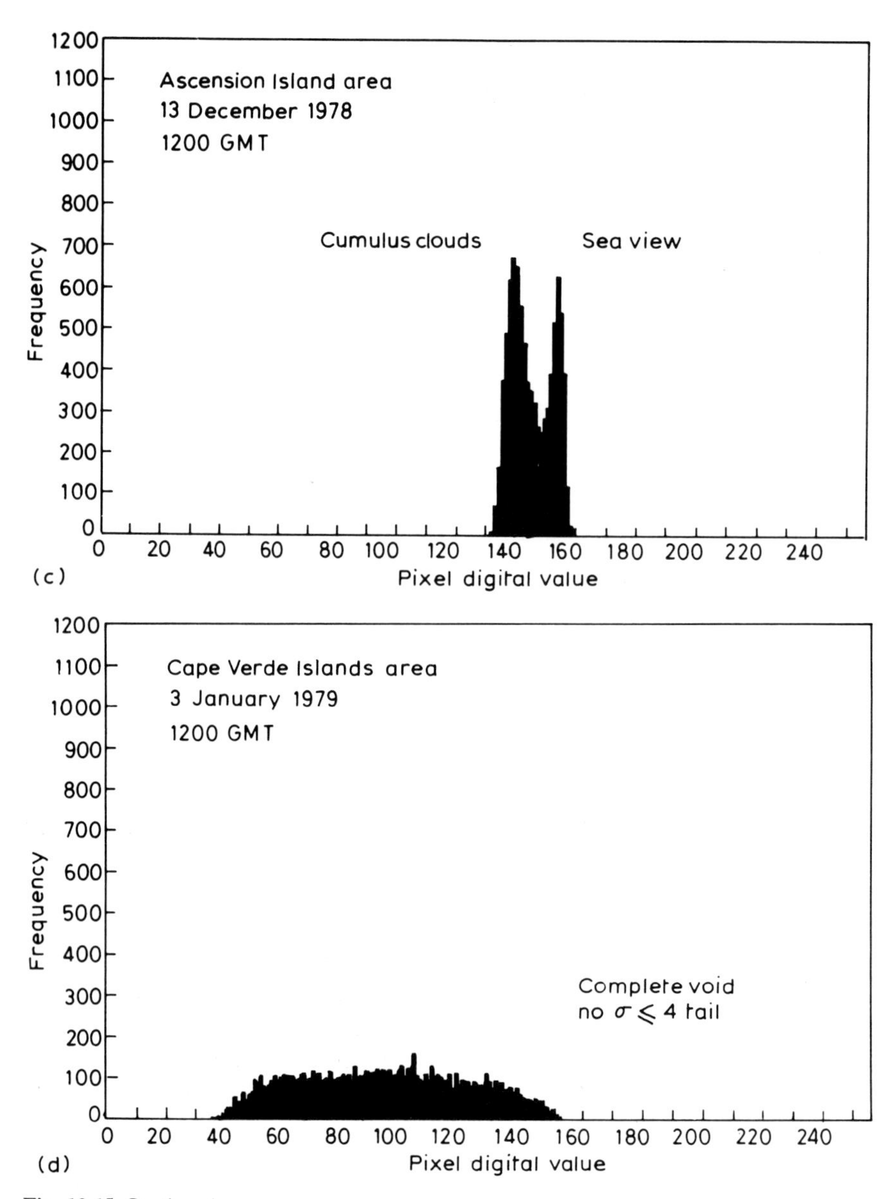

Fig. 12.15 Continued

system as a whole, but also because of the strong economic and human impact of recent El Niño events, e.g. the exceptionally strong El Niño of 1982–1983. This particular event, associated with a surface temperature change of about 1°C over much of the equatorial Pacific, is reckoned to have caused over 1000 deaths plus severe economic damage to South America, Asia, Australia and even Africa, through associated droughts or torrential rains in different places.

Using Nimbus-7 SMMR data as the satellite information source, NASA scientists have compared 'normal' and 'El Niño' years, as illustrated by Plate 10 (colour section) for the January average of 1980 and 1981, and for January 1983, respectively.

To examine first the 'normal year', Plate 10(a) shows the combined monthly averages for January 1980 and 1981 of the SST, surface wind speed, precipitable water, and the latent heat flux derived

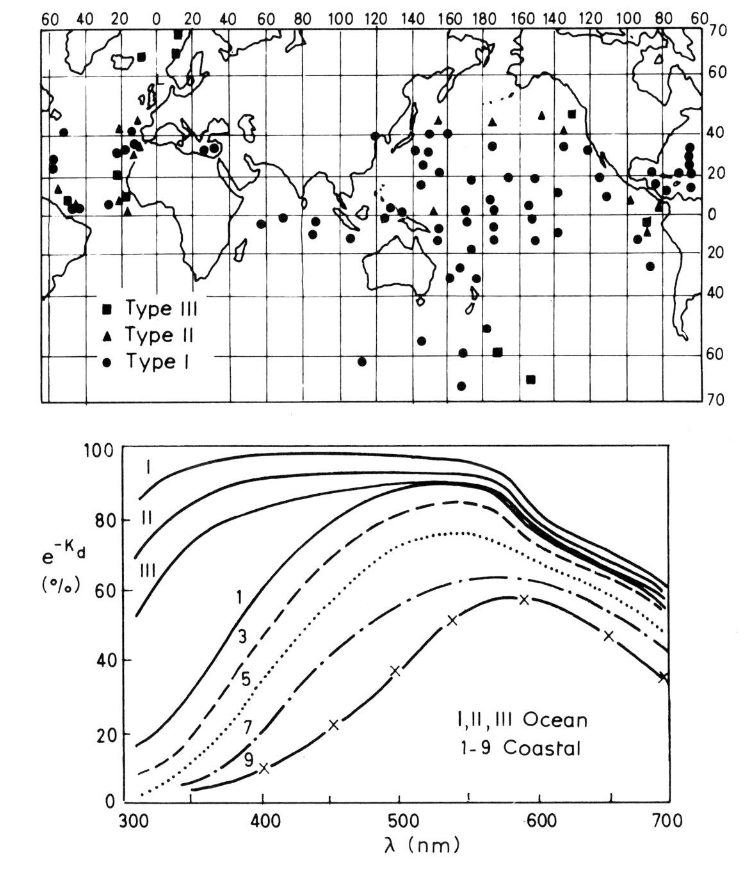

Fig. 12.16 Optical water types classified from aircraft MSS observations. (After Jerlov, 1976; source: Fraysse, 1980.)

from these three parameters. In this normal period, the distributions of these four parameters are closely related. The SST shows that along the Equator and away from the coasts where the SST algorithm breaks down, there is a cold tongue extending from the South American coast that interrupts the warm-water belt of the tropical Pacific. Because of convective lifting of the air and surface evaporation, the regions of large atmospheric water vapour overlie the warm surface water. In contrast, the air over the cold tongue is a region of low atmospheric water vapour. Because of this surface convergence and the rising air, low wind speeds occur in these same regions. Finally, the latent heat flux is approximately proportional to the wind speed.

Later in 1982, one of the most intense El Niño/ Southern Oscillation episodes began. Plate 10(b), which gives the analogous distributions for January 1983, shows the changes that have taken place at the height of El Niño. The SST shows that the cold tongue along the Equator has vanished, owing to the disappearance of coastal upwelling, and that the entire Pacific is warmer. Examination of the winds shows a strengthening of the trade winds north of the Equator and a weakening of the winds to the south. Atmospheric convection above the warm water yields a region of large water-vapour content that extends all along the Equator. Physically, surface observations showed that this change in the water vapour distribution led to large rainfall and flooding in certain of the equatorial islands. Finally, the latent heat flux pattern is strongly altered, with the largest fluxes now lying entirely north of the Equator. The change in these fluxes is dominated by the change in the surface wind patterns.

12.5.2 Water quality and salinity assessments

Turning next to phenomena that are perhaps more significant and certainly more amenable to study at

Table 12.4 Wavebands investigated by the Nimbus 7 CZCS

Band number	Wavelength (nm)	Application
1	443	Multispectral detection of chlorophyll, suspended sediment and Gelbstoffe
2	520	
3	550	
4	670	
5	750	Comparison with Landsat MSS Band 6
6	11.5 (μm)	Ocean surface temperatures mapping

much smaller scales, the *turbidity* or 'murkiness' of water is affected by a wide range of natural and anthropogenic factors, including suspended sediment content, biological activity (e.g. the abundance of plankton), pollution, etc. Entire subdisciplines have grown up to study water quality, especially through reflectance properties of, and light propagation in, water bodies: *optical limnology* deals with freshwater lakes, and *optical oceanography* with oceans and seas. The theory of water optics has been exploited variously in the remote sensing of water quality. Of fundamental importance is the general rule that the spectral distribution of upwelling light from a water body contains quantitative information on the content of suspended and dissolved materials. We may consider two of the many possibilities which depend on this principle.

First, multicategory classifications of oceanic and coastal *water types* have been developed from aircraft multispectral imagery, as exemplified by Fig. 12.16. This case reveals a trend of water clarity extending from the murky inshore waters of high to middle latitudes to the clearest areas of the Earth's tropical oceans. Coastal waters generally are much more turbid than open oceans, not only because of higher concentrations of pollutants but also because of large quantities of so-called 'Gelbstoffe' or 'yellow substance', which is the overall product of dissolved organic matter brought by rivers into these shallow water areas. Since the design of Landsat was optimized for land applications, not for use over oceans and seas, MSS and TM data are of limited use for the mapping and monitoring of water quality types, though MSS Band 5 has proved useful for broad-scale optical wave type determination.

Second, the capacity for useful oceanic application of the Coastal Zone Colour Scanner (CZCS) on Nimbus 7 may be illustrated through reference to monitoring of chlorophyll (phytoplankton) concentrations, as exemplified by Table 12.4. Pre-operational tests suggesting that CZCS visible data should be suitable for estimating chlorophyll through known relationships between chlorophyll concentrations and CZCS band-differences, e.g. the difference between albedos observed in Band 1 minus Band 2 have been amply confirmed by analyses of the CZCS data themselves. Work on this topic is now proceeding with CZCS data, spurred by the knowledge that fishery advisories stand to benefit greatly from successful results, and by the expectation that new sensors of a similar type will soon be operational, e.g. on board the SeaWifs spacecraft (p. 397).

Finally, a word on marine pollution monitoring, a theme of growing international significance at the present time.

Contamination of water bodies by human activities often lends itself readily to investigation by remote sensing techniques since the effluents of commercial and industrial activity often contrast sharply with the waters which receive them. Aircraft have been used in many instances to identify and map discharge patterns into the sea from sewage outfalls and industrial complexes. Oil spillages, both from ships at sea and oil leaks from maritime wells, can be located with particular ease from aircraft (including helicopters, both light and major coastal patrol types) – and potentially from satellites. There are four reasons why this is so:

1. the emissivity for petroleum products is significantly higher than for a calm sea surface;
2. crude oil pollutants have increasing emissivities with increasing gravity;
3. the radiometric response of oil varies little with time of day or the age of the pollutant;
4. it is possible to design reliable microwave systems to map oil pollution.

Already, remote sensing data – particularly from airborne sensors – have been used in courts of law to obtain convictions of those responsible for significant pollution events. Use of remote sens-

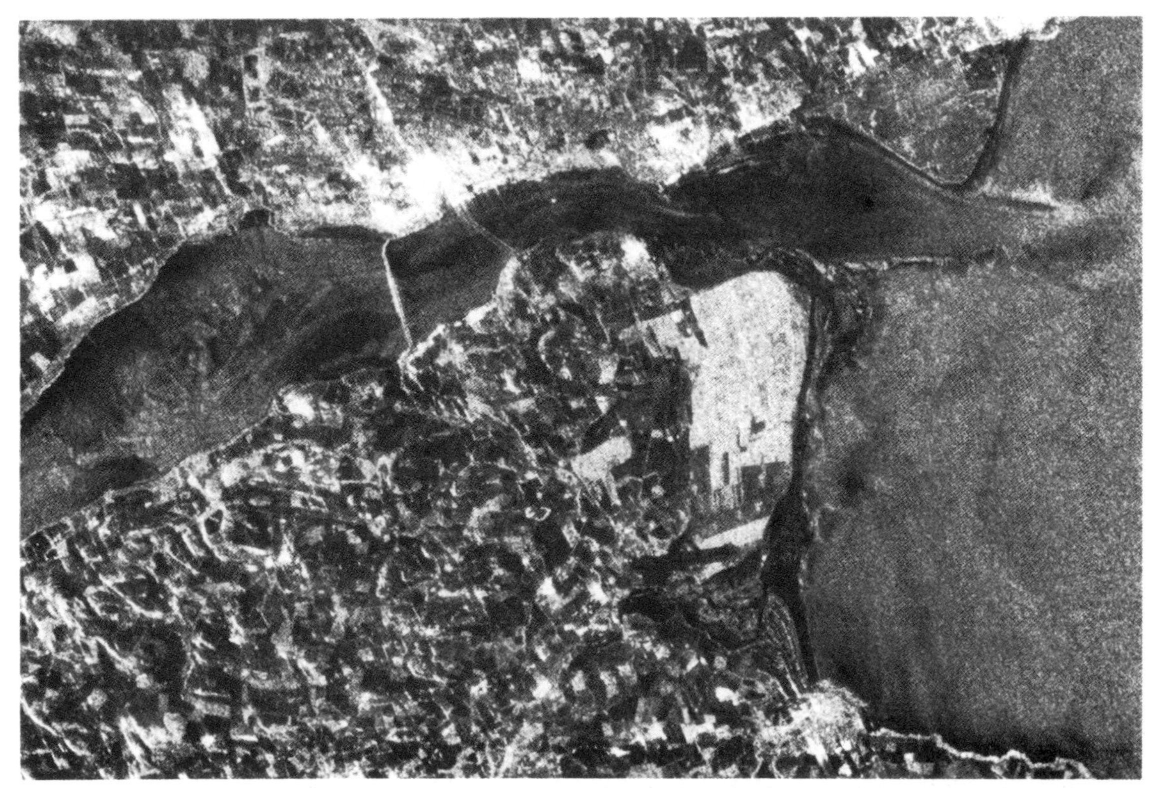

Fig. 12.17 (a) Seasat SAR image of the Tay Estruary, Scotland, 19 August 1978, showing many coastal and land surface features.

Fig. 12.17 (b) Seasat SAR image of Dunkirk, France, 19 August 1978, showing sea surface features related to the topography of the sea bed. (Courtesy, Space Department, RAE, Farnborough; Crown Copyright Reserved).

ing in support of international law is becoming commonplace.

12.5.3 Satellite oceanography: the way ahead

The first purpose-built oceanographic satellite, Seasat 1, was launched into a near-polar circular orbit at an altitude of 800 km on 26 June 1978. Unfortunately the satellite ceased to provide data only 106 days later on 9 October, thus marking the end of its useful life. However, Seasat had already demonstrated the great value of its instrument package, and had returned a rich supply of both land and sea-surface data to Earth. Some researchers have remarked that the early demise of Seasat 1 was fortunate in that it has forced the principal investigators to examine a manageable quantity of data thoroughly, rather than an overwhelming flood of data very partially – which has been the rule, not the exception, with most other satellite data set.

Seasat carried four microwave sensors and an auxillary Visible–Infrared Radiometer (VIRR). The suite of microwave sensors, which gave the spacecraft its all-weather operating capability, included:

1. A radar altimeter looking vertically down to measure wave height and the microtopography of the ocean surface.
2. A synthetic aperture radar (SAR), which scanned a 100 km wide swath located some 250–350 km to the right of the subsatellite track, primarily to measure wave length and direction, to provide data for the study of coastal processes, and to chart ice fields and leads.
3. A scatterometer scanning out to 1000 km on either side of the spacecraft for wind speed and direction assessments.
4. A passive scanning multichannel microwave radiometer (SMMR), which recorded naturally emitted microwave radiation from the target at five frequencies between 6.6 and 37 GHz, primarily for global measurement of sea-surface temperature.

It is generally agreed that the additional primary objectives of the Seasat mission were well met. Many bonuses have also accrued, for example through the realization that much of the SAR imagery over land contains information of potential value for geological, hydrological, glaciological, and even – a useful bonus – some land-use studies. Figure 12.17(a) illustrates the high resolution achieved by the Seasat SAR, and the wealth of detail in the imagery over land and in coastal zones. One exciting and entirely unexpected finding has been that, under suitable conditions of sea-bed topography and tidal flow, seafloor features have visible expressions on the water surface. Figure 12.17(b) is a good example of this effect in the English Channel. Many other interesting phenomena also may be seen in Seasat SAR data, e.g. wave refraction and diffraction in coastal zones, current features, inlet plumes, and wetland characteristics.

Since Seasat, other new satellite/sensor systems have been developed which are of special significance to oceanography. Admittedly MOS-1, the first of Japan's Earth observation programme satellites, was designed with both land and marine applications in mind: its objectives were to establish a basic technology common to both areas of application, as well as to data collection systems. However, the chief emphasis of this family has been on observation of the sea surface through visible, infrared and microwave radiometers. The instrument package on MOS-1 consisted of:

1. A Multispectral Electronic Self-scanning Radiometer (MESSR), a four-channel visible and near infrared radiometer providing a swath width of 200 km and a resolution of 50 km (cf. the Landsat MSS).
2. A Visible and Thermal Infrared Radiometer (VTIR), with one visible and three thermal infrared (one water vapour and two split window) channels, a swath width of 500 km and a resolution of 0.9 km (cf. the NOAA–AVHRR).
3. A Microwave Scanning Radiometer (MSR), a dual-frequency (23.8 and 31.4 GHz) vertically and horizontally polarized imaging system giving a swath width of 317 km and resolutions of 32 and 23 km respectively.

The primary applications of these instruments in oceanography include coastal zone mapping and monitoring, sea-surface temperature monitoring, and sea state (wind speeds near the surface), respectively. Pre-launch simulations of MSR evaluations of sea state are shown in Fig. 12.18.

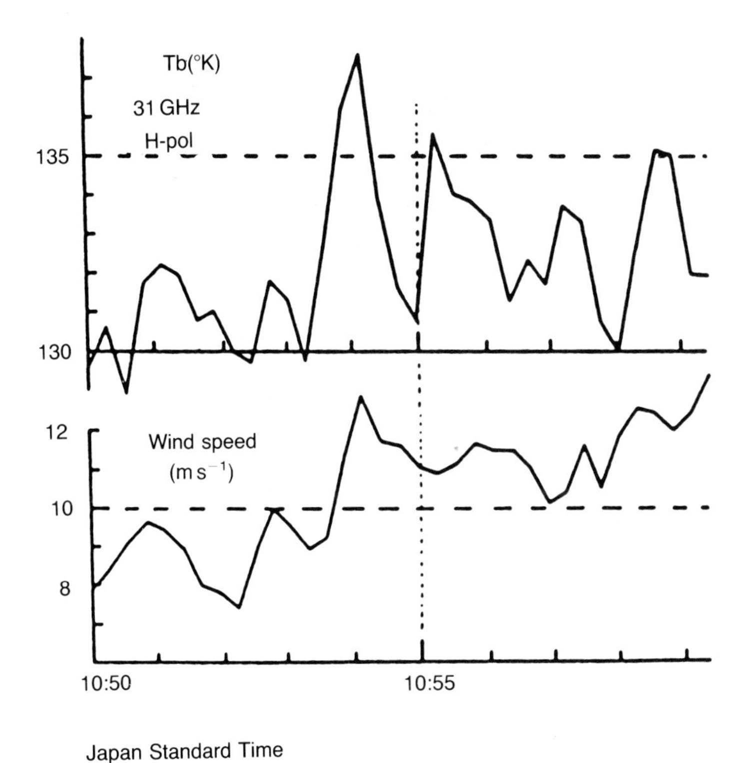

Fig. 12.18 Tb versus wind speed plots from the MSR on MOS-1. Both are 10 min averages for data recorded on 12 November 1982. (Source: Tsenchiya *et al.*, 1987.)

These were based on the expectation that ocean surface roughness and both microwave reflectivity and emissivity are related to one another. One consequence is that, at some frequencies and polarizations, useful relationships are found between microwave brightness temperatures and wind speeds near the surface; some emissions from calm surfaces are strongly polarized, whilst at the same frequencies emissions from less calm surfaces are relatively unpolarized.

These principles have been more thoroughly exploited in relation to data from the wide-angle DMSP–SSM/I sensor, and instantaneous wind speed maps have been prepared both for local regions, and the world's oceans as a whole. Claimed accuracies are of the order of $\pm 2\,\mathrm{ms}^{-1}$, although it seems likely that existing algorithms will benefit from local regional calibration. Clear opportunities exist for such products to be used profitably in a wide range of contexts, from global air–sea interaction modelling at one extreme to fisheries and ship routeing at the other. The same sensor is also proving very useful for sea ice assessment and monitoring, and related practical applications in high-latitude and fringe polar regions.

Last but not least reference must be made to ERS-1, which, after long delays, was launched successfully on 16 July, 1991. It is expected that particular value for oceanography will be derived from its radar altimeter and wind scatterometer, but that its SAR data also will be of significance especially in coastal zones and in areas of sea ice.

As in other Earth-observation application areas, perhaps the most effective oceanographic uses of visible and infrared data, and data from active and passive microwave systems, such as the ERS-I SAR and the MOS and DMSP passive microwave radiometers, will arise from their *combined uses*, the relatively high spatial but low temporal resolution data from the active systems being used for local studies and for calibrating the relatively low spatial but high temporal resolution data from the passive systems, which in turn will be of most value for global applications.

Undoubtedly oceanography will benefit greatly from future dedicated satellites and instrument packages. Indeed, it is not too much to expect that oceanography, potentially, if not presently, the biggest subdivision of hydrology, will undergo a revolutionary growth in scale and practical significance as ocean-monitoring satellites increase our knowledge of maritime conditions by a veritable 'quantum leap': since oceans cover seven-tenths of the global surface, yet have received relatively little attention compared with land areas, no-one could suggest that they do not deserve much more attention in remote sensing than they have yet been given.

13 Soils and landforms

13.1 The nature of soils

Soil is of major importance to mankind because it supports the plants and animals that provide man with food, shelter and clothing. It is also of importance as the foundation material for the buildings in which we live. Because of these quite different reasons for the significance of soils, different concepts of soil have been developed by agriculturalists on the one hand and engineers on the other.

The agriculturalist views soil as a complex entity of minerals, water and air much influencing and influenced by living things. To him or her most of the soil profile is significant for plant growth: the *surface* (A horizon) because it is the seat of biological activity and the principal source of nutrients; the *subsoil* (B horizon) because it affects drainage, soil moisture retention, aeration and root development of plants; and the underlying parent material (C horizon) because it may contain weatherable minerals yielding nutrients, and its texture may affect permeability. Both A and B horizons are important too in respect of their influence on soil temperatures. Deeper parts of the regolith (D horizons) are generally of less importance to plant growth.

The productivity of both natural and agricultural ecosystems depends on maintaining soil quality together with the habitats of beneficial insects and other animals, such as crop pollinators and the predators and parasites of pests. Soil is therefore a crucial part of the life-support system, but disturbance of the natural balance in soil dynamics is now having a major impact on soil fertility in many regions. Soil erosion is a particularly serious problem in tropical countries. For example, in India it is estimated that 6000 million tonnes of soil are lost in this way alone every year from a surface area 800 000 km^2, i.e. about 4 mm in depth over the entire country.

A gradual erosion of the surface layers of soil is a natural feature of the geomorphological process in many regions but in undisturbed ecosystems the soil cover is generally regenerated at about the same rate that it is removed. Under natural vegetated conditions it takes about 1200 years to form about 30 cm of soil. Accelerated soil erosion is often induced by human action following removal of the natural vegetation. Deforestation is particularly damaging in areas of tropical soils, which are often inherently infertile and subject to rapid erosion. Apart from the deleterious loss of soil another problem often occurs as a result of the eroded material collecting as unwanted sedimentation elsewhere, e.g. in lakes or reservoirs.

In contrast to the agriculturalist, the engineer normally views soil merely in terms of unconsolidated sediments and deposits of solid particles derived from the disintegration of rock. Therefore in civil engineering the concept of soil includes all regolith material, and a sharp distinction between rock and soil is no longer made. In fact, the engineer is normally more concerned with what the agriculturalist would term the C and D horizons than with the upper A and B horizons. From the engineering standpoint soil can be regarded as a three-phase system in some state of dynamic equilibrium. The three phases are soil (particulate organic and inorganic material), liquid (consisting of soil solutions containing various salts) and a gas phase containing soil air with changing amounts of oxygen, nitrogen and carbon dioxide. Plant cover and rooting systems are significant as stabilizers of soil conditions, but the engineer has few if any other interests in life within the soil.

Table 13.1 Forms and characteristics of soil water (Source: Curtis, 1977)

Forms of water	Physical characteristics	Biological characteristics
Free (gravitational)	Loosely held at less than 0.5–0.1 bar tension	Undesirable and superfluous, can lead to anaerobic conditions, usually removed in drainage
	Moves in response to film tension and gravitational forces Molecules not strongly oriented	Plant nutrients are leached by this water
Capillary	Tension of water films varies from 0.1–31 bar Moves by film adjustment from thick to thin films (i.e. in response to variations in tension) High tensions may produce orientation of molecules	Desirable, forms the bulk of the water available to plants Functions as the soil solution providing nutrients
Hygroscopic	Tension greater than 31 bar Largely non-liquid Moves mostly in vapour form Water molecules may assume orientation	Not available to plants It is held mostly in the soil colloids

13.2 Conventional mapping of soils

The recognition of soils as organized natural bodies is a development of nineteenth-century science and is generally credited to the Russian scientist Dokuchaiev who first explained the relationships between soil profiles and the environment. It was not until the 1920s, however, that the soil *units* were eventually mapped on the basis of the soil profile using A, B, C horizon notations. With further experience of soil mapping it was recognized that soils were very variable in their properties and several kinds of soil might occur within a mapping unit. Even though a major part of a mapped unit might accord with the definition of a particular soil type, not all of the area would conform. Thus it became usual to recognize that a percentage (up to 15%) of soils within a mapping unit would show variation from the described attributes of the unit. In general, the user wishes within-class variance to be as small as possible, or below a certain threshold value. A number of studies of soil variability have been made. One of the classic studies which focused on variance and intra-class correlations for various topsoil properties in Oxford, England, showed that mechanical properties were acceptable in terms of within-class variance, whereas chemical properties were not.

Studies of soil units by remote sensing methods must take into account the very different concepts of soil held by agriculturalists and engineers, along with the variability of topsoil characteristics, particularly in respect of chemical properties.

The soil categories used by agriculturalists that are significant to remote sensing studies include those of great soil groups, subgroups, soil series and soil phases. The 'great soil group' brings together soils having similar horizons arranged in a similar sequence, e.g. typorthod (podzol), a broad zone of podzolic soils extending across northern Canada and Russia over a variety of rock types. This unit is, therefore, suitable for small scale (1:1 000 000 or less) or regional maps. At medium scales (c. 1:50 000–1:100 000) the 'soil series' category is more commonly used. A soil series consists of a group of soils of similar profiles derived from a particular parent material. The soil series is often subdivided into 'soil phases'. The basis of subdivision into phases may be any characteristic or combination of characteristics potentially significant to our use or management of soils. The most common differentiating attributes for soil phases are slope, degree of erosion, depth of soil, stoniness, salinity, physiographic position and contrasting layers in the substratum.

Soil classifications for engineering purposes

are primarily based on the sieve size of the soil material. Thus the extended Casagrande Soil Classification as used by engineers has, as its major divisions, coarse-grained soils and fine-grained soils, subdivided into gravelly, sandy and fine-grained soils. Finer subdivisions in the classification use criteria such as uniformity of deposit, grading of deposit, plasticity, shrinkage or swelling properties, drainage characteristics and dry density.

13.3 Soil parameters detected by remote sensing

Although remote sensing techniques cannot record the profile characteristics of soils, several useful parameters can be detected thereby. First, the *surface colours* of soils are recorded. These affect the percentage of insolation received at the surface which is reflected back to space. Dry sandy soils may have an albedo percentage of 37% whereas moist black (highly organic) soils have an albedo of about 8% (Table 13.1). Second, *surface roughness* may be observed, e.g. whether created by cultivation of soil or resulting from the inherent soil structure, for both of these affect the scattering of incident radiation.

Third, *soil temperature* can be detected by thermal sensors. A major factor affecting the temperature relations of a soil is its specific heat or *thermal capacity*, i.e. the amount of heat required to raise the temperature of a given substance from 15 to 16°C compared to that required for the same rise in temperature of an equal weight of water. Under field conditions the soil's moisture content more than any other factor determines its thermal capacity. For example, a dry mineral soil may have a specific heat of 0.20 but with a moisture content of 30% the capacity increases to 0.38. It will be evident, therefore, that sensing of *thermal regimes* of soils, i.e. their annual, seasonal, and even their diurnal variations, will provide indirect evidence concerning the likely *moisture variability* in soils.

Fourth, soil moisture can be assessed by remote sensing through its influence on the returns obtained from microwave radar. The related *dielectric properties* of soils have been studied by many researchers: it is clear that the so-called 'dielectric constant' depends mainly on the soil water content. Variations in water content produce substantial changes in soil dielectric properties. As a result many studies of soil moisture/radar relationships are found in the literature and some pertinent key references are given at the end of the book for this chapter. Methods used to measure soil water in the field vary considerably. It may be measured in terms of percentage of dry weight (*gravimetric method*) of the solids forming the soil. This method produces considerable problems in that soils of different organic matter content cannot easily be compared. For this reason measurements as a percentage of the volume of soil (*volumetric method*) are preferred. More recently, however, remote sensing data have been related to the 'available water' in the soil i.e. the water that is available to plants. This water is best described in terms of the tension under which water is held in soil, as shown in Table 13.1. A soil that contains both gravitational and capillary water held at a pressure of about one-third that of the bottom of the atmosphere is said to be at *field capacity*. This is an important measure for the agricultural use of soils, and its relationship to remote sensing data is now being studied.

13.4 Remote sensing of soils and landforms by photographic systems

13.4.1 Conventional air photography

We have seen that engineers and agriculturists adopt different definitions and classifications of soils. As a result, remote sensing techniques are required to serve several purposes based on different definitions of soil and different basic concepts. The first major post-war study of the recognition of soil conditions, made at Purdue University, USA was aimed mainly at identification of soil parent materials and was chiefly of value to those interested in soils engineering. It was based on two assumptions, first, that once photographic characteristics for soils at one site had been determined the identification of similar sites elsewhere on the photographs could be used to identify similar soils; and second, that by study of aerial photographs such features as landforms, surface colours (or tones), erosion, slope, vegetation, land use, micro-relief and drainage patterns could be used to deduce the general character of the soil.

Further development of this latter approach by

the International Training Centre for Aerial Survey in Holland led to a widely used technique for pedological analysis. In this, each element of the landscape is mapped out separately (e.g. slope, vegetation, land use) and maps are prepared from a composite of the element boundaries. The boundaries that were coincident with more than one element of the landscape were given additional weight in the final interpretation of soil boundaries.

The widespread recognition of the close association of soils and landforms promoted much interest in landscape mapping as a means of reconnaissance soil mapping. Research was based on the premise that if detailed characteristics for one terrain unit were known, one could apply them to other analogous terrain units elsewhere. The problem facing terrain analysis was (and continues to be) one of classification and definition of the units to be recognized in aerial photographs. It was necessary to have some means of identifying each terrain unit and to store data relevant to the unit so defined, so that other workers could recognize them elsewhere. Thus a system of hierarchical classification was proposed as shown in Table 13.2.

Table 13.2 Terrain unit classification

Unit	Description
Land element	The simplest part of the landscape; uniform in lithology, form, soil and vegetation
Land facet	Consists of one or more land elements grouped for practical purposes
Land system	Recurrent pattern of genetically linked land facets
Land region	Consists of one or more occurrences of land system local forms which are generally contiguous
Land division	An assemblage of surface forms expressive of a major continental structure (morphotectonic)
Land zone	The world extent of a major climatic type

This 'land system' approach has been widely applied, for example, in Australia by Land Research and Regional Survey teams, in South Africa by engineers, and in Africa by agriculturalists. It is best applied to those areas that display distinctive and well-defined facets of landscape. Where units are ill-defined and/or where cultivation has largely destroyed the natural vegetation cover it is less easily applied.

The development of computing techniques has subsequently led to such investigations of the parametric description of landforms to be placed on an entirely numerical footing. Computer analysis of landform now seems to be within our grasp and future developments of the land system approach are likely to be in the form of *automated cartography* using some form of supervised classification based on knowledge of soil-landscape relationships.

Meanwhile, the accuracy of soil mapping from the air photograph has proved to be very variable. Comparisons of soil maps made from ground observations with photo-interpretation maps have revealed that increasing complexity of geological deposits or increasing importance of properties not closely associated with landforms greatly reduces the accuracy. Also there is increasing awareness that the timing of air photography is critical. Photographs taken at inappropriate times may provide an unsatisfactory and even potentially misleading yield of information. Under European conditions photographs to record soil-tone patterns and soil change should be taken in winter and spring, the best months being March and April (Fig. 13.1(a and b)). Crop patterns which reflect soil boundaries are best recorded in July and early August. These requirements must, however, be fitted into the period when conditions are suitable for aerial survey, so the planning of flights and placing of contracts should be carried out well in advance.

Suitable statistical methods for the testing of the goodness of soil boundaries drawn by photo-interpretation have become necessary following the increasing use of air photography. Multiple correlated attributes of the soil profile can be reduced to a single variate expressing a large proportion of available information by principal component analysis. This variate, the first principal component, can be plotted against distance on a linear transect and approximate positions of maximum (positive) and minimum (negative) slope found by inspection. These positions are then pin-pointed using Student's t statistic. They represent the points where the soil boundaries – the maximum rates of change of soil with respect to distance

Fig. 13.1 Soils on a former estuarine marsh and adjacent upland, near Martin, Lincolnshire: (a) under crops, July 1971; (b) bare soil, April 1971. Note the clear portrayal of the patterns of creeks that formerly traversed the marsh. The tonal variations in the photograph along the creek lines are primarily due to variations in soil texture and moisture content. (Source: Cambridge University Collection; Copyright Reserved.)

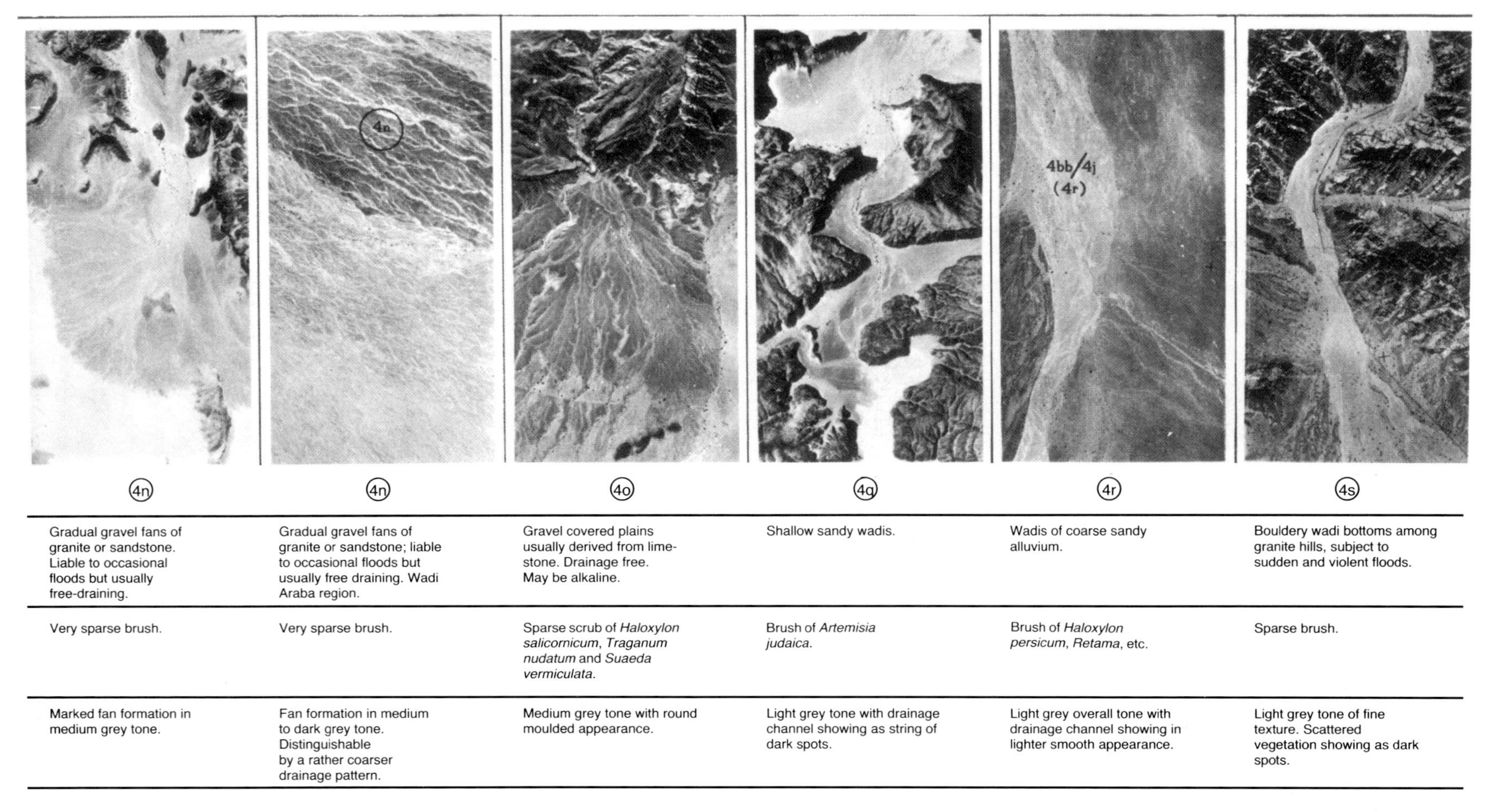

4n	4n	4o	4q	4r	4s
Gradual gravel fans of granite or sandstone. Liable to occasional floods but usually free-draining.	Gradual gravel fans of granite or sandstone; liable to occasional floods but usually free draining. Wadi Araba region.	Gravel covered plains usually derived from lime-stone. Drainage free. May be alkaline.	Shallow sandy wadis.	Wadis of coarse sandy alluvium.	Bouldery wadi bottoms among granite hills, subject to sudden and violent floods.
Very sparse brush.	Very sparse brush.	Sparse scrub of *Haloxylon salicornicum*, *Traganum nudatum* and *Suaeda vermiculata*.	Brush of *Artemisia judaica*.	Brush of *Haloxylon persicum*, *Retama*, etc.	Sparse brush.
Marked fan formation in medium grey tone.	Fan formation in medium to dark grey tone. Distinguishable by a rather coarser drainage pattern.	Medium grey tone with round moulded appearance.	Light grey tone with drainage channel showing as string of dark spots.	Light grey overall tone with drainage channel showing in lighter smooth appearance.	Light grey tone of fine texture. Scattered vegetation showing as dark spots.

Fig. 13.2 A photographic key for soil and range conditions in Jordan. Some mapping categories, e.g. 4n, require more than one photograph to illustrate the appearance of a particular soil and range condition. This key was found to be effective when used in 1990, several years after its inception. (Courtesy, Aerofilms, Hunting Surveys Ltd, Boreham Wood, Herts.)

– cross the transect. Beckett (1974) well reviewed the statistical assessment of resource surveys by airborne remote sensors, particularly in respect of the precision of information and the costs of obtaining data on soil conditions at different scales.

The use of colour photography as an aid in soil mapping only became important after 1960 following advances in technology by film manufacturers. Studies on the use of colour film for engineering soil and landslide studies showed that it gave a saving in interpretation time, often of about 50% in the USA. Research by British and Russian scientists indicated that colour aerial photography yields more data on soil conditions than panchromatic photography. Work has shown, however, that soil colours as recorded in the field and on aerial photography do not correspond exactly in respect of their Munsell colours, indeed overlap of hue, value and chroma is quite frequent. The influence of different soil hues on the optical densities of aerial colour and aerial infrared colour has subsequently been studied by densitometric methods. For example, statistical tests applied to density data obtained from one sample of 12 soils showed that the soils could be separated into two groups on the basis of significant differences in densities. One group consisted of those soils with low chroma (soils grey or neutral in colour) and which were best distinguished by infrared colour. The second group included soils with high chroma (intense soil colour) and which were best distinguished with colour film.

Infrared colour transparencies require suitable techniques for viewing and annotation and suitable light tables and filter frames have been developed as more and more infrared colour has been used. Techniques have also been developed for the production of panchromatic internegatives and positive prints from infrared colour film. There remains, however, the fact that transparencies are less convenient for use in manual photo-interpretation and this may partly explain the somewhat slow adoption of infrared by user agencies despite its considerable information yield in respect of soil conditions and crop vigour. A summary of the advantages and disadvantages of different types of film used in aerial soil surveys is given in Table 4.2.

In order to communicate the photographic characteristics of soil and terrain units to other users, photographic keys of various kinds have been devised. Many are now available, including keys for vegetation, landforms, forest sites, and soil conditions. An example of a photographic key for soil and range conditions in Jordan is shown in Fig. 13.2. The continuing interest in site selection is reflected by recent publications dealing with terrain evaluation, some of which are listed in the further reading recommended for this chapter.

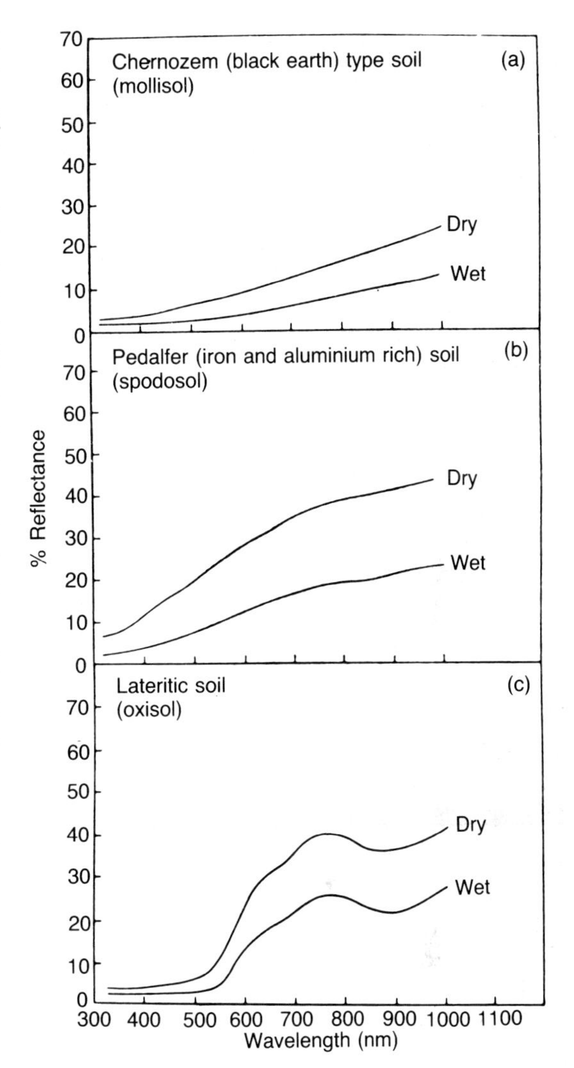

Fig. 13.3 Spectral characteristics of soils. (Source: Condit, 1970.)

13.4.2 Multispectral photography

Since soils are composed of mineral and organic constituents together with varying amounts of soil

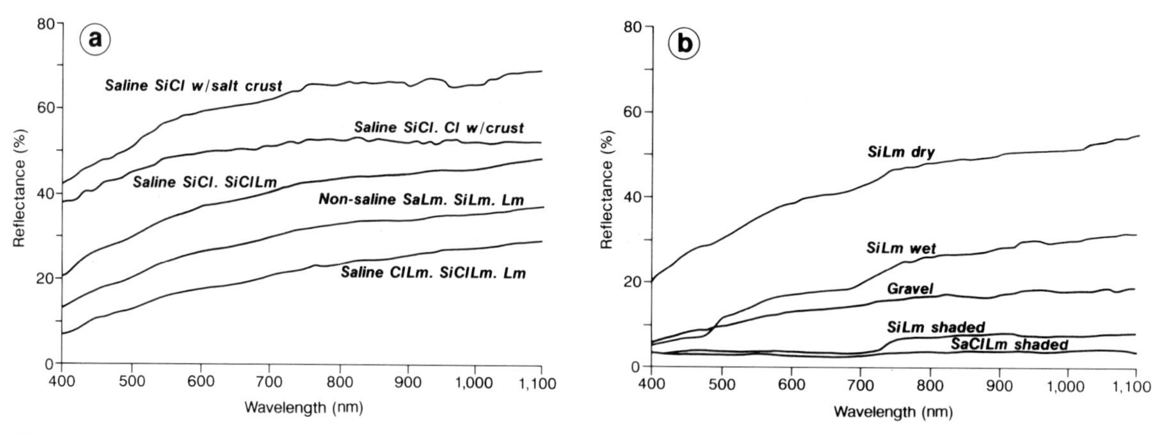

Fig. 13.4 Spectra of sunlit and shaded soils from ground-based studies in Nevada using a 15° for Spectroradiometer. (Source: Satten White and Ponder Henley, 1987).

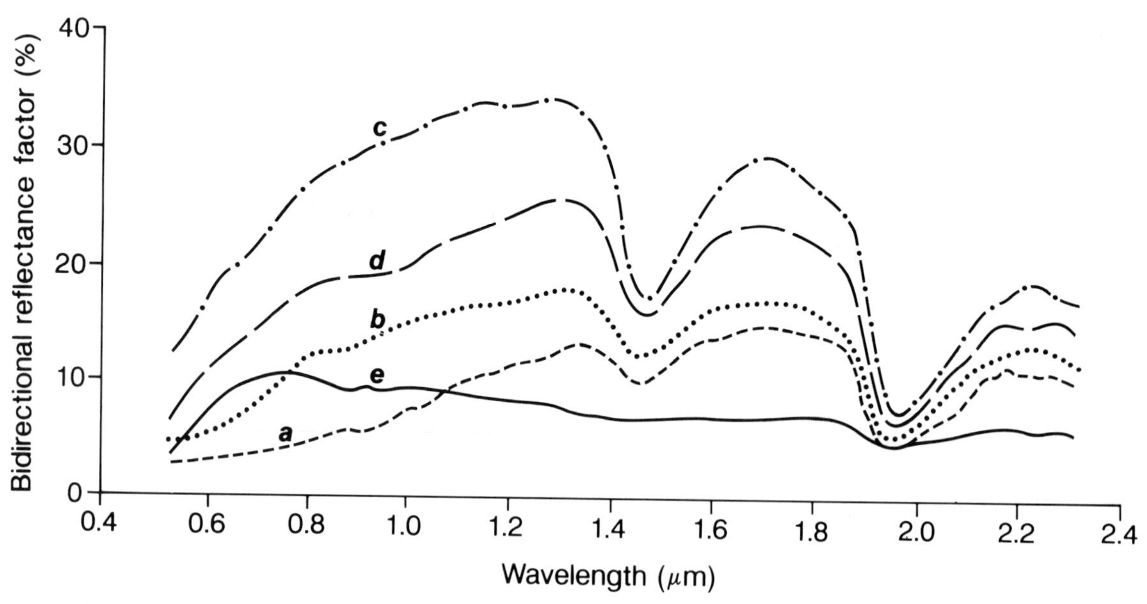

Fig. 13.5 Spectral reflectance curves for five UK soil types; namely organic-dominated (a), organic-affected (b), minimally altered (c), iron-affected (d), and iron-dominated (e). (After Milton and Webb, 1984).

moisture and as these constituents vary in amount and kind, it can be assumed that different soils will show different reflectance characteristics. There was, therefore, early work by soil scientists in the 1950s on the reflectance characteristics of different soil types, particularly in the infrared part of the spectrum. With the increasing use of multispectral aerial cameras and multispectral scanners in more recent years there has been renewed interest in spectral signatures of surface objects, including soils. Recent laboratory measurements within the wavelengths of 0.32–1.0 μm have been made on soils ranging from wet (almost saturated) to dry (oven dried at 43°C) states.

From some 160 sets of curves three general shapes have emerged (Fig. 13.3). Type 1 curve was

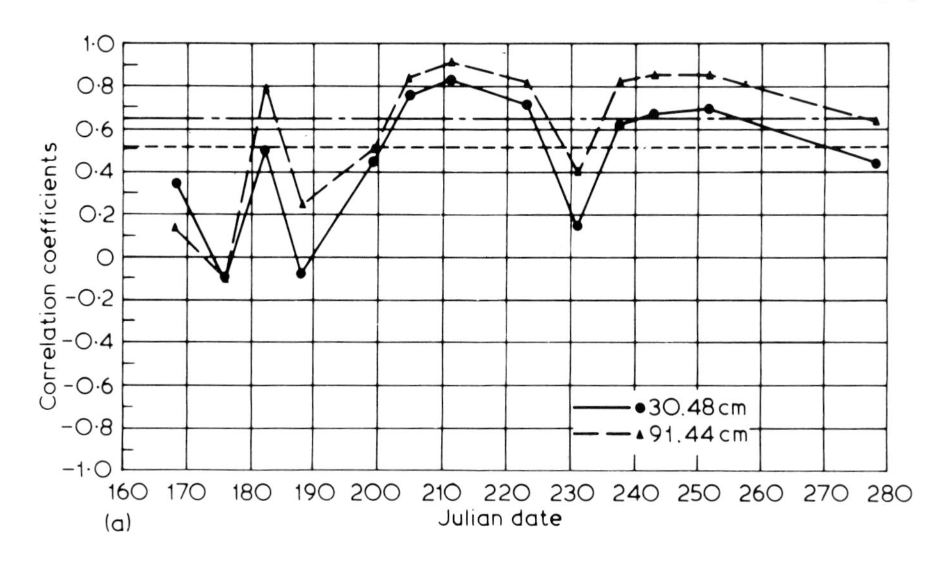

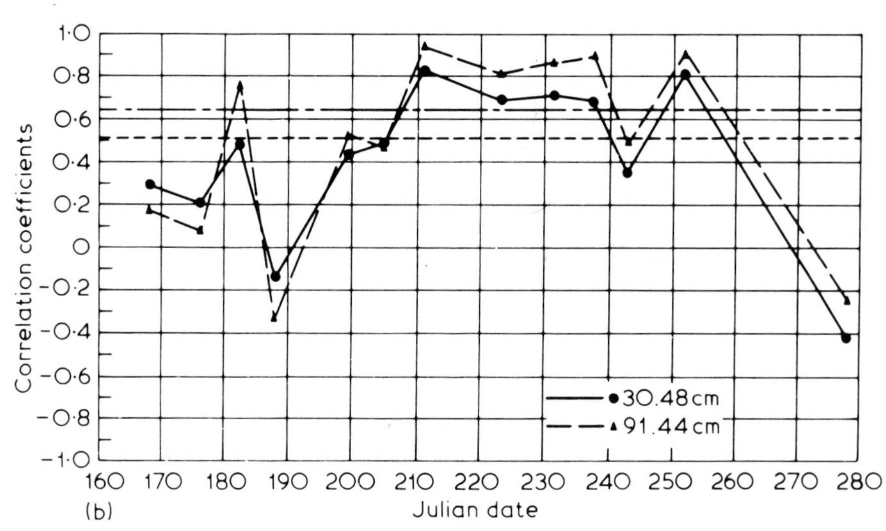

Fig. 13.6 Mission correlation coefficients plotted against time for the available soil moisture in the sorghum versus film density: (a) colour infrared film; (b) green-filtered black and white film. (Source: Werner *et al.*, 1973.)

characteristic of chernozem-like (mollisol) soils, type 2 of pedalfer (spodosol) soils, and type 3 of lateritic (oxisol) soils. An important part of the analysis of these results has shown that reflectance measurements made at only five wavelengths would suffice to predict with sufficient accuracy the other 30 wavelengths measured. It was recommended that the five selected wavelengths should be 0.4, 0.5, 0.64, 0.74 and 0.92 μm. Much more extensive studies of spectral signatures have now been made in the field as well as in the laboratory. Examples can be drawn from work in the USA on soils of Nevada. These spectra have revealed significant differences between saline and non-saline soils, and between gravel and silt loam soils in dry, wet and shaded conditions (Fig. 13.4). Other work in the UK has examined an extended spectrum of soils in the range 400–2500 nm. (Fig. 13.5). Russian studies of the spectral reflectance of sands in the Karakum desert have shown that isopleths drawn

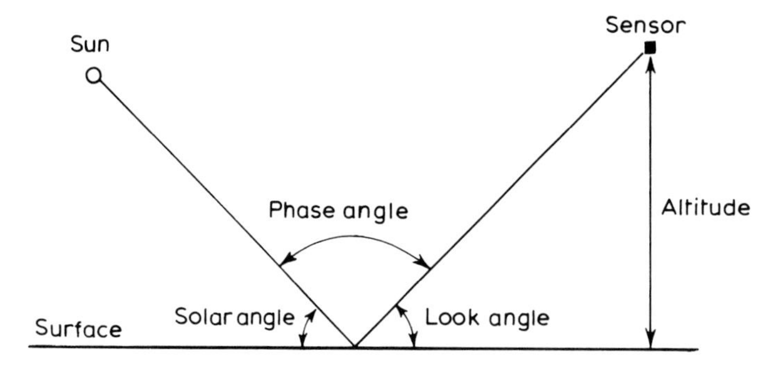

Fig. 13.7 Geometrical relationships and terms for the use of polarized photography. (Source: Curran, 1981.)

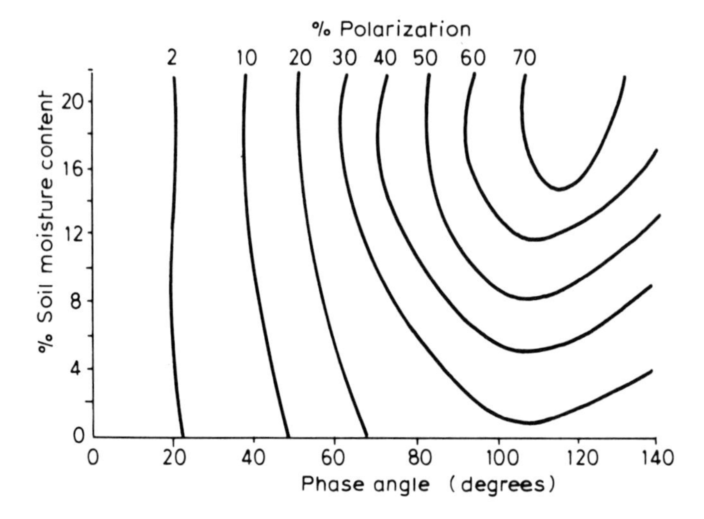

Fig. 13.8 The degree of polarization (PVL) recorded using a photometer–polarimeter in the laboratory for a range of soil moisture contents and phase angles. (Source: Steg and Frost, 1971.)

by trend surface analysis of the spectral data can be used to distinguish three types of sand with differing mineralogical characteristics.

Spectral reflectance data have been examined also for the purpose of evaluating soil moisture and water use by plants through airborne remote sensing methods. Examples of correlation coefficients plotted against time for the available soil moisture in the sorghum versus film density for red filtered (8403 film) and infrared (2424 film) are shown in Fig. 13.6.

13.4.3 Polarized photographic data

Although multispectral sensing for soil moisture assessment has received much attention, less work has been focused on the use of *polarized visible light*. When the unpolarized light waves are reflected from a soil surface the degree of polarization at a high phase angle (Fig. 13.7) is much affected by surface soil moisture. Where free water stands at the surface, light will not be scattered but reflected specularly in one vibrational plane. For rough surfaced soils below saturation the reflectance is diffuse and polarization low, but it will increase as saturation is approached. The intensity of polarized visible light reflected from the surface is related to a very thin surface layer of less than 1 cm. Laboratory studies have shown the relationships between soil moisture content, percentage soil moisture and phase angle (Fig. 13.8).

Early studies revealed that polarization of reflected sunlight provides the basis for a sensitive technique for remote sensing of soil surface conditions. In particular the *polarization percentages* have been shown to be related to both water

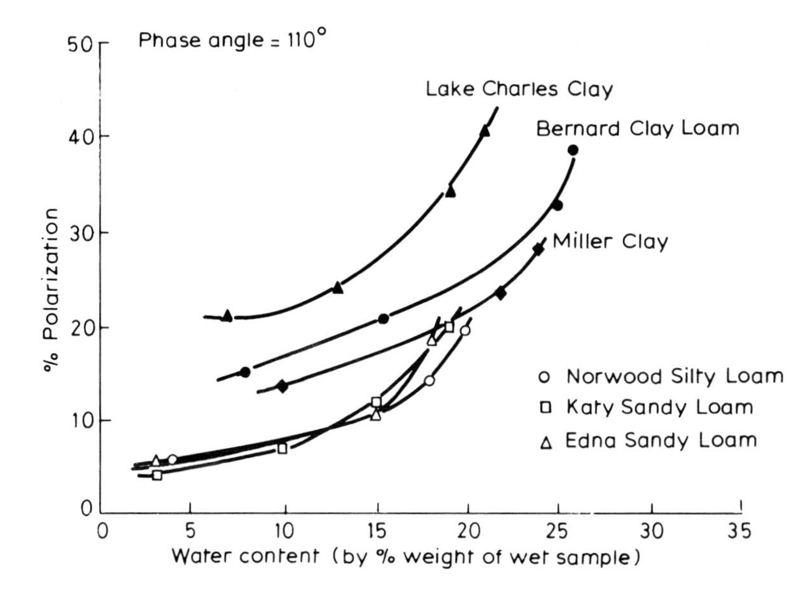

Fig. 13.9 Polarization related to water content and soil type. (Source: Stockhoff and Frost, 1972.)

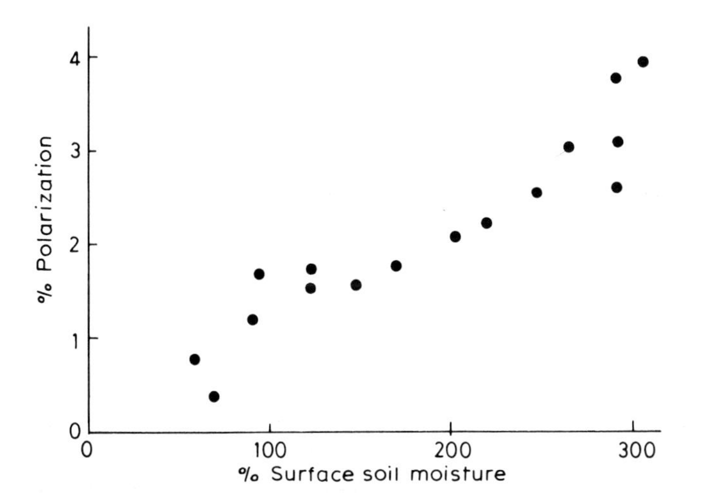

Fig. 13.10 Percentage polarization (PVL) recorded for a range of surface soil moisture contents. The peat soil is recorded from a light aircraft using a camera with polarizing filter. (Source: Curran, 1978.)

content and soil type (Fig. 13.9). The technique seems particularly useful when soils are at maximum moisture retention capacity, or near to this state. However, it should be noted that in some circumstances the degree of polarization is more dependent on soil type than on soil moisture and variations in soil aggregation are particularly important.

Some success has been obtained by analysing data obtained from a light aircraft using a camera equipped with a polarizing filter. Photographs were taken with the focal axis set to face the sun and angled towards the ground surface. Two exposures were made of each scene, both through a polarizing filter on to a panchromatic film. The camera operator can obtain the record of maximum polarization by turning the lines of the filter until they are first parallel with the ground surface and then at right angles to the surface. The densities of the imaged areas then can be measured quantitatively to show percentage polarization (Fig. 13.10) by using a densitometer, and standardiza-

Table 13.3 Western Sudan: land system area percentage error

Land system	Standard map area (km^2)	Landsat map area (km^2)	Difference (km^2)	Error (%)
Qoz	9877	10 190	+313	+3
Baggara	5513	5 228	−285	−5
Basement	6381	6 261	−120	−2
Bahr	70	72	+2	+3

Table 13.4 Western Sudan: soil agricultural potential groupings percentage error (After Parry, 1978)

Soil units–agricultural potential	Ground survey soil units (km^2)	Landsat interpreted soil units (km^2)	Interpreted soil units (% error)
Sandy soils cultivated and suitable for cultivating using traditional hand tillage methods	8 965	9 364	+4
	462	648	+40
	240	271	+13
	405	375	−7
	10 072	10 658	+6
Alluvial soils suitable for mechanized agriculture	2 239	1 413	−37
	1 289	1 756	+36
	1 078	1 258	+17
	44	74	−8
	5 450	5 201	−5
Alluvial soils in major wadis suitable for irrigation	2 310	2 416	+5
	70	72	+3
	2 380	2 488	+5
Basement soils unsuitable for cultivation	1 355	1 630	+20
	2 584	1 774	−31
	3 939	3 404	−14

tion of densities for each scene are made by reference to the density of a marker such as a road or vehicle roof within the scene. Such a technique is, however, limited to very low altitudes (50–100 m) owing to atmospheric effects and to variability in soil areas sampled. Also, such factors as soil particle size, soil albedo and the range of moisture values will affect the correlation between polarized light and moisture content. In brief, the technique is best suited for obtaining an overview of moisture content at small sites where surface roughness is constant and data can be obtained in sunlit conditions.

13.5 Remote sensing of soils and landforms by non-photographic systems

13.5.1 Visible and near-infrared scanner data

Data from all the Landsat satellites have been used extensively in terrain studies. In particular they have been useful in Third World countries where the topographic and thematic mapping base is often poor. The unique, near orthographic perspective of 34 000 km^2 of the Earth's surface provided by each Landsat MSS frame gives good opportunities for soil and land systems to be interpreted. The interpretation of satellite imagery for terrain analysis is essentially an extension of the conventional technique of air-photo interpretation discussed in Chapter 8. It involves evaluation of a number of image characteristics such as tone or colour, texture, size, shape, contrast and shadow. The scale differences between aerial photographs and satellite images pose an initial problem for the interpreter of terrain features, but this can be overcome quickly by training.

Land system and soil boundaries have been interpreted most commonly from 1 : 500 000 and 1 : 250 000 false colour composite prints derived from MSS bands 4, 5 and 7, although some studies have been undertaken at scales of 1 : 100 000 or even less. Clear plastic overlays placed over the image are used as the drawing medium.

In a study of land systems in western Sudan, Hunting Technical Services Ltd compared the 'standard' land system map prepared by field and air-photo interpretation methods with a map

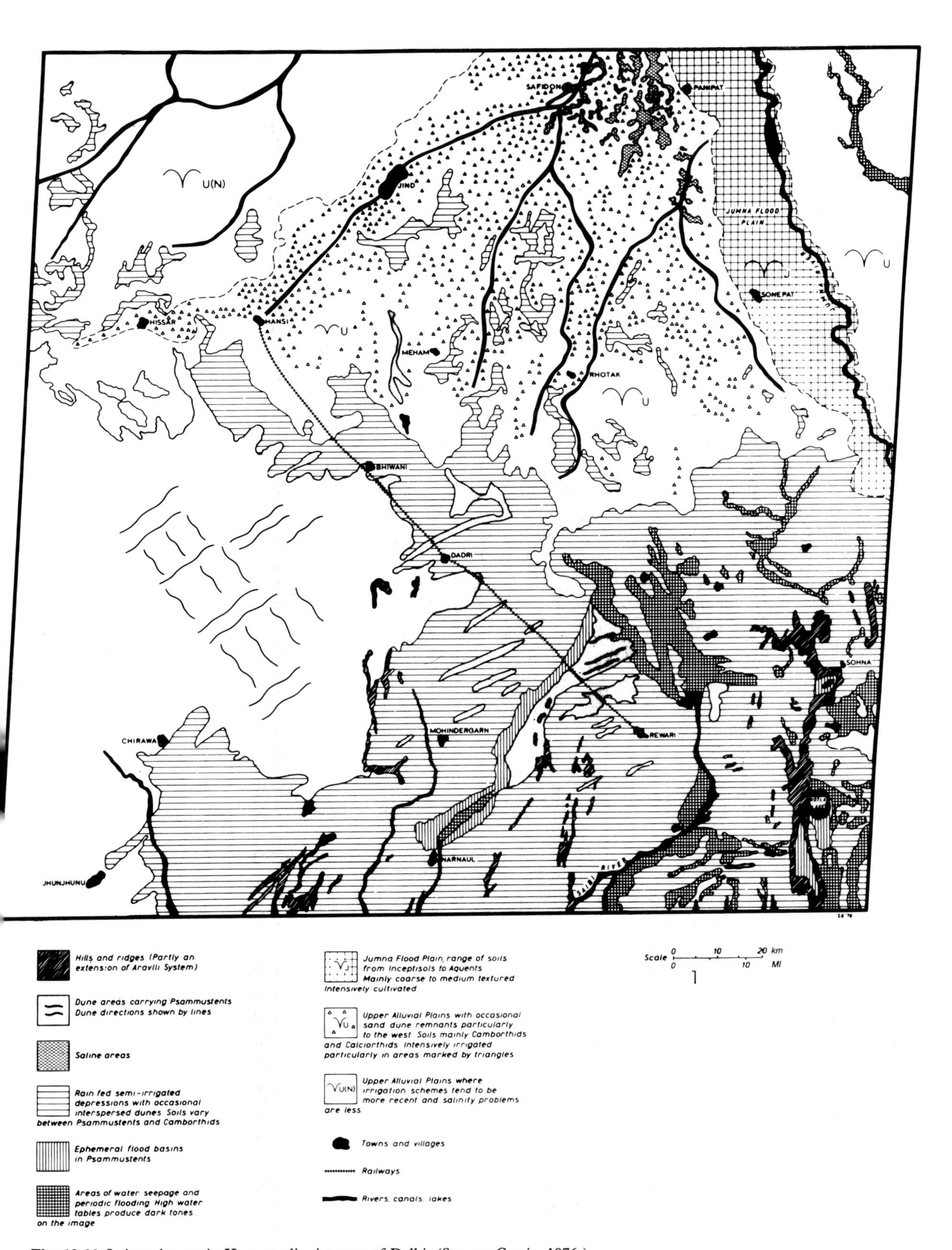

Fig. 13.11 Irrigated areas in Haryana district west of Delhi. (Source: Curtis, 1976.)

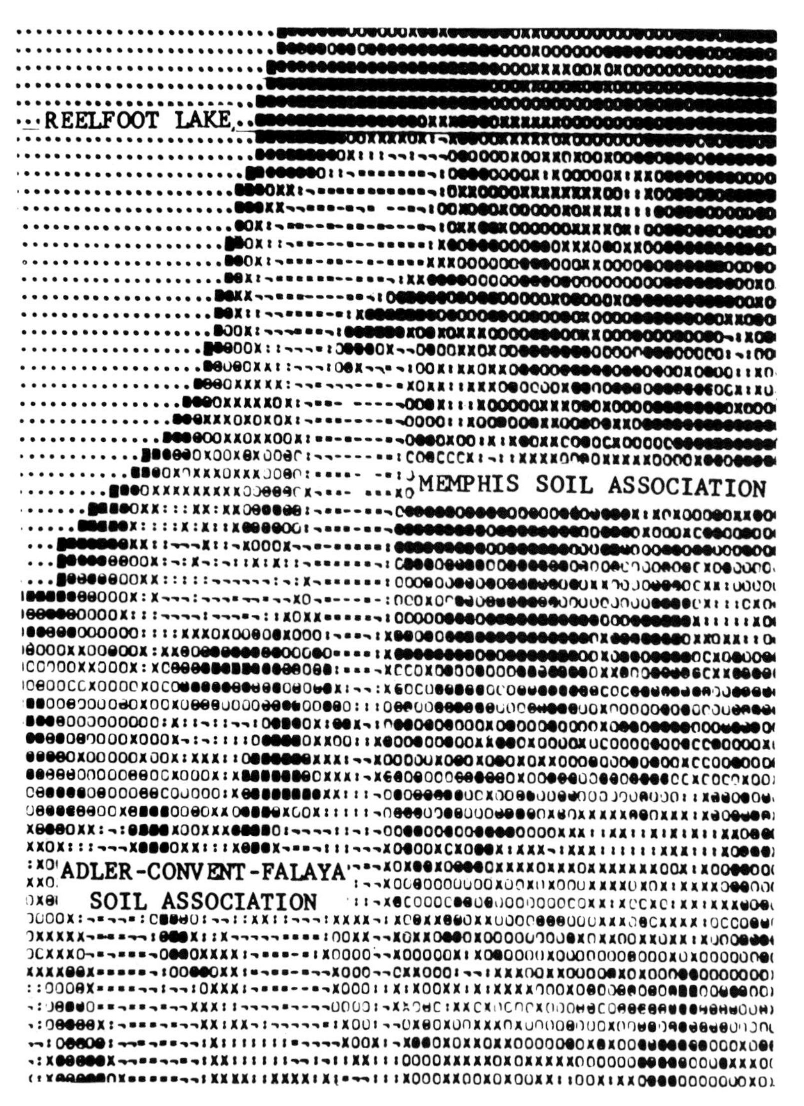

Fig. 13.12 Landsat image and computer print-out of soil association map. (Source: Parks and Bodenheimer, 1973.)

prepared from Landsat 1 imagery. As can be seen in Table 13.3 the analysis showed that no system interpreted from satellite imagery has an areal difference of more than ±5% compared with its equivalent on the 'standard' maps. This level of accuracy was considered adequate for a reconnaissance survey. When soil mapping units were compared for this same area of the western Sudan it was found that there was a greater discrepancy between the 'standard' map and satellite interpretation. For example, the former contained 122 mapping units compared with 102 units derived from satellite interpretation. However, the overall accuracy improved if the soils were grouped according to their agricultural potential (Table 13.4.)

Studies of Landsat MSS data have further showed that in cropland areas useful data can be obtained where bare soils are exposed. However, where significant amounts of vegetation are found the usefulness of remote sensing is seriously lessened. In semi-arid regions with low to moderate amounts of range type vegetation, Landsat digital data have been used to evaluate soil mapping capabilities, e.g. in Colorado, USA. It has been found that some boundaries on soil maps could be

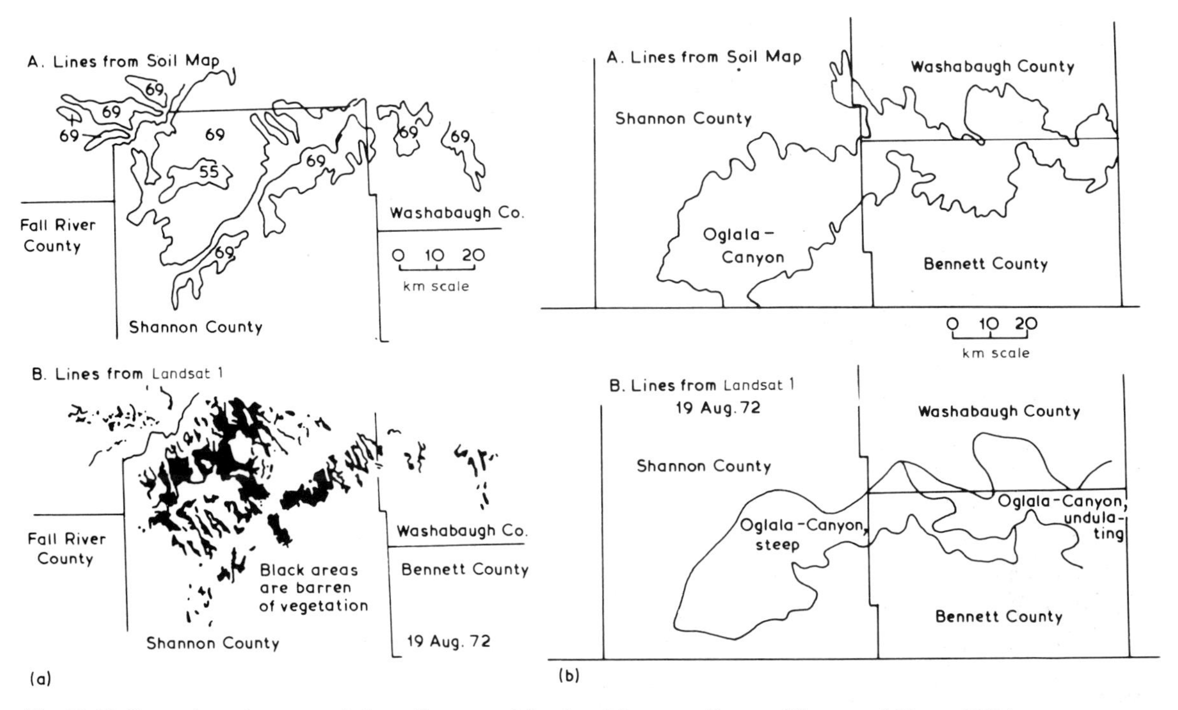

Fig. 13.13 Comparisons between existing soil maps and Landsat 1 imagery. (Source: Weston and Myers, 1973.)

located more accurately if remote sensing computer maps were used in conjunction with conventional soil survey field methods.

Soil mapping from early Landsat data was severely limited by the ground resolution element of the sensors employed, and the mapping units that could be determined were those of major soil associations. Furthermore, the successful application of satellite data has been mostly in semi-arid or subhumid areas of the Third World where terrain units often show close relationships to the soil associations.

North-west India was well covered by Landsat MSS imagery recording seasonal changes in soil state. In one study of the irrigated areas in Haryana District west of Delhi it was possible to identify nine terrain types with different soils and drainage characteristics (Fig. 13.11). The soil groups present ranged from psammustents on dune fields to camborthids and calciorthids in depressions. The interpretation was carried out by examination of a 1 : 250 000 false colour composite with the aid of a monoscopic magnifier. The boundaries were drawn on a transparent overlay. Digital printouts or densitometric plots can, of course, be used as an aid to the mapping of soil associations. For example, in an early study of Obion County, West Tennessee, Landsat-1 imagery was scanned by a high-speed digital scanning microdensitometer using a 25 μm raster. Subsequently the computer printouts were able to separate major soil associations, such as the Memphis occurring at the break between loess soils and the delta soils of the Mississippi floodplain (Fig. 13.12). It was noted even then that water and soil moisture are most effectively imaged in the near infrared: therefore, Band 7 is the most useful MSS channel for such studies.

Separation of small soil associations in intensive row crop agricultural areas has been found to be much more difficult at the scale of Landsat imagery. In this study the reflectance was thought to be at a minimum at low moisture contents (about 2 bars tension) whereas maximum reflectance was obtained at moisture levels slightly below field capacity (1/3 bar tension).

Early studies of Landsat imagery of South Dakota also showed that soil associations can be identified from film colour composites of Bands 4, 5 and 7 when viewed over a light table with magnification. Specific comparisons between boundaries shown on the Landsat composites and soil association maps are shown in Fig. 13.13. It was established that soil association boundaries were mainly distinguished through the identification of boundaries around landscape features. Vegetation differences were also found to assist the placing of boundaries around the soil associations.

The work outlined above was concerned mainly with identification of soils. Other studies have been made, however, to determine whether general terrain units, such as water, exposed rock, grasslands, coniferous forest and lowland marsh areas, can be identified. An example of such a study is that carried out in respect of terrain classification maps of Yellowstone National Park. Training sets of the five categories just described were defined by ground observations. Samples were selected for each of the terrain types and the signature of each type was extracted from the digital record of the Landsat imagery. At first individual signatures (i.e. from one band only) were obtained. Later, signatures from several bands were combined. The optimum channels for classification were then determined by calculating the average probability of misclassification (an estimate of the percentage of points which will be incorrectly classified in a terrain recognition map). The results were as shown in Table 13.5.

Table 13.5 Average probabilities of misclassification for five category recognition (Yellowstone Park)

MSS Bands	Average probability of misclassification
5	0.086
5, 7	0.031
5, 7, 6	0.028
5, 7, 6, 4	0.027

It will be seen that using MSS Bands 5, 6 and 7 was only slightly less accurate than using all four bands. Thus economy of computer effort was possible in subsequent studies of a similar kind.

The subsequent advent of Thematic Mapper and SPOT data at higher resolutions now offers soil surveyors better data for reconnaissance soil mapping, particularly in inaccessible and low cloud-frequency areas of the world. In some arid areas, however, difficulty exists in separating units because of the low reflectance contrast between soil and vegetation and between different plant communities. Most researchers have used ground-based measurements of the spectra of vegetation and soil (Chapter 6) to assist in analyses of TM data. Mean reflectance values for Landsat TM Band 1 (blue 450–520 μm) and TM Band 2 (green 520–600 μm), TM Band 3 (red 630–690 μm), TM Band 4 (near infrared 760–900 μm) are the data commonly used. Band ratios then can be calculated for all TM Band combinations, e.g. through quantifying the so-called Normalized Difference Vegetation Index (NDVI: see also Chapter 16), using the following relationship:

$$\text{NDVI} = \frac{(\text{Band 4} - \text{Band 3})}{(\text{Band 4} + \text{Band 3})} \qquad (13.1)$$

Studies in Nevada, USA have shown that the NIR/Red ratio separates soils and vegetation into three groups, namely *soil and senescent vegetation* (ratios less than 2.3), *grey and yellow green vegetation* (ratios 2.0–7.5) and *green vegetation* (ratios greater than 7.5). It also has been found that NDVI separates the same classes with values of less than 0.3, 0.3–0.7 and greater than 0.7, respectively. Other simple band ratios can be effective discriminants providing there is sufficient reflectance contrast between the two bandpasses, but the overall conclusion must be that such ratios cannot be used consistently for vegetation covers in all situations.

Special note should be taken of the initiatives of the US Geological Survey, the Soil Conservation Service and the Bureau of Land Management. In their work Digital Elevation Model (DEM) data produced by the US Geological Survey together with Landsat MSS data were made available for all areas of continental USA. The agencies combined to develop analytical procedures for ongoing soil surveys. Slope class and aspect class maps were prepared to overlay with 7.5 min quadrangle topographic maps, and quads of corresponding orthophotography. The boundaries of slope classes were interpreted and adjusted with reference to interpretations from topographic maps, aspect maps and spectral class maps. Polygon maps were then created in the laboratory which represented the landscape as an estimate of soil distribution. The resulting map was described as a *soil pre-map*.

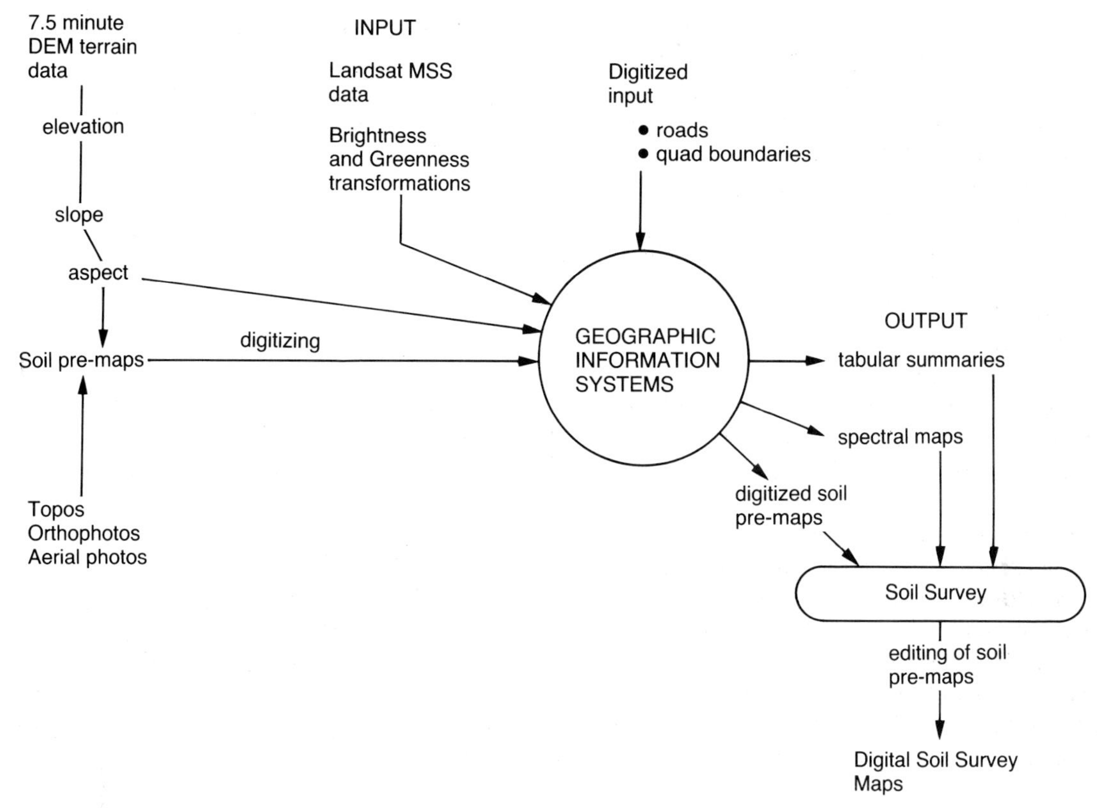

Fig. 13.14 Data processing flow for using Landsat and digital elevation data as inputs to soil surveys, USA. (Courtesy, US Geological Survey.)

The soil pre-map was then digitized and merged with spectral categories derived from classifications of geo-referenced Landsat MSS data and DEM derivatives. This merged data set was then used to create statistical summaries describing each polygon in terms of slope, aspect, elevation, and spectral values. The pre-map units so obtained were used to obtain delineations for the initial soil mapping units, and the statistical characteristics of each polygon provided quantitative data that have been used to develop preliminary definitions of landscape characteristics (Fig. 13.14). Initial soil taxonomy interpretations were developed and further refined through traditional field survey methods.

Landsat data have been used also by Ontario Geological Survey in Canada to map fuel-grade peat lands in that province. Landsat data recorded in mid- to late summer were geographically registered to 1 : 50 000 map formats. Following preliminary visual interpretation of wetland type a selection of field sampling sites was made. Spot-sampling in the field by means of helicopter transport allowed a wide range of *in situ* data to be collected, including, peat depth, peat decomposition rating, pH of surface water, percentage cover by vegetation plant strata, and dominant vegetation.

A supervised classification of the MSS data was then made. This led to the production of colour-coded maps of peatlands, with full geographic referencing. An area of 235 000 km^2 was mapped in this way. Accompanying reports summarize vegetation cover characteristics, peat depth and quality, and site characteristics. The total areas of fuel quality and horticultural quality peatland are estimated for each map sheet and a preliminary estimation of peat values was made.

The first generation of satellite sensors (Landsats 1–3) also have proved useful in mapping coarse geomorphic features, such as rivers, large streams floodplains and terraces. In Arizona, USA, arroyos as narrow as 20–45 m wide have been detected on

Fig. 13.15 Infrared linescan imagery of estuarine lowland near Huntspill River north of Bridgwater. Note dark tones of water and clear representation of field drains by dark lines of lower temperature. (Courtesy, Royal Signals and Radar Establishment, Malvern; Crown Copyright.)

low-contrast and high-contrast Landsat images. The second generation of remote sensing satellites (SPOT and Thematic Mapper) have provided data suitable for much more detailed studies of geomorphic units and classification of soils at the soil series level. Simulated SPOT data obtained by a Daedalus AADS 1268 Digital Multispectral System flown at 6.35 km altitude, have been used to study landscape components in New Mexico. The overall results of the study show that many specific features are well classified (>80%) but poor correlations were found for vegetation areas that are shrub covered or contain fine-grained soils. However, these covered only 3% of the study area.

Second-generation satellites are now proving useful for computer-assisted approaches to drainage pattern classifications. The quantitative study of drainage basins and channel networks first developed by Strahler has led to the evolution of *structural pattern recognition*, which refers to digital image analysis methods that represent drainage patterns in terms of their intrinsic structure. The inherent characteristics of drainage patterns can be summarized as follows:

1. *Dendritic* patterns, with branching, tree-like systems, where tributaries join at an acute angle.
2. *Pinnate* patterns, resembling the dendritic but with secondary tributaries evenly and closely spaced, and parallel.
3. *Parallel* patterns, where tributaries flow nearly parallel to each other.
4. *Trellis* patterns, where primary tributaries are long, straight and parallel. Secondary tributaries are numerous and short.

Table 13.6 Measured and estimated soil moisture and daily evaporation for Grendon Underwood, Bucks, England, 13 September 1977 (Source: Gurney, 1979)

Model	Evaporation (mm)		Soil moisture (%)	
	Estimated	Measured	Estimated	Measured
Bare Soil (Tell-us)	0.85	0.62	29.6	30.1
Grassland (Tergra)	0.53	0.62	49.6	47.7

5. *Rectangular* patterns, with clear right-angle bends in main stream and tributaries.
6. *Angular* patterns, with mixtures of acute, obtuse and right angle junctions.
7. *Radial* patterns, with streams radiating from a centre like spokes of a wheel.

Thus, any drainage pattern consists of a main stream with primary and secondary tributaries (sometimes also tertiary), and different patterns have specific characteristics in terms of the forms and junctions of the component parts. The characteristics can be modelled using features such as *shape*, *angle*, *elongation*, *bifurcation ratio*, and *nodes*. Computer software has been developed for analysis termed the Drainage Pattern Analysis (DPA) system which allows drainage network data from airborne or satellite sensors to be analysed.

13.5.2 Thermal infrared imagery

Although the scanners in the visible and near infrared wavelengths are producing good results there is considerable research into the applications of thermal infrared data for soil studies. The use of thermal infrared for soil temperature studies has been investigated in detail by NASA investigators using aircraft-borne sensors. Here, soil surveys were related to the thermal infrared imagery in a manner which suggested that surface soil temperatures can be indicative of subsurface soil conditions. Such results are to be expected in view of the effects of soil texture, composition, porosity and moisture content on soil temperature regimes in different soils.

Thermal infrared scanning is particularly sensitive to moisture content at the very surface soils. This is particularly the case when surface winds evaporate moisture from the surface layers. In these circumstances the cooling of the surface in exposed areas contrasts with the higher surface temperatures in areas protected from the wind. An example of these conditions is shown on Fig. 13.15 which shows part of an area of Somerset under conditions of 11°C air temperature and 4.5–7.9 m s^{-1} wind strength. This is an area of reclaimed estuarine clay which is under pasture and requires extensive field drainage. The open drainage ditches are readily distinguished by dark tones reflecting the lower temperatures of the water surfaces in relation to the land. The lines of buried field drains also can be detected as a result of the wetter soil conditions along these subsurface channels. The white-toned areas in the lees of the field boundaries reflect the shelter effects of the boundaries and the lower evaporation and hence higher surface temperatures existing in these sheltered places.

The most concentrated space investigation of the relationships of soil temperatures to soil conditions has been through use of the Heat Capacity Mapping Mission, which was developed from the Applications Explorer Mission-A (AEM-A). The HCMM system provided for a circular sun-synchronous orbit at 600 km altitude. The major sensing system was the Heat Capacity Mapping Radiometer (HCMR), which included two bands (0.5–1.1 and 10.5–12.5 μm) and provided a resolution of 500 m with a swath width of 700 km. The HCMM satellite was launched on 26 April 1978. One of the principal investigations of the mission was made by European scientists through the 'Tellus' project, sponsored by the Joint Research Centre of the European Economic Community whose specific objective was to map diurnal soil temperature variations so that, by

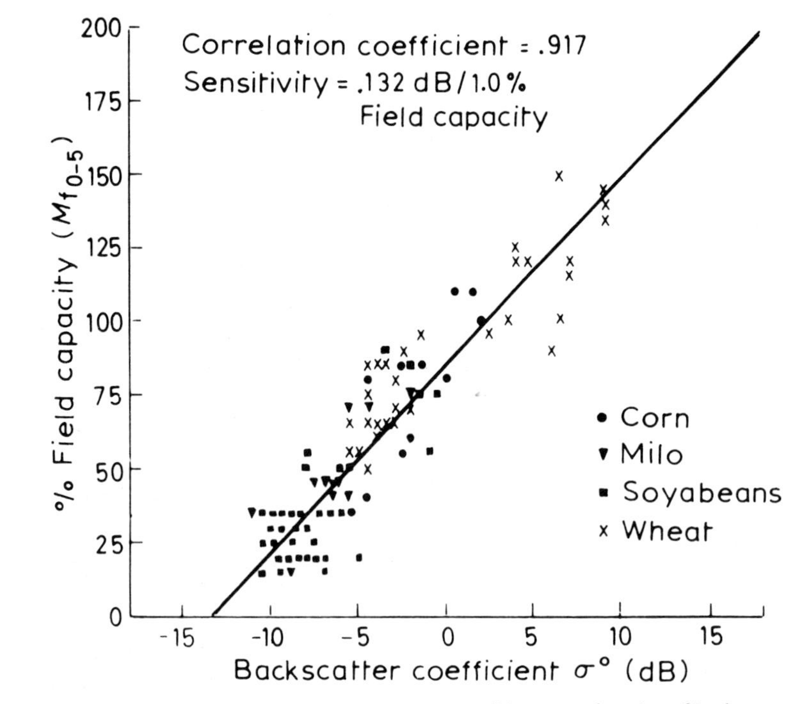

Fig. 13.16 Per cent field capacity in the 0–5 cm soil layer as a function of backscatter coefficient at 4.25 GHz, HH polarization, and 10° incidence angle for corn, milo, soybean and wheat data sets combined. (Source: Ulaby, 1980.)

correlation with ground truth observations, thermal inertia models could be developed for the prediction of variations in soil moisture content of the top soil layers. The principle adopted was that soils of high water content would show lower diurnal variations in surface temperature than would drier soils.

Two digital models were used in the Tellus project. One model, the 'Tergra' model, was for use in grassland areas and used one measurement of the soil surface temperature. The second, the 'TELL-US' model, estimated daily evaporation and thermal inertia by measurement of both day- and night-time temperatures (Chapter 12). Estimated and calculated values for evaporation and soil moisture for a site in Buckinghamshire, England, are shown in Table 13.6.

13.5.3 Microwave radar data

Knowledge of the spatial and temporal distributions of soil water is of economic and scientific significance through many agricultural, hydrological and meteorological applications. In view of the effects of soil moisture on the dielectric properties of surface materials there is growing interest in microwave sensors for the detection of moisture states at the ground surface. Also the dynamic nature of water conditions in the landscape makes the microwave sensors attractive because they can be used on a repetitive basis owing to their relative immunity from atmospheric effects, including cloud cover.

In summary, it can be suggested that the environmental scientist is mainly concerned with four aspects of soil moisture:

1. The *amount of moisture* stored at a given time. This can be expressed best on a volumetric basis or as a percentage of field capacity.
2. The *availability of soil moisture* to the higher plants. The available water is determined by the water tensions occurring at the time of observation.

Table 13.7 Soil moisture regime classes – wetness classes – duration of wet states (Source: Curtis, 1977, after Soil Survey of England and Wales)

Class	
I	The soil profile is not wet within 70 cm depth for more than 30 days[a] in most years[b].
II	The soil profile is wet within 70 cm depth for 30–90 days in most years.
III	The soil profile is wet within 70 cm depth for 90–180 days in most years.
IV	The soil profile is wet within 70 cm depth for more than 180 days but not wet within 40 cm depth for more than 180 days in most years.
V	The soil profile is wet within 40 cm depth for more than 180 days and is usually wet within 70 cm for more than 335 days in most years.
VI	The soil profile is wet within 40 cm depth for more than 335 days in most years.

[a] The number of days specified is not necessarily a continuous period.
[b] 'In most years' is defined as more than 10 out of 20 years.

3. The *movement of water* into and within the soil. This movement is due to positive and negative pressures within the soil.
4. The *periodicity of waterlogging* in a given soil. The Soil Survey of Great Britain classifies soil into drainage categories on the basis of the relationships of soil profile morphology to periods of waterlogging (Table 13.7).

In order to use radar effectively for soil moisture determination, it is not only necessary that the back-scatter coefficient, σ° be dependent on soil moisture, but also that its dynamic range in response to soil moisture should be much larger than its dynamic range to other surface features, such as roughness and vegetation. Much of the initial work to determine the optimum choice of frequency, polarization and angle of incidence of the radar sensor was carried out by the University of Kansas. A Microwave Active Spectrometer was used from a Cherry-Picker platform to acquire a wide range of data beginning in 1972. The multi-year data, containing 190 data sets of bare fields was used to generate an empirical model for back-scatter coefficient as a function of the volumetric moisture content normalized to percentage field capacity at ⅓ bar. This is illustrated by Fig. 13.16.

A model derived by Ulaby, one of the most successful researchers in this field, has been compared with data obtained using a C-Band radar mounted on a crane boom over crops in France (Fig. 13.17).

Since the 1970s many further studies have been made by Ulaby and others of radar response in relation to soil moisture content. Most recently interest has centred upon the relationship between radar returns as a function of the percentage of field capacity. More experiments on different soil types are necessary in order to decide whether percentage field capacity or soil water pressure (pF) is the best soil parameter against which to make radar calibrations.

To date only a limited number of airborne and spaceborne investigations of radar response to soil moisture have been completed. More are required along the lines of the campaigns undertaken in the USA by the University of Kansas and in Europe by means of the SAR 580 missions sponsored jointly by the Joint Research Centre of the EEC and the European Space Agency.

The analysis of aircraft radar response to soil moisture made by the University of Kansas used a site south-east of Colby, Kansas, which afforded flat terrain with relatively uniform soils. It was instrumented with 39 raingauges and three meteorological stations and formed part of a site in the US Department of the Interior High Plains Experiment (HIPLEX). Three airborne radar sensors were operated at frequencies of 1.6, 4.75 and 13.3 GHz from a C-130 aircraft operated by NASA/JCS. The radars collected data at incidence angles of 5–60°. Dual polarization could not be obtained simultaneously. The study showed that the aircraft response to soil moisture is optimum at C-band frequencies and at angles of incidence of 10–20°. Like-polarization (HH) radar response was found to be by vegetation but in other studies has been found to be dependent on the character of row-tillage patterns (Fig. 13.17(a)). Cross-polarization (HV) data were also found to be unaffected by vegetation cover, but were also independent of tillage patterns (Fig. 13.17(b)).

13.5.4 *Further possibilities*

Interest in microwave methods continues to be fuelled by theoretical considerations which suggest

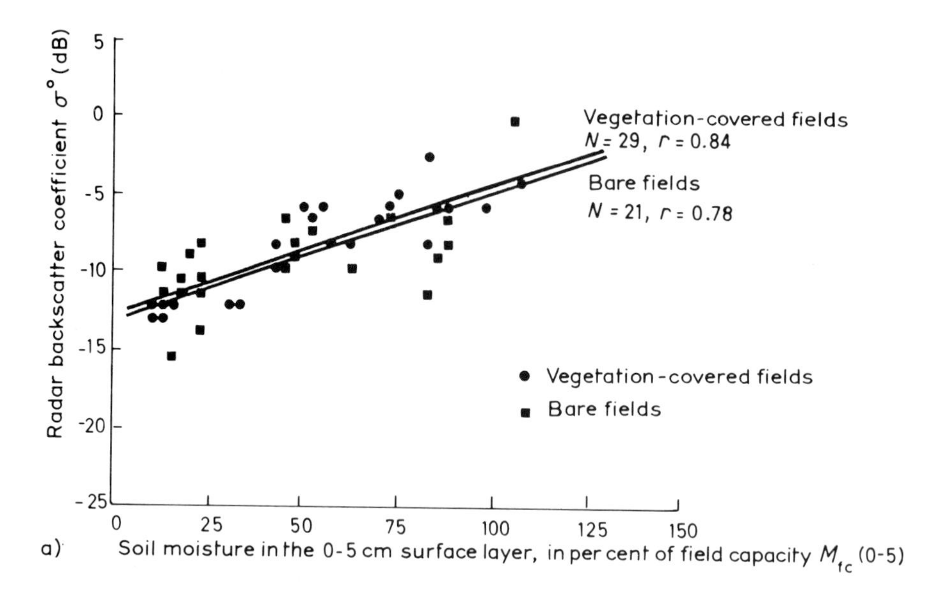

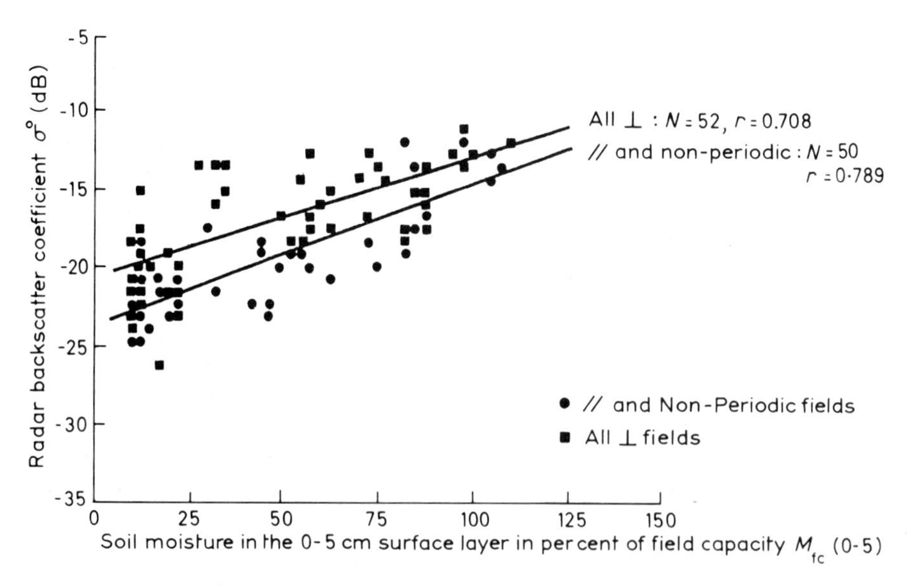

Fig. 13.17 (a) Comparison of radar soil moisture responses to bare and vegetation-covered fields. There is a minimum dependence of the response to vegetation at radar parameters of 4.75 GHz, HH polarization and 10° incidence angle. (b) Aircraft radar cross-polarization response to soil moisture for two categories of agricultural fields. Radar parameters are 4.75 GHz, HV polarization and 15° incidence angle. (Source: Bradley and Ulaby, 1980.)

that some penetration of the soil may be achieved thereby, providing information in respect of all-important subsurface moisture or soil layering. In principle this can be obtained by time domain measurements (Fig. 13.18). For a given range of soil relative dielectric constants and surface roughnesses, approximately one-half of the energy of the incident wave will be reflected from the surface. Some portion of the wave energy that penetrates the surface layer will be reflected back to the observing platform from the first subsurface discontinuity or moisture accumulation gradient.

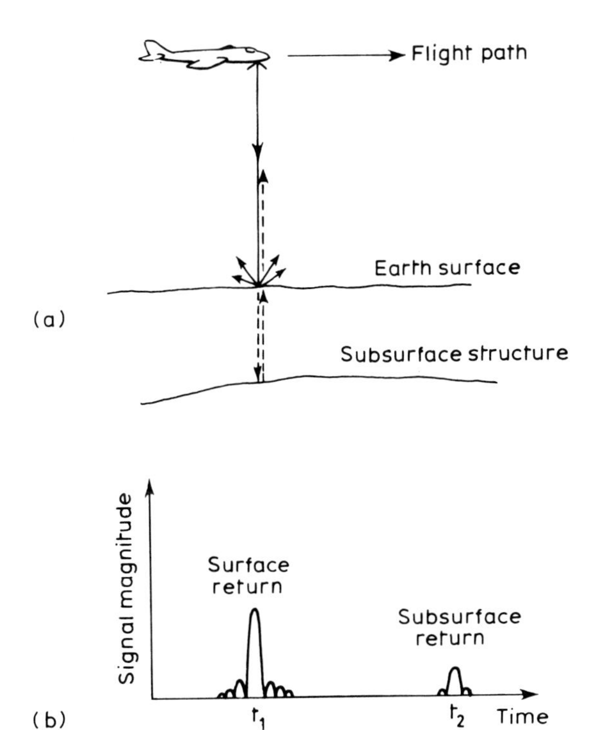

Fig. 13.18 Time domain measurement for subsurface features. (Source: Barringer Research Ltd, 1975.)

The reflected energy from this interface is reduced with respect to the transmitted energy by the transmission loss through the surface layer and the three reflections the wave must undergo. In order to be able to detect the discontinuity, an extremely good range resolution would be required so as to detect the presence of the weak second reflection. This places a requirement on the radar system that the pulse duration be very short, and makes the engineering tasks involved very severe. At the present time there is no evidence that systems are available to achieve this.

More generally, a wide range of new questions and problems have arisen in soil and landform studies, many of which the instrument systems being developed for the EOS platforms are being designed to address. Until now most of the remote sensing research in these germane areas has been directed towards the improved recognition, evaluation, and monitoring of aspects of soils and landforms, the investigations of which began long before satellites were first used for such purposes. The new questions that now demand attention from both conventional and remote sensing soil scientists alike include the following:

1. What and where are the unstable soil conditions that are subject to rapid changes produced by disturbance?
2. Are soils sources or sinks of carbon, nitrogen and other nutrients on a global scale?
3. What are the causes of desertification, and how can these processes be controlled?
4. What are the long-term effects of agriculture and irrigation on soil chemistry and productivity?
5. What is the time evolution of soils, and what can be learned about previous climatic conditions?

Meanwhile, increased attention will be paid to topography from satellites, both because this is interesting in its own right, and also because of its close relationship with soil development. More precise measurements of land elevation will provide an improved basis for studies of soil composition and height above sea-level; slopes determined by stereo-imaging of visible and SAR data will help provide better inputs to digital terrain models and models of surface run-off and landform development; surface roughness measurements of unprecedented detail will help with estimations of surface maturity, soil abundance and particle size.

As in so many areas – perhaps even all – of the natural sciences it is evident that the more we know of soil science, the more we still need to discover.

14 Rocks and mineral resources

14.1 The use of air photo-interpretation in geology

The use of remote sensing techniques in geology is long established through the use of aerial photography in photogeology studies. Sources of further information on the methods used in photogeology are cited in the bibliography at the end of the book.

Aerial photographs have provided a great deal of data for geological studies in all parts of the world. They can be interpreted to give information on the *structure* and *lithology* of rocks. In the study of structural geology such features as bedding, dip, foliation, folding, faulting, and jointing can be observed. Aerial photographs provide evidence of bedding through the occurrence of ridges in the stereomodel and differences in tonal response where beds differ in their mineral constituents. Sometimes certain beds can be recognized by a constant lithological interface which is so distinctive that they can be regarded as marker beds or marker horizons. The dip slopes of rocks often can be recognized more reliably on a stereomodel than on the ground because of the synoptic view of an area of dipping sediments obtained from the air. Accurate measurements of dip can be made by photogrammetric measurements on stereopairs. In regional mapping, however, geologists normally assess the dip by eye and place the dip into categories (e.g. <10°, 10–25°, 25–45°, >45°, <90°) (Figs 14.1 and 14.2).

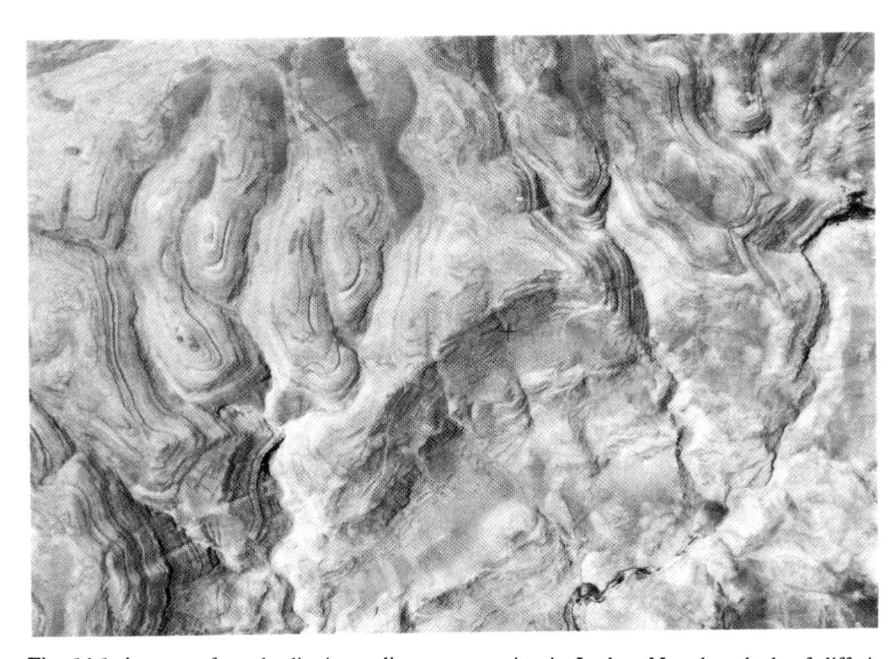

Fig. 14.1 An area of gently dipping sediments occurring in Jordan. Note how beds of differing lithology are characterized by different tones on the photograph. Field systems can be seen on the gentler slopes. (Courtesy, Hunting Surveys Ltd, Boreham Wood, Herts.)

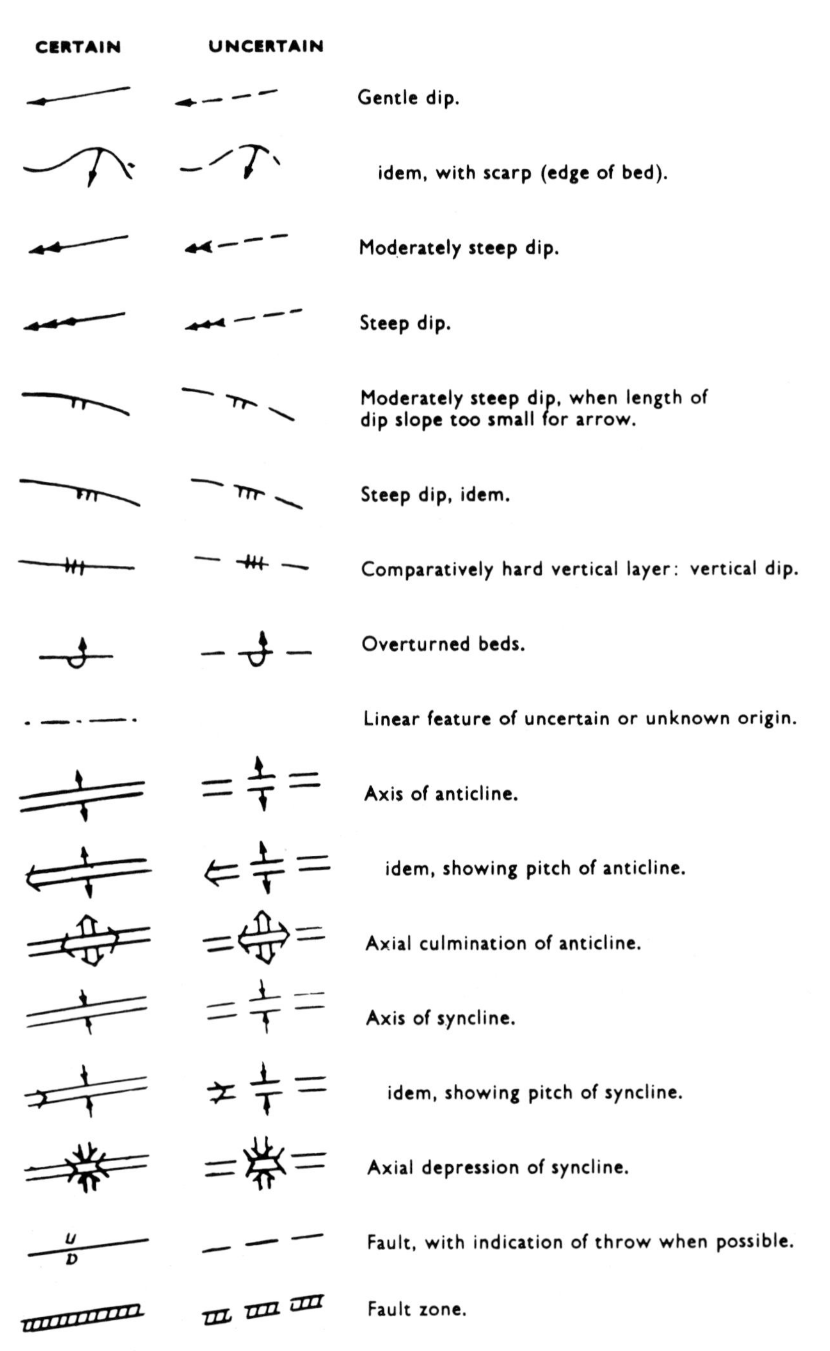

Fig. 14.2 Selected symbols for photo-interpretation in geology. Where the identification of features is 'uncertain', the same symbols are used, but with broken lines as their bases. (Source: Shell Petroleum Company Ltd.)

Any discussion of foliation in rocks is made more difficult by the fact that the term foliation is used in different senses in Britain and America. British geologists distinguish between schistosity and foliation so that for them segregation of minerals into thin layers or folia is foliation whereas parallel orientation of such minerals is referred to as schistosity. On the other hand American geologists give the term foliation wider meaning so that lithological layering, preferred dimensional orientation

Fig. 14.3 Oblique aerial photograph of the Fraser area, Canada (56°45′N, 63°35′W). Note the lineaments in the surface, which are made easily visible by the snow collecting in the lineament depressions. Both short and long lineament features are displayed. (Courtesy, Royal Canadian Air Force.)

of mineral grains and surfaces of physical discontinuity and fissility resulting from localized slip may be included in the term. In this chapter the American usage of the term will be adopted. Lineaments resulting from foliation are generally parallel to one another but are short and do not normally consist of long continuous ridges or valleys (Fig. 14.3).

Folding in rocks is often apparent on air photographs where it would not be noticed in the field by a ground surveyor. The geologist can normally see a representation of the whole fold in the stereomodel so that the axis and plunge of a fold structure can often be seen very readily (Fig. 14.4).

Faulting in rocks occurs where there has been a fracture along which rocks have slipped relative to one another. Generally faults provide fairly straight features and since the fracture is a zone of weathering and weakness the surface manifestation is often one of negative relief. Where mineralization has taken place along a fault line, however, there may be positive rather than negative surface features. In most instances the most reliable evidence of faulting is displacement of bedding along negative linear surface features (Fig. 14.5).

Joints form patterns in rocks which are very similar in photographic appearance to faults, i.e. they often provide fairly straight negative features.

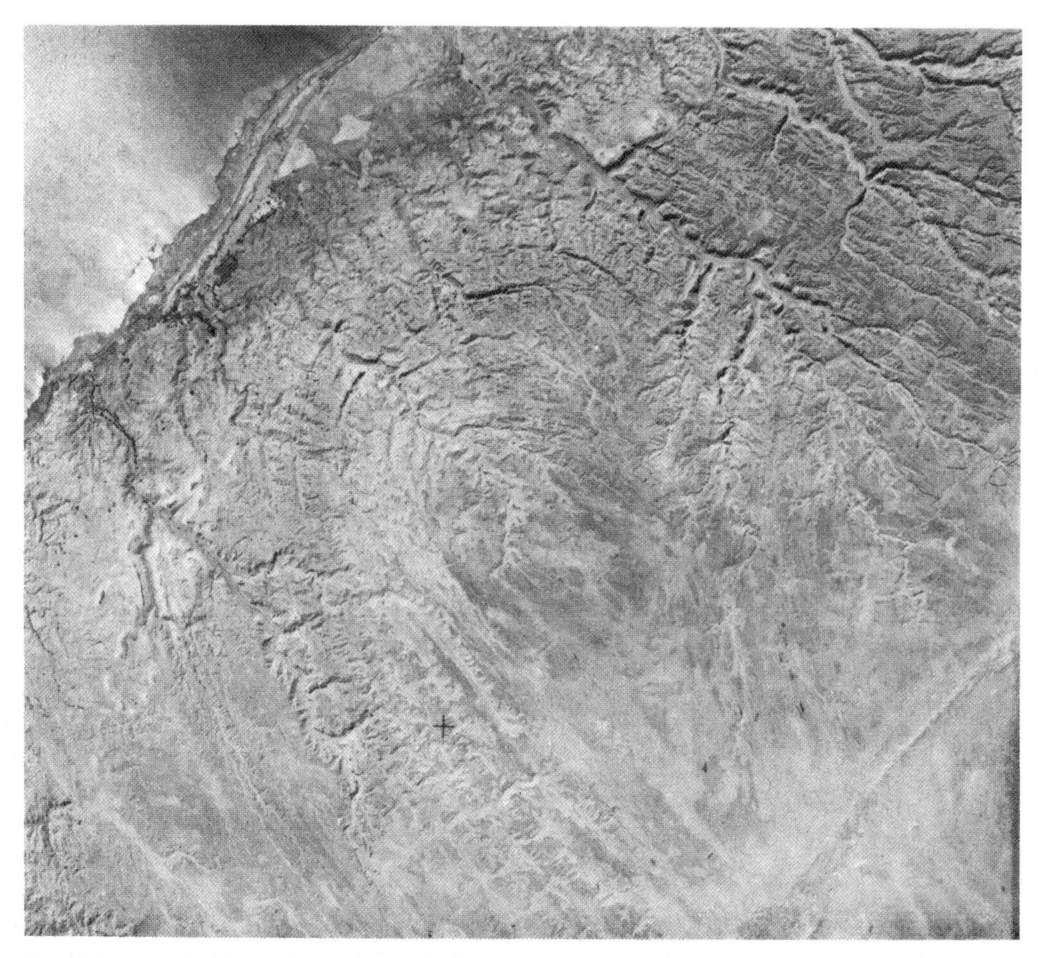

Fig. 14.4 A plunging fold structure near the Dead Sea, Jordan. (Courtesy, Hunting Surveys Ltd, Boreham Wood, Herts.)

The distinction between joints and faults can sometimes be made by careful observation to see if relative movement has taken place in the beds. If relative movement can be seen then the feature can be classified as a fault, and conversely if no movement can be detected it is better to record the feature as a joint. Jointing often plays a part in determining the patterns of river networks (Fig. 14.6). It also can be associated with topographic forms characteristic of granite areas and in such cases the boundaries of granite intrusions often can be plotted with some accuracy.

Whereas structural geology often can be interpreted with considerable certainty from aerial photographs, lithological interpretation is less easy. If the researcher is familiar with a particular field area and has used air-photos extensively it may be possible to recognize rock type with considerable accuracy. However, where an attempt is made to recognize rock types in unfamiliar areas from photo features alone, the task becomes more difficult and uncertain. Various strands of evidence must then be used – in other words the principle of convergence of evidence must be applied. It has been suggested that the following stages may be included in interpretation of lithology and structure:

1. The recognition of the climatic environment, e.g. temperate, tropical rainforest, savannah, desert, etc.
2. The recognition of the erosional environment, e.g. very active, active, inactive.
3. The recognition and annotation of the bedding traces of sediments or metasediments (altered sediments).
4. The recognition and delineation of areas of

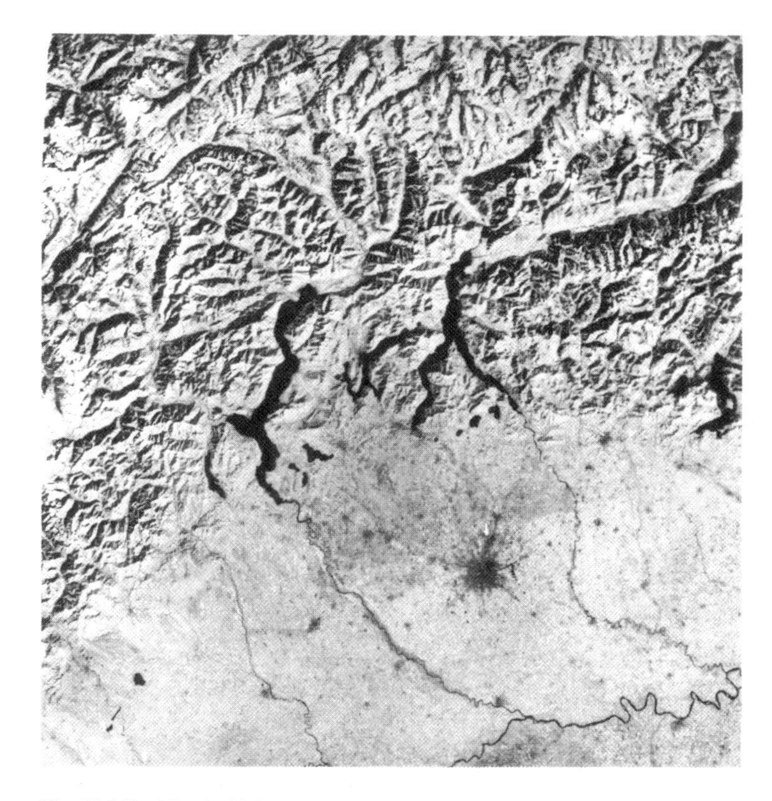

Fig. 14.5 Faulting in Alpine structures fringing North Italian plain as seen in Landsat 1 imagery obtained in Band 7, 7 October 1972. (From Boriani *et al.*, 1974.)

outcrop that do not indicate bedding, e.g. intrusions in horizonally bedded rocks.
5. The recognition and delineation of areas of superficial cover that do not indicate bedding.
6. The re-study of the bedding traces around the noses of folds to determine if possible the approximate position of the axes of folds.
7. The study of lineaments transverse to the bedding traces to determine whether they represent faults, dykes, joints, or combinations of these.

It is important to use a *recognized set of symbols* for annotation of prints in geological studies so that work can be interpreted by co-workers. In Fig. 14.2 the symbols used for photogeological work by the Royal Dutch/Shell Group are shown, and in Fig. 14.7 their use is illustrated.

It has long been recognized that certain plants are characteristic of soils that are found over rocks containing particular minerals. These plants are termed *indicator plants* and they have been used as guides by mineral prospectors seeking the occurrence of ore bodies. There is now a considerable literature dealing with the use of 'geobotanical techniques' for mineral exploration and since vegetation boundaries are often well displayed on air photographs attempts have been made to trace indicator plants on the photographs and thereby outline potential mining areas. These methods have been applied successfully in Africa and Australia and the development of multispectral photography together with infrared colour photography has made such geobotanical studies of growing interest.

The literature dealing with the use of air photography in the study of rocks and mineral resources is now voluminous and most major exploration companies rely heavily on photogeological work. There are also separate photogeological departments in various governmental institutes, e.g. Institute of Geological Sciences, UK.

An example of a restricted working legend used by the Overseas Division of the Institute of Geological Sciences, London, is shown in Fig. 14.8. This set of symbols was used in the interpretation of the geological structure of an area of folded

Fig. 14.6 Faulted terrain in an arid environment in Jordan. Note the sharp changes of direction in the wadi system resulting from the effects of faulting and jointing. (Courtesy, Hunting Surveys Ltd, Boreham Wood, Herts.)

sediments in Australia (Fig. 14.9) for which the final photo-interpretation is also shown in Fig. 14.8.

14.2 The use of non-photographic airborne sensors in geology

14.2.1 Thermal infrared scanning

The development of non-photographic sensors has provided remote sensing data in the infrared (thermal) and microwave regions of the spectrum, which are potentially very important for geological studies. Infrared scanning can be used to differentiate between different strata using data such as those portrayed in Fig. 14.10 from Snowdonia, Wales showing part of a sequence of contorted beds of Palaeozoic slates and sandstones. Infrared has been used also for studies of Italian *volcanoes and volcanic deposits*. These have been overflown with a two-channel Daedalus thermal scanner. The ratio of two thermal channels when plotted can be considered as a relative emissivity map of volcanic materials. It has been found also that such a ratio method can be used for mapping the texture of volcanic materials.

In the Soviet Union airborne infrared images have been used to study *sand desert features* in the Repetek region. Landscape features, such as barchan ridges and sand hills, have been found to show different thermal patterns according to differences in slope steepness, density and composition of the vegetation cover, solar elevation and wind speed and directions. The temperature graduations were found to have a large range from 55°C on the sunlit slopes of barchans to 16°C in the shade of desert bushes.

Another application of thermal scanning arises from the fact that mineral prospectors have long recognized that there are differences in soil temperature over *ore bodies*, therefore thermal infrared linescan techniques are potentially useful for mineral exploration. An example of this is shown in Fig. 14.11 where the microdensitometer scan across the Dugald River Lode in Australia indicates that it is marked by a fairly clearly defined zone of high emission in linescan imagery.

The thermal response of a material to temperature change is termed 'thermal inertia' (see also Chapter 12). The thermal inertias measure the resistance of the rock to changes in temperature and is defined as:

$$(K\rho c)^{1/2} \tag{14.1}$$

where K is the thermal conductivity, ρ the density, and c the specific heat capacity. For most materials

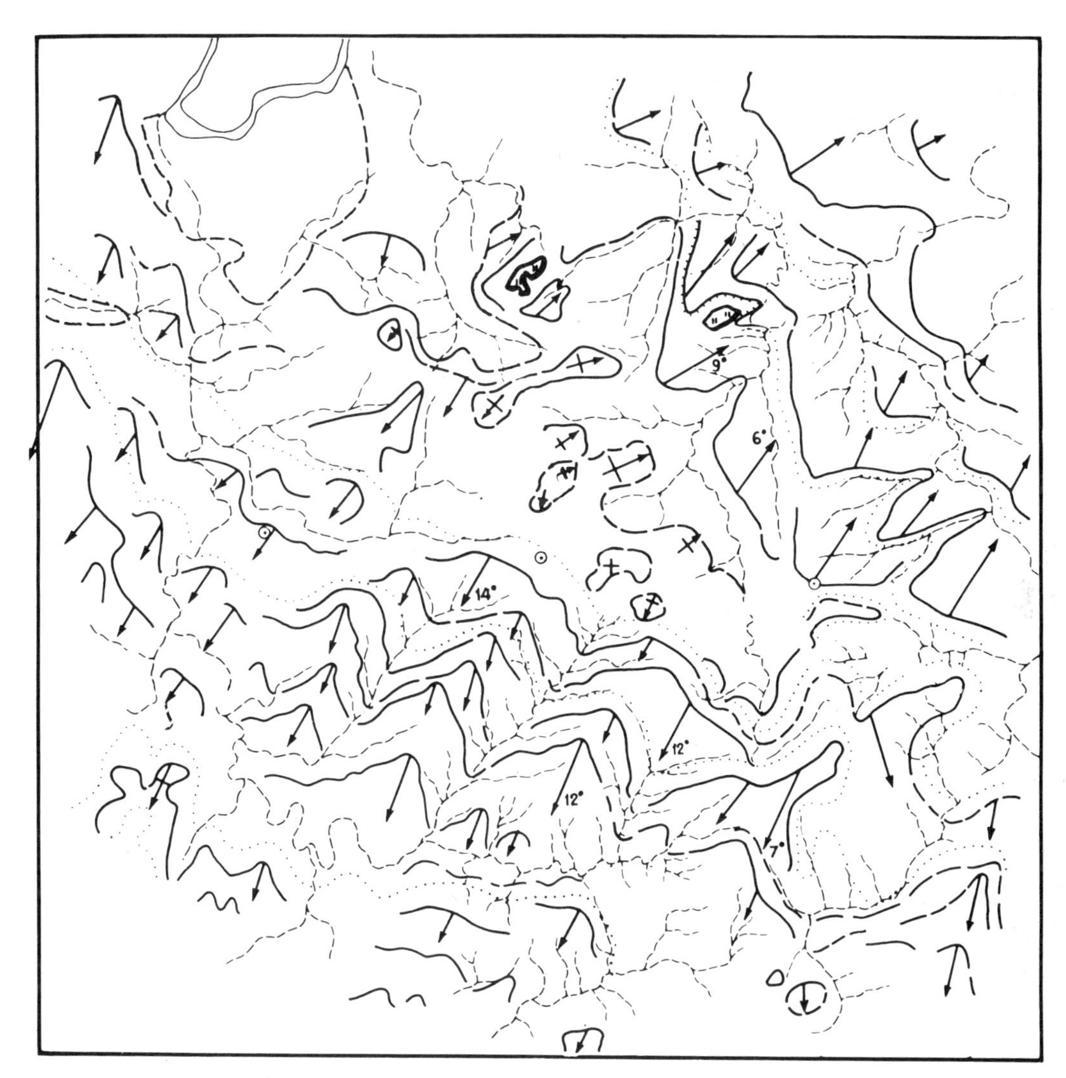

Fig. 14.7 Photogeological interpretation of tropical rainforest areas in New Guinea. Scale of original 1:40 000. (Source: Shell Petroleum Company Ltd.)

thermal inertia increases linearly with increasing density. During a diurnal solar cycle (Fig. 14.12) some surface materials with lower thermal inertia, such as shale, loose pumice or gravel, may reach relatively high temperatures during the day but cool to relatively low temperatures at night. In contrast, rocks or deposits with higher thermal inertia, such as sandstone, basalt or granite, may be cooler in the daytime but warmer at night.

One advantage of the thermal inertia technique over other methods of identifying rocks is that it minimizes the potential masking effects of surface contaminants (e.g. lichens), since the thermal inertia is a volume rather than a surface property. Recent studies have shown that, although there is a significant reduction in contrast of mid-infrared spectral features with increasing lichen cover, the spectral features of granites were still apparent with lichen coverage as high as 71%. Thus lichens do not appear to have a significantly deleterious effect on the thermal inertia estimated from spectral data. Nevertheless, further work needs to be done to fully investigate the usefulness of thermal inertia modelling as a basis for geological mapping.

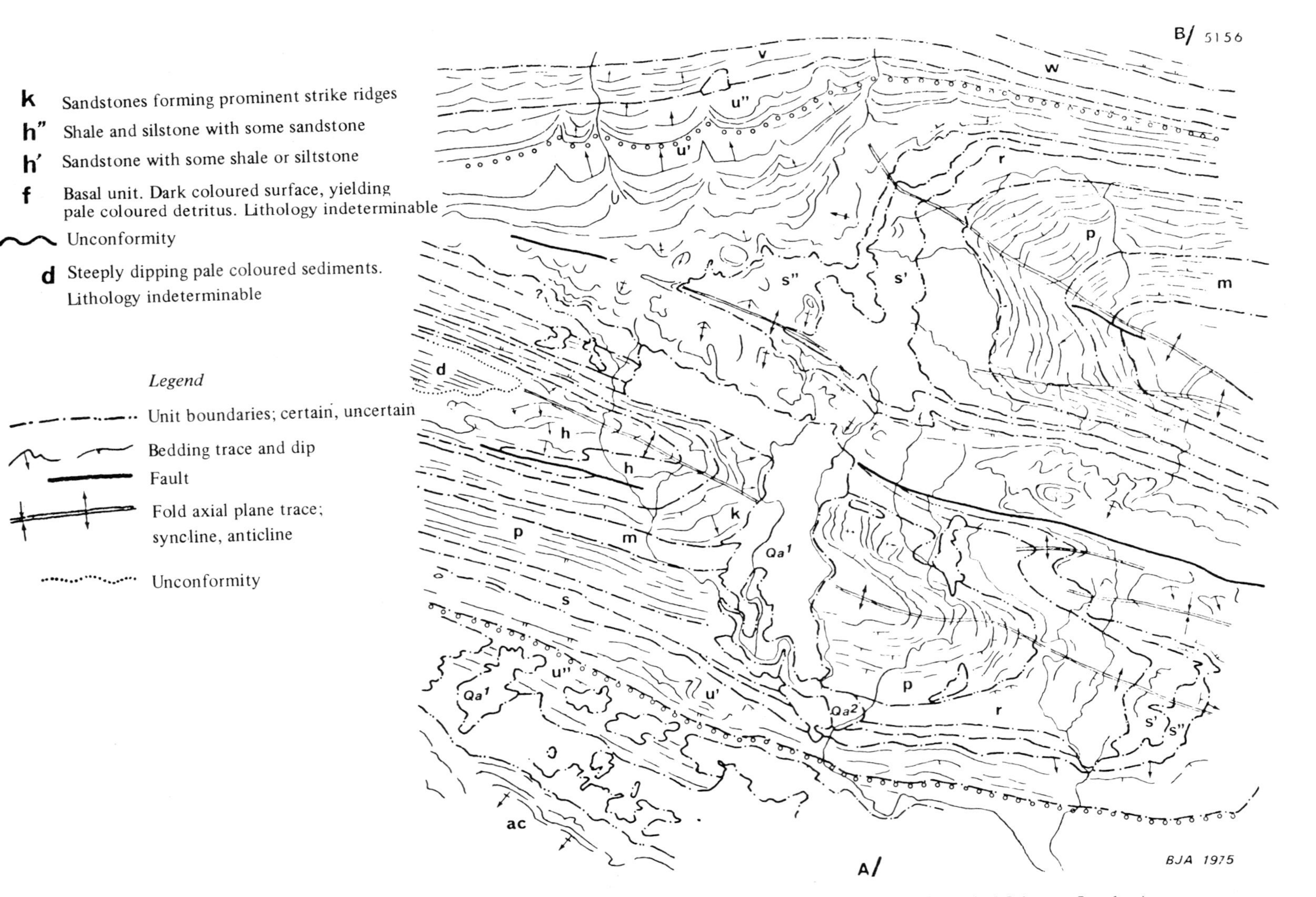

Fig. 14.8 Photogeological map prepared from a vertical photograph of folded sediments in Australia. (Source: institute of Geological Sciences, London.)

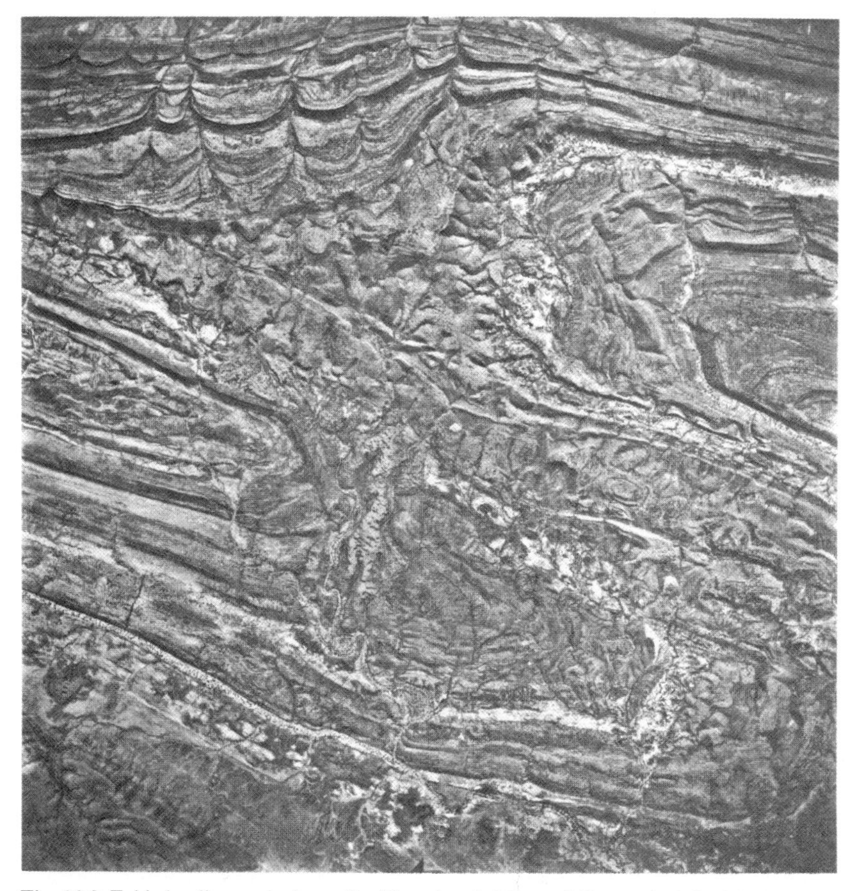

Fig. 14.9 Folded sediments in Australia. Note the pitching anticline on the left of the figure and the dipping sediments on the right. The photogeological map of this area is shown in Fig. 14.8. (Courtesy, E.A. Stephens, Institute of Geological Sciences, London.)

14.2.2 Microwave radar

Perhaps the most extensively used non-photographic system for geological studies is the microwave radar system. Side-looking radar systems are particularly valuable in equatorial and maritime regions where persistent cloud renders photography difficult to acquire. Furthermore the radar system illuminates the ground from the side so that shadowing effects are produced similar to those of low-sun photography (Fig. 15.9). These are well shown in the radar imagery of the Malvern Hills illustrated in Fig. 4.19.

Shape, pattern, tone and texture are used in the interpretation of radar imagery but the side illumination of radar systems places an emphasis on patterns, e.g. lineaments. Tones may vary greatly according to look angle and so must be used with caution. Textures in the imagery mainly reflect the physical form, e.g. roughness of the surface.

Lineaments are often well displayed in radar images (Fig. 12.9) and regional fault and fold patterns can be conspicuously displayed and readily mapped. The orientation of the radar system can, however, give rise to bias in the representation of linear features. Also, where areas are in the radar shadow (Fig. 4.19) ground information is lost.

14.3 Space observations for the study of rocks and mineral resources

14.3.1 Visible and infrared imaging

The advent of the satellite platforms has provided the geologist with a new and wider perspective on geological structures. Thus, for example, photo-

Fig. 14.10 Infrared linescan imagery of a mountainous area of North Wales, including quarries, lakes and settlements. Note the ability of the sensor to detect different rock strata. (Source: Laird, 1977; courtesy, Royal Signals and Radar Establishment, Malvern; Crown Copyright.)

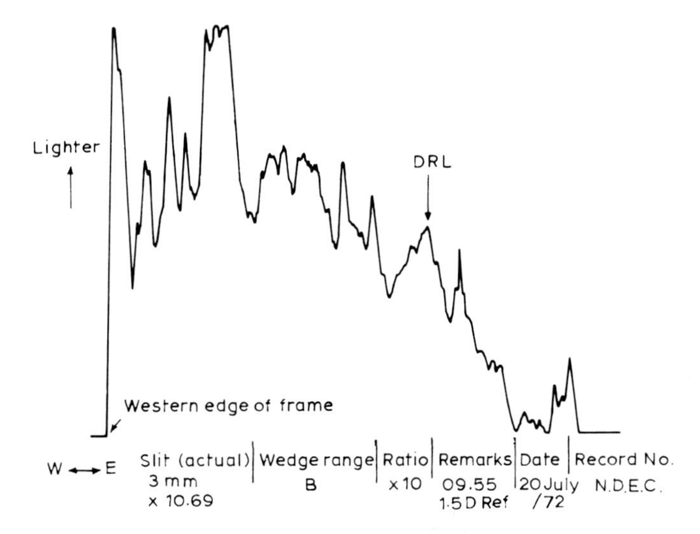

Fig. 14.11 Microdensitometer scan line of the infrared linescan imagery of the Dugald River Lode, Australia. (Source: Custance, in Cole *et al.*, 1974.)

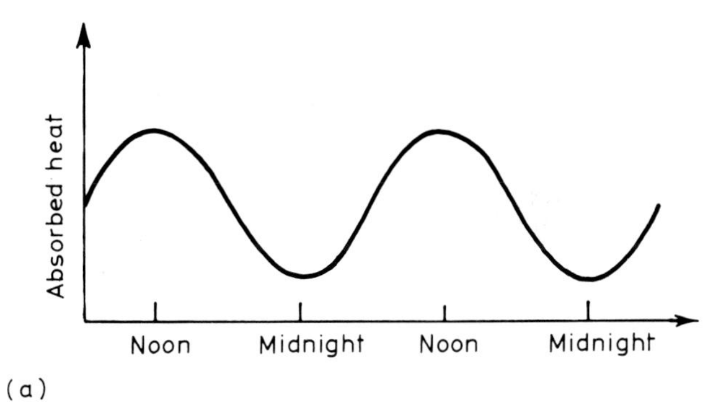

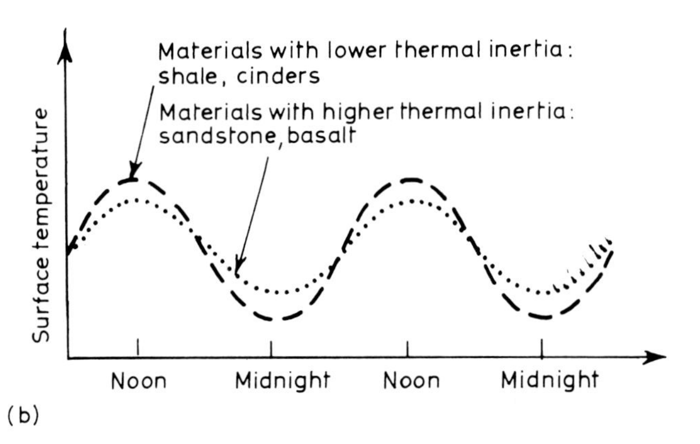

Fig. 14.12 Effect of differences in thermal inertia on surface temperature during diurnal solar cycles. (a) Solar heating cycle. (b) Variations in surface temperature. (Source: Sabins, 1978.)

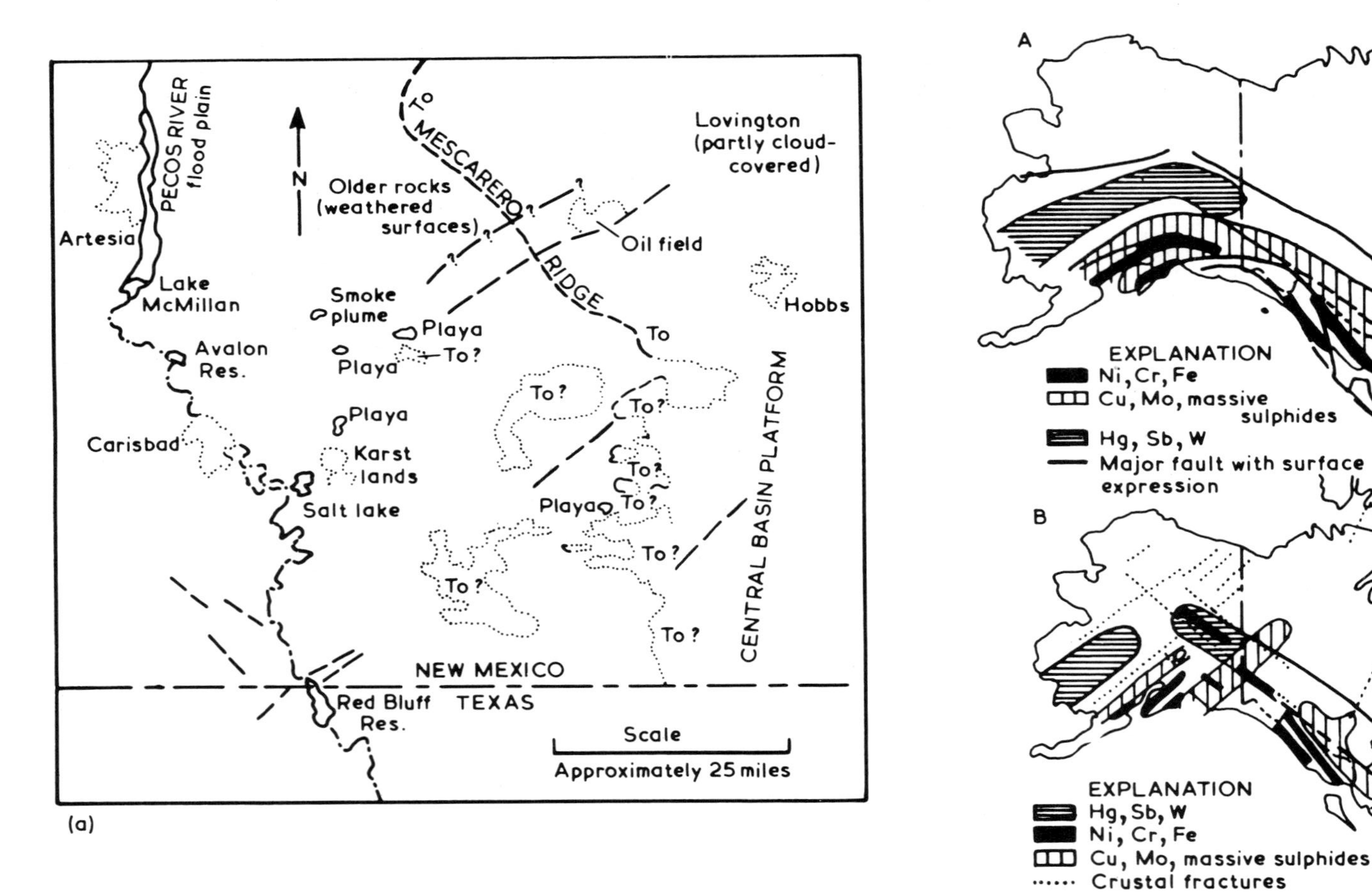

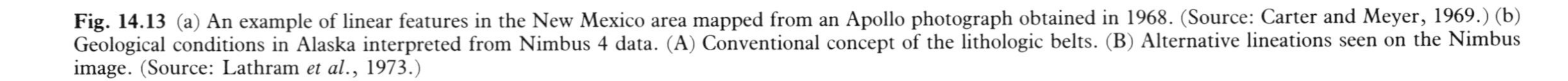

Fig. 14.13 (a) An example of linear features in the New Mexico area mapped from an Apollo photograph obtained in 1968. (Source: Carter and Meyer, 1969.) (b) Geological conditions in Alaska interpreted from Nimbus 4 data. (A) Conventional concept of the lithologic belts. (B) Alternative lineations seen on the Nimbus image. (Source: Lathram *et al.*, 1973.)

graphs of New Mexico obtained from the Apollo Saturn-6 spacecraft launched in April 1968 were examined with interest by geologists. They found a major north-east-trending lineament that was not shown on the most recent geologic map available for the New Mexico area (Fig. 14.13(a)). However, the major impact of the satellite platform has been made by Landsat satellites with their MSS and TM data. Much geological work has been achieved with this data and some selected examples will be discussed below. The general achievements of the Landsat programme have resulted in keen interest on the part of geologists. The most significant results can be listed as follows:

1. The broadest use of Landsat imagery in geologic mapping lies in the construction of regional structural (or tectonic) maps. Major geologic features, contacts between distinctly different, thick rock units, and landform types can be effectively mapped from Landsat images with details comparable to and sometimes superior to mapping by conventional aerial photo or ground methods.
2. Computer-produced geologic maps at scales up to 1:24 000 can be made from Landsat images with surprisingly good accuracy, providing training set data and other supervised methods are applied.
3. Winter imagery, with foliage-free scenes or snow-cover enhancements, provides useful data for mapping in many instances. Such imagery is especially valuable in seasonally vegetated areas and areas with considerable relief.

A number of applications for Landsat geologic studies have been identified. In regions for which only poor quality or outdated maps have been produced Landsat data offers a method of constructing good general maps at small scales. Existing maps also can be checked against Landsat imagery to correct mislocated or omitted rock unit contacts, geologic structures (fold axes for example) and lava flows. Landsat generated maps depicting structural information (particularly lineaments) also have utility in the search for ore deposits, oil accumulations and ground-water zones.

It must be borne in mind, however, that often field checking has been insufficient to verify the accuracy and correctness of Landsat-map data. Also it would appear that no reliable identifications of lithologic types have been made consistently. The band ratio techniques (e.g. MSS 7/5) lead to some enhancement of visual images, which improves the separation of rock types. However, the ranges of ratio values for many common rocks and minerals overlap. Unique spectral signatures have not always been found to occur within the MSS sensing wavelengths. In fact the ratio interval for haematite and serpentine – strikingly different minerals – is almost identical.

It also has been found that individual stratigraphic units are sometimes thinner than the linear resolution capability of Landsat 1 (70–100 m.) Thus remote sensing 'units' are often groupings of stratigraphic units and they sometimes have limited applicability to standard geologic mapping procedures. With these advantages and limitations of Landsat imagery in mind one can examine some selected examples of geologic interpretations in order to illustrate some of the techniques used.

An interesting application of the mapping of lineaments and faults is provided by workers comparing Nimbus and Landsat imagery. Prior to the launch of Landsat 1 a cloud-free image of Alaska was obtained by the Nimbus 4 Image Dissector Camera System. This image showed a set of north-west and north-east trending lineaments which suggested previously unrecognized geologic structures deep in the Earth's crust.

Lineaments and faults on Landsat 1 images have corroborated and added detail to the initial geological trends noted on the Nimbus imagery. These lineaments have been compared to the relation of known mineral deposits and fundamental fractures in the Canadian Cordillera. As a result it has been possible to provide an alternative map to guide mineral resource application in Alaska and western Canada. (Fig. 14.13(b)). The great detail obtainable from Landsat images is also borne out by the detailed mapping of main fault systems revealed by Landsat 1 data for the Italian peninsula, as shown in Fig. 14.14.

Many researchers have noted that snow cover does not obliterate lineaments, indeed major structural features may be accentuated thereby (Plate 14.15). Observations in New England, USA indicate that a heavy blanket of snow (22.5 cm (9 in)) accenturates major structural features whereas a light dusting of snow (2.5 cm (1 in)) accentuates

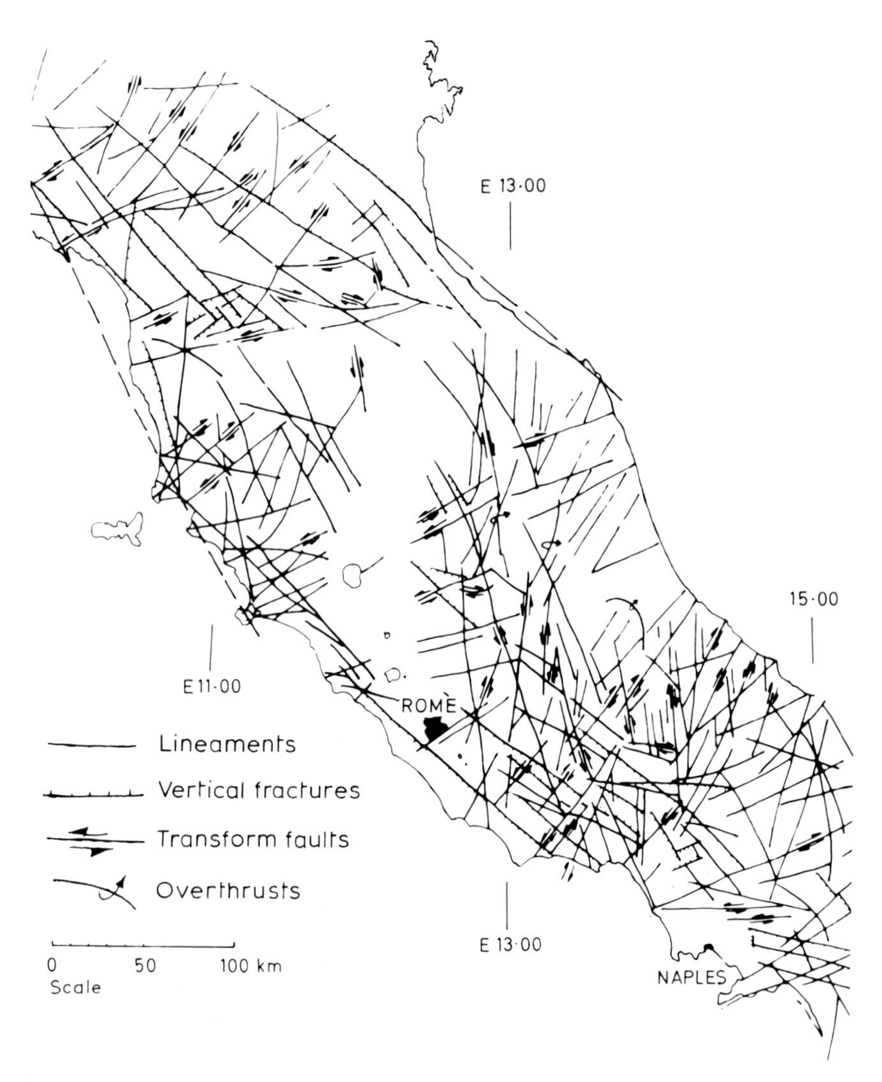

Fig. 14.14 Main fault systems in central Italy, as interpreted from Landsat data. (Source: Bodechtel *et al.*, 1974.)

more subtle topographic expressions. Comparisons of snow-free and snow-covered Landsat 1 images for the same area have shown that snow-covered imagery allows more rapid fracture analysis and provides additional fracture detail.

Some researchers have used the methods of *spectral ratioing* of reflected radiances of selected pairs of Landsat multispectral channels and production of analog ratio images for the mapping of large exposures of iron compounds in Wyoming. The data were combined with laboratory data as training sets wherever possible.

Enhancement techniques were used in which the digital data of one channel were divided by the data from a second channel. For example where channel 7 data were divided by channel 5 data the resultant ratio digital greymap is given the notation R_{75}. When the analog ratio images were printed in colour it was found that green areas represented vegetation, violet areas primarily rock and vegetation, blue areas rock outcrop and red represented iron-rich outcrops. Darkest red was found to occur only in an iron mine area and along pond edges, where muds and tailings were present. Other

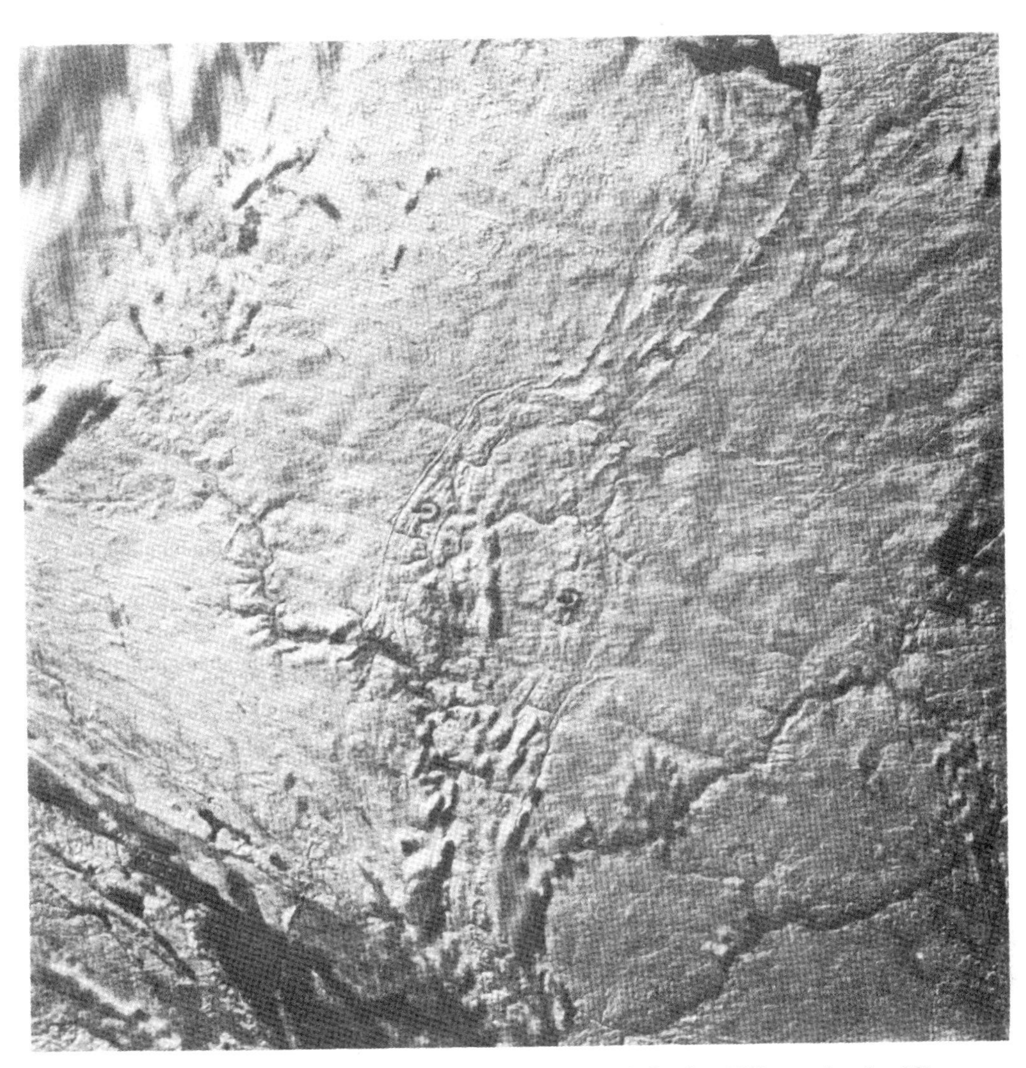

Fig. 14.15 Landsat 1 image of New England, USA.; early winter, 9 October 1972; sun elevation 25°; no geometric corrections; MSS, Band 6. Lineations of quartzite marked G. (Source: Gregory, 1973.)

images were also created by analog ratio images. For example the analog R_{74} showed iron mines as unique dark areas in the scene. These digital ratio and analog ratio maps can be used to focus the attention of the geologist or scientist on areas in the scene where chemical properties are different from those of adjoining or surrounding localities.

Landsat data have been widely used in *mineral exploration*. For example, a training set for digital classification of rock types with potential for copper deposits was derived from a Landsat image of Saindak copper deposit, Pakistan as shown in Fig. 14.16. This training set was given computer symbols (see Table 14.1) and then used to make a digital classification map of Western Pakistan.

In this manner large areas can be examined at a lower cost than would be incurred for conventional field reconnaissance. Subsequent field work can then be selective in its approach and the most promising areas for copper extraction can then be determined.

Substantial improvements in the quality of an image can be achieved by a variety of computer techniques (Chapter 9). These include principal

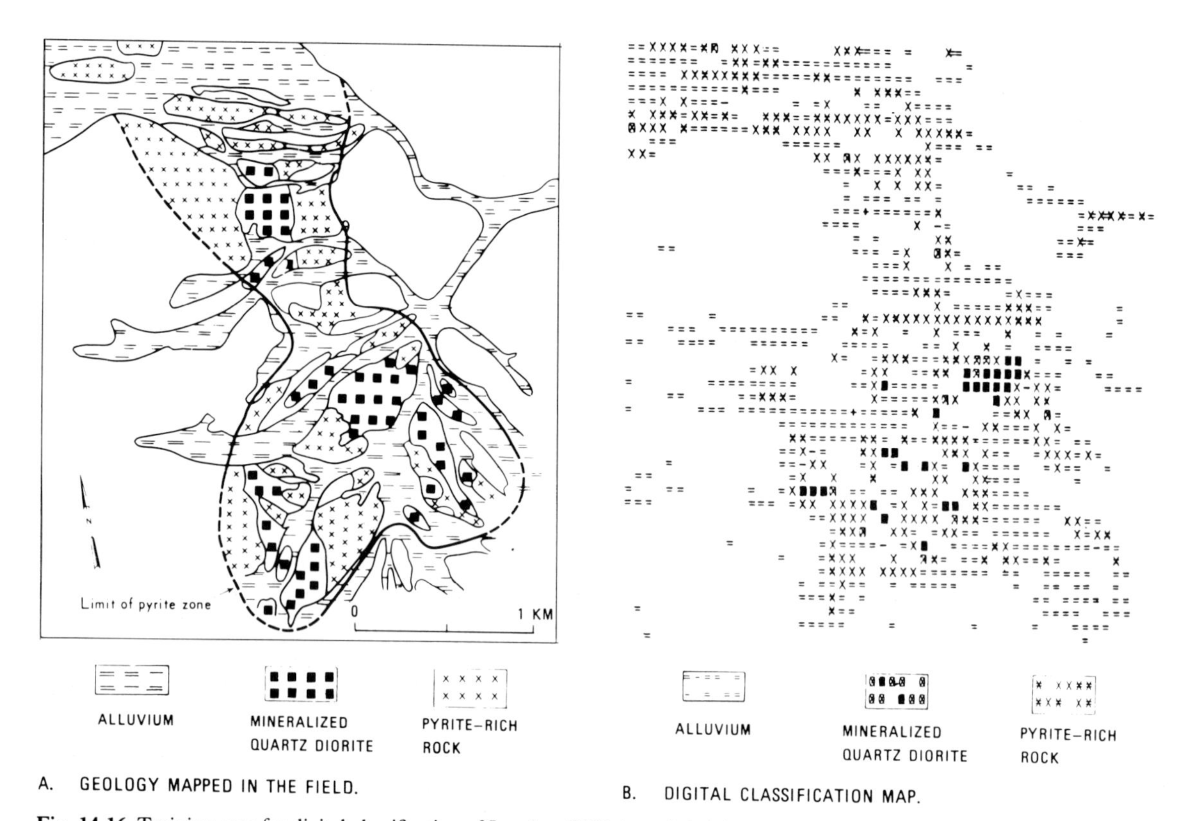

Fig. 14.16 Training area for digital classification of Landsat CCT data. Saindak copper deposit, Pakistan. (Source: Sabins, 1978.)

components analyses and contrast enhancement algorithms.

Automatic pattern recognition by computer analysis can, of course, be applied to geological materials. However, it is necessary to make certain assumptions if such mapping is to be undertaken. First, it must be assumed that subsurface materials will manifest themselves as spectrally separable classes at the Earth's surface. Since subsurface rocks are normally veneered by soil and vegetation it also must be assumed that spectrally separable surface features are correlated with subsurface variations. Secondly, the assumption that lithologic types are naturally segregated into a limited number of discrete compositional and textural categories which can be recognized in some classification system must be made. This assumption is clearly wrong, as it is for many soils. However it does provide a basis for grouping similar lithologies

Table 14.1 Digital classification table for Saindak training area (Source: Sabins, 1978)

Rock type	Classification reliability	Computer symbol
Mineralized quartz diorite	High	0
	Low	Ⓧ
Mineralized pyritic rock	High	X̶
	Low	X
Dry wash alluvium	High	=
	Low	–
Boulder fan	High	+
Eolian sand	High	·
	Low	,
Dark rock outcrops, desert-varnished lag gravels, and black sand		1 # H

into discrete classes. Where intergrades occur, e.g. sandy-shale, the classifier may assign to either sandstone or shale in a random fashion.

Where these automatic recognition techniques have been applied to Landsat images of south-west Colorado a correlation of 89.9% has been obtained with existing geologic maps. It is apparent, however, that extensive ground truth observations and aircraft underflight data are needed before definitive statements can be made concerning the reliability and accuracy of machine-made maps. Nevertheless the vast amount of information now coming from satellite sensors makes it necessary to think in terms of automatic mapping procedures for the future.

14.3.2 Microwave imaging

Radar imagery from aircraft has the useful characteristic that it often accentuates surface irregularities. However, the usefulness of satellite radar in geology and geomorphology is at least equally promising but has been limited so far. Radar imagery from satellite orbits has been obtained from the SIR-A and SIR-B sensors. It has been found that the high depression angle dictated by orbital platforms generally means that imagery with extensive shadow effects – which is typical of low-level airborne data – is mostly absent. Rather, orbital radar imagery is strongly influenced by local incidence angles. Thus topographic detail and lineament differentiation may be less pronounced. As a result the tonal variations that are so often valuable in photogeological image interpretation are often very subdued. In these circumstances *image enhancement* techniques become essential if effective geological interpretation is to be achieved. Research has shown that enhancement by one of the following methods may be of assistance:

1. linear contrast stretch;
2. speckle reduction (nearest neighbour average or mean/median filtered image);
3. directional enhancement using edge filters;
4. principal component analysis for combining more than three look directions.

Orbital radar studies based on SIR-A data, when compared with analyses of Landsat MSS data, have shown that satellite radar facilitates mapping of small-scale features owing to higher resolution and greater sensitivity to surface roughness. However, MSS data do permit the mapping of rock units, owing to varying spectral response from rock surfaces, regolith, or surficial soils (Fig. 14.17). The advent of ERS-1 is expected to provide a better opportunity of assessing radar for geological exploration, although the amount of overland imagery will be quite restricted in the early years of operation.

14.3.3 Terrestrial imaging spectroscopy

A new class of sensor called the *imaging spectrometer* is now under development for geological prospecting. The first images from an Airborne Imaging Spectrometer (AIS) were acquired in late 1982 and they constituted a new class of data requiring new approaches to information handling and extraction. The concept of imaging spectroscopy is shown in Fig. 14.18. Energy reflected from the Earth's surface is dispersed by a spectrometer and is used to form as many as 200 registered spectral images of the scene. The essential advantage of the technique is that each pixel in the scene has associated spectral information sufficient to allow for the reconstruction of a complete spectrum. Thus the diverse narrow-band spectral signatures that are characteristic of most surface materials can be used to identify those materials.

Field-acquired reflectance spectra (Chapter 6) for selected phenomena are shown in Fig. 14.19. Simultaneous imaging in many narrow contiguous spectral bands has required new thinking in respect of sensor design. Opto-mechanical systems, as operated from the Landsat series, are unsuitable but linear array detectors offer prospects for the future. The Modular Opto-electronic Multispectral Scanner (MOMS) flown aboard the Space Shuttle was an experimental two-spectral band instrument. Two approaches to the design of imaging spectrometers, one using *line arrays* and the other using *area arrays*, are shown in Fig. 14.20. Linear arrays for high spatial resolution imaging (IFOV 10–30 m) are only suited to airborne detectors that allow for the sensors to be moved slowly enough for the detector array to provide data at a small fraction

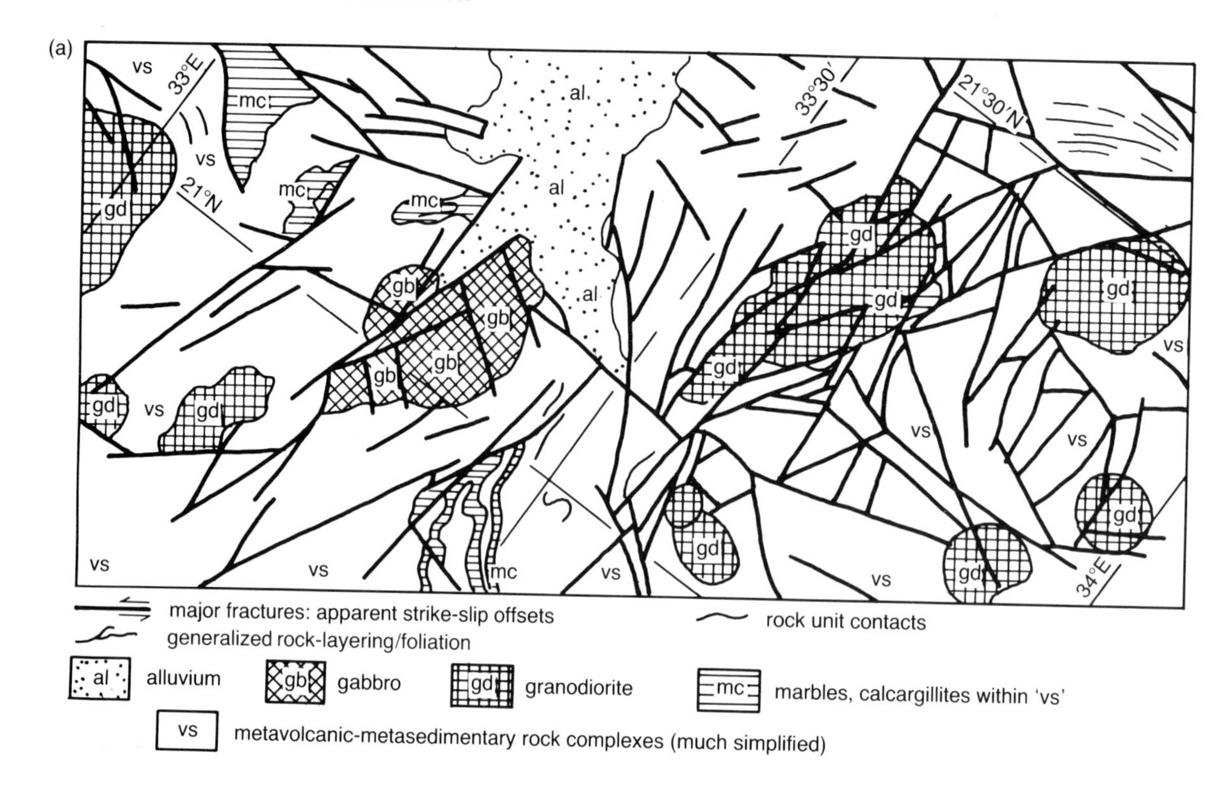

Fig. 14.17 Geological interpretation from (a) MSS imagery of the study area in the Rocky Mountains (53 × 111 km) and (b) SIR-A imagery.

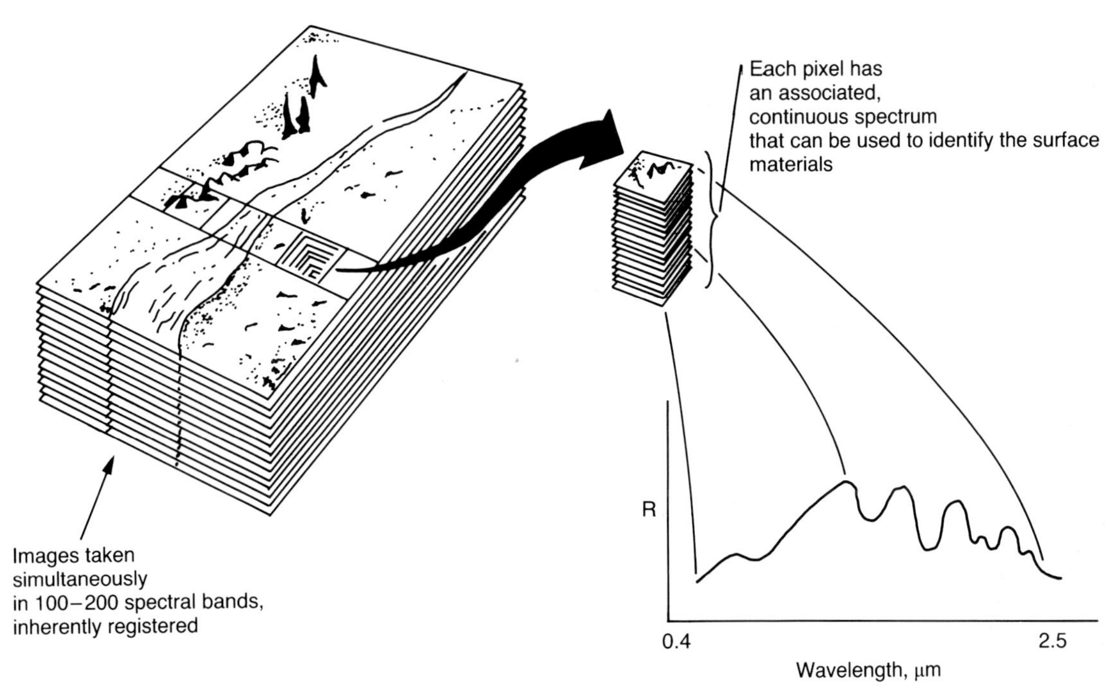

Fig. 14.18 The imaging spectroscopy concept. Two hundred or more images are acquired simultaneously, each in a narrow spectral band. In this manner, a complete reflectance spectrum can be constructed for every pixel in the scene. (Source: Vane and Goetz, 1988.)

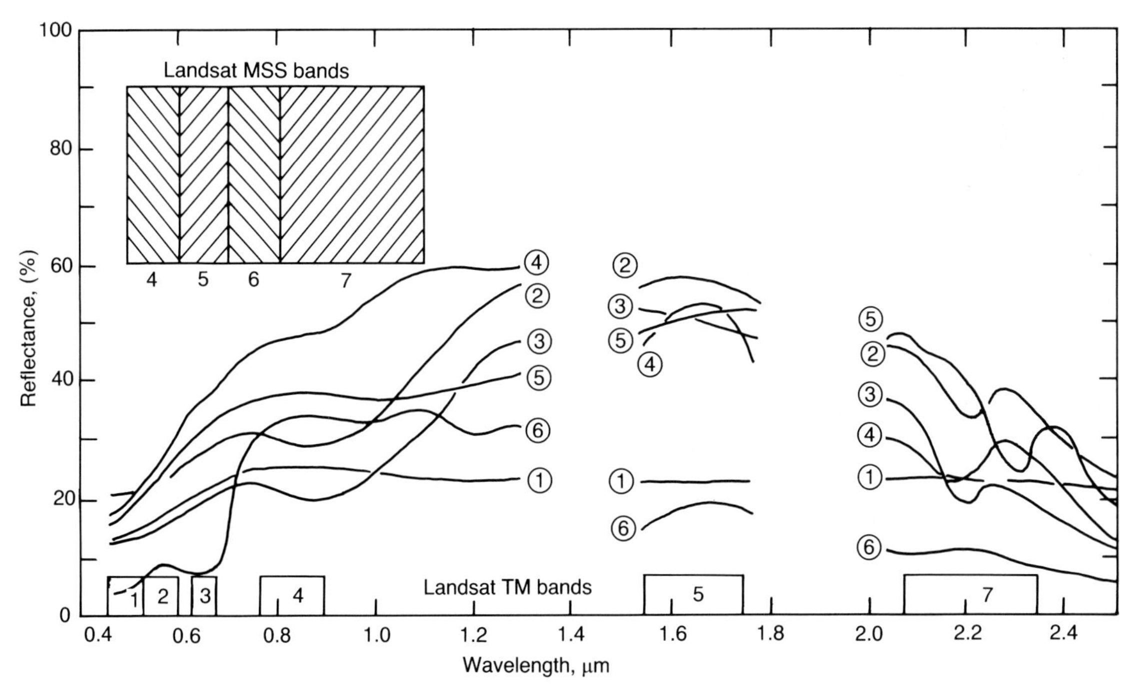

Fig. 14.19 Field-acquired reflectance spectra for (1) unaltered tuff fragments and soil, (2) argillized andesite fragments, (3) silicified dacite, (4) opaline tuff, (5) tan marble, and (6) ponderosa pine. Charge transfer bands that extend from the ultraviolet into the visible portion of the spectrum are responsible for the general increase in reflectance between 0.4 and 0.8 μm. The gaps at 1.4 and 1.9 μm are the result of atmospheric water absorption. (Source: Vane and Goetz, 1988.)

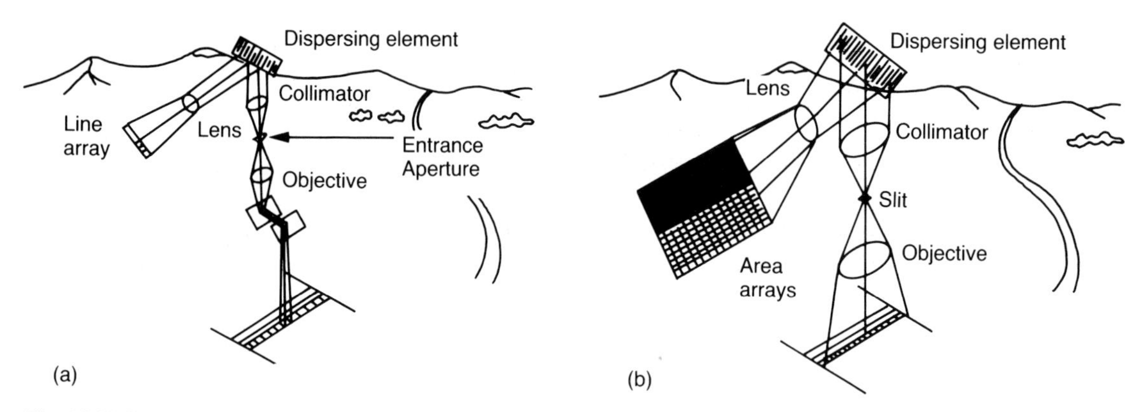

Fig. 14.20 Image acquisition techniques. (a) Whiskbroom imaging spectroscopy with line detector arrays (e.g. AVIRIS) and (b) pushbroom imaging spectroscopy with area detector arrays (e.g. AIS and HIRIS). (Source: Vane and Goetz, 1988.)

Table 14.2 Characteristics of imaging spectrometers planned for the EOS programme (Source: NASA)

	Sensor	
	High Resolution Imaging Spectrometer (HIRIS)	Moderate Resolution Imaging Spectrometer-Nadir (MODIS-N)
Spatial resolution	30 m	1.5 km × 1.5 km or 0.5 km × 0.5 km
Swath width	30 km	1500 km
Viewing area	25 off track +60/−30 in track	Nadir centred
Spectral range	0.4–2.5 μm	0.4–2.3 μm 3 –5 μm 6.7–14.2 μm
Spectral resolution	10 nm	1 km and 0.5 km
Number of bands	196	27 at 1 km 13 at 0.5 km

of the integration time. The Airborne Visible/Infrared Imaging Spectrometer (AVIRIS) has been built for this purpose and began operations in 1987.

AVIRIS covers the entire spectral region of 0.4–2.5 μm in 210 contiguous 10-nm-wide spectral bands. At 20 km altitude (e.g. from a U-2 aircraft) the imagery spans an 11 km swath width and provides a ground pixel size of 20 m. It is anticipated that AVIRIS will become the key sensor for terrestrial imaging spectroscopy for the next decade.

Because of high spacecraft velocities, imaging spectrometers designed for Earth orbit require area arrays of detectors at the focal plane of the spectrometer. The NASA Earth Observing System (EOS), scheduled for launch in the late 1990s is planned to carry two imaging spectrometers – the Moderate Resolution Imaging Spectrometer – Nadir (MODIS-N) and the High Resolution Imaging Spectrometer (HIRIS). The characteristics of these instruments are shown in Table 14.2.

There is likely to be considerable interest in the geological and mineral exploration capabilities of these instruments. The analysis of imaging spectrometer data has been investigated at various Data Analysis Workshops held at the Jet Propulsion Laboratory (JPL) and a software package has been developed by JPL. The package, called Spectral Analysis Manager (SPAM), contains a number of general image processing algorithms together with several that have been designed specifically for the analysis of airborne imaging spectrometer data.

Imaging spectroscopy is still in its infancy but the search is on for techniques for the detection of specific mineral assemblages and particular plant stress situations, which are often indicative of ore bodies. The use of such systems is being hailed a 'quantum leap advance' in remote sensing. Only time will tell whether their significance will be so great, but the early signs are highly promising.

15 *Ecology, conservation and resource management*

15.1 Introduction

There is now a growing awareness that man's use of nature's processes and resources must be adjusted to the limitations and requirements which nature sets for us. Thus twentieth century man has become more interested in the ecological changes which human settlements and industry have brought about in the local and the world environment. In particular he is now concerned with the *conservation of species* in danger of extinction and the *maintenance of ecological balance* in environments which have been altered to suit his economic needs. The important and fundamental fact remains that all animal life, including man, ultimately depends on the plant life, which alone is able to synthesize elements into the form of food.

Wherever conditions are sufficiently favourable the climax vegetation cover consists of forest. In some areas, notably in savanna regions, forest clearance has led to an extension of grassland where forest formerly occurred. A summary of data for the Earth's cover of natural vegetation is shown in Table 15.1 from which it will be seen that approximately 42% of the total land area is potentially forest land, 24% potentially grassland and 34% essentially desert.

Forests and woodlands are significant not only in their own right but also as balancing mechanisms in terms of CO_2 in the Earth's atmosphere. They are valuable in resource terms to both affluent societies in industrial nations and to poor rural communities in developing countries. Products such as timber, sawnwood and panels (for construction, doors, shuttering and furniture), pulpwood (for paper, cartons and rayon), poles, posts, mining timbers and railway track sleepers, fuelwood, fodder, fruits, pharmaceuticals, fibres, resins, gums, dyes, waxes and oils are some of the many items originating in woodlands. The value of annual world production of forest products exceeds \$115 500 million, and international trade is worth about \$40 000 million a year at present prices. Thirty countries (eight of them developing countries) each earn more than \$100 million a year (five more than \$1000 million) from forest products.

Alongside the demand for forest products there is a huge burden placed upon forests and woodlands for fuel and as sites for shifting cultivation. More than 1500 million people in poor countries depend on wood for cooking and keeping warm. This annual consumption of wood is estimated to be more than 1000 million cubic metres. In Africa the contribution of trees to total energy use is as high as 58%, and in South-east Asia and Latin America it is 42% and 20%, respectively. Around one fishing centre in the Sahel region of Africa, where the drying of 40 000 tonnes of fish consumes 130 000 tonnes of wood every year, deforestation extends as far away as 100 km. Fuel wood is now so scarce in the Gambia that gathering it takes 360 woman-days per year per family.

There is now sufficient expertise available concerning forest and woodland management to plan for sustained cropping of woodlands whilst maintaining forest cover and habitats for woodland animals. Unfortunately, owing to over-use and exploitation of woodland areas the world position is now becoming sufficiently grave as to affect even world climates. In addition the habitats of many wild animals are under such pressure as to lead to the virtual extinction of many species.

Permanent pastures (land used for five years or more for herbaceous forage crops) and other grazing lands are generally in areas unsuitable for

Table 15.1 Natural vegetation cover (Source: Roberts and Colwell, 1968)

	Area in square miles	Percentage of total land area
Forests		
Tropical rainforest	3 800 000	7.5
Temperate rainforest	550 000	0.9
Deciduous forest	6 500 000	12.0
Coniferous forest	7 600 000	15.0
Monsoon (dry) forest	2 000 000	3.8
Thorn forest	340 000	0.6
Broad sclerophyll forest	1 180 000	2.1
Total forest	21 970 000	42.0[a]
Grasslands		
High grass savanna	2 800 000	5.3
Tall grass savanna	3 900 000	7.5
Tall grass	1 580 000	3.1
Short grass	1 200 000	2.4
Desert grass savanna	2 300 000	4.3
Mountain grass	790 000	1.4
Total grassland	12 570 000	24.0[a]
Deserts		
Desert shrub and grass	10 600 000	21.0
Salt desert	30 000	—
Hot and dry deserts	2 400 000	4.7
Cold desert (tundra)	4 400 000	8.3
Total desert	17 430 000	34.0

[a] Some areas of potential grazing land occur within areas of forest and desert. A more realistic figure for total land area of potential grazing is 46%.

crops without intensive capital investment. Their productivity is generally low, ranging from 3 to 5 Grazing Livestock Units (GLU) per hectare in Central Europe to 1 GLU per 50–60 ha in Saudi Arabia. Even so grazing lands and forage support most of the world's 3000 million head of domesticated grazing animals and hence most of the world's production of meat and milk.

Mismanagement of grazing lands is widespread. Overstocking has severely degraded grazing lands in Africa's Sahelian and Sudanian zones. In parts of North Africa, the Mediterranean and the Near East it contributes markedly to desertification. Overstocking together with uncontrolled grazing is also a serious problem in the Himalayas and Andes. Even in the upland areas of England and Wales there is evidence of overstocking affecting the natural composition of the rough grazings. In the world as a whole there are many areas where intensive grazing has removed both trees and grass cover, leaving the soil open to the risk of soil erosion.

Many of the extensive grazing areas of Europe are in upland areas of scenic beauty. In consequence they have been designated as 'protected areas' and efforts have been made to protect the natural vegetation by reducing the problem of overstocking.

Most of the critical problems in the management of grazing lands (range resource management) and husbandry of grazing animals occur in remote areas where least is known about native grazing land. In most cases what is needed is better information on the following:

1. acreages of useable grazing land by ecologically appropriate classes;
2. the ecological characteristics of each kind of range (phytosociology, plant succession, present range condition, autecology of the important species);
3. the special management problems associated with different grassland ranges;
4. indices of potential productivity of each kind of range.

Information is needed also on the numbers and kinds of animals that make use of grassland resources. These include not only the farm animals but also wild herbivores.

Wildlife is an important subsistence resource in developing countries, and a significant recreational resource in both developed and developing countries. In parts of Ghana, Zaire and other countries in west and central Africa, for example, up to three-quarters of the animal protein comes from wild animals. The nutritional importance of wild animals and plants for many people in developing countries is often underestimated or ignored.

Ecological studies require detailed biotic and abiotic information, some of which can be obtained only by ground survey, such as chemical data for the environment and organisms concerned. However, remote sensing has provided valuable support for ground sampling methods. It has proved particularly useful for mapping plant associations and estimating the populations of larger animals occupying remote regions.

In the case of plant ecology it may be noted that images from remote sensors are especially affected by variations in the growth forms of plants. Thus the broad vegetation categories of trees (Phanerophytes), shrubs (Chamaephytes) and perennial forms, such as the grasses (Hemicryptophytes), normally can be recognized readily. Also, the habitat of the community being studied can be viewed by examination of the images of the surface of the area. In this way the interrelationships of topography, local climate, soil, surface water, rock formations, human artefacts and vegetation cover can be assessed.

Seasonal changes in plant forms usually produce marked differences in the images obtained by remote sensing. For example, winter and summer images of deciduous forest vary greatly from each other. The term *phenology* has been applied to that branch of science which studies periodic growth stages in the plant and animal world depending upon the climate of any locality. Phenological studies can be assisted greatly by remote sensing observations. Conversely the interpretation of remote sensing data can be aided considerably by applying knowledge of local phenological changes to the study of images obtained at different times of the year.

Although only a minute proportion of the world's plants and animals have been investigated in respect of their medicinal values, medicine depends heavily upon them. It has been estimated that more than 40% of USA prescriptions each year contain a drug of natural origin as sole active ingredient or as one of the main ones. The value of medicines from higher plants in the USA alone has been valued at $3000 million a year and rising. The most important applications of higher plants and animals for medicine are:

1. as constituents used directly as therapeutic agents, e.g. digitoxin, morphine, atropine;
2. as starting materials for drug synthesis, e.g. adrenal cortex and other steroid hormones;
3. as models for drug synthesis, e.g. cocaine, which led to the development of modern local anaesthetics.

Genetic diversity and its preservation is an issue of great importance in respect of world survival. Genetic material presently contained in commercially important crops, trees, livestock, aquatic animals and organisms provides useful qualities for a limited time. They are rarely, if ever, permanent. New genetic material is often necessary in order to breed strains for improved yields, nutritional quality, flavour, durability, pest and disease resistance, responsiveness to different soils and climates, and other qualities. For example, the destruction of European vines by *Phylloxera* in the 1860s was overcome only when it was noted that the native American vine is tolerant of *Phylloxera*. Europe's wine production was saved by grafting European vines on to American rootstocks. This practice continues to the present day.

Even many useful breeds of livestock are at risk. Of the 145 cattle breeds in Europe and the Mediterranean some 115 are threatened with extinction. The genetic pool contained in these rare breeds may yet be valuable. The very rare Wensleydale sheep was used to breed heat-tolerant sheep capable of producing good quality wool in subtropical lands. The Cornish hen, once a rarity, was crossed with other strains to produce the meat-bird now used widely in the broiler industry.

The scale of the loss of domesticated plants and animals is matched in losses of wild species. Some 25 000 plant species and more than 1000 vertebrate species and subspecies are presently threatened with extinction. These figures do not take account of the great number of small species, e.g. molluscs, insects, and corals lost as their habitats are progressively destroyed. Some estimates suggest that from half a million to a million species will be made extinct by the end of the century.

15.2 Reflectance from vegetated surfaces

The reflectance of light from a vegetated ground surface is determined by several factors, which may include leaf geometry, morphology, plant physiology, plant chemistry, soil type, solar angle and climatic conditions. Plant structure, soil background and the surface condition of the plants' reproductive and vegetative parts are particularly important in determining spectral reflectance.

When considering reflectance it is, of course, necessary to bear in mind that the angular nature of reflectance can be described as either hemispherical or directional (Chapter 6). As we have seen the term 'hemispherical' refers to circumstances where radiant flux is measured over a hemisphere whereas

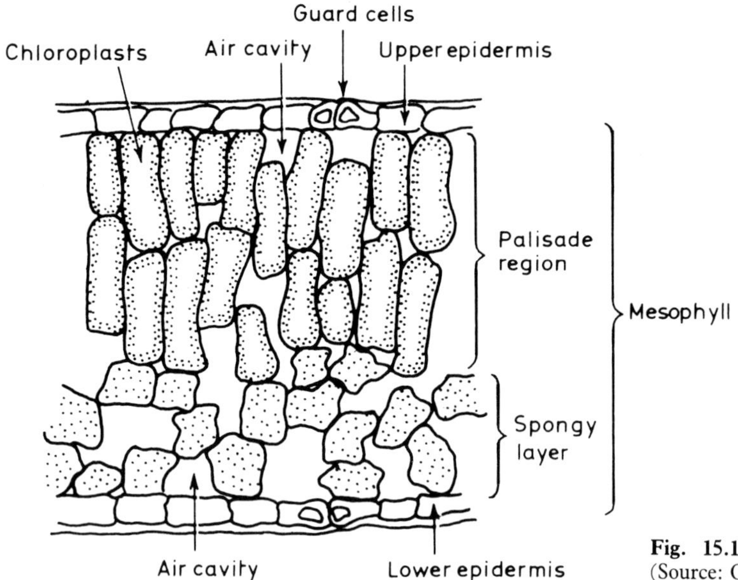

Fig. 15.1 A generalized diagram of leaf structure. (Source: Curtis, 1978.)

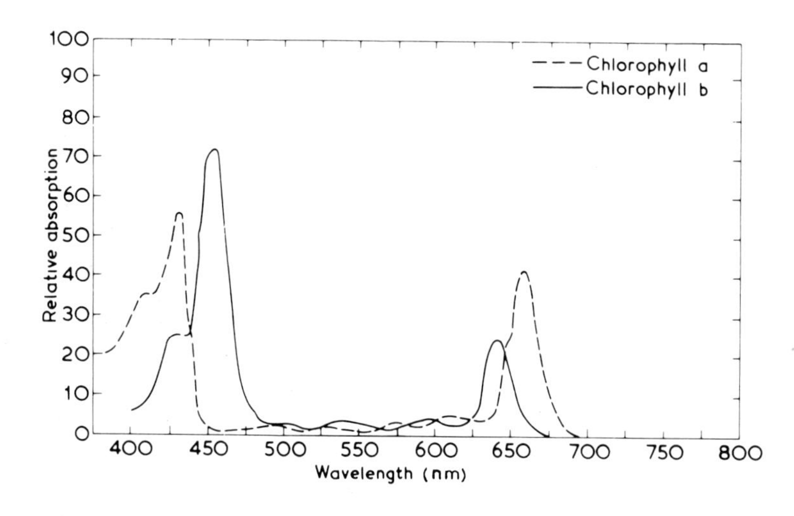

Fig. 15.2 Curves showing relative absorption of chlorophyll *a* and chlorophyll *b* as a function of wavelength. The combined absorption would be a summation of the two curves at each wavelength. (Source: Colwell, 1969.)

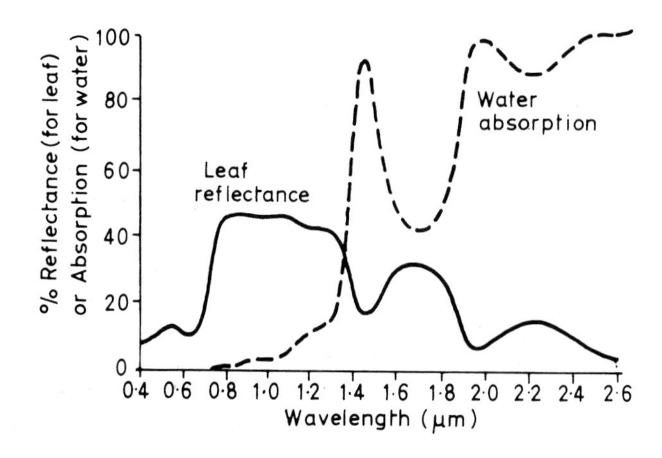

Fig. 15.3 A generalized spectral signature for leaf reflectance. Note the characteristic infrared 'plateau' at 0.75–1.2 μm and the regions of water absorption. (Source: Curtis, 1978.)

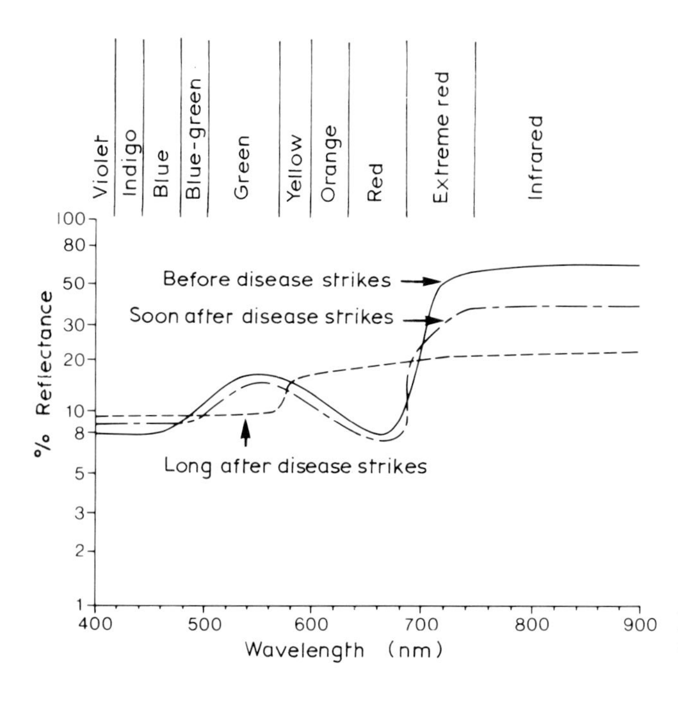

Fig. 15.4 Reflectance characteristics of healthy and diseased leaves. (Source: Colwell, 1969.)

'directional' refers to collection in one direction only. It is convenient, however, to think of reflectance as being 'bidirectional', where the angles of incidence and reflection are directional, as happens with satellite observations on a sunny day. We have discussed the term Bidirectional Reflectance Factor (BRF) in the context of field measurements of spectral response (Chapter 6). The BRF is the ratio between the spectral radiance at a given angle of incidence and the reflectance of a standard reflector at the same angle within the scene. This measure is sometimes termed the *bidirectional reflectance* (BDR) for convenience. In remote sensing studies of vegetation the BDR relates to the vegetation canopy, consisting of a mosaic of leaves, and other plant structures, together with the *soil background*. Before considering the models used for the study of vegetation using remote sensing it will be useful to consider reflectance from a single leaf.

The characteristics of light reflectance and transmittance can be explained mainly on the basis of critical reflection of visible light at the cell wall–air interface of both the palisade and spongy layers of the mesophyll (Fig. 15.1). It has been shown that reflectance increases with an increase in the number of intercellular air spaces. This is because diffused light passes more often from highly refractive hydrated cell walls to lowly refractive intercellular air spaces.

The leaf structure is important in that the upper and lower epidermal layers together with the palisade cells and spongy mesophyll each play a part. Spongy mesophyll is important because it scatters near-infrared light. On the other hand, the lower sides of leaves reflect more light than the upper side and this is thought to be due to the absence of palisade cells on the lower side. Often the leaf becomes more spongy with age with the result that the mature leaf displays less reflectance in the visible bands (about −5%) and more in the infrared (about +15%).

Frequently the different parts of the spectrum are affected differently by variations in plant composition and structure. The 0.5–0.75 μm band is characterized by absorption by pigments consisting mainly of chlorophylls *a* and *b*, carotenes and xanthophylls (Fig. 15.2). The 0.75–1.35 μm band is a region of high reflectance and low absorption which is greatly affected by the internal leaf structure. The 1.35–2.5 μm band is influenced somewhat by internal structure but is more particularly affected by water concentration in the tissue. In general the spectral *transmittance* curves for mature and healthy leaves are similar to their spectral

Table 15.2 Seven of the more commonly used BDR ratios, where G = green, R = red and IR = near infrared

Name	Ratio
Simple subtraction	$IR - R$
Simple division	IR/R
Complex division	$\frac{IR}{R + \text{other wavebands}}$
Simple multiratio (vegetation index of normalized difference)	$\frac{IR - R}{IR + R}$
Complex multiratio (transformed vegetation index or normalized difference)	$\left(\frac{IR - R}{IR + R} + 0.5\right)^{1/2}$
Perpendicular vegetation index (vegetation directional reflectance departure from soil background)	$[(R_{soil} - R_{veg})^2 + (IR_{soil} - IR_{veg})^2]^{1/2}$
Green vegetation index (as above for use with Landsat satellite wavebands)	$-0.29(G) - 0.56(R) + 0.60(IR) + 0.49(IR)$

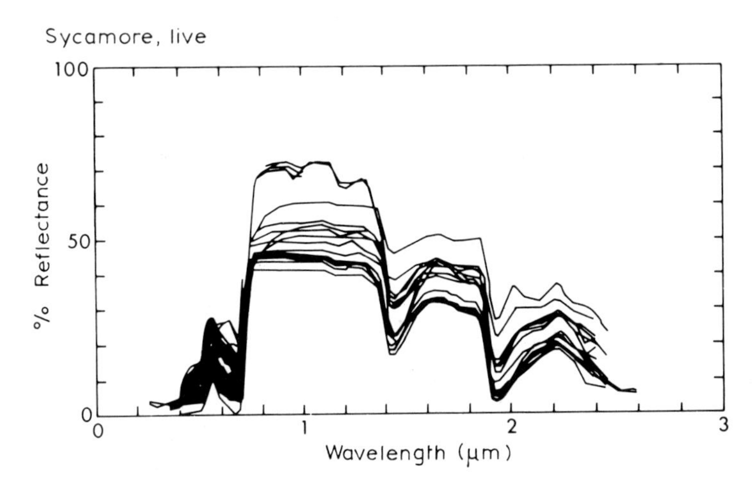

Fig. 15.5 An example of laboratory determinations of spectral reflectance curves. (Source: Leeman *et al.*, 1971.)

reflectance curves for the 0.5–2.5 μm bands but are slightly lower in magnitude (Fig. 15.3).

Leaf senescence occurs as the leaves end their functional life. Most herbaceous annual plants also have a progressive senescence from the older to younger leaves. During leaf senescence, starch, chlorophyll, protein and nucleic acid components are degraded. Thus the familiar autumn colours are caused partly by loss of green chlorophyll and the build up of yellow and orange carotene and red anthocyanin pigments. Usually light reflectance increases markedly in the 0.55 μm (green) band when chlorophyll degradation takes place. Changes in leaf water content usually can be correlated with near-infrared reflectance but, in general, dehydration leads to increased reflectance over the whole range of wavelengths. Leaf senescence, however, leads to decreased infrared reflectance and the infrared plateau at about 0.75 μm is usually reduced considerably.

The reduction in infrared reflectance which characterizes senescence also occurs where disease strikes and the leaves lack their functional condition (Fig. 15.4). It is for this reason that infrared sensors have proved particularly valuable in studies of plant disease. It must be recognized, however, that the resolution capability of the sensor is important. Where there is a non-uniform distribution of dead and dying vegetation along with patches of more healthy vegetation, the classification becomes difficult if the pixel contains a mixture of elements.

Catalogues of the spectral properties of leaves

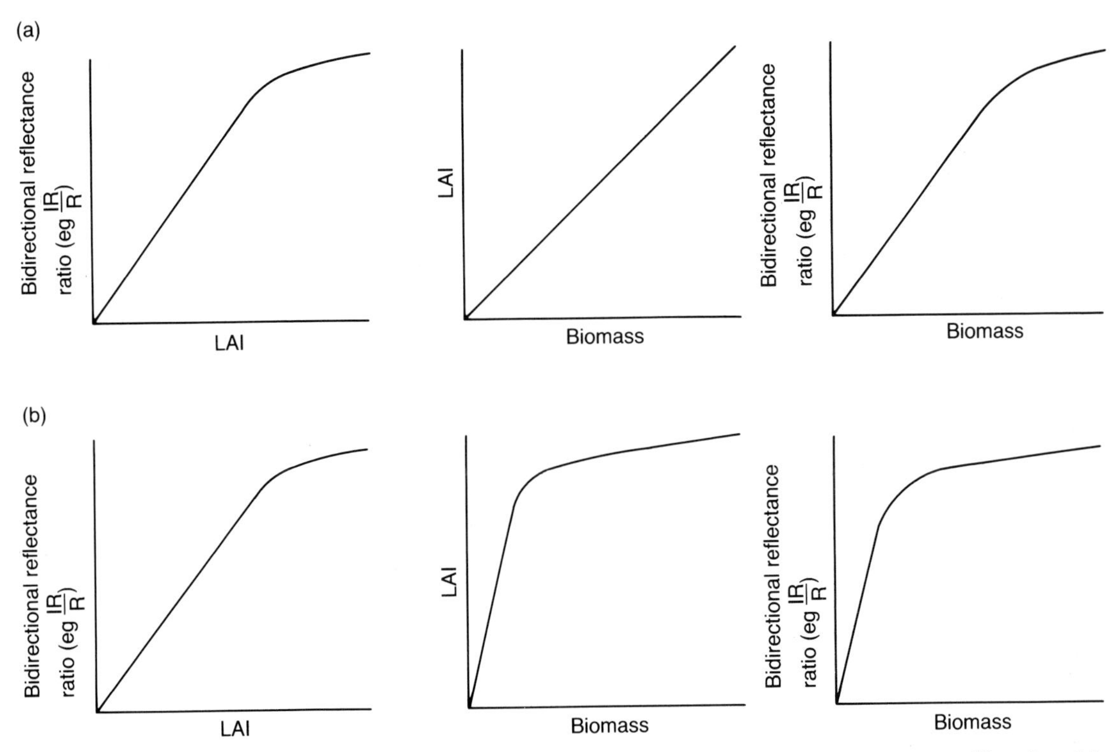

Fig. 15.6 The theoretical relationship between BDR, LAI and biomass for (a) a grass and (b) a tree canopy. Note that while grass canopy biomass can be used as a surrogate for LAI, this is not so for the tree canopy.

have been derived mostly from laboratory determinations (Fig. 15.5) but the field spectral response may be rather different. As a result, analytical models of the natural scene have been developed and tested against real conditions. In many early studies the plant canopy was assumed to be an ideal diffusing medium having a uniform distribution of light scattering and absorbing components. More recent studies have envisaged a layered canopy, the biological components being represented by three, flat, 'Lambertian' plane sections. These planes are mutually orthogonal projections, one horizontal and two vertical.

15.3 Phenological studies

Models of the scattering and absorption by plant canopies requires estimates to be made of the leaf area affecting the response. Such estimates have been made by measuring the cumulative one-sided leaf area per unit ground area projected from the canopy top to a plane at a given distance above ground level. Such measurements are termed the Leaf Area Index (LAI) and can be measured for horizontal and vertical projections.

Vegetation has a characteristically high BDR variation between the red wavebands (strong absorption) and the near infrared bands (strong reflectance). The relationship between a ratio of remotely sensed red and near-infrared BDR and measurement of leaf cover (LAI) has been studied by various researchers (Table 15.2), initially using Landsat MSS data, but increasingly using NOAA–AVHRR data also, as their suitability for lower spatial resolution but broader area and much more frequent vegetation assessments became appreciated (see also sections 13.5.1 and 16.3.3). It is also helpful to note that any estimate of leaf area is likely to be indicative of the weight of plant material (biomass) at the surface. The degree of correlation between BDR and biomass clearly will be dependent on the strength of the BDR/LAI relationships as well as the LAI/biomass relationships. For non-woody monocotyledons, such as the grass species, BDR is almost linearly related to

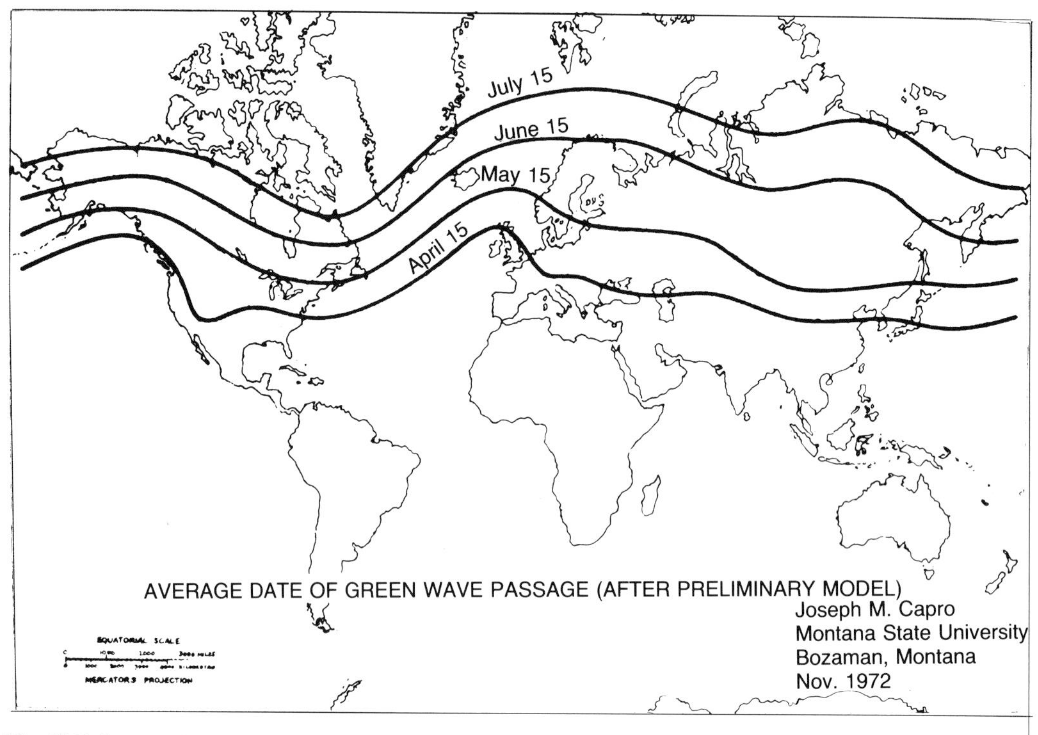

Fig. 15.7 Average date of the Green Wave passage in the Northern Hemisphere. (Source: Dethier *et al.*, 1973.)

biomass (Fig. 15.6). In other species (including woody species) that have high biomass (e.g. maize) the relationships are weak or inconsistent. Nevertheless species-specific studies can be made to establish BDR/biomass relationships. For example, measurements of *Calluna vulgaris* showed that in the low biomass growth phase there was a linear correlation. However, the growth of heather produces marked changes in the canopy morphology with consequent changes in the relationships to reflectance over time. Obviously such variations in response make estimates of biomass for mixed stands of heather much more difficult. Furthermore, although BDR/LAI/biomass relationships exist the strength of the correlations with reflectance may be affected greatly by other factors, such as soil background, presence of senescent leaves, or phenological and man-made changes in the canopy.

In practice such LAI measurements can be correlated with plant stage and vegetation height (Fig. 6.5). Thus modelling of vegetation reflectance is closely concerned with the phenology of the plant types and the amount of biomass characteristic of different stages of growth. These relationships have been exploited in studies of the productivity of different ecosystems by remote sensing. For example, remote sensing data for the shortgrass prairie was used as an input into biosystem models for the IBP (International Biological Programme) Grassland Biome programme in the USA.

The synoptic view provided by satellites can be used effectively in studies of the development of growth stages in vegetation over wide areas. For example, NASA investigators have described a phenology satellite experiment using Landsat data. In their study two phenological sequences were observed:

1. The Green Wave. A succession of images were used to record the geographical progression with time of foliage development (greening stage) as spring conditions extended over wide areas (Fig. 15.7).
2. The Brown Wave. A record of the geographical progression with time of vegetation senescence (maturation of crops, leaf coloration and leaf abscission). This progression plays the analogous role in the autumn to the Green Wave.

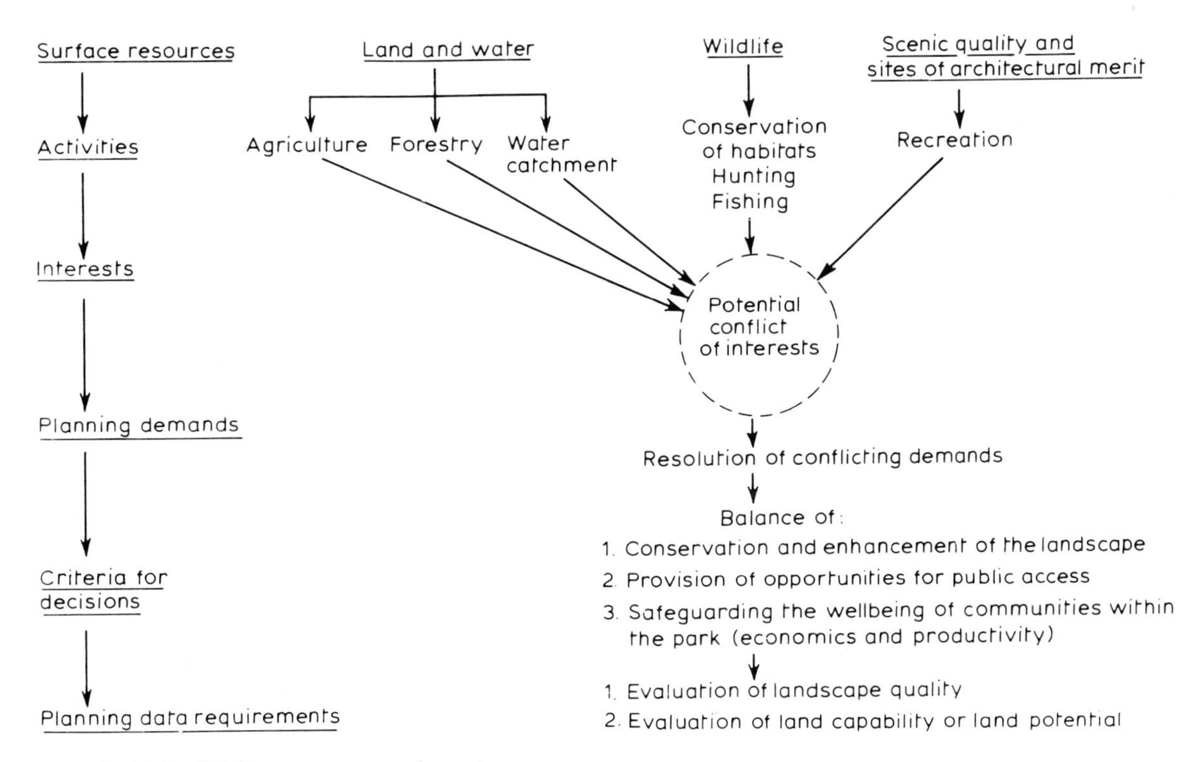

Fig. 15.8 National Park management requirements.

Results from the Phenology Satellite Experiment show that it is possible to develop phenoclimatic models showing the progression of vegetational change during a particular year. For countries with highly developed agricultural or grazing economies such information is very valuable. The LACIE programme (section 16.3.2) was used for determinations of crop status, crop yield and for management planning. Outside the cropped areas the ecological conditions of grassland ranges, particularly the steppe, prairie and savannah regions, are of great significance. The most important use of Landsat imagery for grassland ranges is that of monitoring changes in the forage conditions and development of grazing areas as the season progresses (i.e. Green/Brown Wave effects). It is possible to determine areas and dates when plant growth ceases owing to drought or soil moisture depletion.

15.4 Conservation in National Parks

We can now turn to consider some examples of the application of remote sensing to conservation management. Interest in conservation has grown steadily as the effects of economic exploitation of animal, plant and land resources became increasingly apparent throughout the world. In some measure this interest has been linked with growth in leisure time in the developed nations and the often expressed need for urban man to find places which are relatively remote and retain elements of wildness.

The National Parks in various countries provide areas of refuge for wildlife and recreation for mankind. They differ widely in their characteristics. Some are virtually untouched by human beings and stringent precautions are taken to maintain them as wild areas. Others, such as the National Parks in England and Wales are areas of outstanding landscape and wildlife interest which also support local farming populations. These Parks might aptly be termed *protected landscapes*.

In protected landscapes the preservation of natural beauty and wildlife requires careful management, and sometimes control, of several competing uses of the land and wildlife resources (Fig. 15.8). For good results it is necessary to start with an inventory of existing wildlife and then to

Table 15.3 Classification of landscape features in UK National Parks, based on surface and remote sensing data (Source: Huntings Technical Services Ltd, 1986)

Class	Feature	Subdivision
A. Linear features		
	A1 Hedgerows	
	A2 Fences and insubstantial field boundaries	
	A3 Walls	
	A4 Banks	
	A5 Open ditches	
	A6 Woodland edge	
	A8 Strip woodland	
	A9 Grips	
B. Small or isolated features		
	B1 Individual trees in linear features	
	B2 Individual trees outside linear features	
	B3 Groups of trees, all species	
	B6 Inland water	
C. Wood and forest land		
	C1 Broadleaved high forest	
	C2 Coniferous high forest	
	C3 Mixed high forest	
	C4 Scrub	
	C5 Clear felled/newly planted areas	
D. Moor and heath land		
	D1 Upland heath	
	D2 Upland grass moor	(a) grass moor
		(b) heath
	D3 Bracken	
	D4 Unenclosed lowland areas	(a) rough grassland
		(b) heath
	D6 Upland mosaics	(a) heath/grass
		(b) heath/bracken
		(c) heath/blanket peat
	D7 Eroded areas	(a) peat
		(b) mineral soils
	D8 Coastal heath	
E. Agro-pastoral land		
	E1 Cultivated land	
	E2 Grassland	(a) improved pasture
		(b) rough pasture
F. Water and wetland		
	F1 Open water, coastal	
	F2 Open water, inland	
	F3 Wetland vegetation	(a) peat bog
		(b) freshwater marsh
		(c) saltmarsh
G. Rock and coastal land		
	G2 Bare rock	(a) inland
		(b) coastal
	G3 Other coastal features	(a) dunes
		(b) sand beach
		(c) shingle beach
		(d) mud flats
H. Developed land		
	H1 Built-up land	(a) urban area
		(b) major transport route
	H2 Quarries, mineral workings and derelict land	(a) quarries and mineral
		(b) derelict land
	H3 Isolated rural developments	(a) farmsteads
		(b) other
I. Unclassified		

monitor changes in the natural environment over periods of time.

The important role of remote sensing can be illustrated by the inventory of moorland made by Exmoor National Park where concern had been expressed at the loss of moorland habitats as a result of ploughing of the moorland and its conversion to agricultural use. In order to determine the scale of the problem and to provide an inventory of existing moorland, a Moorland Map has been constructed by interpretation of infrared colour photographs and back-up field work. The vertical infrared colour photography of 1 : 10 000 scale was obtained using a Wild RC-8 camera flying at approximately 1800–2000 m. The photography was examined on light tables using both stereoscopic and monoscopic viewers.

Moorland could easily be separated from permanent and temporary pasture by colour and texture differences (Plate 11, colour section). Of some 70 000 ha of land within the National Park boundary, approximately 19 500 ha were classified as grass and heath moor. In order to obtain a comprehensive record of the surface conditions an uncontrolled black and white photomosaic was constructed from black and white internegatives derived from the original infrared colour photography. The mosaic was printed at a scale of approximately 1 : 25 000 and was subsequently marked up to show boundaries of the areas of moor and heath vegetation. It provides a record of the vegetation in great detail, which enables farmers and landowners to identify small-scale features within fields and blocks of land that they manage. Both the Moorland Map and the photomosaic offer a valuable basis for discussions between farmers and the Exmoor National Park Authority as to the best options for conservation and sheep grazing policies. The map is now widely used by field officers of the National Park, the Agricultural Advisory Service, and farmers and their agents as an acceptable statement of the field conditions existing at the end of 1979.

As a result of that successful application of remote sensing to monitoring of Exmoor moorland, in 1981 the UK government introduced a requirement for the preparation of moorland maps in all National Parks. Subsequently the requirement has been broadened to include other land cover categories, such as woodland and coastal marshes of conservation value.

The role of remote sensing data collection for conservation purposes in National Parks was further emphasized by the acquisition of Thematic Mapper data (Plate 12, colour section) for these areas and by a decision of the UK Department of the Environment (DOE) and Countryside Commission for England and Wales (CCEW) to set up the Monitoring Landscape Change (MLC) Project in 1984. The principal objectives of the project were as follows:

1. to obtain reliable information on the current and past distribution and extent of landscape features of major policy importance;
2. to determine the magnitude of any changes in distribution and extent of the features between specific points of time;
3. to develop a method by which future changes in the extent of features can be monitored.

The MLC project was based on air-photo interpretation, image analysis of Landsat data and field checking for collateral data. The results provided a baseline against which future changes in landscape features may be measured (Table 15.3). The acquisition of SPOT data has given further impetus to a follow-up study. Work in progress seeks to combine remote sensing, map and file data within a geographic information system which can be accessed by land managers.

The Commission of the European Communities (CEC) decided in 1985 to establish an environmental data base for community countries. The existence of good quality information on the state of the environment was seen to be essential for efficient implementation and orientation of the Community environment policy and integration of environmental objectives within other Community policies. The CORINE project (co-ordination, information, environment) included a programme for gathering coherent land-cover data and integration of this data into a geographical information system. An initial feasibility study was completed in 1988 covering the whole of Portugal. The classification showed that the spatial and spectral resolution of TM and SPOT data is sufficient only to provide area estimates for relatively broad cover types, compared with the more varied and detailed information in the MLC classification.

15.5 Resource management: the example of forestry

Remote sensing studies by aerial photography have been used for several decades in the field of forestry. Air photo-interpretation has been used for classification of forest stands and types, survey of mortality and depletion, planning of reforestation, inventory of timber and other forest products and assessment of property taxes.

15.5.1 Tree stand inventories and forest mapping

As well as using photographic tone, texture and colour to identify different tree stands, foresters can make precise measurements of tree height, crown diameter or stand density. Tree height is closely correlated with tree volume and stand volume and can be measured on photographs in a number of ways. Measurement of shadows, measurement of parallax, and measurement of relief displacement on single large-scale vertical or oblique photographs are some of the methods used. Tree shadow measurements are normally made with a micrometer scale consisting of a finely graduated series of short lines, one of which is matched with each shadow visible on the air photograph. Measurement of crown diameter can be made by either micrometer or dot-type crown wedges. The micrometer wedge used normally consists of two converging lines calibrated to read intervening distance to the nearest thousandth of an inch. The dot-type wedge usually consists of a series of dots differing in diameter by 0.0025 in. It is laid alongside the image and moved until the dot that just matches the size of the crown is identified. Alternatively the dot images can be moved over the crown until the appropriate size that covers the crown is found. The accuracy of crown measurement is largely dependent on the scale of the photographs. The error may be about 1 m with either kind of measuring device on photographs of 1 : 12 000 scale.

Transparent dot templets are probably the most widely used area measuring instruments in forest inventories made from aerial photographs. The density of dots in a templet varies from 1 to 65 in^{-2} depending on the intensity of the survey and the scale of the maps on photographs used. The ratio of dots in a given class to dots in the entire tract gives the proportion of the tract occupied by that class.

These elementary devices for the measurement of size and area can now be replaced (at a cost) by automated scanning equipment. If very large areas are to be studied such equipment is essential for data to be available within a reasonable period of time.

Once average height, crown diameter and crown cover are known it is possible to estimate the stand volume of timber from look-up tables. Stand volume tables have been compiled for a variety of species and species groups occurring in specific regions. They usually give average volumes for stands, often in 10-ft height classes, 10% crown-closure classes, and 5-ft crown width classes. Statistical analysis has shown that little volume variation is associated with variation in crown width in some areas. This variable is, therefore, often omitted and volume is classified by stand height and crown closure alone. The accuracy of the estimate will depend on the accuracy of the measurements made from the image and on the quality of the correlations of tree height, diameter and crown cover with volume of timber. In ground surveys the wood volume is estimated from individual tree measurements, usually diameter at breast height (dbh) and the total height of each tree. Breast height is standardized at 140 cm (4.5 ft).

Forest inventory techniques using air photographs normally fall into three basic classes: combined air and ground surveys, aerial surveys using stand volume tables, and aerial surveys using tree volume tables.

Table 15.4 Classification of forests

Symbol	Composition
S	More than 70% by volume of softwood
M	Mixed — 30–70% softwood by volume
H	Less than 30% softwood by volume

Qualitative and quantitative forest estimates often include a combination of descriptive classes of cover type as well as size and density classes. The number of classes and categories may be quite varied according to scale, quality and type of photographs, purpose of the survey, skill of the interpreter, and allowance cost per unit area.

Cover type is usually based on species and species groups that provide relevant information for a particular inventory. The classification could be a simple one of proportion of conifers (softwoods) and broad-leaved deciduous (hardwood) species. For example the classes might be as shown in Table 15.4. Alternatively, detailed species analysis may be used or the concept of utilization may be introduced into cover-type classification. For example in a study of a commercial forest in the northern latitudes the classification might include sawtimber, poles, saplings or reproduction. Species identification is a complex problem of interpretation and various photographic keys have been produced over the years. Table 15.5 is an example of a key for the identification of the northern conifers of New England and the Maritime Provinces of Canada. The key was devised for use with panchromatic photography at scales from about 1 : 6000 to 1 : 16 000. It would require considerable revision for use with colour photography or other scales of images.

At present, the majority of forest inventory is carried out using black and white photography but the usefulness of colour is generally accepted. Infrared colour photography is less commonly used in forestry studies at present but there is evidence to show that it offers many advantages: infrared colour film is always exposed with a yellow filter (Wratten 12 or similar) and is far superior at high altitude where natural colour is often affected by haze. A wildland vegetation and terrain survey in California compared infrared colour at scale of 1 : 120 000 with a survey of the same area using

Table 15.5 Air photograph interpretation key to northern conifers[a] (Source: Hilborn, 1978)

Key	Description	Go to / Species
1(a)	Crowns proportionally large, more or less irregular especially in mature specimens, spreading. Branches often protruding, tops not narrowly conical, usually rounded or flattened	2
1(b)	Crowns smaller, regular, conical to narrowly conical or almost columnar. Tops conical, more pointed, well defined	6
2(a)	Light-toned, irregular crowns, prominent branches or ragged and open crowns	3
2(b)	Darker toned more irregular shaped crowns	4
3(a)	Prominent branches, 'star' shaped in plan. Solitary or in small groups in mixed stands. Pure stands often on sand plains or well-drained sites	*white pine*
3(b)	Ragged, open crowns. Branching sparse, not prominent. In mixed stands solitary or scattered, often taller than other species. Sites well drained	*overmature white spruce*
4(a)	On sand plains or dry sites. Broad, rounded, rough, open crowns	*red pine*
4(b)	On moist or ill-drained sites, rarely in pure stands	5
5(a)	In swamps. Crowns irregular, open or ragged, if conical, blunt-topped. Some specimens may lean precariously (see also 10)	*white cedar*
5(b)	On moist sites. Dense, broadly conical to irregular crowns. Crowns appear fuzzy. Very dark shadows	*eastern hemlock*
6(a)	Mostly in uniform, pure stands on sand plains or excessively dry sites. Crowns small, irregular, pointed. Tones medium	*jack pine*
6(b)	In pure or mixed stands, rarely on excessively dry sites. Crowns conical to columnar	7
7(a)	On moist to boggy sites. Crowns narrowly conical or columnar	8
7(b)	On moist to well-drained sites. Crowns conical, rounded, or pointed-topped	9
8(a)	Crowns deciduous, light-toned especially in autumn, open. In mixed stands often taller than other species	*tamarack*
8(b)	Crowns darker, more dense, narrowly conical to columnar. Stands may be irregular in height	*swamp type black spruce*
9(a)	Crowns very symmetrical, smooth. Height irregular. Tops pointed, rounded, or blunt	10
9(b)	Crowns narrowly conical, not so symmetrical, not noticeably rounded or pointed	11
10(a)	Crowns symmetrical, round-topped, dense, light-toned. Mostly in pure stands and clumps on upland, limestone sites	*white cedar*
10(b)	Crowns symmetrical, pointed in front lighting (blunt in back lighting). Dark-toned shadows	*balsam fir*
11(a)	Crowns narrow conical to columnar, rougher branching than balsam fir. In stands heights may be irregular	*black or white spruce*
11(b)	Crowns more broadly conical, larger. Branching tends to be more prominent	*red spruce*

[a] Red spruce has the broadest, most luxuriant crown, while black spruce has the narrowest. Black spruce will occupy drier and poorer sites. Red and black spruce hybridize making distinction of the eastern spruces very difficult.

black and white at scale 1 : 16 000. The results were comparable but the infrared colour was more efficient because the work was carried out in half the time. This was partly due to less handling of photography – three infrared colour as against 78 at the larger scale in black and white.

In the infrared wavebands there is strong reflectance from deciduous trees but lower reflectance from coniferous trees. Thus both infrared black and white and infrared colour can be used to make broad distinctions between deciduous and coniferous trees (Plate 11 for infrared colour, and Fig. 4.5 for infrared black and white images of trees).

15.5.2 Radar imaging of forests

Radar imagery has been used in mapping forest areas of Nigeria where the weather conditions point to radar as the most practical sensor in many regions. The whole country was imaged on to radar at a scale of 1 : 250 000 within a five month period. Strips of imagery were then formed as mosaics on sheets that coincided with an existing 1 : 250 000 map sheet series of the country. The project showed that radar could be used successfully for forestry. While the more arid northern areas required more field work and presented greater difficulties, the southern high forest areas gave a

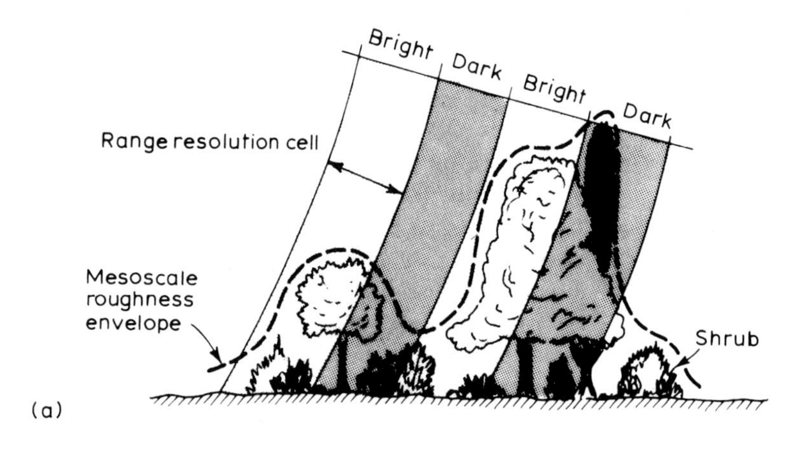

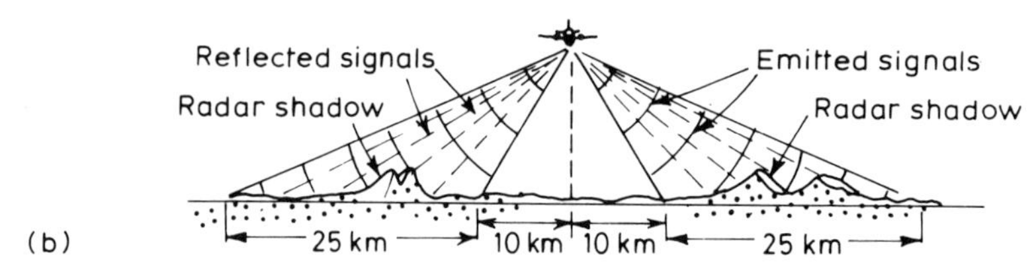

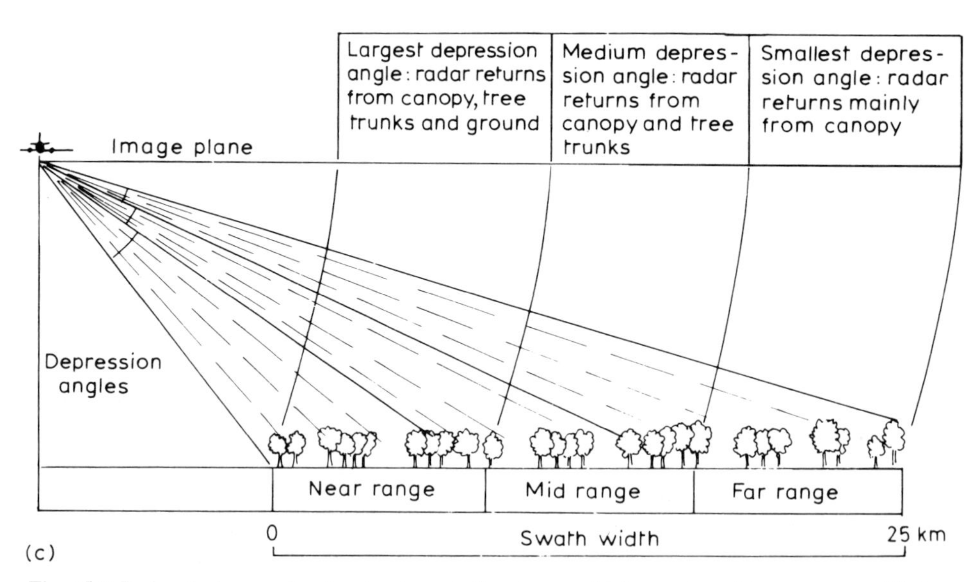

Fig. 15.9 Radar shadows and radar signatures in forest areas. (a) Formation of image texture. (If the amplitude of the canopy roughness on the scale of resolution is large, then there will be very different contributions to the scattering from adjacent resolution cells, giving a corresponding variability in the image brightness.) (b) Image acquisition by SLAR, using two simultaneous beams, showing areas of radar shadow. (c) SLAR signature differences in a uniform vegetation unit on level terrain. (Source: Parry and Trevett, 1979.)

good correlation between interpreted units and ground observations. The survey will enable Nigeria's Department of Forestry to make an inventory of different forest types, primary stratification units, and existing reserves. It also will be able to assess potential for plantation development and firewood areas in heavily cultivated zones.

It will be appreciated that the radar beam is directed at an oblique angle away from the aircraft and in areas of surface roughness this produces a

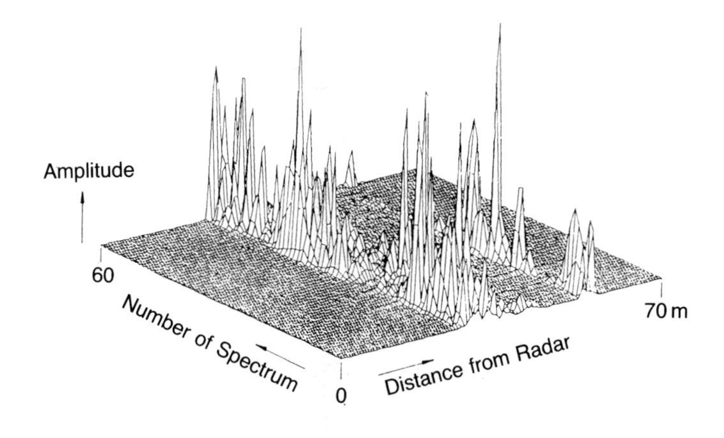

Fig. 15.10 Consecutive radar return spectra for forests in southern Finland. In consecutive radar return spectra the back-scatter power peaks at the minimum and maximum distances from radar correspond to back-scatter from tree tops and the ground, respectively. (Source: Hallikainen *et al.*, 1989.)

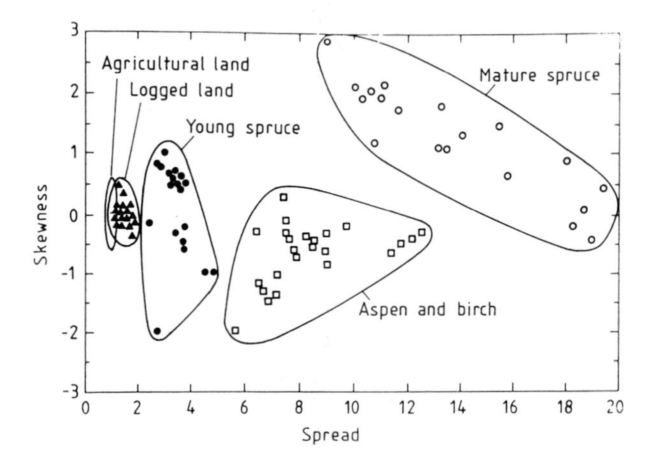

Fig. 15.11 Spread versus skewness for averaged radar return versus range spectra in southern Finland; 5.4 GHz, VV polarization. (Source: Hallikainen *et al.*, 1989.)

radar shadow (Fig. 15.9(a)). In order to counter this, two simultaneous beams are sent out on each side of the aircraft. The eventual flight lines are then arranged so that each area is covered by two 'looks' of imagery, both north and south (Fig. 15.9(b)). Owing to the oblique angle of observation, a further complication is that a uniform vegetation unit may show differences in the signal return according to the range of the image. Care must be taken, therefore, to define and cross reference the signatures for particular areas (Fig. 15.9(c)).

A helicopter-borne non-imaging scatterometer using 5.4 and 9.8 GHz frequencies has been flown over Finnish forests. It used all four (VV, HH, VH, HV) polarizations at both frequencies. The back-scattering coefficient was not useful in discriminating between five forest categories but studies showed that the shape of the radar return versus range reflected the different forest species (Fig. 15.10). Principal component analysis of the spread and skewness of the data has shown that identification of tree species may be possible (Fig. 15.11). The role of space radar for forest mapping will be evaluated as radar data from ERS-1 becomes available. The analysis of such data will be demanding and it is likely to remain in the research rather than the operational domain for some time to come.

15.5.3 Use of satellite data for forestry applications

The use of satellite data in forest resource management has been slow to develop. This is because satellite data is often inferior to air survey data in terms of resolution and interpretability. Also, image analysis using satellite data usually requires computer facilities lacking within forestry organ-

Fig. 15.12 'Fish-bone-like' clearings for small farm settlements leave scars in the Brazilian jungle. Based on Landsat TM imagery (Bands 5, 4 and 3).

izations, whereas air photo-interpretation has an established use in forestry.

Where extensive tracts of forest occur the advantages of scale offered by satellite data become attractive. Therefore, attention has been directed towards optimum methods of analysis. In a French study of Landsat TM data of the Vosges forests good results were obtained only after considerable treatment of the data. In this case the data were geometrically corrected, a digital terrain model was constructed, and radiometric corrections were applied to the data to suppress the effects of slopes on the reflectance of different forest types. Clearly, the investment of so much computer effort may be outside the competence or economic framework of operational forest management unless models can be used year upon year.

The uncertainty of the position in relation to the use of satellite data has been reinforced by studies of Scots pine in Belgium. Landsat TM and SPOT HRV data were used to study diameter, stand basal area, average canopy height and stand volume. It was found that stand density, average canopy height and stand volume could be estimated, but with accuracies ranging only from 50 to 60%. It is likely, therefore, that for relatively small, intensively managed forests typical of many European countries, aerial infrared colour mapping will continue to be preferred for same time to came.

Where there are large, extensively managed areas and the level of detail needed is less stringent there is a role for studies based on combinations of satellite and ground data. At its simplest level the effective use of satellite data to measure *rates of clearance* of the tropical rainforest areas, e.g. in Thailand, Brazil, etc., reveals an application to which space data is well suited: Landsat and even NOAA imagery has been used to monitor the clearance of Amazon rainforests and the fires by which much of this has been accomplished (Fig. 15.12).

It is likely, however, that there will be a general trend towards integration of space data with aerial survey and ground data to provide general information of interest to forest managers. For example, *income taxation of forest* in Finland is based on forest site productivity. Existing methods of delineation of forest stands into taxation stands use ground measurements and air photographs. Studies have now begun in which ground measurements, digital terrain modelling and ground checking are used in the context of Landsat TM imagery. The proportion of correctly classified pixels has been found to vary from 50 to 70%, so other input data were

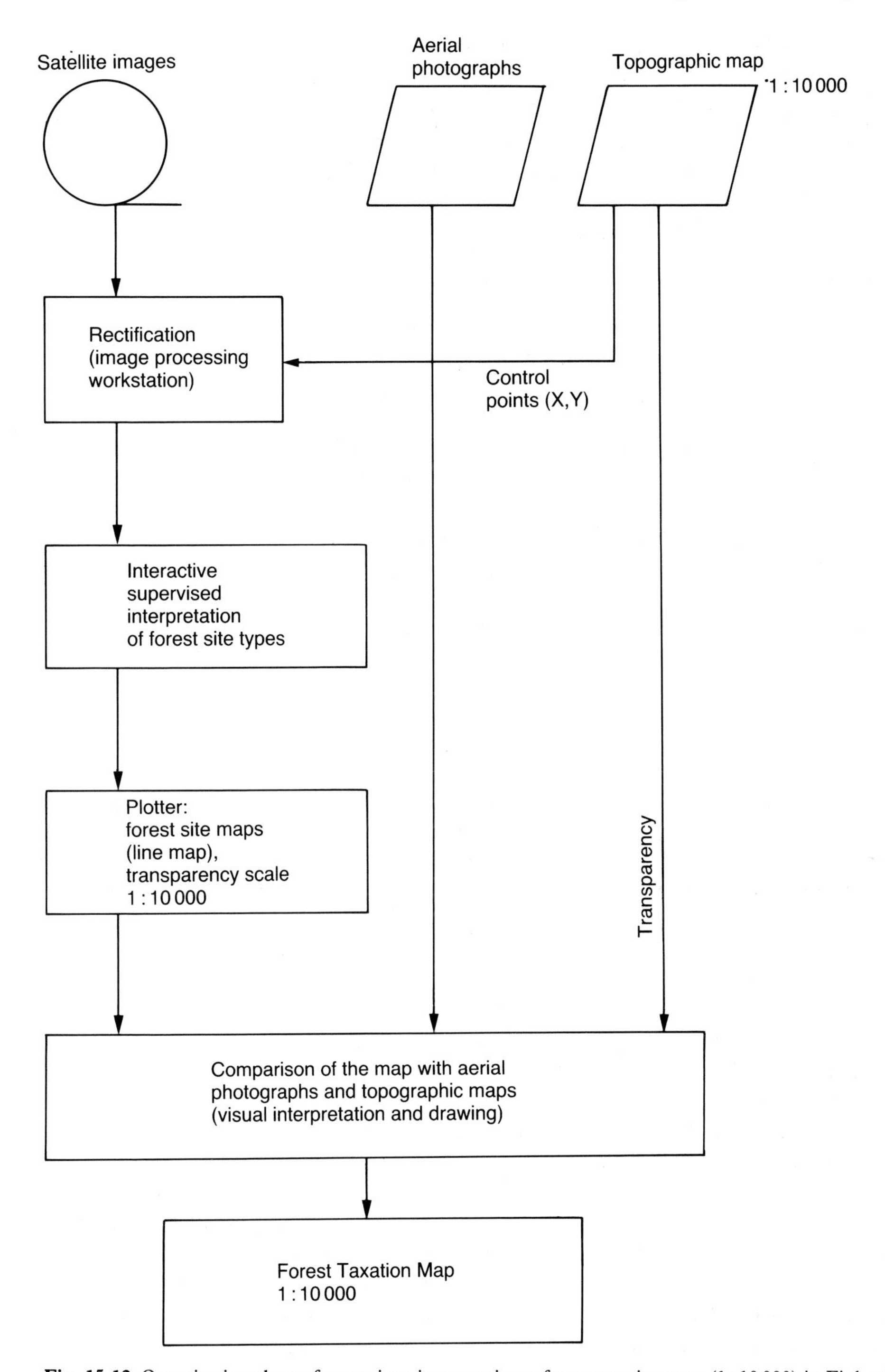

Fig. 15.13 Organization chart of operations in preparing a forest taxation map (1 : 10 000) in Finland.

considered necessary. Thus satellite data is essentially used as *support* following initial aerial survey. The organization of such operations is shown in Fig. 15.13.

15.6 Disease and fire hazards

One of the first indications of disease in a plant is discoloration of the leaves. Similarly, where plants are suffering stress due to attack by pests the changes in leaf structure and colour often provide a guide to the worst affected areas. The detection of disease and insect damage in forest areas can, therefore, be aided by remote sensing using colour photography.

The change in reflectance in the near infrared wavelengths as disease affects a plant can be very considerable (Fig. 15.4). Therefore the use of infrared colour photography for the detection of forest disease is growing. Healthy broad-leaved trees image in red or magenta in infrared colour, whereas dead or dying trees usually image in blue–green. Red autumn leaves appear yellow and yellow autumn foliage appears white. It will be evident that distinct shades of colour can be associated with a particular species at a given season of the year. Any photography for the purpose of disease monitoring should, therefore, be carried out at the period of maximum growth and where regular checks are required the data should be obtained for the same phenological stage.

The role of non-photographic imagery is likely to increase in the next decade. Thermal infrared imagery has already proved useful in forestry because of the moisture and local temperature effects which can be observed. For example, small smouldering fires started by lightning strikes or abandoned camp fires can be detected by thermal infrared detectors on regular aircraft patrols; conventional photography is affected by smoke plumes, but these detectors can identify 'hot spots' when a major fire is under way. Infrared imagery from aircraft or helicopters can help in the efficient deployment of fire crews and on occasions guide them away from dangerous positions where they might be overtaken by the fire.

The US Geological Survey and Bureau of Land Management have used Meteosat AVHRR multitemporal data (1982) to assess the dangers attached to wildland fire fuels. The AVHRR data were used to obtain the normalized difference Green Vegetation Index for four dates. The data were then interpreted using digital terrain data, false colour Landsat images and field knowledge as additional inputs. The land cover/fuel-type map for a 6.4 million acre study area was assessed as having 92% accuracy. The cost of mapping some 42 million acres was about 0.04 cents per acre. As a result of the study the Bureau of Land Management is using these techniques for mapping fire fuel hazards for the entire western USA to support their national fire programmes (Fig. 15.14).

Meanwhile, in Canada the infrared sensor has proved useful in fighting infestations of spruce budworm. This is the most destructive insect in the fir–spruce forests of eastern Canada. Certain localities or epicentres provide optimum climatic conditions for budworm development, but climatic observations from special towers erected in the canopy are expensive and liable to be only partly representative of the overall pattern of climate. As an alternative, thermal sensing of the canopy has been carried out by an airborne sensor providing a series of strip records of temperature variation along a metre transect. In this way the potential epicentres were identified.

15.7 Wildlife studies

The most important uses of remote sensing studies in wildlife management are for making censuses of animal populations and mapping and evaluating the vegetation in the area occupied by the game. Inventories of wildlife populations are necessary for the planning of management programmes and formulation of fishing and hunting regulations. When surveys are made by ground methods animals may be counted twice or perhaps not at all. Aerial surveys minimize such errors by making it possible to cover an area quickly and completely. Therefore, aerial surveys by direct observation were made as early as the 1930s in the USA. The use of aerial photographs developed somewhat later but gradually became important. The animals counted by air-photo methods include antelope, deer, elk, barren-ground caribou, moose, musk oxen, waterfowl, seals, and in certain circumstances fish. It is interesting to note that one of the main

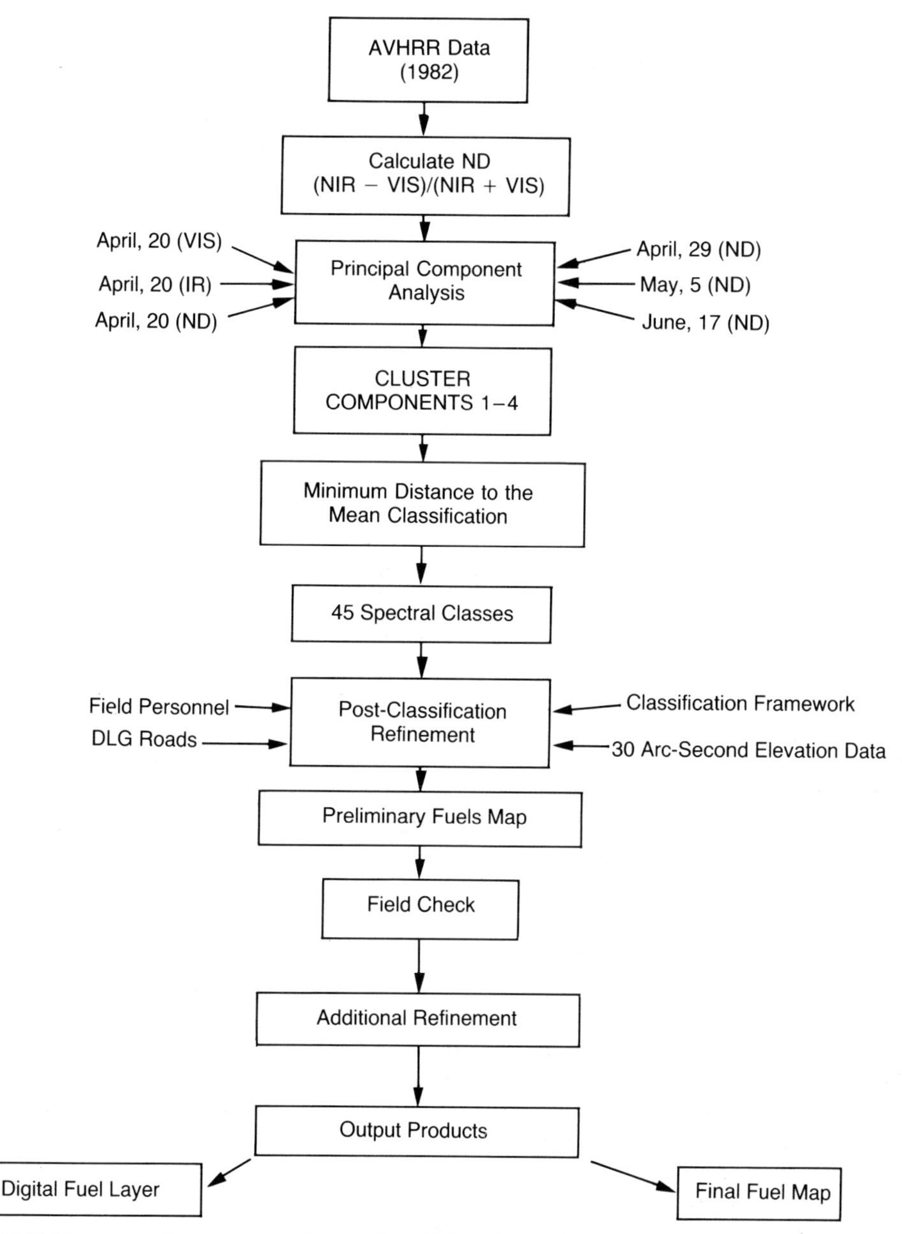

Fig. 15.14 Data processing procedures for mapping wildland fire fuels with AVHRR data, western USA. (Courtesy, US Geological Survey.)

difficulties facing wildlife researchers was the task of counting thousands of animals. This was particularly the case with photographs of waterfowl where some 14 000 geese may be recorded on a single photograph. Various attempts were made to develop counting methods, e.g. using a scanning device, which allowed successive slits to be examined, through which the animals were counted using a binocular microscope. Another method consisted of placing transparent overlays over the print and pricking the image of each animal counted.

It is clear that the development of automatic scanning densitometers allows such tedious (and

Table 15.6 Comparison of image counts (IC) with ground enumeration (GE) of livestock numbers for area sampling units – cultivated stratum (Source: Roberts and Colwell, 1968)

Sampling unit number	Livestock species							
	Cattle		Sheep		Other		All species	
	IC	GE	IC	GE	IC	GE	IC	GE
1	45	47	0	0	0	0	45	47
2	0	0	0	0	0	1	0	1
3	0	0	0	0	0	0	0	0
4	22	25	184	180	0	3	206	208
5	135	182	0	1	8	1	143	184
6	202	255	0	0	15	0	217	255
7	0	0	0	0	0	0	0	0
8	0	0	0	0	0	0	0	0
9	0	0	0	0	0	0	0	0
10	2	0	0	0	0	4	2	4
11	0	0	0	0	0	0	0	0
12	58	61	316	608	6[a]	0	378	669
13	0	0	0	0	0	0	0	0
14	0	0	1373	1000	0	0	1373	1000
15	0	0	0	0	0	0	0	0
16	41	43	0	0	5[a]	1	46	44
Totals	505	613	1871	1789	34	10	2410	2412

[a] Cattle incorrectly identified as horses.

somewhat inaccurate) counting to be superseded. Automatic counting techniques are now readily available for such work. However, before such counting can take place it will be necessary to identify the characteristic spectral responses for different animals. The application of colour and black and white photography to the inventory of livestock has been studied by workers in the USA. A comparison between image counts (IC) and ground enumeration (GE) for an area of about 1000 square miles in Sacramento Valley, California is shown in Table 15.6.

In the great National Parks of East Africa, such as the Serengeti in Tanzania, the success or failure of wildlife conservation is measured in terms of the population figures. Millions of animals are involved and regular censuses of the more common species can be achieved only by aerial survey. The terrain is often semi-arid, travel is slow over land and the wild animals are elusive. Aerial photography has provided the ability to monitor the sizes of the populations of the commoner species to new limits of accuracy.

Two methods have been reported. Total counts, using oblique overlapping photographs, have been made for very gregarious animals such as wildebeeste and buffalo. Other animals showing a more scattered distribution have been counted by sampling techniques. In this method the number of animals is determined for random sample strips of photography of known surface area. The results of a census of wildebeeste in the Serengeti region of Tanzania have been reported as 322 000 (±7%), 381 875 (±6.5%), 334 425 (±7.5%) and 350 000 (±8.5%) for the years 1963, 1965, 1966 and 1967, respectively. The size of these populations underlines the difficulties ecologists would face in making reasonable estimates by means other than remote sensing.

Other counts reported using aerial survey include

those of crocodile and hippopotamus and estimates of the physical condition of large herbivores such as the buffalo.

Infrared scanning techniques are of interest in livestock studies in that they offer a potential means for counting and studying the distribution of animals at night. The body temperatures of animals normally contrast with the surface ground temperature (Fig. 4.12). It is clear that thermal scanning used alongside aerial photographic methods also could be a means of estimating mortality rates. This idea also has been proposed for counting dead seals after culling operations.

Radar studies of migrating swarms of locusts began when they were accidentally seen on radar over the Persian Gulf. It is well known that long distance migrations are made by swarms of sexually immature locusts. Apart from these swarms, however, locusts also can exist in the solitary phase (section 18.6.2) and the solitaries fly by night and rest by day. From field studies using radar by night in the Niger Republic of Africa it has been shown that solitary locusts are able to orient themselves downwind in contrast to the essentially random orientation within a swarm.

Although extensive survey by radar over large areas is not economic at present, locust research workers have suggested that radars placed at strategic points – for example on frequently used migration tracks or regions where control operations are taking place – could enable population estimates to be made. In this way protective measures could be planned (see also Chapter 18).

Preliminary studies made by NRSC, Farnborough, UK recently assessed the usefulness of Landsat Thematic Mapper data to derive baseline information on the area and distribution of upland bird habitats for predictive models of bird populations. Promising correlations between satellite-derived habitat parameters and later data sets gave poor predictive results except for red grouse. Studies by NRSC have also examined the role of Thematic Mapper imagery in mapping suitable wader habitats in peatlands. The study showed that the data could map afforestation of peatlands, which reduces drastically the feeding grounds for large numbers of breeding wetland birds.

We may conclude that the applications of remote sensing methods can do much to monitor the conditions within a wildlife habitat. It is possibly in this area that most assistance will be forthcoming for those concerned with wildlife management. The problem facing many conservators is that measurements of changes in an ecosystem are time consuming and often occur at a few points. Also, the sites where change takes place may be remote and expensive to reach. It is for these reasons that most conservators now include in their budgets some provision for air photography or other imagery. The savings in staff time, together with the value of a permanent record of transient conditions, make remote sensing an economic way of meeting managerial needs.

15.8 Marine conservation

The role of satellite remote sensing in monitoring studies of oceanographic conditions is addressed also in Chapter 12. It also may be noted here that the early studies using the Coastal Zone Colour Scanner (CZCS) enabled scientists to evaluate the use of sensors to study a range of environmental parameters, including:

1. Formation of the yellow substance ('Gelbstoff').
2. Particulate materials in the subsurface layers of the oceans (related to dynamics of phytoplankton). The concentration of suspended matter in the subsurface layers determines the penetration of sunlight in water. In this way suspended matter can be a limiting factor for phytoplankton production.
3. Marine water quality.

The CZCS sensor was launched in 1978 aboard the Nimbus 7 satellite. Since 1977, airborne oil surveillance systems have been developed based on aircraft sensor packages. Sensors used include a passive microwave imager (PMI), multispectral line scanner (IR/UV) with three channels (two infrared and one ultraviolet), side looking radar and high-resolution aerial camera. The passive microwave imager has limited quantifying ability but provides some all-weather capability. Thin oil slicks have a low 'brightness temperature' whereas thicker oil slicks appear radiometrically warm. The multispectral scanner detects thermal differences between oil and water in IR bands, whereas UV detects relative differences between oil and water and will detect very thin slicks that may escape IR detection. Such systems can provide early warning of environmental changes that are damaging to sea organisms. The US Coastguard Aireye system is an example of such an airborne monitoring facility. An Italian system is described in more detail in Chapter 18.

Further reading on aspects of remote sensing for studies in ecology, conservation and resource management is recommended in the bibliography for this chapter at the end of the book.

16 *Land use and crop production*

16.1 Introduction

The world population continues to grow apace (6000 million is forecast for the turn of the century) and it is characterized by high populations in third world countries that are undernourished. Yet, in advanced countries, such as Europe and the USA, there is a problem of surplus production. One of the most pressing problems facing society is how to feed a hungry world without causing permanent environmental damage by intensive farming methods unsuited to the area concerned. In order to achieve this it is necessary to improve our knowledge concerning crop distributions and crop yields.

In the field of agriculture the main need is for early information on crop conditions and areas to allow for efficient management of the farming industry, and for planned distribution and marketing of produce. In an age when there is substantial (often ecologically damaging) government intervention in the form of support prices and subsidies there is a need for national and international data concerning the following:

1. crop production;
2. animal production;
3. farm structures (average size of fields, access, irrigation facilities, farm buildings).

Crop production is a function of crop area × yield (productivity per unit area). Thus there are three fundamental aspects to the production of information, namely crop identification, area measurement, and yield prediction.

Furthermore, in order to be useful the data must be delivered to the authorities in a timely manner. For example, the area estimates in Europe are needed in December (winter cereals within plus or minus 4%) and May (spring cereals plus or minus 3%). Yield estimates for cereals are required in August. Thus early acreage determination of main crops, early yield predictions, information on plant diseases and pest attacks, quantification of the effects of moisture deficiency or waterlogging and continuous updating of inventories are some of the requirements for land use planning in modern agriculture.

At present agriculture dominates the land use of most countries, although forestry may play an important role in some. Nowadays agriculture is increasingly seen to be competing for land with the claims of recreation, mineral extraction, housing, industry and waste disposal sites as well as forestry. The Earth's surface consists partly of natural features such as vegetation, snow and ice (sometimes termed 'land cover') and features resulting from human activities ('land use'). Any inventory showing areas devoted to different purposes should be compared with the total land area concerned. Therefore, care must be taken to account for land cover areas (e.g. streams, moorland, etc.) even though the main purpose may be to establish areas of a particular land use, such as crops. A land use and land cover classification suitable for use with remotely sensed data is shown in Table 16.1. Information relevant to many of the 'land cover' categories can be found in sections of the book dealing with water, ecological studies, resource management and the built environment.

Agricultural support absorbs 70% of the European Community's budget and all decisions on important matters, such as prices, export and import regulations, levels of aid, etc., are made by the Community authorities at the European level. Europe is, however, frequently dependent upon releases from the US Department of Agriculture for global information. In the light of this situation

Table 16.1 Land use (man-made) and land cover (natural or semi-natural) classification system for use with remote sensor data (Source: Anderson *et al.*, 1976)

Level I	Level II
1 Urban or built-up land	11 Residential
	12 Commercial and services
	13 Industrial
	14 Transportation, communications, and utilities
	15 Industrial and commercial complexes
	16 Mixed urban or built-up land
	17 Other urban or built-up land
2 Agricultural land	21 Cropland and pasture
	22 Orchards, groves, vineyards, nurseries, and ornamental horticultural areas
	23 Confined feeding operations
	24 Other agricultural land
3 Rangeland	31 Herbaceous rangeland
	32 Shrub and brush rangeland
	33 Mixed rangeland
4 Forest land	41 Deciduous forest land
	42 Evergreen forest land
	43 Mixed forest land
5 Water	51 Streams and canals
	52 Lakes
	53 Reservoirs
	54 Bays and estuaries
6 Wetland	61 Forested wetland
	62 Non-forested wetland
7 Barren land	71 Dry salt flats
	72 Beaches
	73 Sandy areas other than beaches
	74 Bare exposed rock
	75 Strip mines. Quarries, and gravel pits
	76 Transitional areas
	77 Mixed barren land
8 Tundra	81 Shrub and brush tundra
	82 Herbaceous tundra
	83 Bare ground tundra
	84 Wet tundra
	85 Mixed tundra
9 Perennial snow or ice	91 Perennial snowfields
	92 Glaciers

the Commission has set up a 10-year project aimed at the introduction of remote sensing in the European Community's statistics system for agriculture. This will form part of an Advanced Agriculture Information System (Fig. 16.1). This system will be based on data from high-resolution sensors (SPOT, Thematic Mapper) and low-resolution sensors (AVHRR) together with meteorological data for agro-meteorological models. The ultimate objective will be to use remote sensing techniques in a complementary manner alongside conventional data gathering. An outline of the desired development of the project is shown in Table 16.2. Much of other work in Europe is directed from the Joint Research Centre of the EEC at Ispra, Italy. The JRC Pilot Project for remote sensing applied to agricultural statistics is engaged in a 10-year programme directed towards regional inventories, monitoring of vegetation and production indices, yield forecast models, rapid estimates of areas and potential yields, at a European scale. In the case of regional inventories, test sites totalling some 20 000 km^2 have been set up in Germany, France, Italy, Greece and Spain. The availability of SPOT and TM data was not a limiting factor for these chosen sites, although one would expect severe difficulties in cloudier territories in Europe, such as the UK. The methodology for integration of ground and satellite data is complex and JRC has developed special software programmes to ensure that results can be obtained in a timely manner.

As might be expected, it has been found that within the matrices established there are important areas of confusion between the crops of interest and other land uses such as forest or urban areas, and/or other crops under study. It is recognized that this is often the result of real spectral confusion. Two strategies are under investigation to improve the accuracy of identification:

1. Grouping crops together to form a more homogeneous class for better spectral recognition and then splitting these group categories by ground information only.
2. Masking of water, forest and urban areas by means of visual interpretation of images of previous years.

At the end of the first two years of the JRC Pilot Project it has been found that in half of the 13 crop categories surveyed the area estimates were more

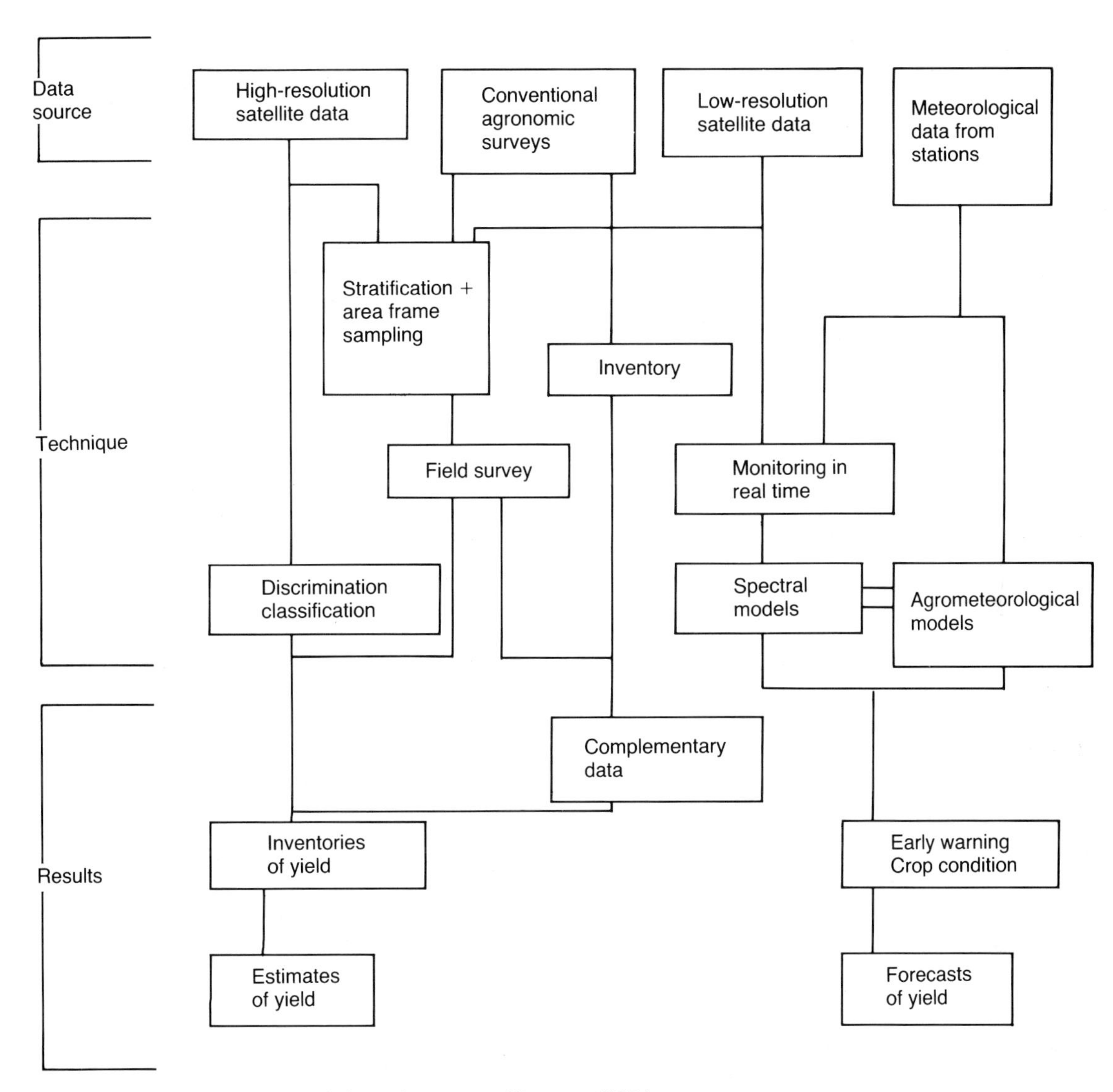

Fig. 16.1 Advanced agricultural information system. (Courtesy, EEC.)

than 5% in error. In the case of yield estimates correlations of better than 0.5 were found in several crops. The project staff hope that in the remaining years of the programme it will be possible to refine the methodology so that it can be used operationally. Clearly the work illustrates the need for satellite and ground survey data to be seen to be complementary to each other. Satellite data alone cannot fulfil the established task. It must be recognized that the operational use of such a methodology will depend on the cost benefits and accuracies which accrue from the use of remote sensing. Where there is a fundamental need for ground survey there will be those who will wish to use ground methods alone if satellite technology appears too expensive. However, the cost savings for EEC agricultural budgets which may devolve from improved accuracies in crop data may amount to some millions of ECUs and so the stakes are high.

16.2 Aerial photographic surveys

Early examples of the use of aerial photography in land use studies include experimentation with air-photo cover in Northern Rhodesia (Zimbabwe) in the mid-1930s. By the 1940s there were air-photo

Table 16.2 Development of EEC Agriculture Project (Source: Contzen, 1989)

Action	Main orientation	Geographic localization	Input	Output
Regional inventories (acreage)	Semi-operational	Three to four selected administrative regions	High-resolution satellite data	Precise regional inventories (plus localized statistics)
Vegetation conditions and yield indicators	Semi-operational plus research	Selected regions and sampling of sites, then all of Europe	Low-resolution satellite data (mainly AVHRR)	Vegetation monitoring, alarm for man crops
Models of yield prediction	Research		Low-resolution satellite data High-resolution satellite data Meteorological data	Yield prediction
European rapid estimates of acreages and potential yield	Semi-operational	Sampling of some 40 or 50 sites throughout Europe	High-resolution satellite data	Estimates of surface and prediction of yield at European level for main crops
Advanced agriculture information system	Semi-operational and research	Sampling of some 40 or 50 sites throughout Europe	All data available	Integration of preceding methods with conventional ones
Area frame sampling associated surveys	Support for other actions	Wherever it is needed, mainly on three to four regions and 40–50 sites	Satellite imagery to build area frame	Support documents or data for other actions
Long-term research	Research		New satellite data (microwaves) New analysis methods (expert systems, GIS)	Improved inventories and yield prediction

interpretation studies of agriculture and forestry in many areas of the world. These early studies were limited to black and white photography, which was becoming readily available in vertical line overlap form.

Photo-interpretation keys were developed for various crops for particular agricultural regions. A useful review of these early techniques in mapping agricultural land use from air-photos is given in the Manual of Photographic Interpretation published by the American Society for Photogrammetry in 1960. At this time features such as patterns of fields, tone, texture, shape and size of the cropped areas were used as aids to interpretation. Further evidence was sought where photos were available for different seasons. Changes in the photographic appearance of particular areas often could be related to conditions of tillage, crop growth, crop rotation or conservation measures. As in all manual interpretation work the best results were obtained by a combination of field knowledge and careful examination of the images. The same principles apply today in the visual interpretation of both aerial photography and satellite pictures.

At a very early stage American and British research workers recognized that aerial photography was useful for the detection and mapping of crop diseases and pest infestations. However, the development of colour, infrared colour and multispectral photography has added further impetus to this area of research in the last decade. Diseases such as potato blight, take-all disease in cereals, leaf spot in sugar beet and yellow dwarf disease in barley have been observed and their effects on crop

Fig. 16.2 (a) Stem nematode infection in lucerne showing as dark patches. Background differential growth is due to soil variations.

conditions noted. Likewise the effects of nematode attacks (eelworm) and soil fungi may be detected and affect the spectral response of a particular crop (Fig. 16.2 (a and b)).

Where crops have been affected by disease or adverse environmental conditions dead portions of the crop can be identified readily by using infrared colour photography. On this emulsion, healthy plants image in red whereas diseased and dead portions assume different colours. The colours of diseased plants often range from salmon pink to dark brown depending on the severity of the attack. Dead portions image in green or bluish grey.

Crop discrimination using infrared colour photography has been studied closely and it has been found that the percentage accuracy depends on time of year, location and environment. For example, in North Eastern Kansas in late July all crops were found to image red. In Britain, however, it is noted that the quality of grass pasture often can be detected by variations in the intensity of the red hue, and in some cases this is a reflection of the fertilizer treatment the field has received (Plate 11, colour section).

16.3 Multispectral sensing of crops

16.3.1 Basic research

In view of the different spectral responses observed for different plant types, a great deal of effort has been made to measure the spectral reflectance characteristics of a number of crops. A major

Fig. 16.2 cont. (b) Patches of stunted growth owing to barley yellow dwarf virus infection differentially invaded by sooty mould fungus (*Cladosporium* spp.). (Courtesy, Ministry of Agriculture, Fisheries and Food, Cambridge.)

source of data concerning crop signatures was obtained under the Air Force Target Signatures Measurement Program in the USA. These were mostly measurements in the visible wavelengths, and much of the data was derived from laboratory investigations. Examples of the signatures obtained for certain crops are given in Fig. 16.3. The spectral response of a field crop depends partly, however, on the layering within the crop and so modelling of crop canopies also is important (Chapter 15).

The principal objective now being pursued is that of automatic recognition of land use categories and crops by means of computer analyses of spectral reflectance data. Such analyses can be applied to either airborne sensor data or satellite data.

There are broadly two approaches which may be adopted in mapping from spectral data. On the one hand a *training set* of data can be obtained from matching known crops to sets of data from particular sensor systems. This training set then can be entered into the computer so that the computer classifies all similar data as the crop in question. Such classifications are termed *supervised* classifications. Alternatively, it is possible to adopt an *unsupervised* classification approach. In this case the remote sensing data are grouped and mapped by clustering algorithms (see also Chapter 9).

The sequence of operation for crop surveys includes data preprocessing, training set selection, and classification by statistical pattern recognition. In the case of aircraft data a similar sequence is necessary as, for example, was used in the Laboratory for Agricultural Remote Sensing (LARS) method at Purdue University (Fig. 16.4).

Data preprocessing includes radiometric cor-

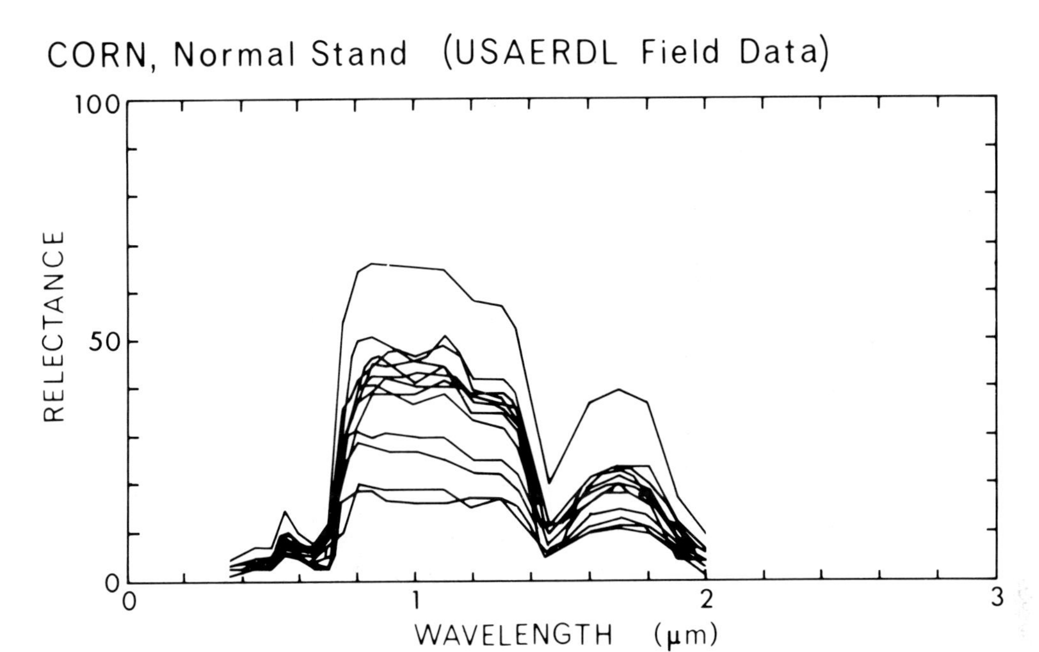

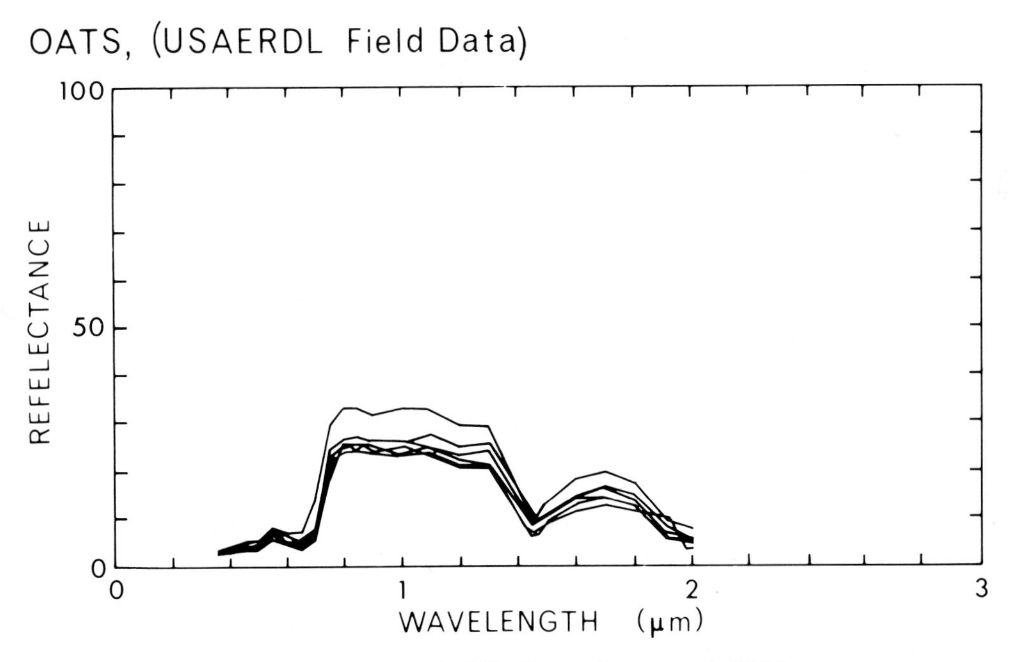

Fig. 16.3 Spectral signatures of corn, oats and alfalfa. (Source: Leeman *et al.*, 1971.)

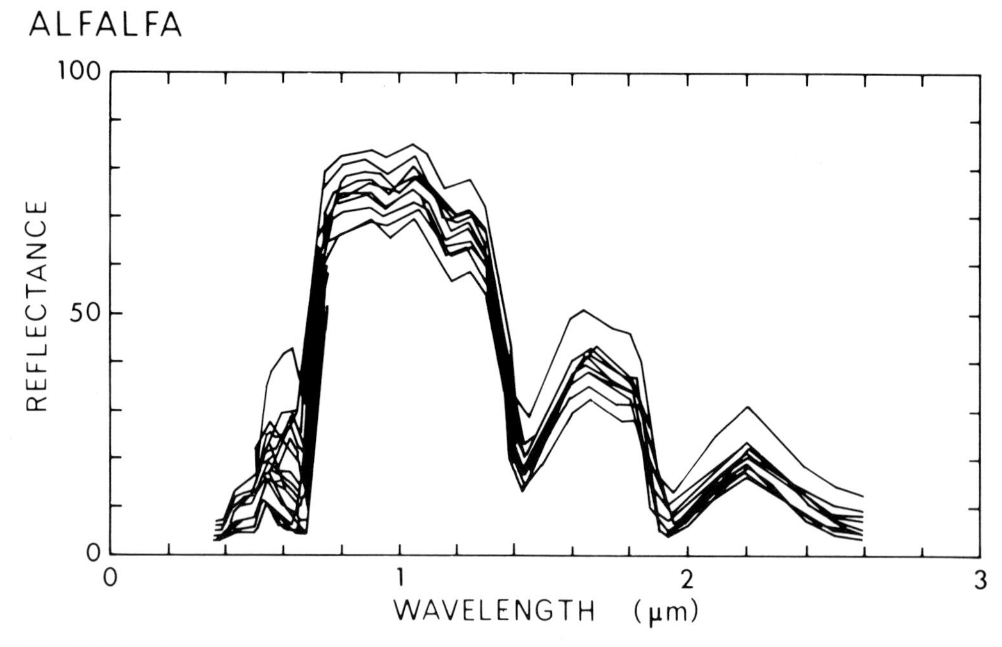

Fig. 16.3 Continued

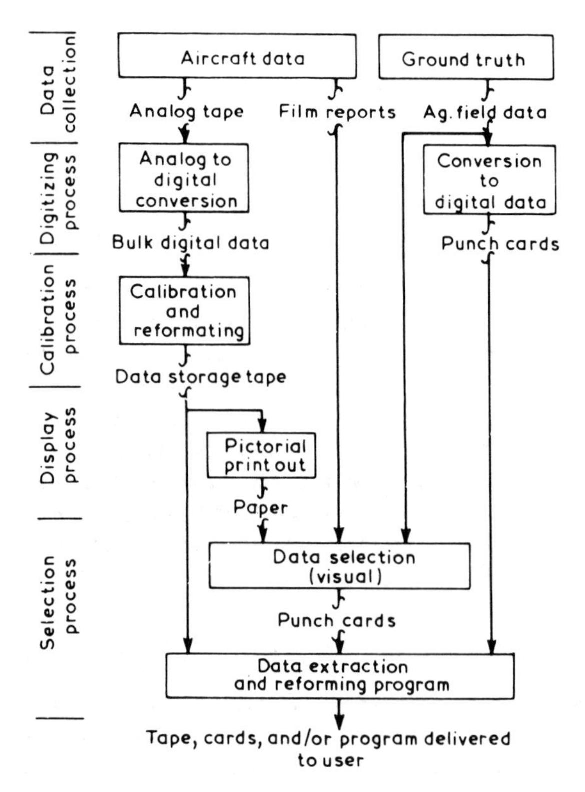

Fig. 16.4 LARS data flow system for crop studies.

rection and registration of each picture element in geometric coincidence. The training sample selection phase aims to determine the separable classes and subclasses in a given data set. Pattern analysis systems have been developed that allow several methods of class selection to be employed as appropriate. Statistics can be computed for up to 30 wavelength bands and printed out in the form of histograms, correlation matrices, and coincident one sigma spectral plots. One of these types of output can then be used to group areas having similar spectral responses. Another method of class separation consists of using clustering techniques to group image points such that the overall variance of the resultant sets is minimized.

The fields for which ground truth data are available are then identified by line and column coordinates so that they can be entered together with multispectral data into the computer. Histograms and statistics are then computed and printed for each field. An adequate statistical sample (>30) for each crop type is necessary at this stage. Samples of data from each of the classes identified by the statistical process are then used for training the pattern classifier, e.g. the histograms for barley can be used as training sets in respect of the automatic classification of barley.

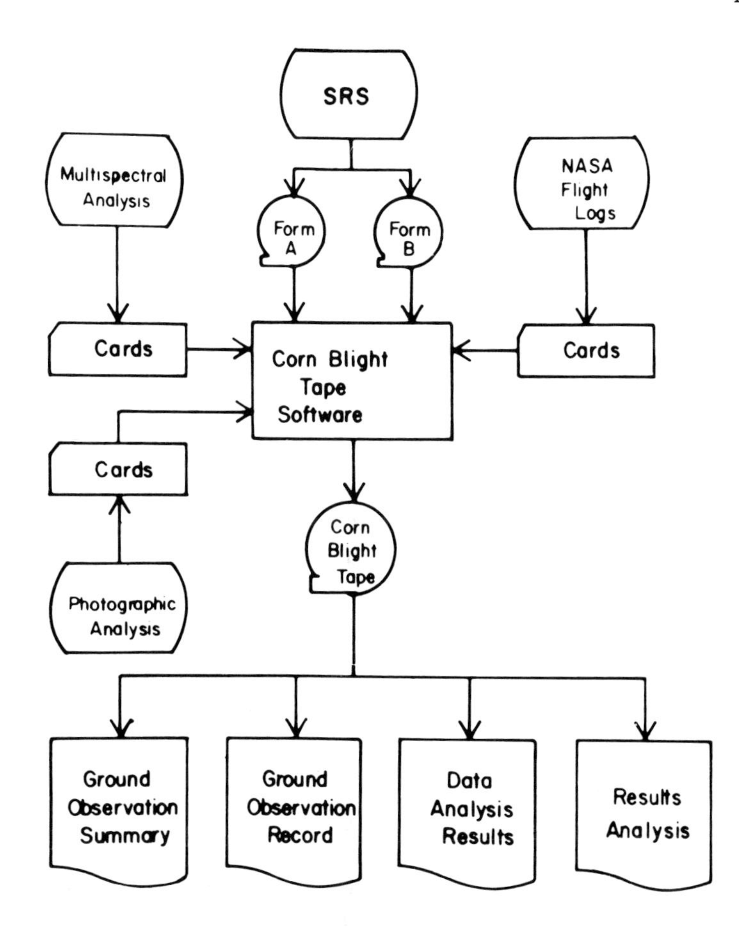

Fig. 16.5 Merging of ground data, flight logs and analysis results in 1971, Corn Blight Watch Experiment. (Source: MacDonald *et al.*, 1973.)

In the pattern recognition stage two methods can be used to classify the multispectral imagery. One method classifies each image point into one of the defined classes. Alternatively an entire field can be classified as one decision. The latter technique, termed 'per field classification', has the advantage of speed. However, it demands that the field co-ordinates must be known and fed into the classifier before any classification can be performed. Classification by the 'per point' method is time consuming since every resolution element in the image is classified separately. However, it can classify any area without the field boundaries being specified.

Following the pattern recognition phase, which produces automatic classification of the land use, it is necessary to evaluate the classification accuracy quantitatively. A large number of test fields are necessary for this purpose. These are located in the computer classification and the ground data is compared with the computer result.

Supervised classification techniques of the type described above have certain drawbacks. There is the very considerable difficulty that spectral signatures show high variability and supervised techniques generally demand that reference signatures be collected directly from a training area lying within the survey area or nearby. The unsupervised classification technique avoids this difficulty by not requiring reference signatures in the data processing phase. Unsupervised techniques will group the multispectral data into a number of classes based on the same intrinsic similarity within each class. The meaning of each class in terms of land use category is then obtained after data processing by checking a small area belonging to each class.

One of the most extensive crop studies so far made by remote sensing techniques, using supervised classification based on ground data, is that of the Corn Blight Watch Experiment, 1971. The corn (maize) leaf blight is caused by the fungus *Helminthosporium maydis* and is widespread in maize growing areas in tropical parts of the world.

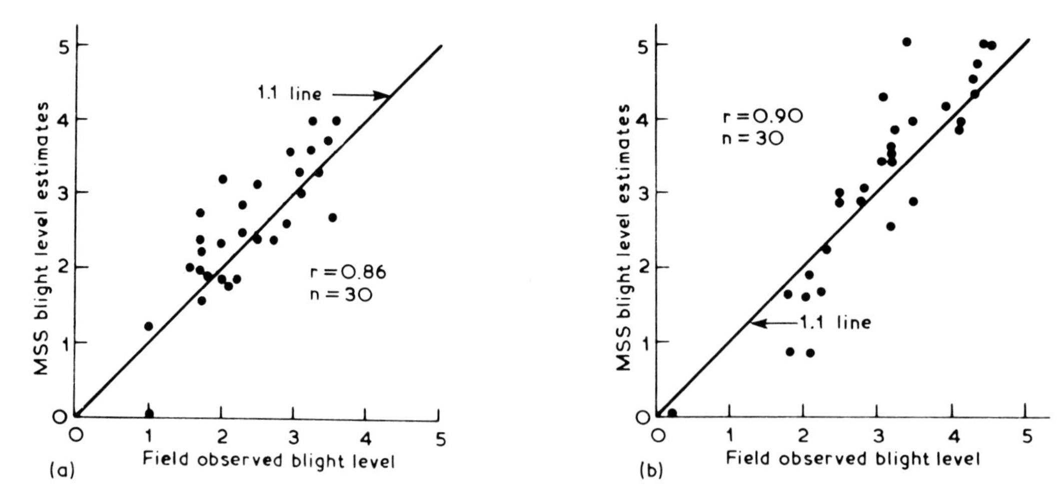

Fig. 16.6 Correlation of field observation and machine-assisted analysis of multispectral scanner data estimates of segment average blight severity levels, 23 August (left) and 6 September (right). (Source: MacDonald *et al.*, 1973.)

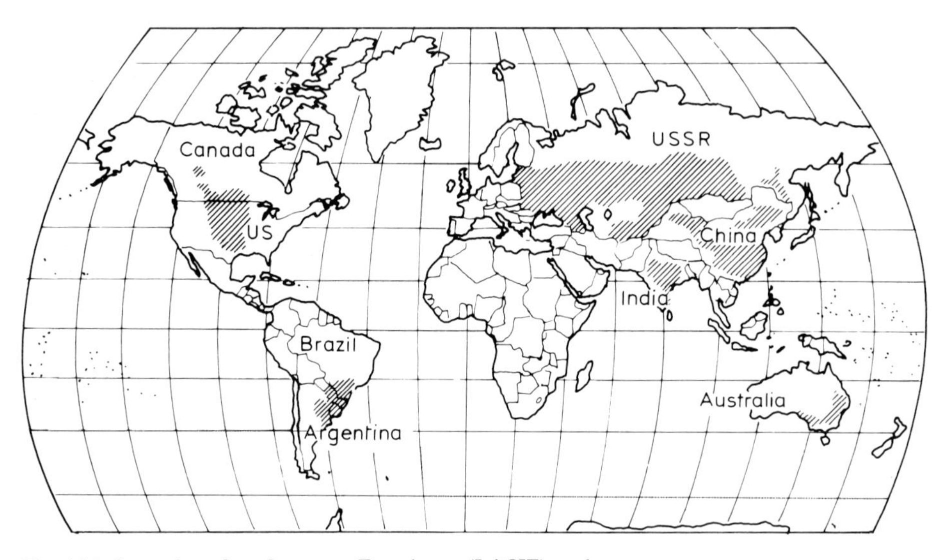

Fig. 16.7 Large Area Crop Inventory Experiment (LACIE) study areas.

Until 1969 corn leaf blight was a minor problem in the USA, but in 1970 there was extensive corn blight as a result of the development of a new race of fungus. As a result, yields are thought to have dropped by about 700 million bushels in 1970.

In 1971 two aircraft collected colour infrared photography and 12-channel multispectral scanner imagery. The ground data, flight logs and the image analysis results were combined and analysed to record the results of the corn blight survey (Fig. 16.5). The interpretation phase of this study required extensive correlations of field observation, photo-interpretation and automatic analysis of MSS data. Five categories of corn blight severity were used in the field. The best correlations were achieved when the blight symptoms were well established. It also appeared that the correlation between MSS analysis and field data became better at later dates in the study, thus emphasizing the importance of phenological conditions in such studies (Fig. 16.6). This important early study concluded that neither manual interpretation of small-scale photography nor computer-made analysis of MSS data gave adequate detection of

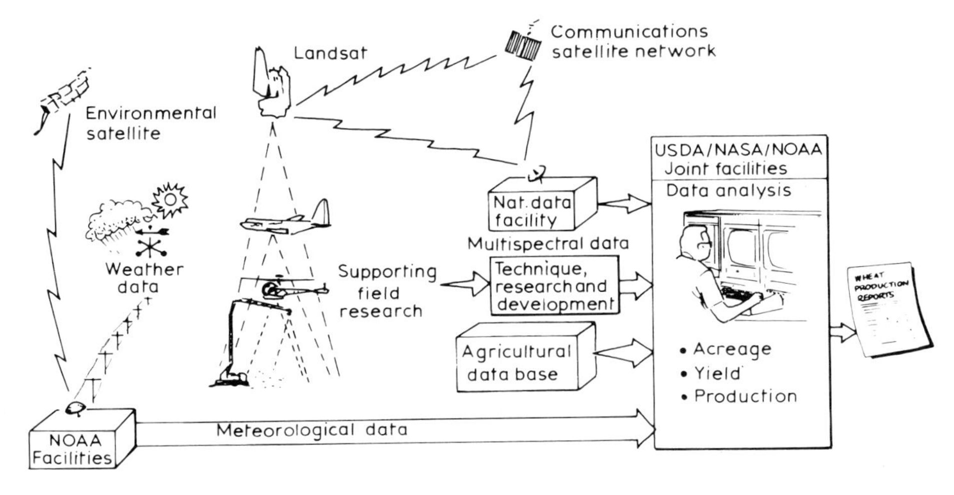

Fig. 16.8 The LACIE system data flow.

corn blight during early stages of infection. Analysis of the data did, however, permit the detection of outbreaks of moderate to severe infection levels.

16.3.2 Integrated studies in crop monitoring using multispectral data

It is useful to bear in mind developments in the USA over the past three decades towards a crop monitoring system. The Agricultural Board of the National Research Council set up its Committee on Remote Sensing for Agricultural Purposes in 1961. In the 1960s the multispectral concept was developed and initial feasibility experiments were carried out. Improved airborne multispectral sensors were developed in the mid-1960s and the development of decision algorithms to analyse multispectral data soon followed. Thereafter there was a long period (1966–1972) when controlled field experiments were carried out together with over-flights with an airborne scanner. Then in 1969 a Landsat 1 multispectral scanner was flown on Apollo 9. Subsequently the corn blight watch and the launch of Landsat 1 in 1972 gave added impetus to both analytical techniques and concepts. As a result in 1974 the pioneering LACIE (Large Area Crop Inventory Experiment) programme was initiated to test the technology and know-how developed by applying it to the assessment of crop production over several of the most important agricultural regions of the world (Fig. 16.7). The objectives of the LACIE experiment were to:

1. Evaluate the applicability of space technology to the global monitoring of agricultural crops.
2. Integrate the necessary components into an experimental system to permit the acquisition and analysis of the required volume of data in a timely manner.
3. Conduct the experiment in a quasi-operational manner to evaluate the technology under conditions representative of any future operational missions.
4. Assess the suitability of the 'first-generation' technology and identify areas requiring improvement and suggested means of development.
5. Meet an accuracy goal such that the estimates at harvest should, on average, be within ±10% of the true country production 90% of the time.
6. A timeliness goal be set such that Landsat data should be reduced to acreage information within 14 days after acquisition.

The LACIE study was focused on monitoring wheat production in the selected regions shown in Fig. 16.7 and the LACIE technology was evaluated over a nine-State USA 'yardstick', together with exploratory studies in India, China, Australia, Argentina and Brazil. The LACIE system data flow supporting the programme is shown in Fig. 16.8.

In order to estimate wheat production a country was subdivided into areas (strata) where yield and prevalence of wheat were fairly uniform. As shown in Fig. 16.9 the yield and the areal extent of wheat within each stratum were estimated by independent

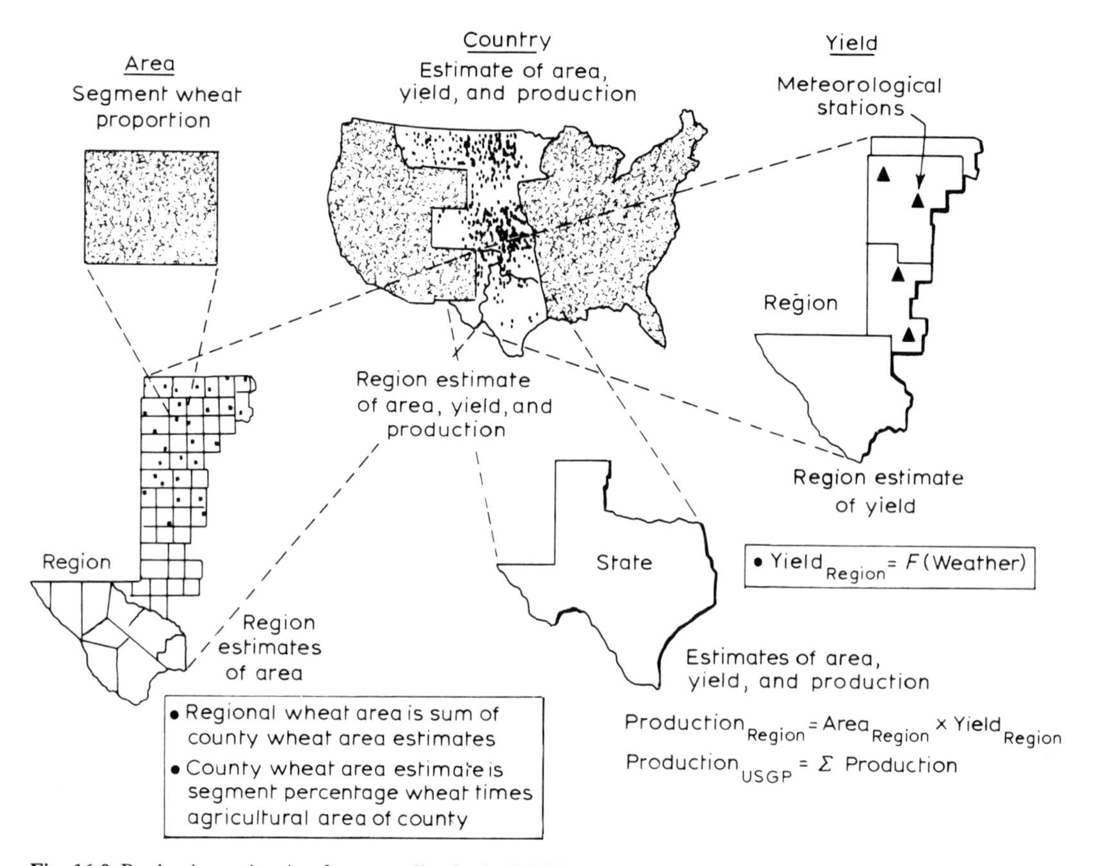

Fig. 16.9 Production estimation from sampling in the LACIE study.

methods and then multiplied together to obtain production at the stratum level. The production estimates in each stratum were then added to obtain the production estimates at other geographical or political levels. In addition, area and yield were aggregated to determine wheat area and yield at other hierarchical levels within the country.

The crop identification and mensuration of USA 'yardstick' areas was carried out using a stratified random sampling strategy based on 5 × 6 nautical mile segments randomly allocated to strata according to the 1969 Census of wheat growing areas. The basic sample unit was termed a LACIE image and consisted of 196 picture elements (pixels) by 117 scan lines for each of the four MSS channels. These segments were extracted from full-frame Landsat images and four acquisitions of data representing four different stages of wheat growth were taken to classify wheat from non-wheat. These four channels (N) were merged into a multitemporal (4N) image and then classified by a sum-of-likelihoods classifier. The LACIE classifier was dependent on the concept of analogous areas and analyst/interpreter techniques.

In the period 1974–1977 the algorithms used were reviewed to allow for improved stratification of the inventory region and different acquisitions of data were programmed. Despite processing problems estimates were made of the USSR wheat production by the end of 1977. These were compared with the US Department of Agriculture estimates made by the Foreign Agricultural Service (FAS) (Fig. 16.10).

An investigation of the Landsat data and the yield model response at subregional levels showed that drought conditions observable by Landsat were accurately reflected by reduced yield estimates in the affected regions. The LACIE estimates met the accuracy goal at harvest and even achieved this in its forecasts made 1.5–2 months before harvest. It was noted, however, that in small field areas there was greater confusion in the identification of

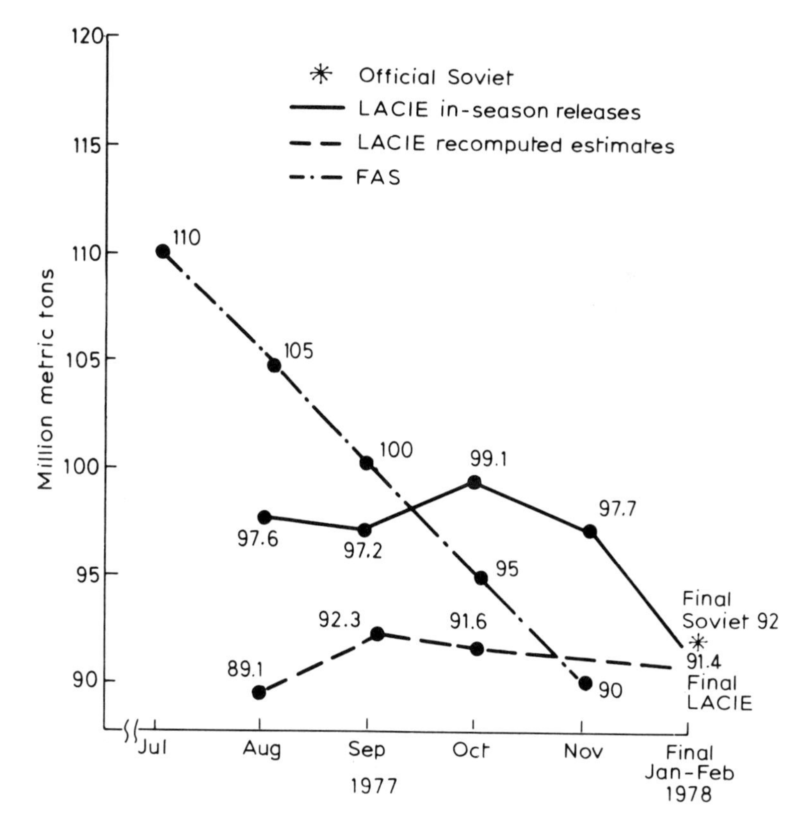

Fig. 16.10 LACIE estimates of 1977 Soviet wheat production.

crops and a resulting underestimation of spring small grains.

As noted above, the LACIE experiment revealed that the accuracy of estimate of cropland acreage decreases where field sizes become smaller. This is a fundamental problem when crop identification is being attempted in Europe where field sizes may be less than 1 ha in some cases. The crux of the matter is that from space altitudes many of the ground resolution elements are individually composed of a mixture of crop categories. Thus many of the picture elements generated by multispectral sensors are not characteristic of any one crop but relate to a mixture.

Morrison, in 1977, expressed a NASA view of the prospects for a Worldwide Crop Information System by writing 'Present satellites do not have the resolution necessary for providing worldwide crop data in field situations. For this we have to wait for the Thematic Mapper'.

Following the LACIE programme the USA introduced the AgRISTARS (Agriculture and Resources Inventory Surveys through Aerospace Remote Sensing) project in 1980. Its primary purpose was to assess the feasibility of integrating aircraft and satellite data into the data collection systems used by the US Department of Agriculture (USDA). The programme led to much progress, especially in the use of NOAA and geostationary environmental satellite data in support of crop monitoring, particularly in respect of meteorological data useful for crop forecasting models. Unfortunately budgetary problems had reduced the programme substantially by the mid-1980s, and although TM data were available to AgRISTARS investigators after 1982, these data, supporting improved crop recognition, were never as well integrated as those from LACIE. Today, even with both TM and SPOT data available, Morrison's view still holds good, especially in areas of small fields and mixed crops of similar types.

16.3.3 Remaining problems in multispectral sensing

Europe is one area of particular difficulty, where the problem of *field size* is reflected in the small size

Table 16.3 Number and area of agricultural holdings with 1 ha arable area and over by size groups in the EEC (Source: Eurosat)

Size group	Quantity	Area (ha)
1–5 hectare	2.429×10^6	6.14×10^6
5–10 hectare	1.067×10^6	7.67×10^6
10–20 hectare	1.067×10^6	15.14×10^6
20–50 hectare	0.845×10^6	25.32×10^6
>50 hectare	0.2961×10^6	34.48×10^6

The % of the holdings with ≥20 hectare is 54%

of the agricultural holdings (Table 16.3). It will be evident that about 46% of holdings are less than 20 ha in size. In these small holdings several crops often divide the farm into parcels smaller than 1 ha. It can be estimated that correct recognition of a 0.5 ha parcel would require a pixel size of no more than 25 × 25 m. Furthermore, in Europe the cloud cover makes it difficult to obtain more than two repetitive scenes using visible sensors. Present experience indicates that to achieve an accuracy of 95% in acreage estimation, using only two multispectral scenes, would require a ground resolution of 20 m.

It is also being recognized increasingly that the problem of *modelling* the radiance of crop canopy is, in many ways, more complex than that of modelling a complete cover of natural vegetation. As the crop grows different plant characteristics are developed and simultaneously the amount of underlying soil visible to the sensor changes through the growing season. As we have noted, remote sensing devices detect radiance primarily from the canopy but additional complicating factors must be borne in mind. The reflectance characteristics are dependent on a number of variables which change during the growing season, including:

1. optical properties of stems and reproductive structures within the canopy;
2. leaf area indices (LAI);
3. canopy densities;
4. leaf orientations and shapes;
5. foliage height distributions;
6. transmittance of canopy components;
7. the amounts of soil background viewed and their reflectance contribution.

With regard to optical properties we have noted already, in Chapter 14, the relationships between chlorophyll and the red/infrared portions of the spectrum. As chlorophyll changes with maturation of the crop so does the wavelength of the reflectance. The sharp change in leaf reflectance over the wavelengths 0.68–0.75 μm has been called the 'red edge'. Experiments have shown that chlorophyll content, background material returns, multiple leaf stacking, and leaf water content affect the maximum slope of the 'red edge'. It has been concluded that:

1. Shifts in the wavelength of the 'red edge' are related directly to leaf chlorophyll content. An *increase* in chlorophyll is marked by a shift of the 'red edge' to *longer wavelengths*.
2. Provided the 'red edge' is measurable it can be used to distinguish *live* vegetation from other background targets.
3. 'Red edge' measurements over sufficiently narrow wavelength intervals can be used as early indicators of *crop stress*.

For crop canopy studies based on MSS data it has been found useful (as explained in Chapter 13) to use Normalized Difference Vegetation Index (NDVI) values using MSS band 5 (red) and MSS band 7 (infrared), i.e.

$$\text{NDVI} = \frac{\text{IR} - \text{R}}{\text{IR} + \text{R}} \tag{16.1}$$

Early in the modelling of crop canopies by such a method the *brightness* and *greenness* of a crop were tackled by seeking *orthogonal indices* using a linear function derived from four Landsat MSS bands. The new axes that were developed comprised what was termed the 'Tasselled Cap'. This transformation rotated the MSS data so that the majority of the information was contained in two components of the physical scene, i.e. brightness and greenness.

Subsequently the Tasselled Cap concept was extended to six-band Thematic Mapper data. In this case the rotation was adjusted so that three dimensions were defined which broadly accorded with soils, vegetation, and a transitional zone that might be associated with conditions of canopy wetness and/or soil moisture.

Meanwhile other vegetation index models were developing, including the Perpendicular Vegetation

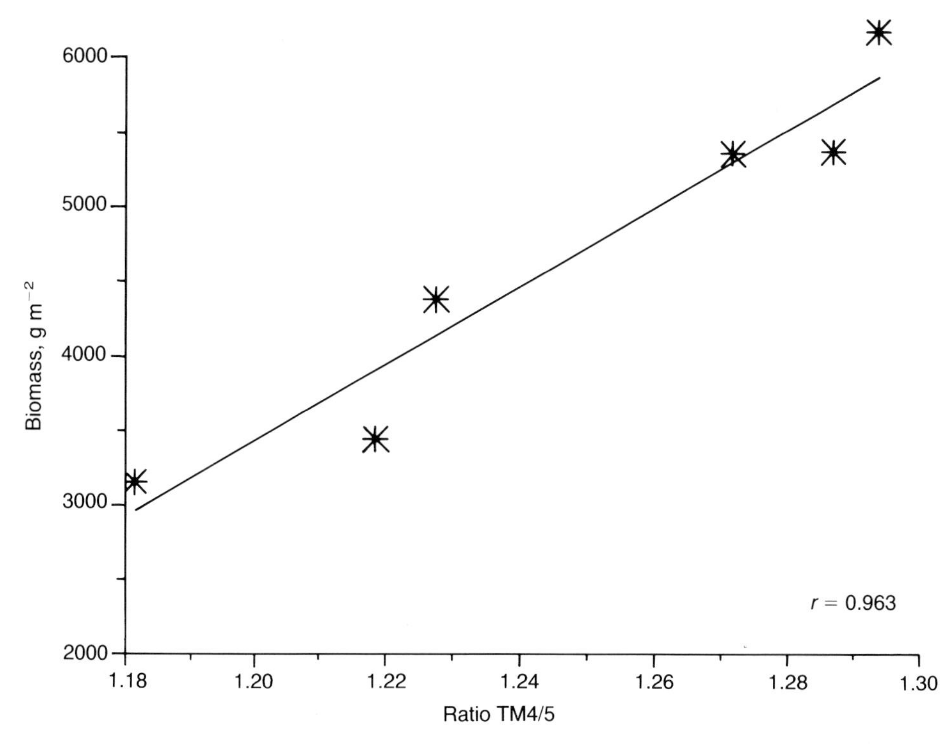

Fig. 16.11 Functional relationship between TM ratio (4/5) and biomass of corn based on regression analysis of data from 27 June, 20 July and 14 August 1986, Freiberg, Germany. (Source: Mauser, 1988.)

Index (PVI). We have already noted that as vegetation grows the red reflectance decreases and the near-infrared increases. The perpendicular (or orthogonal) distance of a *vegetation spectral plot* to the *soil line* is, therefore, a measure of greenness. The Perpendicular Vegetation Index defines for each pixel the distance that the crop reflectance is from the plane of soil reflectance in the feature space defined by the near infrared and visible wavebands. One may calculate PVI as follows:

$$PVI = [(R_s - R_v)^2 + (NIR_s - NIR_v)^2]^{1/2} \qquad (16.2)$$

where R is the Red waveband reflectance, NIR the near infrared waveband reflectance, and s and v the soil and vegetation respectively.

A positive PVI indicates vegetation, zero indicates bare ground and a negative value indicates water.

The work of the Joint Research Centre, Ispra, has shown that *monotemporal* crop classifications of Landsat 5 TM data contain a high level of spectral confusion. Furthermore, it has been found that *multitemporal* data do not always provide sufficient information for classification. However, Tasselled Cap transformations have been found to present efficient compromises in many cases, although precise geometric correction has been found to be necessary.

At present the task of modelling crops effectively remains formidable, although considerable progress has been made in the generation of soil adjusted line data. Recent results are tantalizingly near to achieving operational accuracy under favourable circumstance but there is a need for sensors with 20 m ground resolution with visible and infrared data. It is for this reason that there is pressure for SPOT 3 to include infrared channels. However, it has been shown that land cover/crop classifications are affected by the sensor view angle. In general, highest classification accuracies are obtained at view angles close to those at which the training sample data were obtained. Thus, care needs to be given to the use of SPOT data which are obtained at different viewing angles. Summarizing

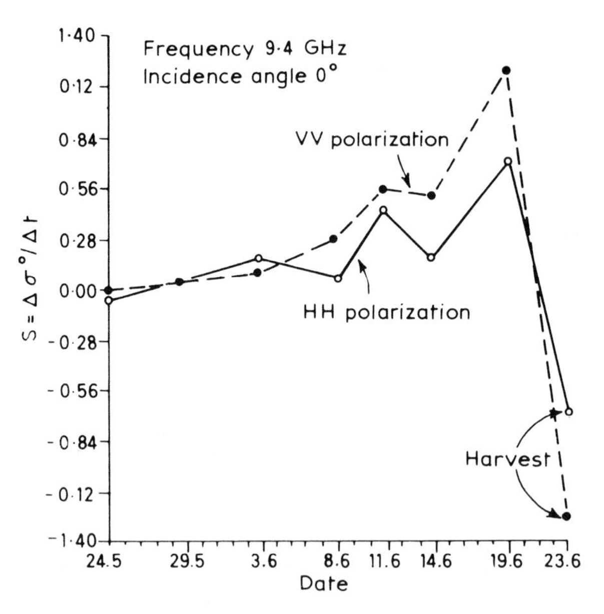

Fig. 16.12 Changes in radar response as a wheat crop matures in Kansas. The period of observation extends from 24 May to 23 June 1974. (Source: Bush and Ulaby, 1975.)

the position so far as MSS data is concerned in crop analysis the following conclusions may be drawn:

1. The crop calendar is a very important guide as to which data sets are likely to be most useful and should serve as a basis for ordering data.
2. Multitemporal image analysis is likely to improve classification accuracy compared with monotemporal studies.
3. Image ratios and transformations of data are useful in improving crop classification accuracy. In addition they can remove redundant data.
4. Field size, topography and variability of crops within a field can have a marked effect on classification accuracy.
5. SPOT data are likely to be improved by the addition of a mid-infrared channel but the user must be aware that the different viewing angles obtained by SPOT can have an impact on classification accuracies.

The importance of including infrared data in crop studies has been borne out by many studies. Thematic Mapper data have presented advantages in this respect. For example, studies in the Upper Rhine near Freiburg, which were carried out in relation to the AGRISAR radar campaign of 1986, used TM data for estimates of biomass. Promising functional relationships between multitemporal ratioed TM bands 4 and 5 and the recorded biomass of corn were obtained (Fig. 16.11).

In conclusion, it is important to note the growing trend towards multiparameter 'knowledge based systems', in which data such as altitude, temperature, and existing map information can be used in the classification stage. Also, one may note the increasing interest in 'all-weather systems' such as radar for those areas where multitemporal data is difficult to obtain because of frequent and/or persistent cloud cover.

16.4 Radar sensing of crops

Two aspects of crop quality lend themselves to radar monitoring: moisture status and the topology of the crop. A strong correlation between the radar back-scattering coefficient and plant moisture content was found by workers at the University of Kansas using a 9.4 GHz radar system mounted on a platform 26 m above the ground. Linear regression analyses of the radar back-scattering coefficient (σ°) on plant moisture provided good correlation of about 0.9 and a slope of -0.275 dB per cent plant moisture with nadir observation. Also it was found that σ° undergoes rapid variations shortly before and after wheat is harvested (Fig. 16.12). These variations suggest that radar could be used for estimating wheat maturity and for monitoring the progress of harvest.

In parts of Europe there may be only a few days in the year when conditions are ideal for Landsat MSS observations. In these circumstances the use of radar has many attractions. In cloudy regions throughout the world it may be the only sensor capable of providing data for time-discriminant analysis. Evaluation of radar has taken place in the USA, Canada and Europe in the last decade and the following decade will see continued emphasis on radar studies. Perhaps the most promising avenues of research have been concerned with ratioing polarized radar returns at different grazing angles and using the data in time-discriminant analysis (Fig. 16.13). Much of the work of an experimental nature has been carried out under the auspices of the Agriculture and Resources Inventory Surveys Through Aerospace Remote Sensing (AgRISTARS) programme, the successor to the LACIE project in the USA.

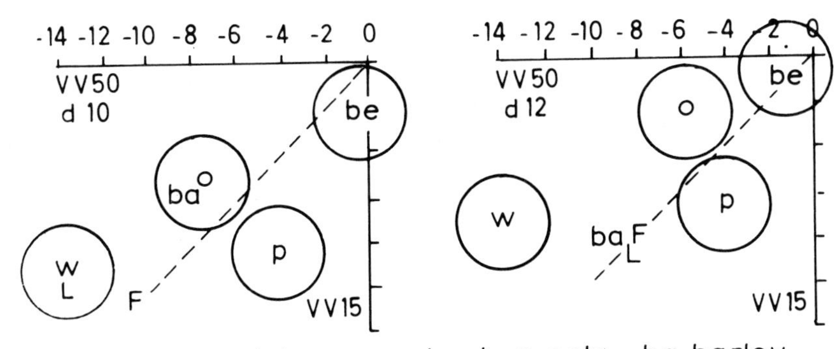

be beets, p potatoes, w wheat, o oats, ba barley, L *Lolium*, F *Festuca*

Fig. 16.13 An example of the discrimination of different crops by a ground-based X-band radar observing at different angles of incidence (VV 50°: VV 15° on day 10 (d10) and day 12 (d12)) of a series of observations. The circles are drawn at 2 decibels radius and VV indicates vertical polarization. (After Kasteren and Smit, 1977.)

In Europe the AGRISAR campaign of 1986 obtained SAR images over various sites using the French VARAN-S system mounted on a B-17 aircraft. The campaign, which was organized by JRC, Ispra, aimed to provide multipolarized X-band SAR data for various agricultural regions. Temporal values for back-scatter were obtained from different cereals. An example from a test site in the Upper Rhine near Freiburg shows that winter and summer cereals can be distinguished. The X-band SAR images that were used were of different polarizations and dates 27 June (VV), 17 July (HH), and 13 August (VV). For cereals, separation was best in June and July (the crops had been harvested by the August date). Other crops studies (Fig. 16.14) were best separated by means of July data. Although present SAR technology and analytical methods are far behind those used for optical data there are indications that crop types can be distinguished using multitemporal X-band data. The analytical processes are lengthy and complex so that an operational system seems to be some distance away.

16.5 Crop yield modelling using infrared sensing

Whilst thermal infrared sensors operating from aircraft can provide high-resolution imagery capable of distinguishing hedges, trees and animals (Plate 4.6), the satellite sensors have more limited ground resolution. The Ground Resolution Element (GRE) of the Heat Capacity Mapping Mission (HCMM) was about 500 m, whereas Band 6 (10.40–12.50) on Landsats 4 and 5 provided a GRE of 120 m. As a result of this, low-resolution thermal infrared data are of little use in crop identification. Such data become important, however, when used in crop yield (productivity per unit area) models.

The two primary environmental determinants of crop yield are *temperature* and *moisture*. The technology required to remotely measure surface temperatures of soil and crops is well developed. Thus the crucial issue for the production of crop yields by remote sensing data remains the assessment of available soil moisture (see also Chapter 13). Temperature measurements may, however, provide valuable information and there have been efforts to model the relationships between air temperature, leaf temperature and crop yield. The underlying theory is that moisture deficiency in the crop leads to an increase of leaf temperature above air temperature. Thus the term 'stress degree day' (SDD) has been coined to indicate those days when the crop is suffering from a shortage of water, which will depress yield. The final yield of the crop (Y) can be deemed to be linearly related to the total SDDs accumulated over a given critical period. Thus a general least-squares regression equation (intercept α and β) can be expressed as:

$$Y = \alpha - \beta \left(\sum_{i=b}^{e} \mathrm{SDD}_i\right) \qquad (16.3)$$

where SDD_i is the mid-afternoon (1400 hours) value of (leaf temperature) minus T_a (air tempera-

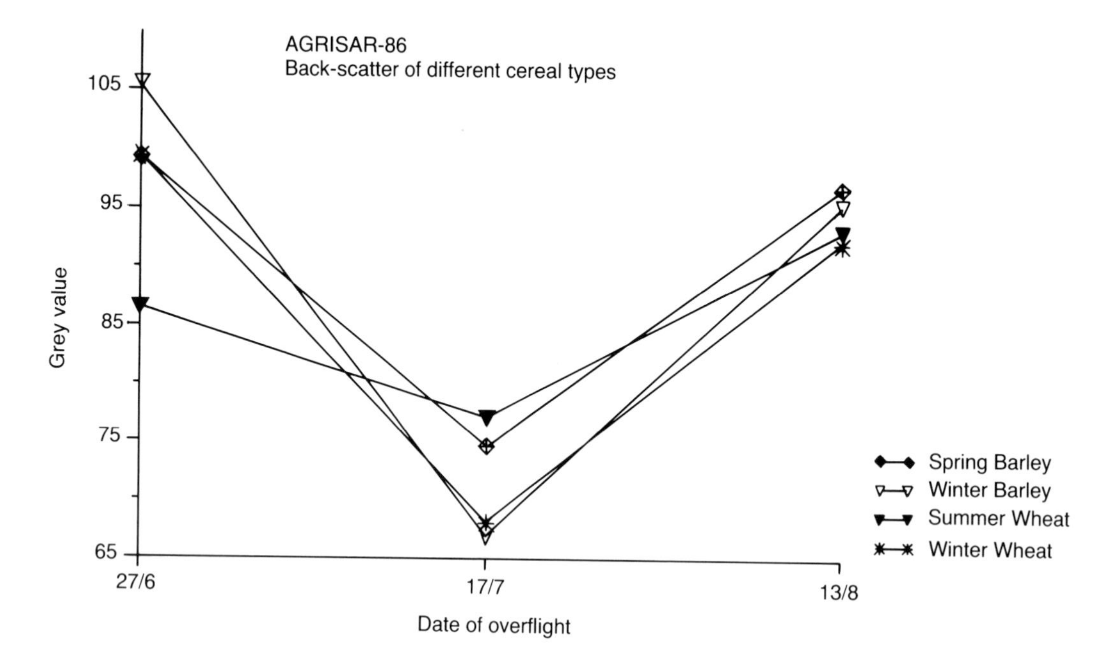

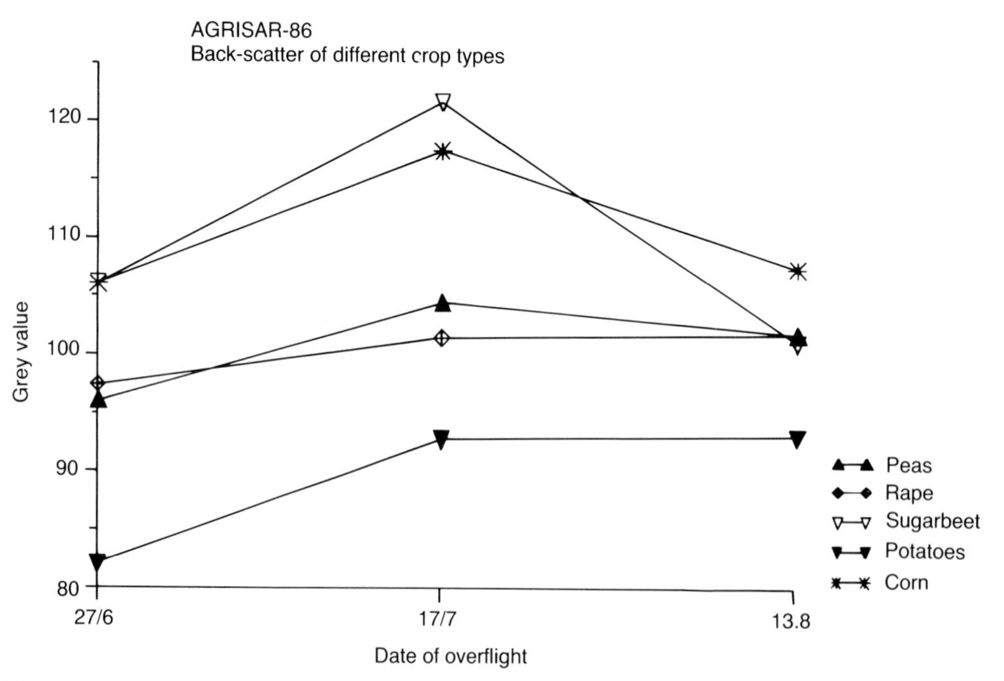

Fig. 16.14 Grey value profile of the AGRISAR data for (a) different cereals in the test site and (b) non-cereal cover types. Grey values are in arbitrary power units. (After Mauser, 1988.)

ture) on day i, b is the day on which summation begins, and e is the day on which summation ends.

It was first found that the best time over which to sum daily values of the SDD measurements is the period from the first appearance of the awns to the time when the head produces no more dry matter. This period can be determined by a discontinuity in the albedo data between first head emergence

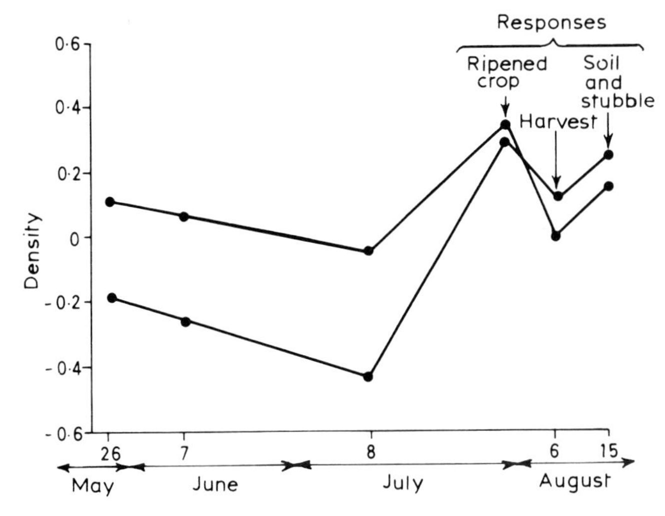

Fig. 16.15 Changes in reflectance of visible light during the growth of spring barley (variety Vada) on soils of the Sherborne Series, Badminton, Gloucestershire. (After Curtis, 1978.)

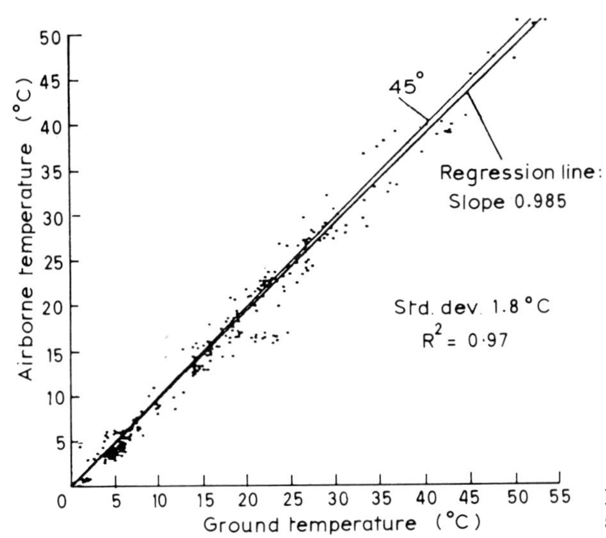

Fig. 16.16 Airborne versus ground-measured temperatures for an entire wheat growing season. (Source: Millard *et al.*, 1979.)

and maturity of the crop. For example, the changes in reflectance of a spring barley crop in England (Fig. 16.15) clearly reflect crop maturity. Albedo measurements during the early stages of crop growth could also provide estimates of the green leaf area index at the time of head appearance. In this way the potential for head growth could be estimated from albedo measurements and then the growth during the crucial period of head growth could be monitored by SDD values obtained by thermal sensing.

There is good reason to believe that temperatures of the canopy of the crop can be measured effectively. For example, a study of a wheat crop at Davis, California, showed that the correlation between temperatures measured on the ground with a PRT-5 Radiation Thermometer (10.5–12.5 μm bandpass) could be correlated closely with airborne measurements at 300 m altitude using a Texas Instruments R-25 infrared line scanner in the same bandwidth (Fig. 16.16).

Attempts have been made recently to combine the stress degree day concept (moisture factor) with the classical growing degree day (GDD) in order to improve crop yield modelling. The growing degree day can be defined as:

$$\text{GDD} = \frac{T_{\max} + T_{\min}}{2} - T_b \qquad (16.4)$$

where T_{max} and T_{min} are the daily maximum and minimum air temperatures, and T_b is the base temperature (c. 5.5°C) below which physiological activity is inhibited.

Measurement of the climatic factors of GDD plus calculation of daylight minutes in the period between emergence of the crop and the appearance of heads and awns are now used as additional inputs for the estimation of crop yield.

The sensors on the Thematic Mapper and SPOT satellites which are available for the study of land use represent a considerable advance on those existing on satellites in the 1970s. The growing experience of remote sensing experts will doubtless refine still further the radiometric characteristics of future satellites. Improved interpretation and classification should be possible as radiometric bands as narrow as 0.04 μm become practicable. The addition of Bands 5, 6 and 7 in the near and middle infrared wavelengths has already provided a greater range of relevant data, which are gradually being assessed in crop studies.

16.6 Land use in the past

Aerial photographs have been used widely in studies of earlier civilizations and their settlement patterns; in fact air photography is now an accredited, almost veteran, aspect of the archaeological method. Nevertheless, its potential and limitations are still not widely appreciated.

In practice oblique photographs are widely used to illustrate the features of archaeological sites because they afford the advantages of a view comparable to that which could be obtained from a hill top in a hand picked position at close quarters. The vertical air photograph serves a different purpose in that it is the main tool for mapping features and detecting their planimetric relationships. When overlapping verticals are studied stereoscopically they offer many advantages in that the stereomodel enables the viewer to see the physiographic positions in which archaeological remains are found.

When air photographs are used to study former land use patterns there are two principal lines of evidence. First, there are vegetation markings, which mainly occur within cropped land. Where buried features exist beneath the surface the soil may be shallow, e.g. over a buried wall. In this case the rooting depth of crops will be restricted and at times of water deficiency (usually late summer) the crops may show yellowing of the leaves or stunted growth. Conversely the buried feature may be in the nature of a ditch. Under these circumstances the soil over the ditch may be more moisture retentive and higher in nutrients. In consequence the crop may be more luxuriant and greener along the ditch line. Such features are sometimes referred to as positive (stronger growth) or negative (weaker growth) crop marks. They are usually best seen when the crop is mature and when the climatic conditions have produced moisture stress in the soil.

The direction of photography is of crucial importance in archaeological survey. Differences of contrast as a result of photography taken in different directions relative to sun azimuth can be striking (Fig. 16.17).

The second type of evidence of prehistoric features consists of soil markings. These occur where the soil is bare or has only the thinnest covering of vegetation. The disturbed soil along field boundaries or foundations is different in structure, texture and colour from that of the undisturbed adjoining soil. These differences are often sufficient to induce different reflectances from the surface, which can be detected by remote sensors. In England some of the greatest contrasts have been seen on chalk and limestone soils. Usually soil markings are most clearly detected after a period of exposure to wind and rain rather than immediately after ploughing.

Some archaeological sites can be identified best as a result of transient surface climatic conditions. For example, dark bands of bare earth may reveal where snow or frost has melted first above the filling of buried ditches. A similar effect is sometimes seen where the lines of buried foundations (or perhaps more probably of trenches where the foundations have been dug out and robbed for later buildings) are picked out by early melting of hoar frost on grass.

The patterns of ancient field systems are often incorporated into present-day field patterns. It is not always easy to distinguish old field boundaries from maps but aerial photography can often provide startling pictorial evidence for such fea-

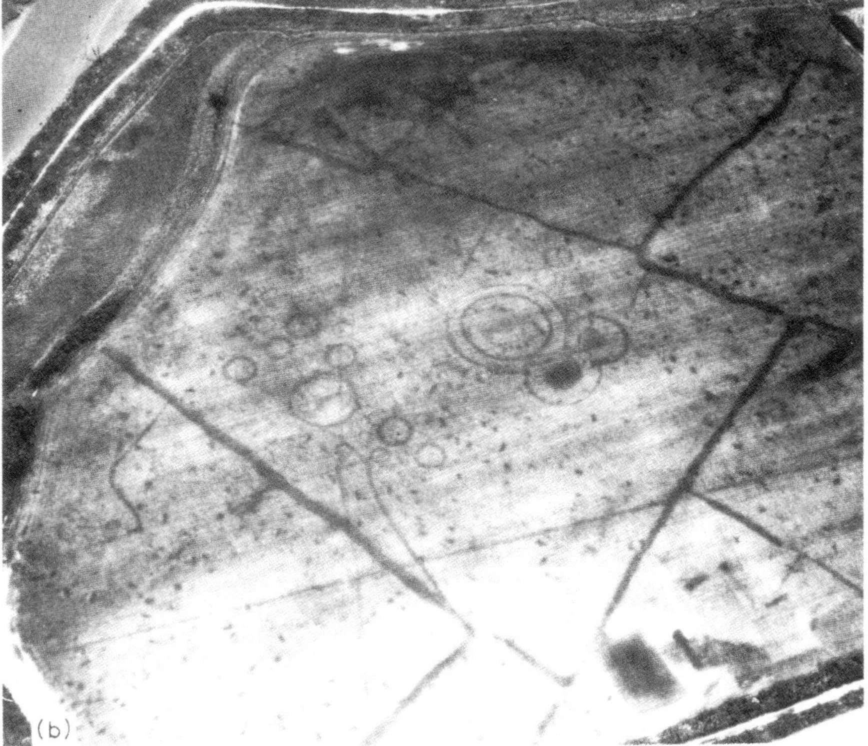

Fig. 16.17 Crop marks of ring ditches at Lawford (Essex), June 1970: (a) looking into the sun; (b) with the sun. Tonal contrasts are entirely due to direction of view since photographs were taken within a minute of each other. (Source: University of Cambridge; Copyright Reserved.)

Fig. 16.18 A pattern of medieval fields in ridge and furrow near Husbands Bosworth, Leicestershire. Note how the sun angle casts shadows that serve to emphasize small changes in surface relief. (Source: University of Cambridge Collection; Copyright Reserved.)

tures. For example the Roman field systems often can be observed in great detail. The Roman method of land partition was to allocate blocks of land which were usually in the form of squares (the centuria quadrata) of 710 m size. High-altitude photographs show the grid pattern of centuriation well and in some cases lower altitude observations allow detection of earlier Greek systems to be identified. Similarly the patterns of ancient strip fields can be mapped from remote sensing data (Fig. 16.18).

In this brief discussion of the use of remote sensing for the study of prehistoric land use attention has been focused on the photographic sensors. However, the use of infrared systems, on both aircraft and satellites, has also provided useful information of an archaeological nature, the latter particularly in broad-area reconnaissance-type studies of areas not yet investigated in detail on the ground. Also there is growing interest in the potential of airborne active sensor systems for such purposes, e.g. using synthetic aperture radar designed originally by Jet Propulsion Laboratories in the USA for imaging the surface of Venus. When flown at an altitude of about 7 km over parts of Belize and Guatemala it was able to penetrate the dense jungle canopy to reveal widely distributed patterns of intensive ancient agriculture, leading to a new perspective on classic Maya civilization. All the major satellite-borne imaging systems discussed elsewhere in this book have yielded data of interest to archaeologists, although their interpretation for such purposes is still in its infancy. Imagery from the Landsat MSS and TM sensors, SPOT, the Metric Camera Large Format Camera and MOMS, and SIR-A and SIR-B have all begun to support new archaeological research in areas such as the San Juan Basin in New Mexico (archaeological site densities), parts of Alaska (areas of potential archaeological interest in advance of petroleum exploration), Iraq (ancient Mesopotamian canal systems), Egypt (relations between ancient sites and megalithologic units), and the eastern Sahara (buried river systems and associated settlement zones).

Satellite archaeologists anticipate particularly keenly the data expected from the Space Shuttle's SIR-C radar, which will experiment with multiple polarizations and frequencies. Analysed on interactive image processing systems and combined with other types of archaeological data, both traditional and new, such advances should greatly expand our knowledge and understanding of our heritage not only in respect of rural settlements and land use, but also of the built environment, which is the subject of the next chapter.

17 *The built environment*

17.1 General considerations

We remarked in the opening chapter that our environment is partly natural and partly of our own making. Although we have already considered some aspects of our imprint on the surface of the Earth, this imprint is most complete in what we may call the 'built environment'. Whether our focus of attention is a single structure in a rural setting, or some great conurbation, it is here that we find the most unnatural features in the world – but also some of the most important for us all, because it is here that we live, and from here that we organize our use or manipulation of our surroundings.

Since the scales of unitary features in the built environment are smaller than most of those in the natural environment which satellites have been designed to identify, satellite remote sensing is still, at present, less helpful in this connection than the more traditional remote sensing approaches of air-photo interpretation and photogrammetry. However, we shall see that significant use has been made already of Landsat and SPOT imagery for analysis and assessment of elements and patterns within the fabric of urban and industrial areas. At even more detailed spatial scales, the relatively recent recognition that spectral reflectance curves differ from area to area within urban areas (Fig. 17.1) is opening up some important new uses of multispectral airborne sensor data, both in terms of

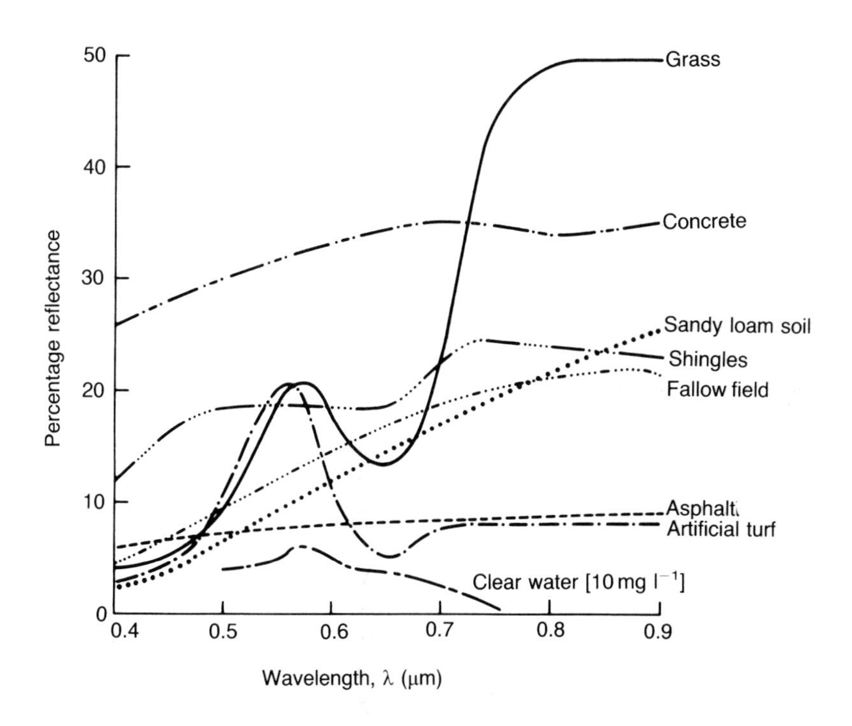

Fig. 17.1 Spectral reflectance curves for selected urban phenomena. The spectra are averages of results discussed in numerous studies. In all cases, data were obtained using a spectroradiometer. (After Jensen, in American Society of Photogrammetry, 1983.)

Fig. 17.2 Bristol airport and surrounding features, 20 April 1975. (Courtesy, County of Avon.)

urban land use mapping, and the evaluation of very local patterns and changes in urban climates, with numerous practical applications. Doubtless, both Earth resources, and even environmental, satellites will play an increasing part in monitoring of the built environment in the future, with growing practical significance for such disciplines as urban geography, urban and rural planning, and transport studies.

This chapter is divided into five parts: the first three are concerned with built environments of increasing degrees of artificiality, namely in rural, urban and industrial contexts; the fourth is a different kind of overview concerned with population, its distribution, and its change; in the fifth, references will be made to the ways in which remote sensing can be applied beneficially in civil engineering, i.e. in helping to bring built structures into being, and sustain them safely.

17.2 Rural structures

Since we have already discussed rural land use in Chapter 15, and much of this involves open (i.e. unbuilt) countryside, our present focus must be on settlements themselves, plus some specialized and often quite distinctive aggregates of features, e.g. airports (Fig. 17.2). As in other contexts in the present chapter, remote sensing is a valuable tool for rural survey purposes because it permits relatively cheap, rapid and repetitive mapping even in areas to which access on the ground might be difficult or dangerous. It is convenient to differentiate between a number of contrasting geometrical components of the rural landscape. These are:

1. point features, e.g. dwellings, storehouses and animal sheds;
2. linear features, e.g. field boundaries, pipes and power lines, roads, railways and canals;
3. area features, e.g. settlements (hamlets, villages and small towns), airports, military establishments, large schools, hospitals and penal institutes.

In the case of point features remote sensing surveys may be carried out as aids to topographic mapping. If repeated later, air photography or Earth resources satellite imagery may be analysed to reveal rural change, perhaps for interpretation in economic terms. In this respect a relevant question may be this: Is there evidence of changing rural wealth, e.g. through increasing dereliction of homes and out-houses, or through new or improved rural structures?

Fig. 17.3 A clear example of an administrative boundary (the USA–Canadian border), which is revealed in Landsat imagery through associated differences in land use and farming practice, except where topography overrules. To the north of the border lie the grassy central Great Plains of Alberta and Saskatchewan; to the south, the agricultural lands of Montana, where small grains are grown by strip farming methods to reduce soil erosion. In the west the forested mountains of Sweetgrass Arch appear more 'Canadian' than 'American'. (Courtesy, Earthsat Corporation, Bethesda, Maryland.)

In the case of linear features, such as field boundaries, different sizes or patterns of fields may be found in relation to ethnic and even national differences because of differences in associated cultures and rural practices (Fig. 17.3). Or field patterns may evidence spatial variations in terrain and/or soil characteristics. Significant changes in field boundaries may result from the implementation of land reform policies, especially after a change of government or for more complex agro-economic reasons, such as those responsible for the EEC's 'set aside' policy initiated in the late 1980s and destined to have a growing influence on the use of rural land across much of Western Europe. Remote sensing surveys provide the most convenient means whereby checks can be made on the progress, and effects on the landscape, of any widespread policies of agrarian reform.

The use of remote sensing in respect of existing communication lines (roads, railways and canals) is more restricted, for most principal routes have always been mapped in detail and usually with considerable precision. Furthermore, alterations to these networks are usually controlled by governments or corporations answerable to the general public. Therefore appropriate changes can be incorporated into topographic and route system maps without recourse to additional data sources. However, remote sensing is fulfilling an increasingly vital role in respect of topographic surveys for new road and rail construction projects. Such surveys are widespread, and increasing, especially in developing countries where there is a particularly high premium on a combination of survey accuracy and speed. Even in developed countries, multispectral air photographs and infrared linescan data have long been used by commercial remote sensing survey companies to appraise ground conditions for road and rail planning purposes.

Using remotely sensed data in conjunction with established *terrain evaluation* procedures it is possible to assess many factors affecting the cost of alternative alignments and types of construction methods. For example, in conjunction with ground truth from selected sites, the evaluation of surface conditions in key localities can be extended along the proposed construction line(s), taking special account of features having practical significance (e.g. slope stability, weathering, erosion, hydrology, etc.). Associated inventories of indigenous materials for road or railroad construction also may be vitally important if the final alignments are to represent the most advantageous and economic routes permitted by the terrain.

The use of aerial photographs and even satellite imagery in supporting public utility projects too is more than a future dream. For example, aerial photographs have been used successfully for many years in the development of water distribution services: such photographs provide quick indications of the numbers of premises to be served, the distances involved, and types of land use in the service area. Sewage collection is also efficiently planned thereby. Remote sensing provides perhaps the most practical and economic method whereby gradients can be assessed, the layout of pipelines planned, and pumping stations and force mains appropriately sited. Photogrammetric surveys are always conducted in the USA and other developed countries for the laying of long-distance pipelines and high voltage transmission lines. In the UK, as elsewhere, the electricity industry regularly checks the condition of major power lines by helicopter-borne infrared linescan. Microwave transmission towers also can be sited best by such means, and their heights selected to meet the necessary requirements for line-of-sight arrangements.

Lastly, in this section reference must be paid to area features, most commonly in the form of small rural community settlements. These may be mapped from air photographs, e.g. in exploratory surveys of regions as yet unexplored on the ground (e.g. in tropical rain forest regions such as Amazonia and New Guinea), prior to initial attempts to contact primitive tribes. Similar surveys may be repeated to build up realistic pictures of movements of rural populations still practising shifting cultivation, e.g. in humid tropical forest environments. Radar imagery may be a useful supplement to visible and infrared photography for such purposes, for there is a high incidence of cloud cover in the tropics, and it has been shown that radar images of forested areas may possess internal textural variations that result from a 'slash and burn' economy. Indeed, radar is being seen increasingly as a valuable system for delimiting and tracing many artificial features in the rural environment. In rural areas, many frequently recurrent characteristics of elements of the built environment (e.g. flatness, sharpness of corners, and unusual material composition) combine to yield stronger radar echoes from artificial structures than from more natural features of the landscape.

The use of radar can be particularly effective in rural areas when the principles of *polarization* and *cross-polarization* are exploited (see also Chapter 4). We have already seen that all electromagnetic waves are *polarized*; that is to say, once propagated, they continue to move at a given angle measured against a standard plane of reference – unless some outside force or object changes that angle. When radar waves are transmitted horizontally or vertically and are received at the same angle or polarization, they are termed *like-polarized*. If they are received at a different angle, they are said to be *cross-polarized* (section 4.4).

In general, cross-polarized signals produce grainier images than the like-polarized, thus limiting the usefulness of the former. However, under certain circumstances some objects in the rural environment are revealed more clearly in the cross-polarized imagery. For example, detecting and tracing communication nets is performed most easily, completely and accurately using cross-polarized imagery if the net traverses the flight path. On like-polarized imagery the best results are obtained when communication net components are parallel to the direction of flight. Therefore, the most efficient system seems to be one in which both like- and cross-polarized components can be assessed together. The most useful and obvious applications of such a system are in mapping in developing countries. However, it is also possible that, with suitable improvements and refinements to such schemes, urban road systems may be investigated with special reference to their surface materials, an application of considerable practical value to many developed countries also. It is to urban areas that we may now turn our attention in greater detail.

17.3 Urban areas

Whilst urban centres afford the greatest concentrations of facilities for living and working, they also present man with a wide range of attendant problems. These most notably include *internal* ('intra-urban') problems concerned with the provision of acceptable residences, transportation networks, and places of employment, in conditions of suitable freshness and cleanliness, and *external* ('inter-urban') problems concerned with the maintenance of acceptable relationships between one city and another, and between the cities as a group and the intervening countryside. It has been suggested that many of the contemporary problems of cities may be traced to:

1. the reasons for city foundation and growth, and the different histories of city evolution;
2. the ongoing, competitive and conflicting processes of arrangement and rearrangement of land use and functional areas within cities;
3. the scales, complexities, and patterns of concentrations of intra-urban activities.

Today there is widespread interest in the development of 'urban information systems' for the collation, interrelation, and practical utilization of urban data from a wide variety of different sources. Urban geographers and planners already recognize remote sensing data as a vital part of the total information pool. Here remote sensing data contribute to two classes of information in particular. These are as follows:

1. Static phenomena. These include such things as city size; the number, pattern and capacity of roads; building sizes and types; and the characteristics of types of neighbourhoods (e.g. industrial, residential, commercial).
2. Dynamic phenomena. These include variables that cannot be observed directly, either because they change so rapidly, or because they are not physically visible, e.g. population statistics, (aggregated) traffic flow patterns data, and socio-economic conditions.

We will now illustrate and exemplify the value of remote sensing as a source of valuable data for urban information systems by reference to a number of different aspects of the urban environment.

17.3.1 Urban spatial structure and setting

Here the interest focuses on the way in which cities are laid out, and how they are related both to each other and any smaller intervening communities. Cities can be analysed in terms of hierarchies, which take account of the rank orders of cities, and the distances between them on the ground. Small-scale satellite imagery can be used to establish general hierarchical relationships between cities and towns. For example, night-time visible imagery from the low-light intensifier camera system on some of the early Defense Meteorological Satellite Program (DMSP) weather satellites of the early 1970s have been interpreted in this way over areas of subcontinental scale. More recently, major physically related features of contrast between cities and their surroundings have been investigated through analyses of satellite infrared images, for example the 'heat island' patterns that develop with the growth of cities as increasingly well-known features of urban climates.

17.3.2 Urban subregions and the internal structures of cities

Here the purpose is to identify, classify, and evaluate important component features and areas within each city in order to understand better their roles, functions and associated activities. Common steps in analysing the subregions of a city may be summarized as follows:

1. *Locating* the major features, structures and areas with basically similar characteristics and appearances. These may include the Central Business District, interior and suburban residential communities, inner and suburban commercial centres, industrial complexes, transportation hubs, etc. Air-photo interpretation has long proved useful in these respects. Today, increasing use is being made of airborne radar, and Landsat and SPOT image data for such purposes.
2. *Classifying* major features, structures and areas

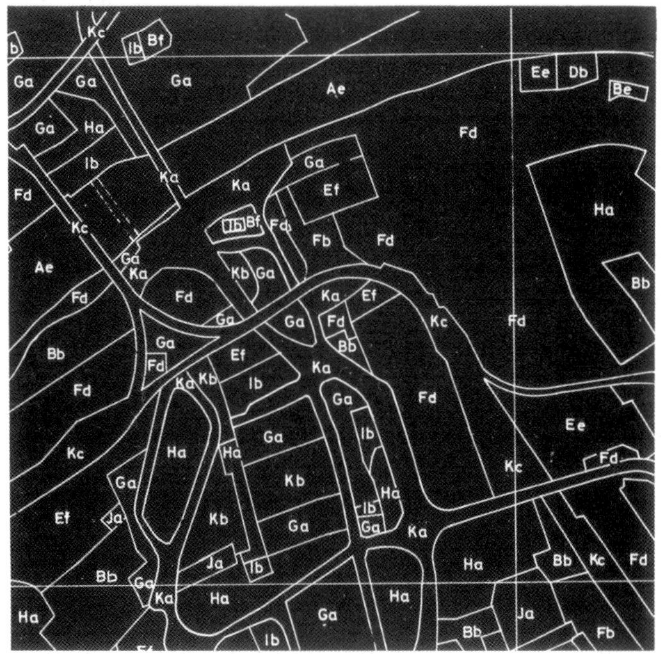

Fig. 17.4 An example of aerial photograph interpretation of an urban area: the centre of Newcastle astride the River Tyne. (Courtesy, Fairey Surveys.)

into a suitable number of categories for systematic mapping. Figure 17.4 exemplifies urban mapping from aerial photographs. Supervised computer classifications based on Landsat MSS data are being developed and applied widely at the present time, as illustrated by Fig. 17.5, and increasingly today, in conjunction with urban geographic information systems (GIS).

3. *Counting and/or measuring* important static and physically visible attributes of the cityscape

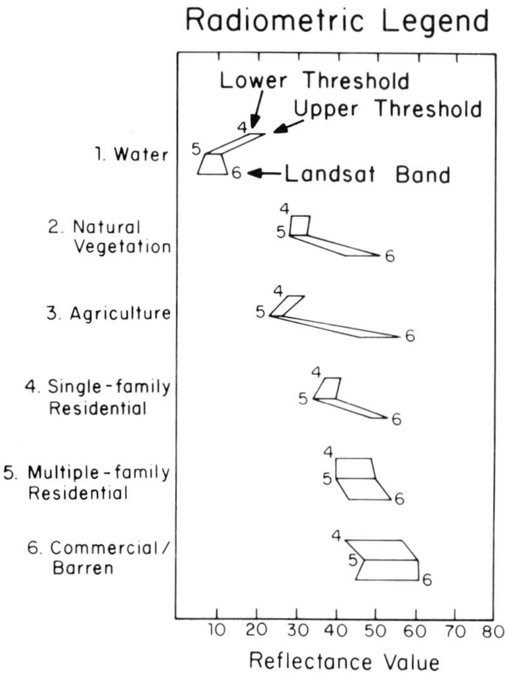

Fig. 17.5 Radiometric interpretive legend developed using training statistics obtained from the 4 May 1974 Landsat image of Goleta, California. Six classes of land use are investigated. (Source: Jensen, 1979.)

(e.g. size and area of subregions and subareas, numbers and dimensions of buildings and transportation systems elements).

4. *Generating and collating data* on dynamic and other urban phenomena for which the image data analyses are inferential, not direct. Such information 'surrogates' are used, for example, in large-scale air-photo and radar image data analyses for the differentiation of residential areas in terms of socio-economic class, population densities, and associated quality of life.
5. *Deriving information and calculating indices* to represent dynamic urban features for which clear visible evidence exists, e.g. car parking distributions and habits, availability of land for development, etc.

17.3.3 Transportation systems

Aerial photography has been used successfully for many years in the field of transport studies. In the USA, for example, a survey of major highway organizations has revealed that nearly three-quarters of them have made use of aerial surveys in highway planning, although only one-quarter have used them 'extensively'. Four types of studies have been undertaken, namely:

1. Road planning. Photogrammetric analyses of air photographs have provided engineers with both qualitative and quantitative data, especially in the early stages of highway routing and design. As we saw earlier, in some countries sources of road-building materials have been located from the air.
2. Traffic studies. Aerial photography has been used to pinpoint areas and causes of traffic congestion, and to provide information on traffic flow, both for future road design purposes and immediate ameliorative action. Closed-circuit television is used on some urban motorways, for example on the Chiswick Flyover in West London, to assist speedy breakdown and recovery operations and help maintain free movement of traffic. Remote camera systems are being used increasingly to record traffic offences and identify offenders both at important junctions and on busy sections of motorways.
3. Parking assessments. Aerial photography can reveal where the heaviest concentrations of automobiles tend to build up, and where additional parking spaces might be most beneficial.
4. Highway inspection. Air surveys are a quick and convenient means of assessing the states of road surfaces that may be in need of repair. In winter the successfulness of snow removal may be adjusted in a similar way.

Although present data from Landsat and SPOT are generally rather coarse for transportation studies, the applicability of spacecraft observing systems to transportation geography and linkage analysis has been carefully assessed. The types of transportation facts ideally required have been identified as follows:

1. Network information. Ideally this necessitates infrared and colour photographic systems, chemical devices and multispectral sensors, capable of providing data with resolutions of between 1 and 10 m. From such data, maps of the physical linkages and terminal facilities in a network could be compiled. Pattern analyses by

computer processes could be carried out once the information had been transformed into matrix and vector form.
2. Flow phenomena and associated problems. Here infrared and panchromatic photography and radar could be used, giving an optimal resolution of 0.5 m. Matters commensurate to investigation would include origin–destination patterns and daily 'tidal' flows of traffic as a whole. These would have useful applications in road design and traffic control.
3. Transport/land use interrelationships. These necessitate infrared, colour and panchromatic photography, and chemical sensing devices particularly sensitive to phosphorus and nitrogen. The optimal resolution would be 1 m. Matters such as the relationships between land use intensity and distance between road and rail links, land use capability, and the positions of areas within their larger economic regions should be amenable to study.

Clearly the ranges and resolutions of remotely sensed data generally available at present are not yet adequate for the achievement of such aims. Nevertheless the list indicates something of the potential of the satellite in transportation studies – as well as some of the hopes and aspirations of would-be data users, which are important stimuli for further advances in remote sensing technology, science, and operations.

17.4 Industrial complexes

In view of the great economic, social, and political significance of manufacturing and extractive industries, and the dominant role they play in many urban areas, we may consider these in more detail, both for their own sakes and in order to exemplify in greater detail methods of application of remote sensing data to the built environment. Three aspects of the industrial components of urban areas may be elucidated most usefully by aircraft or satellite imagery. These are:

1. The present location of industry, and different types of industrial land.
2. Heat loss and the spread of atmospheric and water pollutants away from an industrial complex.
3. The opportunities that exist for the establishment of new industries, i.e. for the redevelopment of land under existing industrial use.

Historically, the analysis of industrial activity from remote sensing imagery can be traced back to World War 2, when air photography was used for the identification of tactical military targets and the selection of military objectives for aerial attack. More recently, various types of airborne and spaceborne sensor data have been used to support the offensive actions of the Coalition Forces in the Gulf War of 1990–1991. Today, the analysis of remotely sensed imagery of industrial areas has many peaceful uses also, not only in mapping industrial areas and their changes through time, but also in associated activities, such as updating inventories of stock-piled materials.

Aerial photography is still the most commonly used remote sensing imagery for industrial analysis: usually high-resolution data are essential for the adequate interpretation and assessment of features of the industrial landscape. Aerial photography of industrial areas is frequently recorded at scales larger than 1 : 10 000, even in some cases larger than 1 : 5 000. Colour film is sometimes used in preference to panchromatic film on account of the higher definition it provides. Often photo-interpretation keys for use in industrial areas are very detailed and technical; their compilation is a highly skilled business, requiring intimate knowledge of industrial structures and processes.

Stock-pile inventories may be made more accurately and expeditiously from air photographs than from ground where raw materials or finished goods, such as coal, ores and other minerals, pulpwood and lumber are accumulated in the open air. Neatly stacked materials can be assessed very accurately, albeit only from air photographs at scales of 1 : 1000 or better. Where stockpiles are of less regular shapes more variance is to be expected in the periodic estimates. However, some experts claim to be accurate within 2%, and assert that the lowest cost ground estimates have an average accuracy of only 15%. The same experts emphasize that air survey costs can be substantially below those of surveys on the ground. Since both methods follow essentially the same set of procedures, they can be used together or interchangeably. It is commonplace to:

1. map the main stockpile with closely spaced contours (often only 1 or 2 m apart);
2. map the tops of the piles (whose areal dimensions are smaller) in greater detail;
3. determine stockpile volumes by planimetering areas between successive contours and multiplying by the depth of material;
4. convert the cubic unit volumes into weights, using tables for specific minerals;
5. adjust the initial estimates of weights using compaction factors to allow for settling of material in the pile(s).

Similar kinds of measurements can be made in support of extractive industries, e.g. of the amounts of material removed from an open-cast mine through selected periods of time. The resulting data can be of great value to mining companies, and, potentially, to governments seeking an accurate, low-cost means of monitoring mineral industries for taxation on a weight basis. Prompted by the major fluctuations in global oil prices that have taken place in recent years much interest has been shown in the use of infrared linescan imagery for the mapping and assessment of heat loss from buildings (Plate 13, colour section). Poorly insulated buildings are often quite clearly evident even in the raw infrared image products. By thermographic mapping, values can be placed upon the rates of heat-loss through every structure covered by a survey. Many commercial remote sensing firms today provide airborne heat-loss survey services. Indeed, these are widely used not only by industrialists, but also by bodies such as central and local government, housing authorities, hospital management boards, etc.

When we broaden the discussion to embrace the whole field of environment pollution monitoring, it emerges that a wide range of remote sensing techniques have now been tested in the search for efficient methods to measure the concentration and spread of pollutants. Examples include:

1. Visible waveband photography. This has been used for such studies as the behaviour of chimney plumes and the appearance and intensification of urban, grassland and forest fires.
2. Infrared imagery. This has been used as the basis for studies as diverse as the monitoring of heat-loss from buildings, forest fire detection, stubble burning, and the delineation of warm effluents in rivers, lakes or the sea.
3. Microwave techniques. These are increasingly being used to map oil spills in coastal and deeper waters.
4. Multispectral processing techniques. These can be used to assess water quality through the different levels of light penetration in selected regions of the thermal emission spectrum.

In general, it has been suggested that remote sensing of such aspects of environmental quality in and around urban centres and elsewhere may have two significant contributions to make: first, through the provision of a more complete spatial inventory of areas of atmospheric and water pollution to a single standard throughout a nation, or indeed, the world; and second through the determination of regional, national and global burdens of pollution.

Obviously the wider of these developments can be contemplated only with satellites, not aircraft, as the sensor platforms. Experience with Earth resources satellites has produced encouraging results in the following contexts:

1. Air pollution. Satellite remote sensing can detect particles emanating from both point (e.g. industrial) and mobile (e.g. aircraft) sources.
2. Water pollution. Large-scale patterns of turbidity in rivers and oceans can be mapped using the types and resolutions of data supplied by these satellites. Water pollution from domestic, municipal and industrial sources also can be observed and differentiated.
3. Land pollution. Large-scale landscape problems can be identified, including strip mines, tailing piles and dereliction. Using the SPOT system, even quite small areas of solid waste disposal can be identified.

We turn lastly to the social needs of redeveloping old industrial and related land. Much old industrial land is rather poor in quality, often as a direct result of its exploitation for industrial purposes. In many cases it is not sufficient just to know the extent of such areas. The precise nature and quality of the land may be equally vital in influencing the new use to which it might be put.

In the UK, for example, studies of derelict industrial land have been made in the West Riding of Yorkshire (Table 17.1), using aerial photo-

Table 17.1 An air-photo interpretation scheme for the identification and mapping of derelict land in the UK using panchromatic photographs (Source: Bush and Collins, 1974)

No. in air-photo key	Code in derelict land key	Brief description of derelict item
1r	A1ia	Ridge tip
1f	A1ia	Low flat tip
1c	A1ia	Conical coal tip
2	A1ib	Coal dump
3	A1if	Degraded land above ground level associated with coal mining
4	A1ig	Degraded land above ground level peripheral to a coal mine
5	A1ij	Coal sludge above ground level
6	A2ia	A tip of domestic refuse
7	B1ic	Open-cast coal workings
8	B1ih	Open-cast coal workings not yet 'excavations'
9	B1iic	A dry brick clay quarry
10	B1iid	A wet brick clay quarry
11	B1viic	A dry sand and gravel excavation
12	B1viid	A wet sand and gravel excavation
13	B1viie	Degraded land below ground level, resulting from sand and gravel workings, but partially restored
14	B1viih	Sand and gravel workings not yet 'pits'
15	C1ib	Coal dump site at ground level
16	C1ie	Degraded land at ground level, resulting from coal workings
17	C1if	Degraded land at ground level, associated with coal mining
18	C1ig	Degraded land at ground level, peripheral to a coal mine
19	C1ij	Coal sludge at ground level
20	C1viif	Degraded land at ground level, associated with sand and gravel workings
21	C1viij	Sand and gravel sludge at ground level
22	C3if	Degraded land at ground level, associated with a brickworks
23	C3vj	Power station waste at ground level
23a	B1i/3v	Power station waste used to fill a coal excavation
23b	B1vii/3v	Power station waste used to fill a sand and gravel excavation
24	C3vij	Sewage sludge at ground level
25	C4iiif	Railway dereliction at ground level
26	D3i	A disused brickworks
27	11i	A coal mine
28	11vii	A sand and gravel works
29	13i	A brickworks
30	13v	A power station
31	13vi	A sewage works

graphy at a scale of 1 : 10 500. Field checks revealed that a very high degree of accuracy was achieved in the identification of different types of derelict areas within the study area, which measured 200 km^2. The detailed stereoscopic examination of the prints took approximately 15 h. In contrast 15 employee weeks would probably have been required for a field survey to collect and map the same information. There is much scope for such studies in the industrialized world, especially where land reclamation and redevelopment is proposed.

17.5 Demography and social change

17.5.1 Population studies

An important use of remote sensing data today is in *population estimation*. In developing countries, such

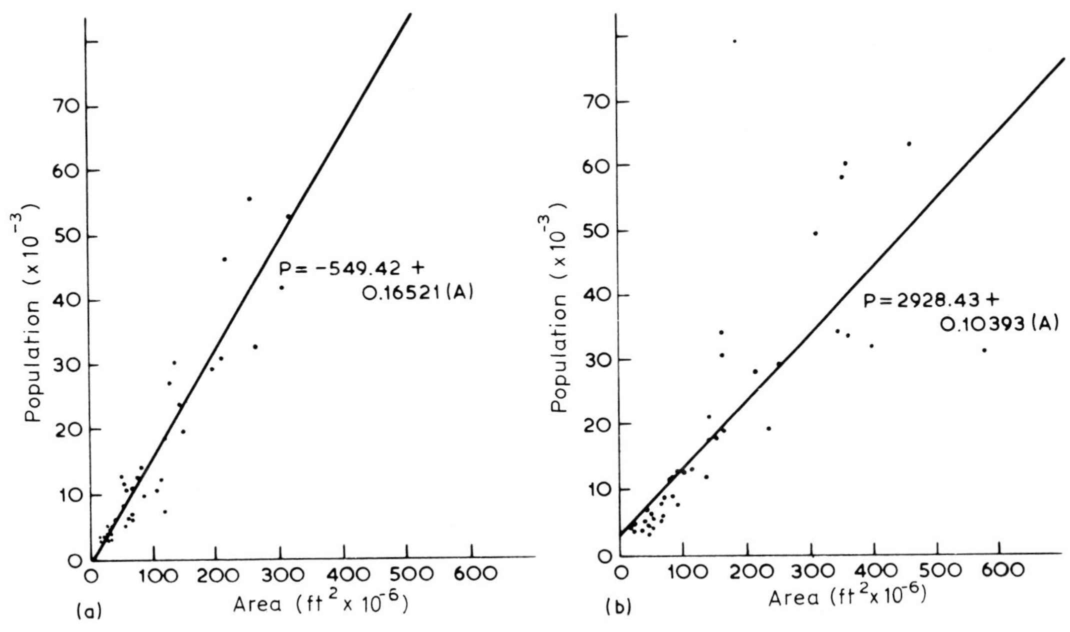

Fig. 17.6 Observed relationships between population and area of towns in the Tennessee River Valley, in (a) 1953 and (b) 1963. (Source: Holz *et al.*, 1969.)

as India and Nigeria, where full censuses may be difficult to organize, use has been made of air photographs and Landsat imagery for classification of both rural and urban settlements in terms of types and densities of dwellings so that more complete estimates of population can be made. In the developed world, it has been shown that remote sensing data can be used to provide valuable information on population changes in intercensal periods. A pioneering study of this kind was undertaken in the 1960s in the Tennessee River Valley region for which aerial photographs were available for 1953 and 1963. Four hypotheses were treated:

1. the population of an urban area is positively related to the number of links it has with other urban areas;
2. the population of an urban area is positively related to the population of the nearest larger urban area;
3. the population of an urban area is inversely related to the distance to the nearest urban area;
4. the population of an urban area is proportional to the observable area of occupied space of such a population.

These hypotheses were tested using stepwise linear regression. It emerged that, except for urban areas, the order of the independent variables differed from time to time, the values of the various coefficients differed significantly, and for one variable in particular (the distance to the nearest larger urban area) reversals of the direction or sign of the coefficient occurred. However, notwithstanding these complications, the multiple correlation coefficients between actual and estimated population for 40 selected central places were 0.95 for 1953 and 0.88 for 1963. These indicated that over 91 and 77%, respectively, of the variations in population of the urban areas is explained on the bases of the selected independent variables. Because of the high correlation between urban area and population (Fig. 17.6) the inclusion of the remaining independent variables adds little to the explanation of the variables that remained.

Other studies in the USA have attempted to estimate population densities and totals for small areas within larger urban centres, using multispectral air photographs and ground truth from sample areas. One such model was applied to Washington, DC in 1970. It was formulated as:

$$Y = f(x_1, x_2, \ldots, x_n) \qquad (17.1)$$

where Y represents housing unit counts or population (1970 tract statistics) and $x_1 \ldots x_n$ are

Table 17.2 A comparison of housing areas in Austin, Texas, based on selected environmental criteria amenable to identification on remote sensing imagery (Source: Davis *et al.*, 1973)

Criteria	Low-income areas	Middle-income areas
House size (ft^2)	380–1220 (average 731)	110–1560 (average 1305)
Placement of house and lot (distance from street in feet)	12–42 (average 27)	34–45 (average 40)
Potential landholding per housing unit (ft^2)	5670	7500
Building density (%)	14.2	17.7
Average lot size and frontage (ft^2)	4337	7376
Image, pattern, and texture	Narrow frontage Lack of uniformity Irregular	Wide frontage Uniform Regular
Houses with driveway (%)	8.3	97.0
Houses with garage (%)	3.0	97.0
Number of visible automobiles per house	0.20	0.76
Unpaved street (%)	65	0
Street width (ft)	12–24 (average 18)	24–31 (average 28)
Quality of curbing	Generally lacking	Intact
Quality of vegetation	Not as vigorous	Cultivated, vigorous
Housekeeping	Presence of debris	Lack of debris
House orientation to street	Short side	Long side
City block pattern	Irregular, dead end twisting streets	Regular
Vacant lots per city block	0.6	0
Proximity to manufacturing and retail activity	Close to manufacturing Remote from shopping outlets	Remote from manufacturing Close to retail outlets

imagery derived variables, such as the number of single-family structures, the number of multiple-family structures, distance from the central business district, etc. Relationships between x and Y were developed by multiple regression analysis.

In practice, residential land use was identified on 1 : 50 000 scale aerial photographs, and the residential land area was computed. A block-by-block count was then conducted to establish the number of dwelling units per structure according to a four-category classification (single-family housing units, 2–5, 6–14, and more than 15 housing units). Multiple regression coefficients of 0.65 and 0.54 were obtained for central city and suburban tracts, respectively. These were encouraging, but pointed to the need for a number of improvements before such a scheme became operational. These included the provision of remote sensing imagery of a type in which tree cover poses less of a problem for the location and classification of housing units, plus a greater range of ground truth data for the formulation of the model and the checking of its output.

We may conclude that remote sensing, even at a coarse scale, can clearly provide valuable information concerning population levels and changes in most areas of the world. The application of aerial photography to dwelling-unit and population estimation would appear to have a particular potential in developing countries where rates of urban and population growth are especially great. Demographic data in such countries are frequently less than adequate for planning purposes, since national censuses are difficult to undertake and often yield inaccurate results.

A partly related question of considerable significance to government and municipal authorities is that of *housing quality*. For example, concerted attacks on the problems of urban poverty neighbourhoods cannot be planned – still less carried out – until such areas have been defined, and located on the ground. Conventional methods of amassing data on housing quality distributions are extremely time-consuming. Once again remote sensing techniques may provide new and better data than surveys on the ground, while providing such data more quickly and frequently.

Table 17.2 provides a comparison of housing areas in Austin, Texas, based on selected environmental criteria. Most of these features may be assessed from remote sensing imagery under normal viewing conditions. We may appraise housing quality, study certain socio-economic aspects of a neighbourhood, and estimate the level and distribution of family income on the basis of such a range of urban characteristics.

17.5.2 Changes in the built environment

We have seen how remote sensing systems may be exploited for both rural and urban land use detection; without doubt, remote sensing data are of greatest value when analysed for *landscape change*. It should be noted, too, that because such data are unselective (given a particular image type with its attendant spectral and spatial resolutions), the remote sensing view of a built environment reveals it as a single system, organically related to the open land within and around it. Since cities are the most powerful built environments we may conclude this section with reference to a range of methods and projects designed to monitor their growth and changing impacts on their own environs. Although remote sensing data can be analysed manually to provide rough estimates of urban growth and associated landscape change there is a trend towards the development of increasingly sophisticated automatic interpretation procedures for such purposes.

One early and ambitious project to evaluate the usefulness of photography from high-altitude aircraft and Earth-orbiting satellites for urban land use change detection was the Census Cities Project. This was organized by NASA in conjunction with the Geographic Applications Program (GAP) of the US Geological Service (USGS). Originally 26 cities were named as test cases, and the US Air Force Weather Service and NASA's Manned Spacecraft Center acquired multispectral, high-altitude photography from 20 of them. The year chosen for this survey was 1970, the time of the 10-year US Census. The basic remote sensing data were colour infrared photographs. This film type is especially valuable for use over towns on account of its high haze-penetration capability. A further useful characteristic of this type of film is that it distinguishes clearly between vegetation (reddish in colour) and cultural features (which have a blue appearance on the film). This makes it especially useful where cities have a strong urban structure/woodland mix. The ultimate goal of the Census Cities Project was the production of an Atlas of Urban and Regional Change, accommodating many types of data presentation, including photomosaics, conventional maps, computer-printed maps, tabulated data and text. In a pilot study for the city of Boston, a 24-category land-use classification was used. The minimum cell size was about 10 acres. The land use data were computer-processed to make them compatible with the 1970 census data. Consequently they can be retrieved either singly, or in selected combinations, by the census tracts.

On a smaller scale, Landsat imagery has been processed to give land use maps of urban changes at scales from 1 : 250 000 upwards. One such programme brought together the Planning Intelligence Department of the UK Department of the Environment (DOE), and the Image Analysis Group at the UK Atomic Energy Authority (UKAEA) at Harwell, Berks. For planning purposes, land use information is needed especially around urban areas, but, at present, there is still no single source of comprehensive data in the UK which can provide up-to-date information of this kind on a national scale. Consequently studies have been undertaken to investigate the potential of Landsat as a suitable information source. Using imagery obtained from the European receiving station at Fucino, Italy, a NASA software package has been further developed for quick and efficient geometrical rectification. Each Landsat scene is computer processed using a multicategory supervised classification in multidimensional measurement space. To verify the suitability of Landsat MSS data for monitoring urban growth, four test areas (each about 35 km^2) covering a wide range of conditions were examined in detail. The accuracy of computer classification of the Landsat data was checked against ground truth obtained from aerial photography and Ordnance Survey maps. Taking Northampton, an important English Midlands

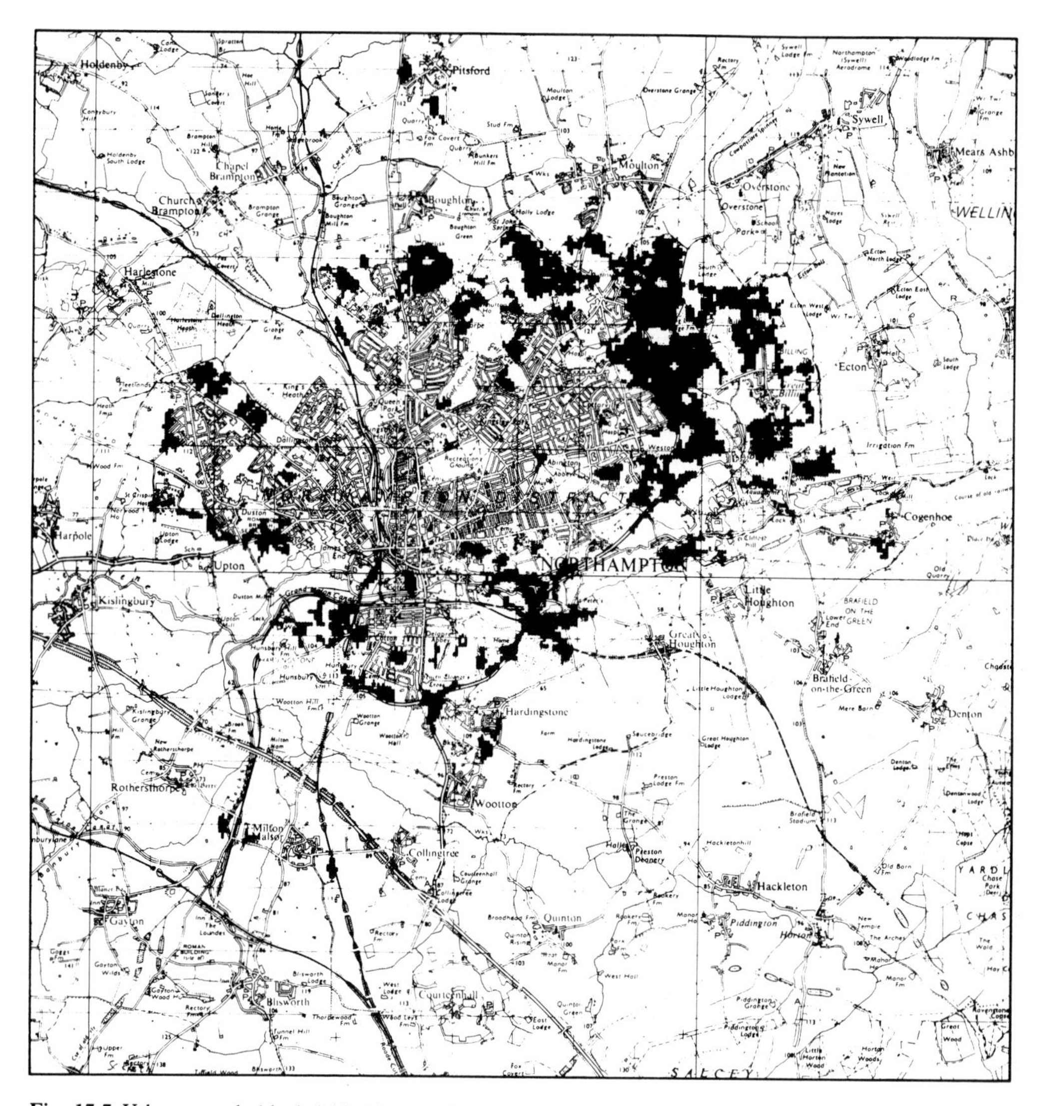

Fig. 17.7 Urban growth (shaded black) around the English Midlands town of Northampton, beyond the 1969 developed area boundary, as determined from Landsat (1975) imagery. (Courtesy, UK Atomic Energy Authority and the Department of the Environment.)

town (population 130 000) as an example, over 90% of new development that took place from 1969 to 1975 was accurately identified (Fig. 17.7).

More recently still it has been shown that urban areas can be classified and mapped with even greater success using the higher spatial resolution data from the Landsat TM and SPOT multispectral or panchromatic systems. However, it is clear that greater accuracies would be achieved if ground survey classifications were better suited to the characteristics of the satellite data; traditional land use classifications cannot always be applied very successfully to either supervised or unsupervised classifications of satellite imagery.

However, information obtained from satellite studies can be compared and integrated easily with digital data of other kinds (e.g. from population censuses) for many central and local governmental planning applications, and is increasingly being used in support of commercial planning and development, e.g. in relation to the selection of sites for new supermarkets, filling stations, even fast-food outlets and many more.

Turning finally to the change-detection poten-

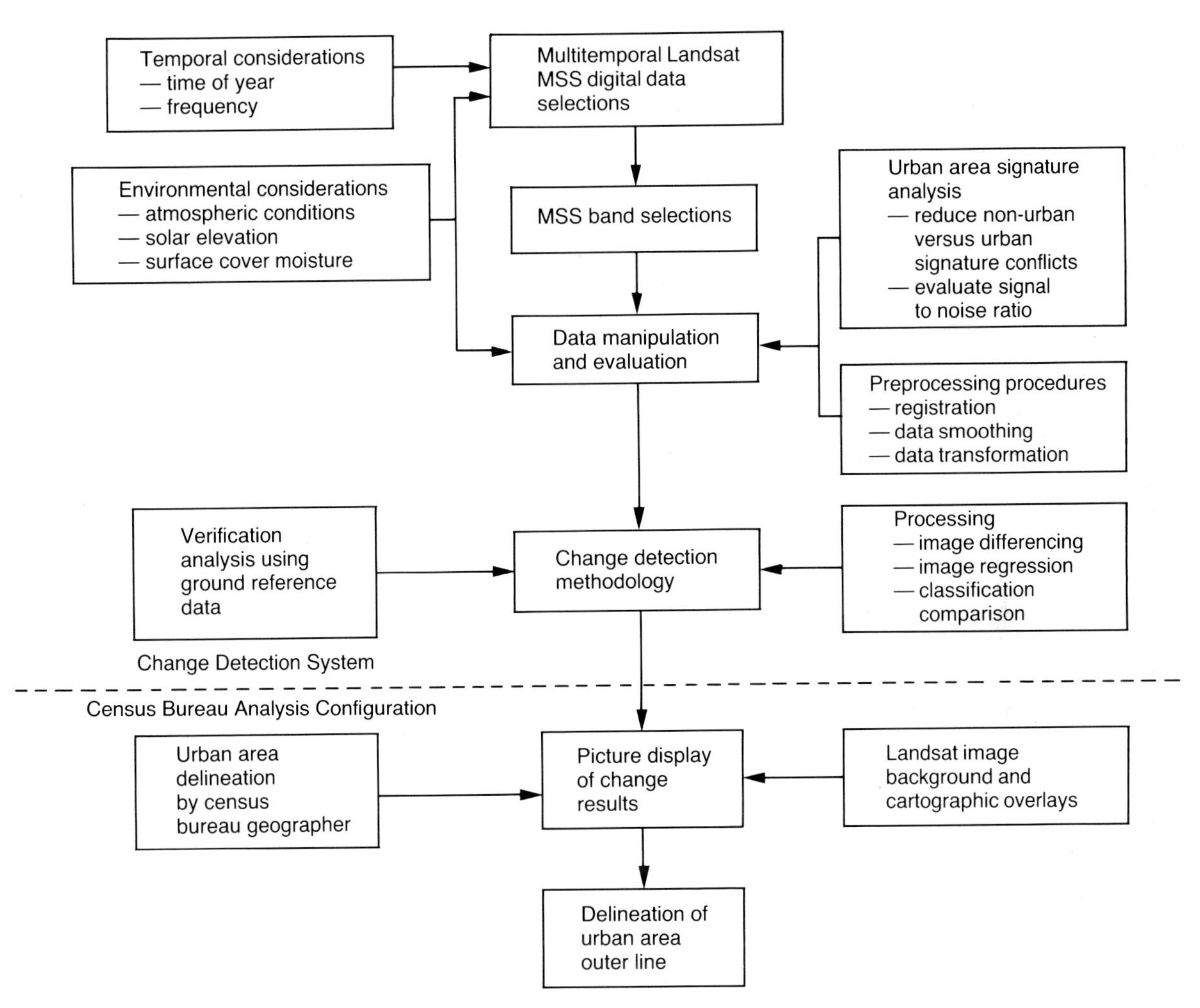

Fig. 17.8 Flow diagram of an urban change detection system for use in urban area delimitation. (Source: Toll, 1980.)

tialities of remote sensing data, studies of general user requirements for land use data in the urban environment confirm that, *for different types of user, different scales of analyses are necessary, and, therefore, different types of remote sensing data must be used.* For example, if urban land-use information is required only on a grid square or block basis, then a relatively low-resolution system will provide adequate information and be advantageous in eliminating the finer details of urban morphology, which may then be regarded as a form of picture 'noise'. Aerial photography or satellite imagery, e.g. from Landsat TM or SPOT, supporting maps prepared at a scale as small as 1 : 100 000 may then suffice. On the other hand, if details of land use at the plot or parcel level are sought, then much larger-scale aerial photography is necessary, supplemented by additional types of information.

Governments are now beginning to set up formal procedures for the operational inclusion of remote sensing data into urban and regional land use change-detection schemes. The form such a scheme may take is suggested by Fig. 17.8. Remote sensing is vital to it, since this provides the 'status report' information involved in the 'update' section of the scheme in operational use. Clearly, frequent snapshot data from high-altitude platforms should increase the dynamic element in studies of urban land use and morphology, with resulting benefits to the whole population through more economic and rational control, and/or development of the complex urban/rural systems of which the modern nation is composed.

17.6 Remote sensing for civil engineering

Although we have noted some civil engineering applications of remote sensing in passing in the earlier sections of this chapter, this activity is so important in the planning and development of the built environment that it deserves specific mention on its own. For nearly 50 years the interpretation of aerial photographs has been an indispensable technique for accomplishing a wide variety of engineering projects. In recent years newer remote sensing systems, such as infrared linescans, radars, and satellites, have become increasingly important sources of information of value to the civil engineer, not only through the *mapping* information they provide for relatively static features of land areas and coastal zones, but also through their *monitoring* capabilities with respect to weather, ice and snow, water currents, lakes and rivers, and other relatively dynamic phenomena.

Although each engineering project is more or less unique, the chief functions and basic methods of image utilization are similar in all projects. These may be summarized as follows:

1. Preliminary planning:
   (a) listing of the engineering survey purpose(s) and the prime factors which must therefore be evaluated;
   (b) review the literature and the existing remotely sensed data;
   (c) plan for the new image data/results acquisition.
2. Data collection:
   (a) perform preliminary field investigations to determine reflectance, emittance and other key properties of natural materials and critical terrain features, to help select types of data required, and the tools and methods of analyses;
   (b) select best weather conditions for data collection;
   (c) obtain imagery as planned;
   (d) collect collateral data simultaneous with aircraft/satellite overpasses.
3. Data analysis:
   (a) merge and display the remote sensing data for the whole area to provide an overview of the study region;
   (b) review the collateral data to help identify pertinent features;
   (c) develop a regional concept of the study area based on the above, and a growing understanding of the features of significance;
   (d) perform detailed analysis of the remote sensing data, e.g. with respect to topography, drainage, erosion, vegetation, and cultural patterns;
   (e) develop a classification relevant to the project and define the basic units to be mapped and/or monitored;
   (f) delineate the basic units classified in the data set(s).
4. Field verification:
   (a) select representative sampling locations, and routes of travel;
   (b) perform field verification, bringing remote sensing data and/or products into the field for detailed correlations;
   (c) collect samples to help refine knowledge of basic surface properties;
   (d) complete laboratory testing of field samples.
5. Final analysis and presentation of results:
   (a) this uses field verification data to correct and complete the original analysis and prepare the final product;
   (b) prepare final report.

The most frequent and important types of civil engineering studies which have used remote sensing data and so followed the above sequence include:

1. Terrain analyses. Almost every major construction project is controlled by the natural terrain: it is rare that some adjustment of terrain is not required, e.g. through cutting, infilling, banking, draining or reclaiming. The cost of moving the natural materials is usually the largest single cost in any engineering construction project. The main value of remote sensing imagery in this respect is that it provides a wealth of detail in the form of a spatially continuous model of the landscape, often able to reveal features not evident to the naked eye.
2. Mapping 'landform engineering soil characteristics', especially soil texture, drainage conditions, and slope categories. For many

Table 17.3 Types and scales of remote sensors for different stages of site investigation

Stage of investigation	Type of sensor	Range of scale
Large area evaluation	Air-photo mosaics and imagery two-dimensional (Landsat) and three-dimensional (SPOT) viewing Earth resources satellite	1 000 000 to 1 : 80 000
Selection and evaluation of alternative sites	High-altitude black and white panchromatic aerial photography (stereoscopic viewing)	1 : 80 000 to 1 : 40 000
Intermediate stage terrain examination and evaluation of selected site	Black and white panchromatic and/or colour infrared aerial photography (stereoscopic viewing)	1 : 40 000 to 1 : 20 000
Detailed terrain classification and mapping followed by site location	Conventional black and white panchromatic, colour, or colour infrared aerial photography (stereoscopic viewing), plus radar surveys	1 : 20 000 to 1 : 10 000

engineering studies air-photos may even be used as the mapping bases in preference to existing topographic or soils maps.

3. Construction material inventories, especially inventories of granular materials (either naturally occurring unconsolidated deposits, or consolidated deposits suitable for crushing). Air photos have been used for many years to identify and assess such deposits; more recently both colour infrared and Landsat MSS data have been used for the same purposes.
4. Regional materials inventories. Regional inventories of construction materials have been prepared from Landsat data to distinguish between major type categories, e.g. organic, alluvial, littoral and glacial deposits, gravel pits, Precambrian rock, Palaeozoic rock and lineaments at scales of 1 : 100 000, supplemented by colour infrared images from airborne surveys for key areas at scales of down to about 1 : 25 000.
5. Site investigations, whether for drains and reservoirs, bridge and pipeline crossings of rivers, airstrips and airports, staging and docking sites, industrial, town or recreational sites, etc. Within the last two decades interpretation of conventional (panchromatic) air photographs has been increasingly complemented by recourse to colour and colour infrared air-photos, airborne infrared linescan and radar imagery, plus Landsat and SPOT data, through a hierarchy of scales (Table 17.3). In future the use of imaging spectrometers may be expected to become important too.
6. Water resources engineering. Perhaps more than any other area of engineering this demands information of an environmental monitoring as well as an Earth resources mapping nature: many types of remotely sensed data are directly applicable, e.g. through watershed analysis, surface water mapping, snow quantification, flood damage quantification and even ground water assessment. Increasingly efforts are being made to bring these and *in situ* and collateral data sets together in a geo-based information system. The same is true in river, lake and coastal water environments, although here data from other specialized sensors, including the Nimbus-7 CZCS, the forthcoming SeaWifs ocean colour sensor and the ERS-1 SAR, Altimeter and even Scatterometer, become important.
7. Transportation facility planning and construction, involving such matters as highway planning, route location, traffic surveys, railroad construction and maintenance, slope stabilization, etc. In these contexts satellite imagery may predominate in developing countries, but airborne data in developed countries.
8. Litigation and claims. Areas in which remote sensing imagery (mainly from aircraft) have been used extensively for such purposes have included *condemnation cases* to help determine

the value of land and appurtenances; *damage suits*, e.g. as a result of a highway accident; and *environmental impact hearings*, e.g. in relation to the construction of some new building(s) or complex of buildings and their anticipated effects on topography and landscape. In such contexts remote sensing data are important because they are permanent and unbiased records of all natural and cultural elements, as revealed by their intrinsic characteristics.

The only possible conclusion to this chapter is that remote sensing applications relating to the evaluation, management and planning of the built environment are amongst the most directly and obviously cost-effective of any of the applications covered by this book. Remote sensing is capable of delivering information that is often *more abundant* and *timely* than that which is otherwise available.

18 Hazards and disasters

18.1 Definitions

We live in an age of growing sensitivity to environmental traumas. As population densities continue to increase, living standards and their expectations rise, and the value of human property chronically inflates, so the impacts of abnormal and extreme events in the environment become ever more serious and costly. Today environmental traumas are both numerous and very varied, as Table 18.1 exemplifies. Many of these events are *natural*; more and more, though, are *anthropogenic*, i.e. man-induced, for example acts of war and terrorism. Some anthropogenic traumas are caused unconsciously or indirectly by man, for example landslides resulting from slope destabilization by the excavation of road or rail cuttings, undesirable soil changes resulting from vegetation clearance and subsequent agricultural activities, and climate changes related to increased emissions of greenhouse gases into the atmosphere, and damage to the ozone layer induced by chlorofluorocarbons (CFCs), as discussed in Chapter 1. However, natural traumas probably still constitute the biggest threats to us and our economy. It is on these that we will concentrate most.

First we must define our terms.

1. A *hazard* is a condition (natural or anthropogenic) of the environment which can exert an adverse influence on human life, property or activity. Since we have to a large extent geared ourselves to the normal state of our environment, the most significant hazards usually accompany environmental states or phenomena which are liable to high degrees of variance about the mean.
2. A *disaster* is a serious, damaging effect on human life, property or activity which results from the impact of a hazard that has exceeded its critical level(s).

It was estimated in 1978 that the cost to the global economy of natural disasters was at least \$US 40 billion. Table 18.1 lists those known to have resulted in economic losses of more than 1% to national economies between 1960 and 1987. It should be noted that Table 18.1 refers only to 'sudden impact' disasters: many others of similar or greater severity occurred during the same period, but these were 'creeping' disasters, i.e. slowly developing and/or chronic situations (e.g. drought-induced famine) whose costs (e.g. through increased food imports, loss of productive manpower, etc.) are much more difficult to estimate. Today the cost of natural disasters to the world economy probably exceeds \$100 billion – and is on an ever-rising trend. Nearly one-half of these astonishing totals involve expenditure on *disaster prevention*, including the monitoring of potentially serious environmental hazards. Clearly, whilst hazard monitoring and disaster assessment procedures will play an increasingly vital role on the world stage in the late twentieth century every effort must be made to ensure that they are carried out not only as efficiently, but also as economically, as possible. Remote sensing is highly promising in both respects.

18.2 Disaster assessment

Today most large nations, and many international agencies and organizations, are specially equipped to respond to disaster situations. The United Nations itself has a practical interest in such matters, spearheaded by UNDRO (the United Nations Disaster Relief Organization), with head-

Table 18.1 List of major natural disasters, which caused economic losses of more than 1% of GNP, 1960–1987 (Source: Galli de Paratesi, in Barrett *et al.*, 1991)

Country	Event	Date	Deaths	Loss (10^6 $)	GNP (10^9 $)
Morocco	Earthquake	2–60	13 100	120	12
Chile	Earthquake	5–60	3 000	800	17
Yugoslavia	Earthquake	7–63	1 070	600	45
Philippines	Typhoon	11–64	58	600	32
Italy	Earthquake	5–76	978	3 600	352
Peru	Earthquake	5–70	67 000	500	17
Nicaragua	Earthquake	12–72	5 000	800	3
Honduras	Hurricane	9–74	8 000	540	3
Guatemala	Earthquake	2–76	22 778	1 100	9
Italy	Earthquake	5–76	978	3 600	352
China	Earthquake	7–76	242 000	5 600	280
Romania	Earthquake	3–77	1 581	800	51
Yugoslavia	Earthquake	4–79	131	2 700	45
Caribbean/USA	Hurricane	8–79	1 400	2 000	?
Algeria	Earthquake	10–80	2 590	3 000	47
Italy	Earthquake	11–80	3 114	10 000	352
Greece	Earthquake	2–81	25	920	33
Yemen	Earthquake	12–82	3 000	90	4
Peru/Ecuador	Floods	4–83	500	700	27
Fiji	Cyclone	3–83	7	85	1
Colombia	Earthquake	3–83	250	380	35
Chile	Earthquake	3–85	200	1 200	17
Bangladesh	Cyclone	5–85	11 000	?	?
Mexico	Earthquake	9–85	10 000	4 000	136
Colombia	Volcano	11–85	23 000	230	35
El Salvador	Earthquake	10–86	1 000	1 500	4
Iran	Floods	12–86	424	1 560	90
Vanuatu	Typhoon	2–87	50	200	0.1
Ecuador	Earthquake	3–87	1 000	700	10
Bangladesh	Floods	9–87	1 600	1 300	12

quarters in Geneva, Switzerland, supported by several other UN Agencies, plus numerous international and national relief bodies. Once a disaster has occurred, the priorities for action include *assessment* of the damage caused, *relief* for the affected population, and *planning* of measures necessary to return life to normal in the disaster area. The use of remote sensing in such connections is already well established. Aircraft surveillance of stricken areas is often the most rapid and convenient way of building up a realistic picture of the scale and the intensity of a disaster situation. Air survey and photogrammetry is often useful in the formulation of plans to facilitate safe reoccupation and reuse of the affected region. Figure 18.1 illustrates the value of aircraft remote sensing in disaster situations. Often this is the only way of obtaining a quick overview of the troubled zone, for surface communications may be severely affected.

As the range and ability of sensors for remote sensing has improved, so the range of disasters that can be evaluated by airborne systems has expanded too. By way of example, reference can be made to the special development programme undertaken by the Institute of Machinery, Faculty of Engineer-

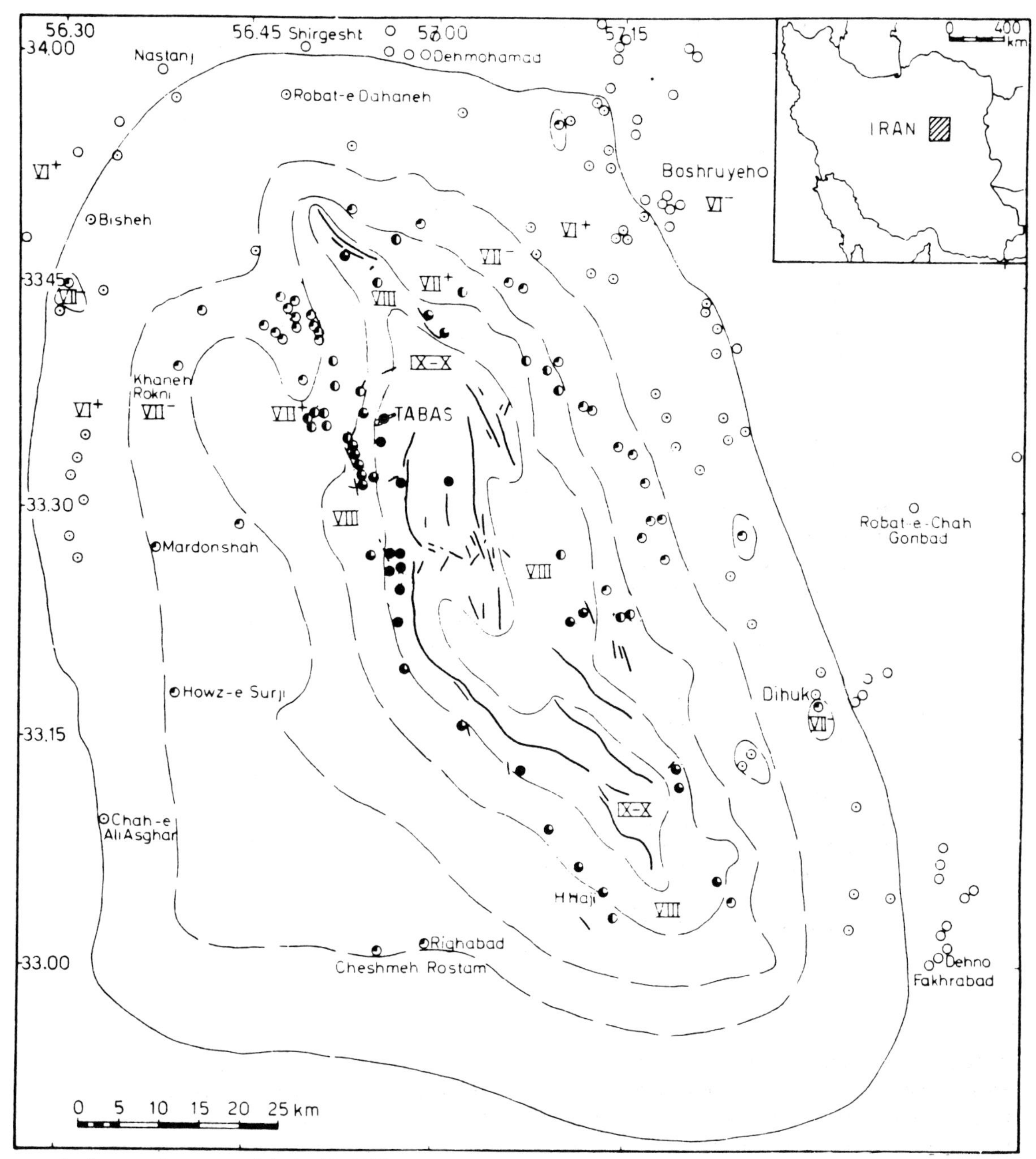

Fig. 18.1 Isoseismal map and aircraft-assessed damage distribution in the epicentral region of Tabas-e-Golshan (Iran) earthquake of 16 September 1978. (Source: Berberian, 1978.)

ing, University of Catania, Sicily, in relation to disastrous and potentially disastrous events in southern Europe and the Mediterranean Sea.

The *platforms* they have developed or exploited for such uses include:

1. fixed platforms (ground-based masts and towers);
2. fixed-wing aircraft, e.g. twin-engined Piaggio P.166-DL3;
3. helicopters, e.g. Agusta A-109A, Agusta Bell AB-204B, AB-212 and AB-412 and Agusta Sikorski SH-3D;
4. naval patrol craft;
5. Earth-orbiting satellites, including Landsat, SPOT and NOAA.

The *sensors* they have used include:

1. photographic, e.g. Aerial Camera System (ACS) and Vinten 618;
2. ultraviolet and infrared scanners, e.g. Daedalus AA 2000 Bispectral scanner;
3. visible and near infrared scanners, e.g. Landsat MSS and TM and Daedalus DS1268 Multispectral scanner;
4. thermal infrared scanners, e.g. FLIR Systems 2500 A/F;
5. lasers;
6. radar;
7. satellite-borne visible, infrared, multispectral and passive microwave radiometers.

Several of these sensors have been combined with an airborne platform in the University of Catania's helicopter-based 'Pelican System'. This system includes provisions for multispectral photography (including simultaneous black-and-white infrared, colour infrared, colour and ultraviolet imaging of the same scene), thermal infrared imaging, and digital image processing.

The system has been designed to monitor and protect the marine environment, which has been the primary concern of the Catania laboratory. Specific applications of the package have included the following:

1. oil spills from ship collisions;
2. illegal discharges from ships;
3. oil slicks, thermal and other types of marine pollution;
4. urban and industrial discharges into rivers and coastal waters.

A recent cost analysis of major oil spills, reported by the International Tanker Owners Pollution Federation, indicates that costs of clean-up operations after tanker accidents range from about $US 5000 to $US 30 000 per metric ton of oil spilled. Additionally claims for damage to the environment following major disasters of this kind may reach many tens of millions of dollars. It has been estimated that a regular use of multisensor remote sensing techniques during oil-spill response phases should reduce clean-up costs by at least 15–20% – whilst reducing general damage to flora, fauna and the environment as a whole by a very significant extent.

Although there is increasing interest in the potential of satellite surveillance for disaster assessment, that potential is sometimes difficult to realize. For example, present civilian weather satellites lack both the high spatial resolution capability, which is often required for such purposes, and the all-weather (microwave) capability, which would enable them to penetrate cloud cover and reveal the situation on the ground irrespective of the state of the sky. Further, present Earth resource satellite types lack the high temporal frequency of imaging which is generally necessary for the evaluation of disasters: typically, these strike with great suddenness and coincide relatively infrequently with the availability of suitable Earth resources satellite data for the area(s) in question. Perhaps the best hope for the serious use of satellites in disaster assessment in the foreseeable future involves the notion of *steerable high-resolution sensor systems on geostationary satellites*, if these could be pointed, on demand, towards any area of key significance. The polar-orbiting French SPOT satellite is equipped with a steerable sensor (Chapter 5), but operates in a low non-geostationary orbit, and therefore lacks the very frequent revisit capability that disaster assessment really demands.

18.3 Hazard monitoring

It is clear, then, that satellites have a much more immediate and extensive part to play in hazard monitoring – and the international community appreciates the overwhelming logic of improving hazard monitoring in order to reduce the impact of hazards when they 'go critical' and a disaster threatens. At the same time, of course, it is clear

Table 18.2 Hazards in the north-eastern USA, which are routinely assessed by US Weather Bureau offices from the evidence of geostationary satellite images (After Wasserman, 1977; from Barrett and Hamilton, 1980)

Type of Hazard	Applicability of satellite data analysis	Key satellite evidence or indications
1. Heavy rain	Direct	Significant, slow-moving convective activity within 100 km of area of interest, or upslope flow of moisture of marine origin
2. Flash flooding	Indirect	Heavy rain, as in (1)
3. River flooding	Indirect	Heavy rain as in (1)
4. Heavy snow	Direct	Significant precipitation within 100 km of area of interest, and/or upslope flow of moisture of marine origin Areas of heaviest snow indicated by imagery
5. Freezing precipitation	Direct	Liquid precipitation in or approaching the area of interest if the surface temperature is near or below freezing, or expected to fall below freezing
6. Poor visibility (highway, aviation, marine)	Direct	Fog and stratus over the area of responsibility, and possible advection of thick fog into a critical area
7. High winds (other than those associated with severe convection)	Direct	Movement and change in intensity of a storm that may produce high winds
8. Severe thunderstorms and tornadoes (high winds, hail, lightning, heavy showers)	Direct	Significant convective activity within 100 km of area of interest or a triggering mechanism approaching an unstable area
9. Low-cloud ceilings at aviation terminals	Direct	Development and change in areal coverage of low clouds
10. Strong low-level wind-shear	Direct	The presence of mountain wave clouds, or boundaries of sharp discontinuities in wind flow
11. Aircraft turbulence (clear air or mountain wave)	Indirect	Cloud features as in (10)
12. Aircraft icing	Direct	Movement of clouds associated with aircraft icing reports
13. Coastal or lake flooding	Direct	Movement and change in intensity of storms with which flooding may be associated
14. High coastal or lake waves	Direct	The presence of intense convection cells over marine areas, and movement and change in intensity of storms that may cause high waves
15. Frost, freeze and cold wave	Direct	Night-time cloud-free areas favourable for radiational cooling Clues to falling temperatures in cloud-free areas in enhanced infrared images
16. Heat wave	—	—
17. Air pollution potential	—	—
18. Fine weather potential	—	—
19. Beach erosion potential	Direct	Movement and change in intensity of a storm that could cause beach erosion

that, whereas aircraft are better than satellites for disaster assessment, they are much less efficient and cost-effective than satellites for hazard monitoring. Aircraft operations are notoriously expensive, ranging at present from some $US 500–10 000 per hour in the air, and are totally inappropriate for assessment of widespread conditions, e.g. snowfall: for these conditions in particular,

Table 18.3 Examples of agricultural hazards which are amenable to monitoring from satellites, classified in terms of the time periods required for them to become effective. (After Howard *et al.*, 1978)

Class of hazard	Types of hazard
Short-term impact (hours to weeks)	Severe storms and hurricanes Flash floods Forest fires Wind Frost Heatwaves Hailstorms
Short-term cumulative impact (weeks to years)	Flooding from prolonged rainfall Drought Ice and snow cover Adverse synoptic weather Persistent cloud cover Insect infestations Disease
Long-term impact (years to centuries)	Climatic change Soil degradation Desertification Pollution

satellite mapping and monitoring is highly advantageous (Plate 14, colour section). Although a few specially equipped aircraft are maintained worldwide for certain key applications, for example hurricane and severe weather monitoring in some tropical and subtropical regions, and for ice monitoring in high latitudes, the cost factor is a very effective limit on the use of aircraft in hazard monitoring as a whole.

There are two different types of approach to the use of satellites for natural hazard monitoring:

1. the *regional* approach, exemplified by Table 18.2;
2. the *thematic* approach, as illustrated by Table 18.3.

Amongst the more common natural hazards, earthquakes and volcanic eruptions are, at least potentially, as dramatic as any in their surface impacts. Of these, the former do not readily lend themselves to monitoring by remote sensing, but the latter do. Ash clouds, e.g. from Mount Patubo in the Philippines in 1991, have been monitored on NOAA satellite imagery, and lava flows and environmental damage have been mapped and assessed after volcanic episodes have ended in many cases, e.g. in the case of Mount St Helens in the north-west of the USA after its spectacular eruptions in 1980. However, the majority of natural hazards are not of geological, but of *atmospheric*, origin. Fortunately these lend themselves rather readily to satellite remote sensing. Consequently, we will focus much of our attention in the remainder of this chapter on the use of satellites in monitoring hazards directly or indirectly related to the state and behaviour of the atmosphere.

Authorities in the hazard monitoring field have established that the most frequent natural disasters are associated with flooding, whether slow or sudden in developing. Some 60% of all natural disasters are said to be of this type. On the other hand, the most frequent stimuli of disastrous loss of life are severe tropical storms and hurricanes (alternatively known as tropical cyclones or typhoons). In view of the significance of such events, and the detailed attention which they have been accorded in remote sensing circles, we may profitably consider severe convective storms (the instigators of most flash floods) and hurricanes in greatest detail, to illustrate the value of satellites in hazard monitoring. Our third case study will involve insect infestations; interesting and varied research has been undertaken and is in progress in respect to several insect pests. Globally these are reckoned to destroy or damage one-third of all food crop production, besides inflicting many diseases on both human beings and their livestock. Fortunately our ability to monitor and control some large insect populations has already been improved significantly using satellite data.

18.4 Severe convective storms

Certain parts of the world are susceptible to severe convective storms, which can cause considerable loss of life and damage to property through associated high wind speeds, short duration high-intensity rains and even tornadoes – the most destructive atmospheric phenomena on the Earth, judged by the fury of their impact in relation to their size. The USA, open in summer to very warm, moist, convectively unstable air from the Gulf of Mexico, has a very significant severe-storm prob-

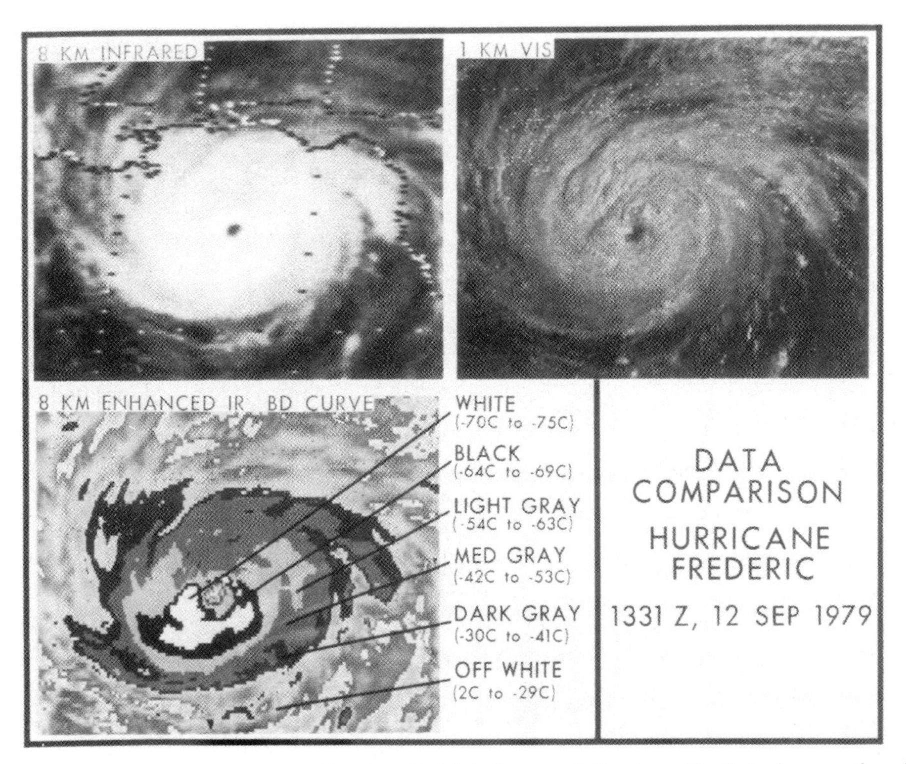

Fig. 18.2 Comparison of raw infrared and visible images of Hurricane Frederic from an American geostationary satellite at 1331 GMT, 12 September 1979, with an enhanced infrared product in which radiation brightnesses have been translated into temperatures of radiating surfaces. These can be interpreted in terms of heights above sea-level (colder cloud tops have higher radiating tops). Enhanced products are often physically more informative than unenhanced images. (Courtesy, NOAA.)

lem. This is magnified by the high property values in this advanced nation. Convectional activity over the coterminous USA is of special concern to the US Weather Bureau, who make intensive use of geostationary imagery from GOES satellites in support of conventional data for monitoring severe convective situations. Of special interest is the work of the Synopic Analysis Branch of NOAA/ NESDIS, who began supporting the Quantitative Precipitation Branch (QPB) of the National Weather Service as early as 1977 with estimates of severe storm precipitation derived from satellite data by the so-called 'Scofield–Oliver technique' (Chapter 11). There are three particular reasons why such an approach is deemed essential in a country well equipped with surface weather-observing stations, including a good weather-radar network:

1. During storms normal communications may be interrupted, and/or conventional weather stations damaged, so that the normal *in situ* data flow is impaired.
2. Satellites provide a more complete areal view than the conventional observing network can give. This is often of special significance in short-term forecasting of severe weather activity.
3. Much of the USA is susceptible to flash-flooding if high-intensity rains occur. Since the exact location and path of movement of a severe convective storm may be very significant in terms of the river basins it may affect, the conventional station network is often not close enough to give the spatial detail desired in such cases.

An appreciation of the scale of the flash-flood threat can be gained from the following examples, all of which occurred during one period of a little over 12 months in 1976–1977:

1. The Big Thompson Canyon flood in Colorado, 31 July 1976. 305 mm (12 in) of rainfall, in parts in only 4 h, caused 139 deaths and an estimated $35 million damage.
2. The Johnstown (Pennsylvania) flood, 19–20

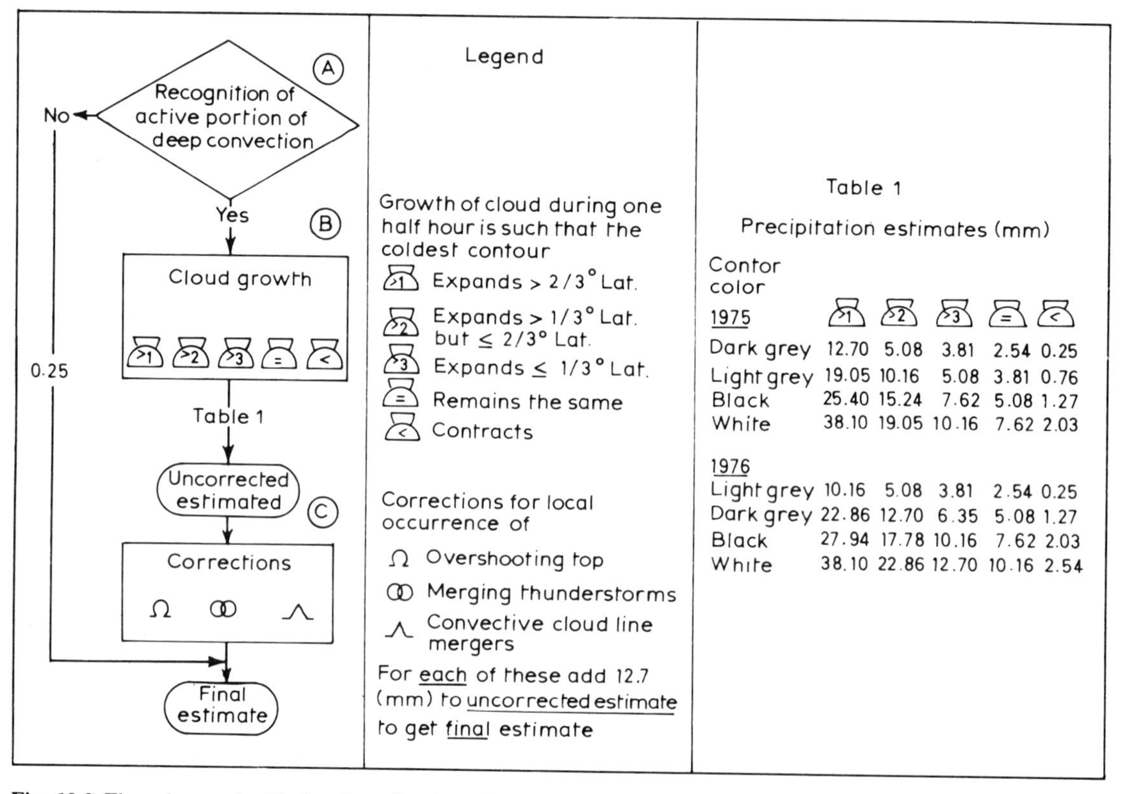

Contor color	>1	>2	>3	=	<
1975					
Dark grey	12.70	5.08	3.81	2.54	0.25
Light grey	19.05	10.16	5.08	3.81	0.76
Black	25.40	15.24	7.62	5.08	1.27
White	38.10	19.05	10.16	7.62	2.03
1976					
Light grey	10.16	5.08	3.81	2.54	0.25
Dark grey	22.86	12.70	6.35	5.08	1.27
Black	27.94	17.78	10.16	7.62	2.03
White	38.10	22.86	12.70	10.16	2.54

Fig. 18.3 Flow chart and table for the estimation of point rainfall from geosynchronous satellite imagery for South America. (Source: Ingraham and Amarocho, 1977.)

July 1977. 305 mm (12 in) of rainfall in 10 h led to 77 deaths and some $200 million damage. The same community had previously suffered from one of the worst flash-floods in USA history: 220 known deaths in 1889.

3. The Kansas City flood, 12 September 1977. 203 mm (8 in) of rain in 6 h resulted in 25 deaths, and an estimated damage to property of $90 million.

Although originally designed as a manual method, nowadays the Scofield–Oliver technique involves a person–machine mix on the purpose-built IFFA (Interactive Flash-flood Analysis) system. Through this, visible, infrared, and enhanced infrared images are analysed via a selected decision-tree interpretation scheme, to establish suitable rainfall estimates for sensitive areas, assessed on a small grid-square basis (down to a few square kilometres). Of special significance is the enhancement process (Fig. 18.2), which contours cold convective cloud tops in terms of temperature (or height). Warmer portions of the imaged areas are displayed by a nearly linear relationship between radiation temperature and picture brightness (dark for warm, white for cold), up to the edges of cumulonimbus anvils (at about −30°C). Within the anvils lower temperatures are represented by a further set of grey shades from black to white. Reduced to its essentials, the Scofield–Oliver method is comprised of three stages. In these the analyst seeks to:

1. Identify the active portions of each powerful convective system.
2. Generate an initial rainfall estimate from the indications of the enhanced infrared imagery, plus knowledge of the typical performance of such convective systems locally.
3. Make allowance for particularly heavy rainfall through the identification of synoptic or sub-synoptic ‘amplifiers’. This involves examining the life history of the event in question, and searching for evidence of the following:

(a) quasi-stationary cumulonimbus systems, and/or such systems regenerating *in situ*;
(b) cold, rapidly expanding anvils;
(c) cumulonimbus mergers;
(d) merging lines of cumulonimbus;
(e) overshooting tops (rapid upward growths of cumulonimbus towers through cumulonimbus anvils).

Since the original Scofield–Oliver technique was formulated, much effort and accumulated experience also has been put into developing and improving it, and most particularly into broadening the general approach to types of storms other than summertime convective outbreaks. Today distinctive variants of the technique are available for winter convective storms, winter (snow-producing) mid-latitude depressions, tropical cyclones (hurricanes), and 'warm top' convective storms from which heavy rains may fall despite an absence of the very low cloud-top temperatures characteristically associated with the heavier rainfalls.

Although this type of approach is complex, and even when implemented on an interactive computer system depends heavily on the skill of the analyst, its operational use in the USA has yielded many valuable forecasts in potential severe weather flash-flood situations. Simpler versions have developed for use by the hydrological offices of other countries, e.g. in Central and South America (Fig. 18.3), whilst under the auspices of the WMO widespread transfer of the full NOAA/NESDIS technique and technology is now being actively encouraged.

18.5 Hurricanes

In the previous section we considered the use of weather satellite data in the monitoring of severe weather, which is organized characteristically at the meso-scale (10–100 km in diameter). Here we consider much larger, more strongly organized and more severe macro-scale weather systems capable of inflicting great damage to human life and property, not only *directly* through high intensities of wind and rain, but also *indirectly* through associated surface effects, especially in low-lying coastal zones. Synoptic-scale revolving or rotating tropical vortices (Fig. 18.2) – variously known as tropical cyclones, hurricanes, or typhoons in different regions of the world – are legendary for their destructive potential. Adjudged by the scale and breadth of the impact they often inflict on some tropical and subtropical lands, this is the most destructive type of weather phenomena on Earth.

Because the hurricane is so significant a threat to human life and property, but also so clearly defined in terms of its vortex of dense, organized convective cloud, the first weather satellites were conceived largely as much-needed monitors of these storms, most of whose lives are spent over remote oceans and seas whence supplies of conventional weather data are very sparse. The monitoring of hurricanes from satellites has been so overwhelmingly successful that no significant tropical storm or hurricanes has gone unseen since the first operational polar-orbiting weather satellite system was inaugurated by two ESSA satellites in February 1966. Concerted efforts have been made, especially in the USA, to interpret weather satellite imagery as fully as possible to improve both the monitoring of existing hurricanes and the forecasting of their likely movements and changes in the future. Currently, satellite imagery are used to:

1. monitor hurricane *tracks*;
2. assess hurricane *intensities*, and estimate their *maximum wind speeds*;
3. identify *areas of influence* of each hurricane;
4. estimate associated *volumes of rain*, although schemes for this are the most tentative of the four.

It is instructive to consider each of these in more detail.

18.5.1 Hurricane tracks

In some cases it is easy to identify the centre of a hurricane because a well-defined, relatively cloud-free 'eye' is plainly evident in or near the middle of the dense clouds of the vortex (Fig. 18.2). However, on many occasions, especially in sub- or post-mature situations, centre location is difficult because no eye is visible and/or the vortical cloud is fragmentary in appearance. Here it is necessary to distinguish between the *central features* (CF) of the hurricane cloud system and peripheral *banding features* (BF), which are arcs of cloud tending to point or spiral towards the centre of the storm circulation in the lower troposphere.

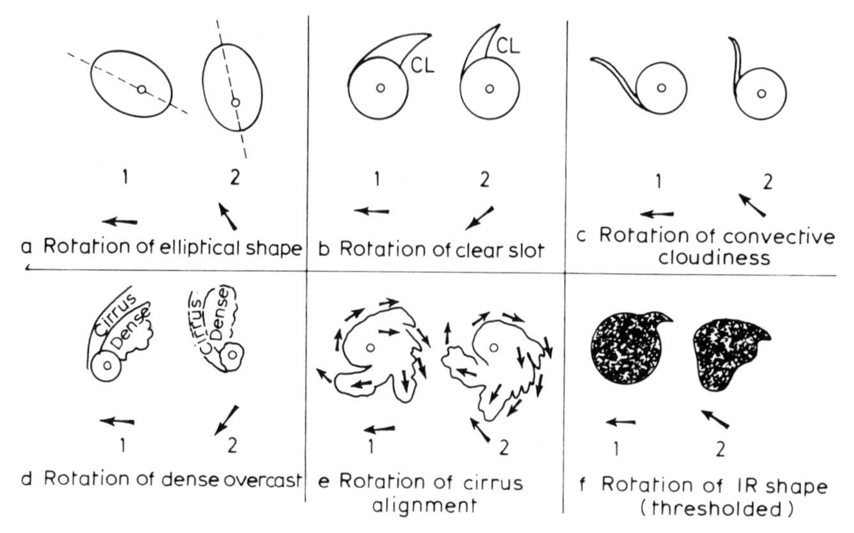

Fig. 18.4 Characteristic changes in cloud patterns associated with directional changes of motion of tropical cyclones. (Source: WMO, 1979.)

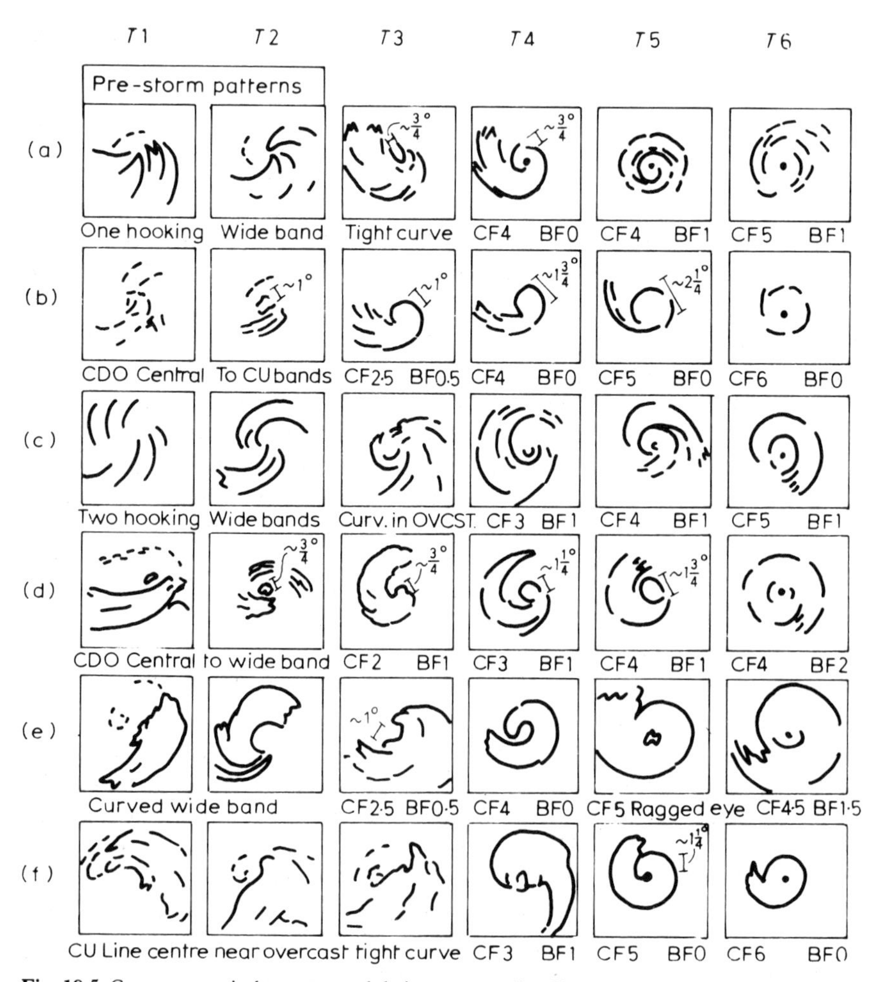

Fig. 18.5 Common tropical patterns and their corresponding T-numbers. The T-number shown must be adjusted for cloud systems displaying unusual size. The patterns may be rotated to fit particular cyclone pictures. (Source: WMO, 1977.)

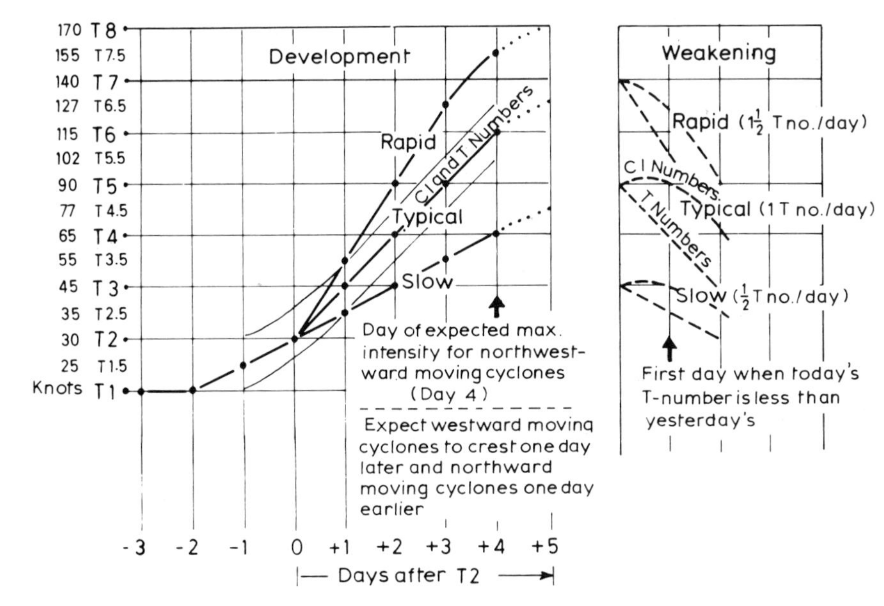

Fig. 18.6 Current intensity change curves of the tropical cyclone. The shaded area surrounding the typical curve represents 'intensity' as a zone, one T-number in width. (Source: WMO, 1977.)

Through careful tracing of cloud arcs, a convergence centre can be established with reasonable confidence in many cases. Results compiled from many hurricane seasons show that tracking is usually correct to within ¼–½° (28–55 km) for the location of the centre of each storm. Only occasionally – in respect of particularly ill-organized storms – are centre locations 1–2° or more astray. For forecasting future movements of the storms, recognition of significant changes of vortical cloudiness is often profitable (Fig. 18.4).

18.5.2 Hurricane intensities

Early classifications of tropical storms and hurricanes have revealed that, through considerations of their sizes and stages of development, good estimates could be made of sustained maximum wind speeds (MWS). This work led to the development of highly detailed classification schemes, which can be used both diagnostically and predictively. This is illustrated by Fig. 18.5, which categorizes common tropical cyclone patterns in terms of 'T-numbers' that are largely (but not entirely) determined by the sum of values allotted to CF and BF characteristics. Comparisons with surface and aircraft observations have shown that *current intensity (CI) numbers* (T-numbers modified to take account of influential meteorological factors not directly evidenced by the cloud features alone) can be interpreted diagnostically in terms of MWS, as exemplified by Fig. 18.6. The T-numbers alone can then be used predictively to give forecasts of hurricane wind speeds for up to a few days ahead, once it has become clear whether the storm in question is following a rapid, normal or slow development curve. Verification tests have indicated that assessments of hurricane intensities expressed in terms of MWS are 90% correct within two T-number categories (i.e. ±1.0 T) which, translated into wind speeds, is ±15 knots. This is reasonable, remembering that hurricane winds often substantially exceed 100 knots.

18.5.3 Areas of hurricane influence

Here the aim is to establish the corridor likely to be affected by strong winds and heavy rain, especially as a hurricane makes a landfall and tracks inland. For this purpose attention is paid to the central dense overcast (CDO) in conjunction with the predicted path of the storm. The CDO is of greatest significance for likely rainfall, but banding features also may have to be taken into account in order to delimit the total regions of possible wind damage.

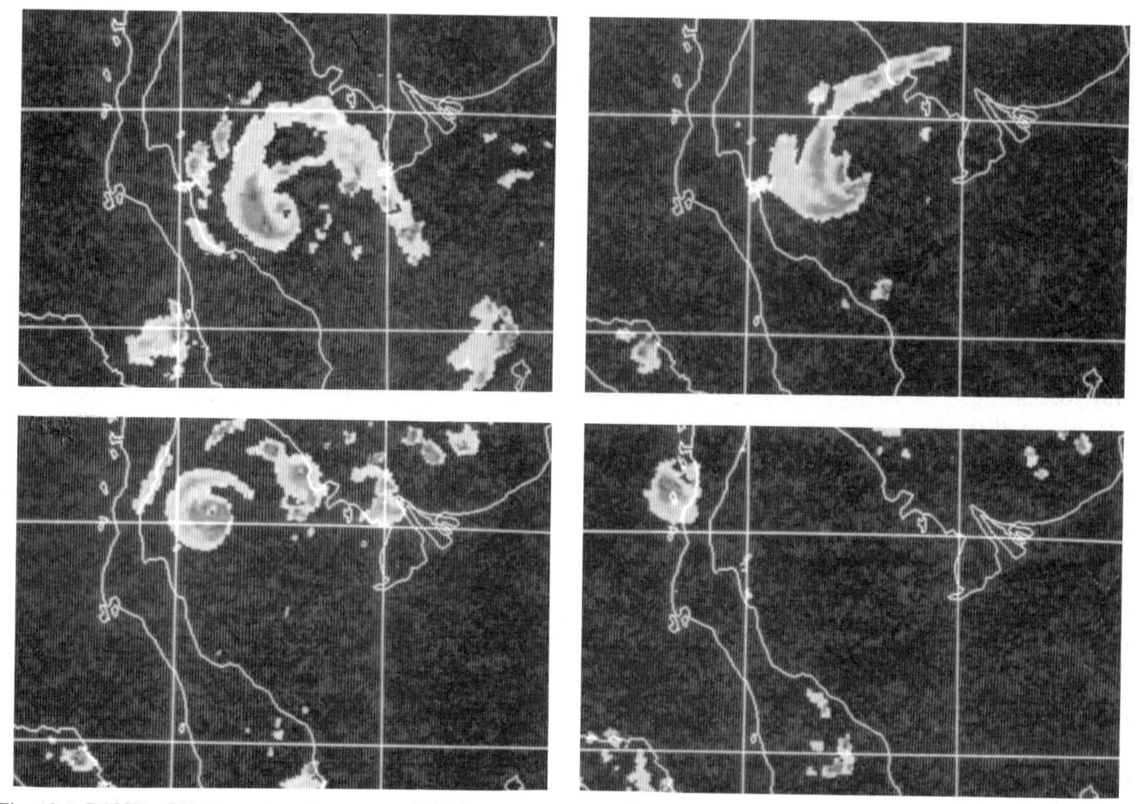

Fig. 18.7 DMSP – SSM/1 passive microwave rainfall intensity images of typhoon Gay over the Gulf of Thailand each day from 1–4 November 1989. (Courtesy, Somsri Huntrakul).

18.5.4 Hurricane rainfall

This has been particularly difficult to estimate accurately, owing both to the extreme rainfall intensities that hurricanes may bring and the propensity for hurricane rainfall to vary spatially to a high degree. Experience with infrared images of hurricanes in the American sectors of the North Atlantic and North Pacific Oceans has indicated that the following rainfall rates may be assumed as representative of a transect marked by the storm centre path:

1. Wall cloud area (around the eye, within 20 nautical miles of the storm centre): 51 mm h^{-1} (±25 mm).
2. Inner central dense overcast (CDO) area (within 50 nautical miles radius of the storm centre, adjustable on the evidence of enhanced infrared imagery): 25 mm h^{-1} (±12 mm).
3. Outer CDO area (radius defined by the outer edge of the CDO): 1–2 mm h^{-1}. If embedded convective bands are present these rates are expected to be higher, between those of 2 and 3.

At the time of writing new efforts are being made to estimate instantaneous rainfall rates for hurricanes more confidently by using passive microwave image data from the US military satellite family, DMSP (Chapters 5 and 10). The early results are promising, not least because these data reveal much more directly the areas of rainfall embedded in the CDO and BF clouds, as shown by Fig. 18.7. They are also thought provoking, because most of the storms so far analysed in this way seem to have had rainfall patterns different from those that existing hurricane models (largely based on visible and infrared imagery) suggested, emphasizing the need to treat each such storm individually.

Calibration of the passive microwave algorithm outputs of rainfall rates is proceeding, but slowly because radar data are required for this purpose, and severe hurricane occurrences in areas served by suitable (digital) radars are rather infrequent. A further challenge stems from the fact that passive microwave data are only available at present from low-orbiting satellites, which cannot provide the temporal detail that fast-changing hurricanes sometimes demand: thus one of today's key research problems is how best to combine the physically more direct but spatially less detailed, and temporally less frequent, evidence of rainfall afforded by the low-orbit passive microwave data with the physically less direct, but higher spatial and temporal resolution, infrared data from the high-orbit geostationary satellites.

However, we must conclude that satellites are already vital tools for use in hurricane monitoring and forecasting generally, remembering that this use of satellite data alone has justified and will continue to justify the operation of environmental satellite systems for many countries in the tropical and subtropical world.

18.6 Insect infestations

Insects are highly significant to the human race, for, as we have seen, it has been estimated that much of the annual food production of the world is rendered inedible because of damage by insects; insects are significant to us also because of the many diseases which they transmit, or more directly cause.

Researchers in remote sensing have long pursued the possibility that conditions conducive to upsurges of several major insect pests might be monitored from aircraft. Perhaps the best examples are from the mosquito family, whose breeding habitats have been extensively defined using multiband aerial photography. More recently, attention to the potential of satellite remote sensing for improved pest monitoring and control has broadened greatly although the apparent promise of satellites for such purposes has, perhaps, still to be generally fulfilled. In principle at least this work has involved a variety of insect pests, influenced by various environmental parameters and characteristics. Here we will consider in more detail projects relating to three pests which seem especially amenable to monitoring from satellite altitudes, and discuss ways in which operational success in respect of each might be achieved.

18.6.1 *The screwworm* (Cochliomyia hominovorax)

This is a devastating cattle pest of Central America and southern North America, being endemic in Mexico, and expanding into the USA in summer. Its adult stage is temperature controlled, the fly being especially sensitive to both very hot and cold weather (for this reason there are no overwintering populations in the USA). The distinctive features of the species are that:

1. the larvae feed only on living tissue;
2. the population density is relatively small;
3. the female flies each mate only once in their lifetime.

Point 3 renders the fly vulnerable to the introduction of sterile males, which therefore can be used as an effective control measure. Without this, annual losses in the USA could exceed $400 million.

In an experimental Screwworm Eradication Program of NASA, high-resolution infrared data from NOAA satellites were processed in conjunction with correction factors for atmospheric attenuation to give daily maps of air temperatures near the ground in northern Mexico and the southern USA, registered to a common geographic grid. Longer term temperature maps and maps of degree-days were also produced. These were judged to have been successful, sterile flies having been released by aircraft over significant areas of Screwworm activity. Unfortunately, as project staff reported 'The logistics of handling images on computer tapes turned out to be a major problem, and the extensive processing required to digitize, register, calibrate, and produce the various products was expensive'. For such a scheme to be implemented widely further model and programme development seems necessary, especially in developing countries where the need is greatest – computer costs, however, have fallen dramatically. With the recent spread of the screwworm to Libya in North Africa (in which continent it was previously unknown) the need for successful control measures is even greater than before.

18.6.2 *The desert locust* (Schistocerca gregaria)

The desert locust is both a traditional scourge of, and a continuing threat to, the agricultural economies of many countries stretching across the desert and semi-desert belt of the Old World, from western Africa through the Middle East to south-central Asia. To counter and contain this threat

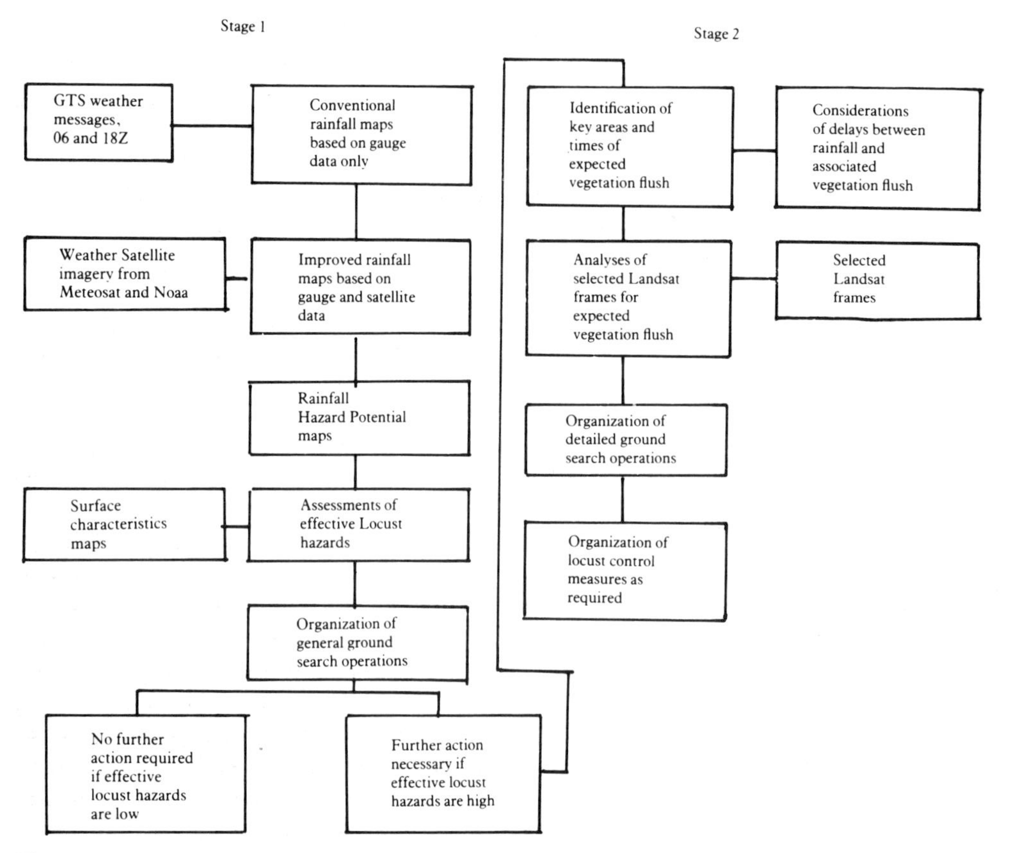

Fig. 18.8 Flow diagram for desert locust monitoring involving satellite (Meteosat, NOAA and Landsat) imagery. This scheme is being adopted for use by the Desert Locust Commission and FAO, beginning with North-west Africa.

a multinational organization, the Desert Locust Commission, has been set up. This meets annually in the headquarters of FAO in Rome, which helps co-ordinate the work of the area Desert Locust Control Commissions, and organizes research into new methods of survey and control. In particular, detailed satellite research began in 1975, focused initially on the area of the North-West African DLCC (covering Algeria, Libya, Morocco, and Tunisia), with the aim of developing an operational method for satellite-improved desert locust monitoring and control (Fig. 18.8).

The desert locust is ideally suited to the exploitation of favourable conditions for breeding as and when they arise. In periods of adversity there is a basal population of locusts spread thinly throughout the affected zone. Some of these 'solitary locusts' become concentrated in rain-fed areas, due largely to the vacuuming effect of converging winds. Here, the solitary locusts change their shape, colour and pattern of behaviour: their feared 'gregarious phase' begins.

The desert locust is capable of considerable migratory movement in search of food and breeding grounds. In suitable areas its reproductive rate achieves geometrical proportions. Suitable conditions for breeding include:

1. a sandy-silty soil for easy egg-laying;
2. sufficient soil moisture in the top layer of the soil to encourage eggs to hatch;
3. green vegetation for the immature locusts to feed on after hatching.

Since 2 and 3 are determined principally by the distribution of rainfall events, *improved monitoring of rainfall* is a primary objective of a satellite-assisted desert locust control operation. However, 1 is determined by desert morphology and pedology, and both 2 and 3 are also influenced by surface characteristics. Consequently Landsat and SPOT imagery have been used for the compilation of *surface characteristic maps* of locust-affected areas, which are much more detailed than existing topographic maps.

In the FAO/DLC research scheme of the late 1970s and early 1980s (Fig. 18.8) weather satellite imagery were used along with surface weather station data to provide the best possible maps of rainfall for periods of 12 h and upwards. Such maps are interpretable in terms of related 'locust hazard levels', for two important practical applications:

1. The identification of areas where significant rain is thought to have fallen in areas whose surface characteristics might be conducive to vegetation flush and locust breeding; post-rainfall Earth resources satellite imagery from such areas can be analysed to reveal the extent and density of any significant growth of vegetation. This should then be inspected in more detail from the air or on the ground.
2. The planning of routes to be followed by ground inspection teams, so that these can be directed more efficiently and economically than before to areas of greatest locust breeding hazard.

Figure 18.9 illustrates results obtained from the operational research programme in its North-West African test area. Since vegetation flush post-dates rainfall by periods from a few hours in spring to weeks or months in autumn and winter, and the locust life-cycle from egg-laying to the first flight of a new generation spreads over about 21 days, there was ample time for the collection, analysis and interpretation of the satellite data and the planning of detailed 'search and destroy' operations before local population dynamics could become critical for the development of plagues of the desert locust.

The project summarized above is a good model of ways in which remote sensing and *in situ* observations can be combined for greatest possible effectiveness, efficiency and economy of pest monitoring and control programmes, and has since been built upon for present operational strategies. However, modifications to the original approach were necessary before an operational version could be implemented, primarily because of the high costs of obtaining and analysing data from the Earth resources satellites, Landsat or SPOT. Fortunately, significant cost savings in implementing such an approach now can be made because of the serendipitous discovery in the early 1980s that data from bands 1 and 2 on the NOAA–AVHRR sensor could be ratioed to provide useful 'vegetation indices', as explained in Chapters 13 and 16. These indices, which provide evidence of *photosynthetic activity* – and even, in areas of grasses and other perennial plants, of *biomass*, and its seasonal changes – can be generated weekly at spatial resolutions down to nominally 1 km. Although even a 1-km spatial resolution may be too coarse to reveal every small stand of post-rainfall vegetation of significance in desert locust monitoring, it has been amply demonstrated that vegetation changes in many key wadis and other areas are clearly evidenced thereby. Consequently, current FAO/DLC operations over Africa follow the general course of Fig. 18.8, but rely mainly on NOAA vegetation index maps rather than Landsat imagery to identify and assess vegetation in the areas – if any – which are deemed likely to support desert locust breeding at any given time, and use an objective rainfall estimation method instead of the original manual approach. Modified in these ways the basic model seems well-suited to applications in many parts of the world.

18.6.3 The armyworm (Spodoptera exempta)

This is the larval stage of a migratory moth which inflicts considerable damage each year to agricultural crops, especially in East Africa anywhere from the Red Sea to Mozambique. Whilst the seasonal movements of the adult moths are generally well known and moderately predictable, many uncertainties remain; it is not without good cause that this pest is colloquially known as the 'mystery worm'. In particular there is uncertainty as to:

1. the exact length of the breeding cycle;
2. the distribution and cause of the quite strongly localized breeding areas that lead to the destructive infestations of armyworm larvae;
3. the ability of individual moths to contribute significantly to the long-distance seasonal migrations that the populations are known to make.

Research investigating evidence of important mesoscale convection within the ITCZ, as imaged by Meteosat, has been undertaken jointly by the

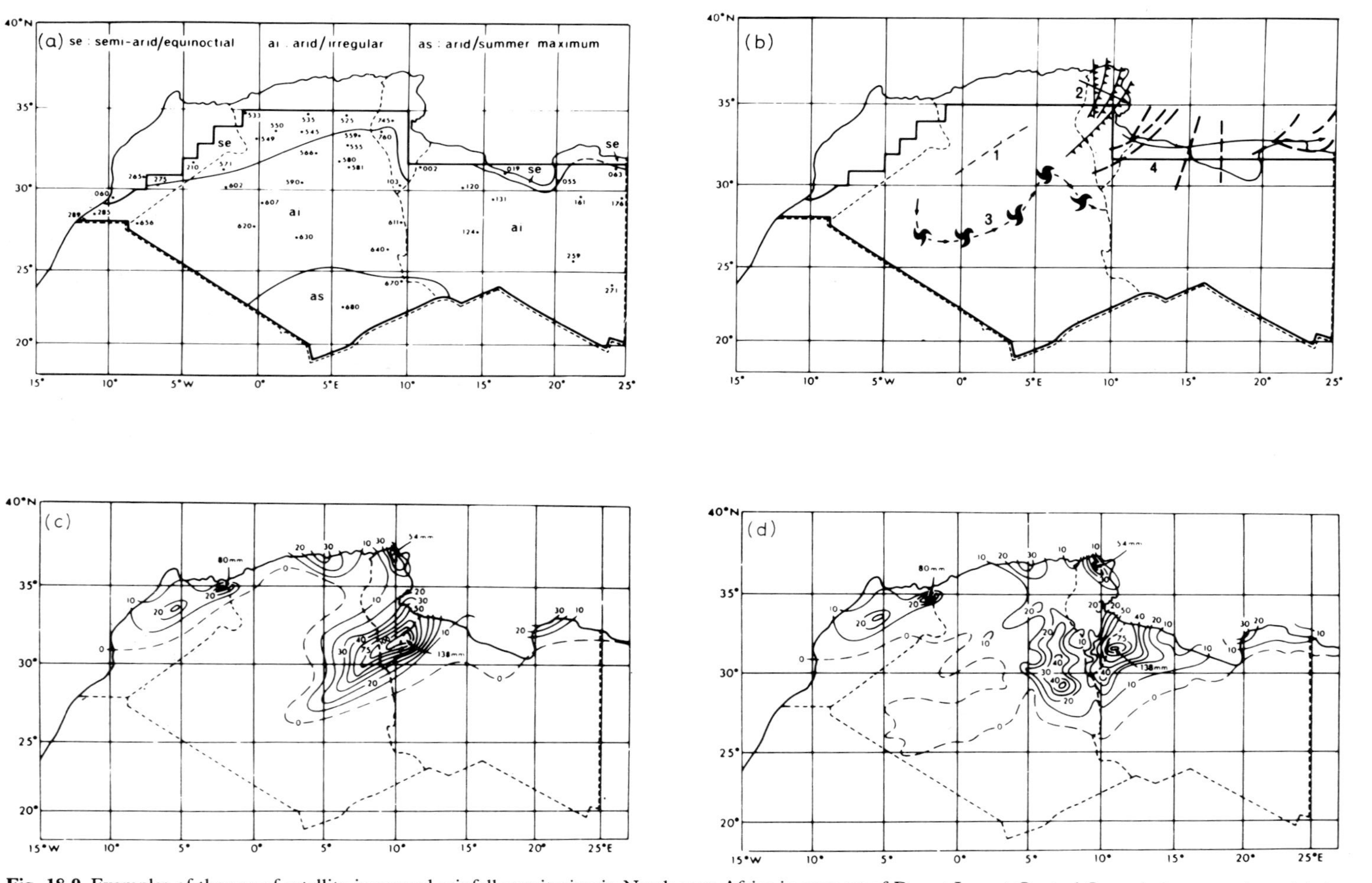

Fig. 18.9 Examples of the use of satellite-improved rainfall monitoring in North-west Africa in support of Desert Locust Control Commission operations: (a) the scatter of GTS rainfall stations, and morphoclimatic regionalization for satellite rainfall estimation, (b) plots of significant rain-cloud systems from satellite evidence, (c) rainfall analysis from gauge data only, and (d) rainfall analysis from gauge plus satellite data, all for 29 March to 4 April 1979.

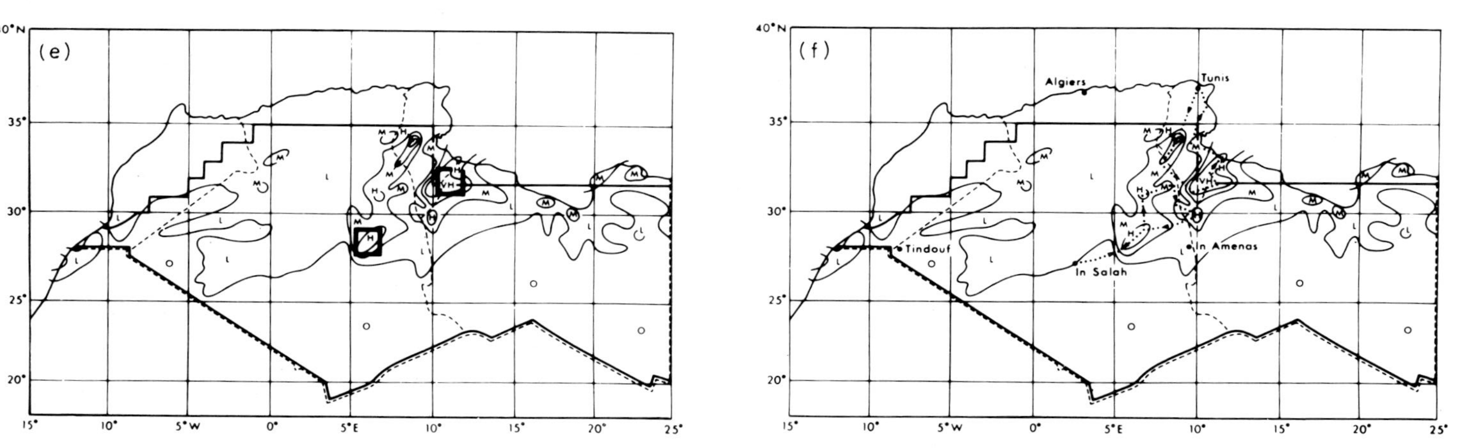

Fig. 18.9 cont. (e) Selection of Landsat frames for post-rainfall vegetation assesment, and (f) selection of preferred routes for ground inspection teams, both on the basis of 'hazard maps' (zero, low, medium, high and very high locust hazard) drawn from satellite-assisted rainfall analysis for spring 1977. (After Barrett, 1981.)

University of Bristol and the former Centre for Overseas Pest Research in London, England (now part of ODNRI, the Overseas Development Research Institute in Chatham, Kent) to test these hypotheses. Ground truth on moth populations and larval infestations was supplied by government organizations in countries of East Africa. The aim was to provide a satellite-improved forecasting scheme for the armyworm; if achieved, this could be almost as valuable as the satellite-improved scheme for desert locust monitoring and control, as outlined in section 18.6.2.

The principal hypothesis was that plagues of armyworm larvae may develop following massive cumulonimbus convection in areas of large moth populations: strong wind convergences may concentrate the moths, which are beaten down by heavy rain and so lay their eggs in areas favourable for vigorous vegetation growth. The principal conclusion from detailed studies of moth concentrations and enhanced infrared Meteosat cloud imagery was that some degree of relationship of the postulated type did seem to occur. However, it was found that the moth concentrations were apparently *peripheral* to major convective clouds, not usually *beneath* them, and associations with selected convective clouds or cloud clusters were almost impossible to predict because the ground data on moth and larvae concentrations were spatially very patchy. This emphasizes the *reliance of remote sensing model and algorithm development upon good collateral data*. At least in cases like the one discussed above special campaigns to provide the necessary *in situ* data must be organized for short periods of time. Subsequently, when adequately developed, remote sensing techniques for pest monitoring and control may be operated with relatively little field data support.

Undoubtedly the use of satellite remote sensing for pest monitoring and control is still very much in its infancy. However, good foundations have been lain, and as the science develops, as satellite data reception and processing costs fall – and as the difficulties of employing staff to engage in arduous surveillance activities on the ground continue to increase – so the use of satellite remote sensing for such purposes will surely grow. Two developments seem particularly probable: a great increase in the geographical areas over which such techniques are routinely applied, and a significant increase in the range of the insect pests involved, e.g. to tsetse flies in central Africa, and sandflies in Africa's semi-arid zones. Directly or indirectly much of the world community stands to benefit therefrom.

18.7 Future developments

Weather satellite data have been an integral part of the data pool for operational meteorology since the mid-1960s. The addition of geostationary satellites, especially since the mid-1970s, when the first SMS/GOES satellites were put into orbit, has greatly enhanced the contribution made by data from space. Stimulated to a large degree by research undertaken by NOAA, especially in its Satellite Applications Laboratory and associated Satellite Research Laboratory, American meteorological centres at both national and regional levels have come to make more thorough use of satellite data in weather forecasting and associated environmental assessment than any other organizations throughout the world.

As early as 1977 it was the considered opinion of a staff member of one US National Weather Service regional office – the Eastern Region Headquarters, Garden City, NY – that 'Satellite pictures, especially when used together with other types of observations, such as radar, are now one of the most important aids we have in detecting hazardous weather in its earliest stages of development'. Even then set procedures existed for the duty forecaster to consider the value of the available satellite data in the preparation of a wide range of special advisories or 'nowcasts' to warn of current or expected weather and weather-related hazards (Table 18.2).

More specialized, but perhaps even more important nowcasts and warnings based largely on satellite data analyses are issued by offices such as the Hurricane Forecasting Center in Miami, Florida, and an increasing number of other national weather offices in areas susceptible to severe weather.

Increasingly, both environmental and Earth resources satellite data are being used in a widening range of other environmental contexts too, especially related to agricultural activity. Crop hazards can all too easily lead to shortages of food, increased food prices, and even famine. Longer term results of satellite monitoring of crop areas

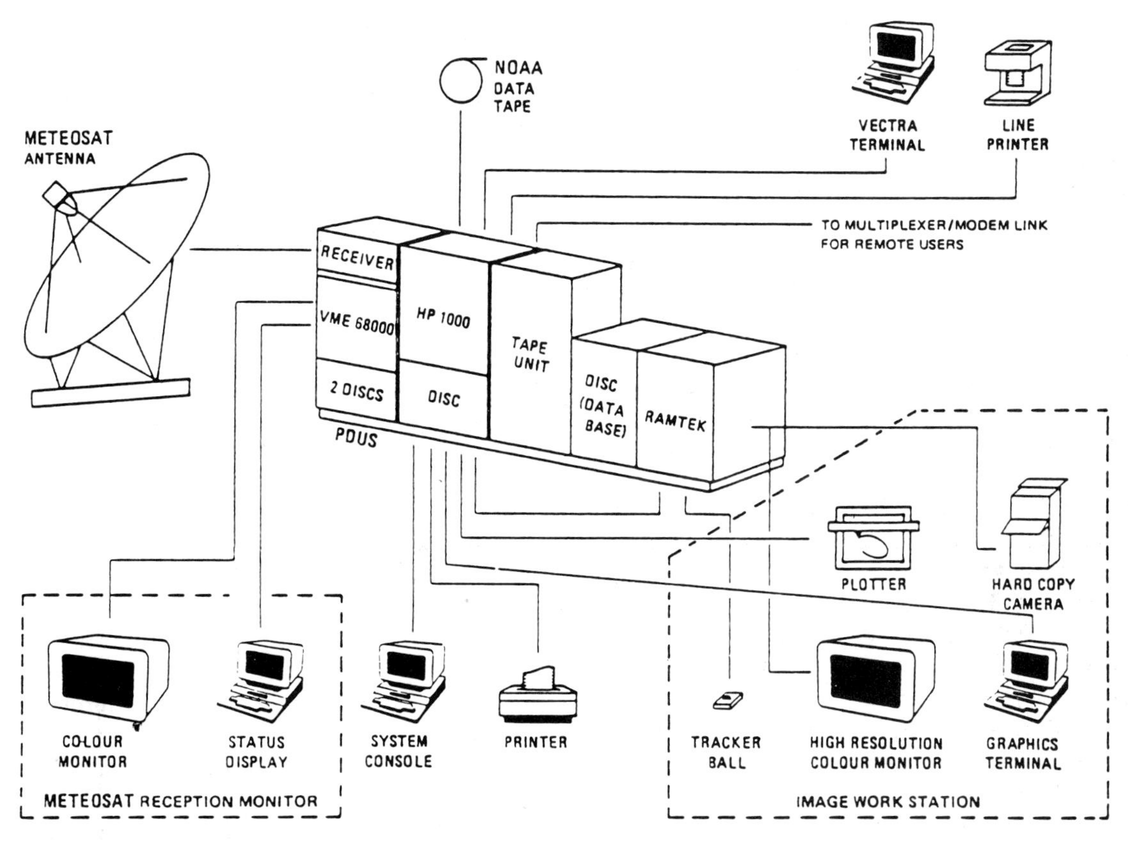

Fig. 18.10 The ARTEMIS system configuration. (Courtesy, FAO.)

and yields (discussed in Chapter 16) may even involve changes in the agricultural policies or internal and external politics of nations or groups of nations.

The Food and Agriculture Organization of the United Nations, with its headquarters in Rome, Italy, is vitally concerned with the global availability of food crops, and crop shortages. Unfortunately the intelligence required for international technical assistance and relief activities is often not available from conventional (ground) sources as quickly as it needs to be if major human disasters are to be averted. However, the satellite, equipped to observe changing conditions and events frequently and very widely over the surface of the Earth is helping greatly, both in the forecasting of crop harvest volumes and the detection and monitoring of acute hazards threatening to agricultural practices and production. The FAO has sought to respond to this challenge, especially through its ambitious ARTEMIS system (see Fig. 18.10). The value of such programmes is recognized also by the commercial world, for which a growing number of meteorlogical bureaux and consultancy firms generate – and frequently update – information on the areas, and conditions of major crops.

Even at smaller scales, including those of local regions, cities, towns, rural districts, rivers and coastal waters, increasing uses are being made of satellite data to improve hazard monitoring as 'distributed systems' of satellite data collection and analysis become possible. Below these scales, and for key instances in time, airborne and even seaborne systems come into play more and more to evaluate disasters and assist with relief operations.

Certainly there is still much unexploited potential in remote sensing, not only for the more efficient and economic monitoring of environmental hazards but also for the increased avoidance or mitigation of natural disasters. What is needed now is a concerted research effort to establish the optimum techniques of data analysis with such ends in mind, and a strengthening of those organizational structures that would foster such approaches, in addition to, or in place of, the existing procedures where these are inadequate to meet the present needs. Perhaps the two events in the 1980s which, more than any others, have focused the attention of the world community on the *needs* for satellite monitoring of hazards and evaluations of associated disasters, and the present *opportunities* for an associated greatly improved watch on the world at all scales from the global to the local were the recognition of the 'ozone holes' in the polar stratospheres, and the nuclear power station catastrophe at Chernobyl in the USSR. In both cases satellite remote sensing has played a key role in helping to establish the dimensions of the associated threats to our planet. Thus it is particularly fortuitous that there is the present trend towards cheaper and more widely distributed systems for remote sensing, for these will help spread the benefits of remote sensing to more areas and people than may be expected through centralized, international systems, helping us to escape some of the worst excesses of environmental behaviour, however stimulated. However, it will be important to ensure that distributed remote sensing systems are networked so that, when necessary, the broader regional pictures can be seen and acted upon. Also, it will be necessary to exercise adequate quality control over local operations to ensure that public warnings and advisories are as suitable as possible: to 'cry wolf' inappropriately can be almost as bad as never to issue possible trauma alerts at all.

19 *Problems and prospects*

19.1 Satellite remote sensing as a global pursuit

We saw in Chapter 1 how remote sensing is a natural facility of mankind, but one which benefits greatly from technological assistance. We saw also how technologically assisted remote sensing effectively took off with the development of aircraft in the early twentieth century. However, there can be no doubt that the greatest single fillip to the growth of environmental remote sensing to the level we find today has been the development of satellite remote sensing platform/sensor combinations. The 'space age' can be said to have begun on 4 October 1957 when Sputnik 1, the Russian automatic satellite, circled the Earth sending back locational bleeps. The 'satellite remote sensing age' may be said to have begun on 1 April 1960 when TIROS 1, the prototype meteorological satellite, was placed in orbit. Since then there has been a continuing and rapid rate of advance in space technology, at first mainly as a result of work in the Soviet Union and the USA, but more recently supplemented by the endeavours of other individual nations, of which Japan is by far the most prominent example, as well as by groups of nations such as those subscribing to the European Space Agency.

Thus, the Russian programme provided the first man in space (Gagarin) and the first space walker (Leonov) whilst the American programme included the exciting Apollo moon landing project, and is aiming to pay great and detailed attention to planet Earth through its ambitious Earth Observation System (EOS) programme (Fig. 19.1). Meanwhile, Japan has developed its own satellite launching capability, and operates successful Earth observation satellites in both low-altitude and geostationary orbits, promising to become a world leader in satellite remote sensing in the twenty-first century. Not to be outdone, the European Space Agency, on behalf of more than a dozen countries in Europe plus Canada is developing a rational programme of space exploration and Earth exploration, which, like the Japanese programme, will partly serve its own regional interests as well as those of the world community as a whole.

Another feature of the more recent past has been the advance of other nations with respect to Earth observation from space. An increasingly long list of countries, already including Canada, China, India, Pakistan, and Brazil have developed, or are developing, their own systems for monitoring the Earth from orbital altitudes. At the same time there has been a very rapid rise in the number of local reception stations capable of acquiring information direct from satellites, even in full resolution digital form. Whereas when the second edition of this book was published in the early 1980s there were relatively few facilities around the globe that had such an ability, today there are many, although this is much more true of environmental satellite data reception facilities than Earth resources satellite reception facilities; cheap stations are technically as feasible in the latter case as in the former but are not encouraged by the major satellite operators. Ground stations for the capture of data from geostationary weather satellites (e.g. Meteosat) are particularly cheap. Reception facilities for near-polar-orbiting satellites (e.g. those of the NOAA family) also are widespread, but less so, because the requirement for a mobile antenna has added significantly to the costs. In this respect, however, advances in ground receiving station design are now proceeding rapidly, and polar-orbiting satellite receivers are becoming substantially cheaper than before, especially those using fixed-horn or manually operated antennae instead of automatic antennae.

As ground station designs improve further, and

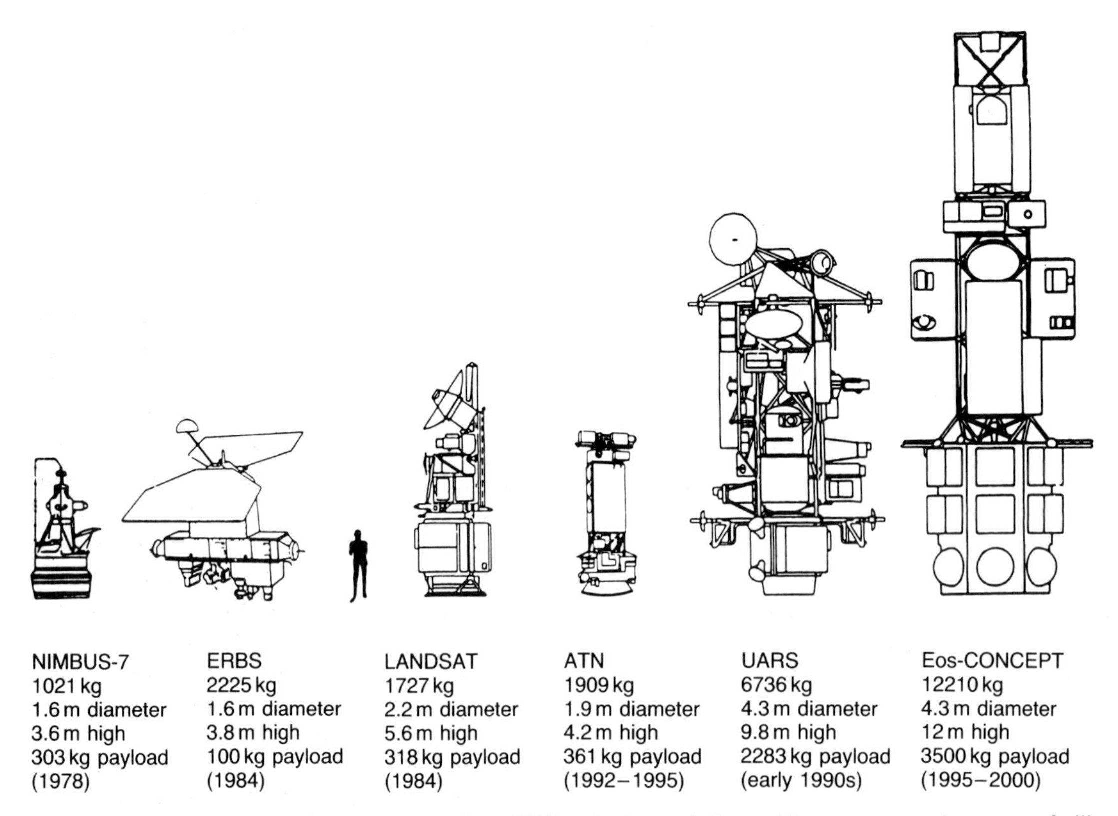

Fig. 19.1 Scale drawing comparing one concept for an EOS payload on a platform with current-generation spacecraft, illustrating stages in the evolution of environmental satellite sizes, complexities and payloads. (Courtesy, NASA.)

associated facilities become more easily affordable than before, other constraints on the widespread use of satellite data need to be relaxed, e.g. the present encryption of signals from some (military) environmental satellites, and the use of narrow bandwidth transmitters by which the data are transferred from some other satellites to the ground. Meanwhile, as suggested above, access to Earth resources satellite data has been almost uniformly restricted by the insistence of the satellite operators to issue few ground station licences, partly because of the security implications of their much higher resolution data, but mostly because of the higher market values of their images and the greater hopes held out by the operators for commercial profits from their space business.

So far as security issues are concerned, there are several apparent sources of difficulty. For example, many sensors and sensing analysis and interpretation systems are classified as secret, and are therefore available to military users only. In some cases security classification rulings have seriously inhibited the exploitation of new satellites or sensors for peaceful purposes. The following are some of the ways in which such restrictions have become evident:

1. Through restrictions on the availability of data and equipment for present programmes.
2. Through restrictions on the education of scientific and engineering personnel who could use new tools and new data with considerable benefits in civilian application areas.
3. Through obstructions to the formulation and development of similar research in the civilian domain.
4. Through the erection of barriers between scientists and engineers concerned with application areas that demand higher resolution data (spatial, spectral or temporal) than those presently available to the civilian user community.
5. Through limitations on the free flow of technical

information within the world community, often resulting in undesirable duplication of effort and incomplete reporting of important research experience.

6. Through artificial restrictions on the development of civilian markets so that system and component costs remain artificially high.

In these areas of difficulty there is clearly a problem of balancing potential gains to the civilian community against potential losses in respect of national and international security. The related judgements are never easy to make, but it seems clear that the classification of remote sensing devices and data as secret should be kept under constant review.

Implicit in these questions concerning the accessibility – or otherwise – of data from satellite remote sensing platforms is the administrative and legal problem of whether 'open sky' observations are permissible or not. For many years there have been differences expressed in the Outer Space Committee (OSC) of the United Nations as to a variety of difficult issues of international law, including:

1. Whether countries have the right to obtain remote sensing imagery of or from any other country.
2. With whom ownership of the satellite imagery obtained over a particular country resides.
3. Whether imagery over a particular country may be made available to neighbouring – perhaps unfriendly – countries against the wishes of the first.

Fortunately for those who believe that open access to all satellite data is desirable, not only are data from western satellites available with a very high degree of freedom, but increasingly access to imagery obtained by the Soviet Union has recently become much more free than before. Perhaps understandably though, most countries restrict access to high-resolution imagery over specially sensitive areas. Whether this still makes sense in an era in which intelligence services have very sophisticated and varied means of obtaining 'target' information from satellites is increasingly open to question.

19.2 Remote sensing as a public service or a commercial activity

Relating to the discussion in the preceding section is another increasingly important question, as to whether satellite remote sensing should be primarily a public service, or a commercial activity. During the last decade there have been concerted attempts by some governments, notably of the USA and France, to ensure that the costs of satellite remote sensing should involve the private as well as the public sectors. Fortunately or unfortunately such efforts have only met with limited success. In the USA it is now recognized that considerable public subsidies must be made available to ensure that programmes such as Landsat will continue, whilst recognizing that there is some role for commercial companies in respect of the marketing of data and products therefrom. In France the experience has been broadly similar, for although a stricter policy has been applied to the marketing of SPOT data than has been the case with Landsat, it has become clear that the income from the sale of data, plus royalties relating to the development of subsequent 'value-added' products, can in no way pay for an entire remote sensing operation from the design and development of a spacecraft, through to its launching and everyday operation.

Studies in the UK have suggested that the day may come when there will be a market for the design, development and manufacture and operation of a satellite or satellites with capabilities and costs attractive to individual companies or groups of companies on a purely commercial basis, but that the basic costs must be significantly reduced before this kind of scenario is likely to become real. In the first instance it would seem likely that multinational companies, e.g. in the oil exploration business, might be the first to feel that their own satellites could be cost-effective. Others believe that, looking further into the future, families of very low-cost satellites ('lightsats') modelled on the 'brilliant pebbles' concept developed by the USA in respect of its 'Star Wars' programme represent a feasible way forward. However, much has to be achieved before such a development may be expected to become widespread.

A further issue relating to all of the above is that transfer of remote sensing to the private sector may negatively affect the benefits of satellite remote sensing to the world community as a whole, not least because developing countries stand to benefit most from a situation within which relatively low-cost access is possible to data from space because in this case the primary costs of establishing the remote sensing systems have been borne by others, without being passed on to the end-user. More broadly it is clear that relatively open access to data from space is much more certain to encourage a continuing increase in the rate of penetration of applied remote sensing into a wide range of areas and monitoring programmes, whereas increased commercialization of space would at least temporarily slow down the rate of such growth.

19.3 Handling large data sets

Great problems are encountered today in the field of data processing as a result of the vast and increasing quantities of data that can be gathered by modern remote sensing systems. For example, the Landsat multispectral scanner produced 15 megabits per second of data, whilst the Thematic Mapper generates over 10^4 megabits per second, and much higher data rates may be expected by the end of the century from the larger space platforms of the Earth Observation System. The acquisition and processing of digital image data therefore demands very significant, and increasingly large, computer resources. For example, a user of Earth observation satellite data who wishes to undertake pans, zooms and three-dimensional displays of satellite data requires a workstation of at least 20 MIPS performance, with some other applications requiring much higher figures even than these. In such cases *memory management* problems are also increasing, especially because data compression algorithms to reduce memory requirements are not very readily available.

Fortunately there have been rapid advances in electronic computing since the first Earth observation satellite was launched in 1960, and this is an accelerating trend. Significant increases have been achieved recently in the capacity and operating speeds of computers suitable for handling remote sensing data sets, and at the same time unit computing costs have come down quite rapidly. A very recent trend of particular significance in these respects is the development of efficient software packages that free the analyst from the earlier requirements for specialized hardware-dependent systems. At the same time, there has been a trend towards the development of *distributed computer networks* through which operators at a number of workstations can function simultaneously on a particular data set, whereas previously only one might have been able to function at a time. However, there are still difficulties in sending large images and attendant voluminous data sets around such networks, owing to limited bandwidths. Developments in communications technology will be required to alleviate this kind of restriction.

In Chapters 1, 7 and elsewhere in this book it has been noted that an important development in the last decade has been in the growth of geographic information systems, through which many different types of spatially distributed sets of data may be organized and intercompared. Today, remote sensing data inputs are seen as increasingly important for modern GIS, and it is certain that the GIS of the future will increasingly capitalize upon the more dynamic types of data which satellite remote sensing systems – both Earth resources and environmental – are able to provide.

In all these areas, however, there have been growing complaints from the remote sensing user community concerning the plethora of available media for data analysis and display, the data formats, and the software packages available for data analysis and interpretation: *incompatibility* between systems in different laboratories is much more common than the *compatibility* that we need for increased co-operation within the remote sensing community. It is certainly increasingly necessary for *standards* to be agreed so that systems can become more compatible with each other, and so that the results and processing packages from different sources can be much more readily interchanged.

19.4 Archiving and distribution problems

Of course, the wish to handle large data sets does not always or only relate to data that have been obtained directly from the satellites themselves, but often in relation to such data that have been obtained at some central facility. Issues related to this mode of operation include *archiving* and *distribution*. The long-term storage of data and their

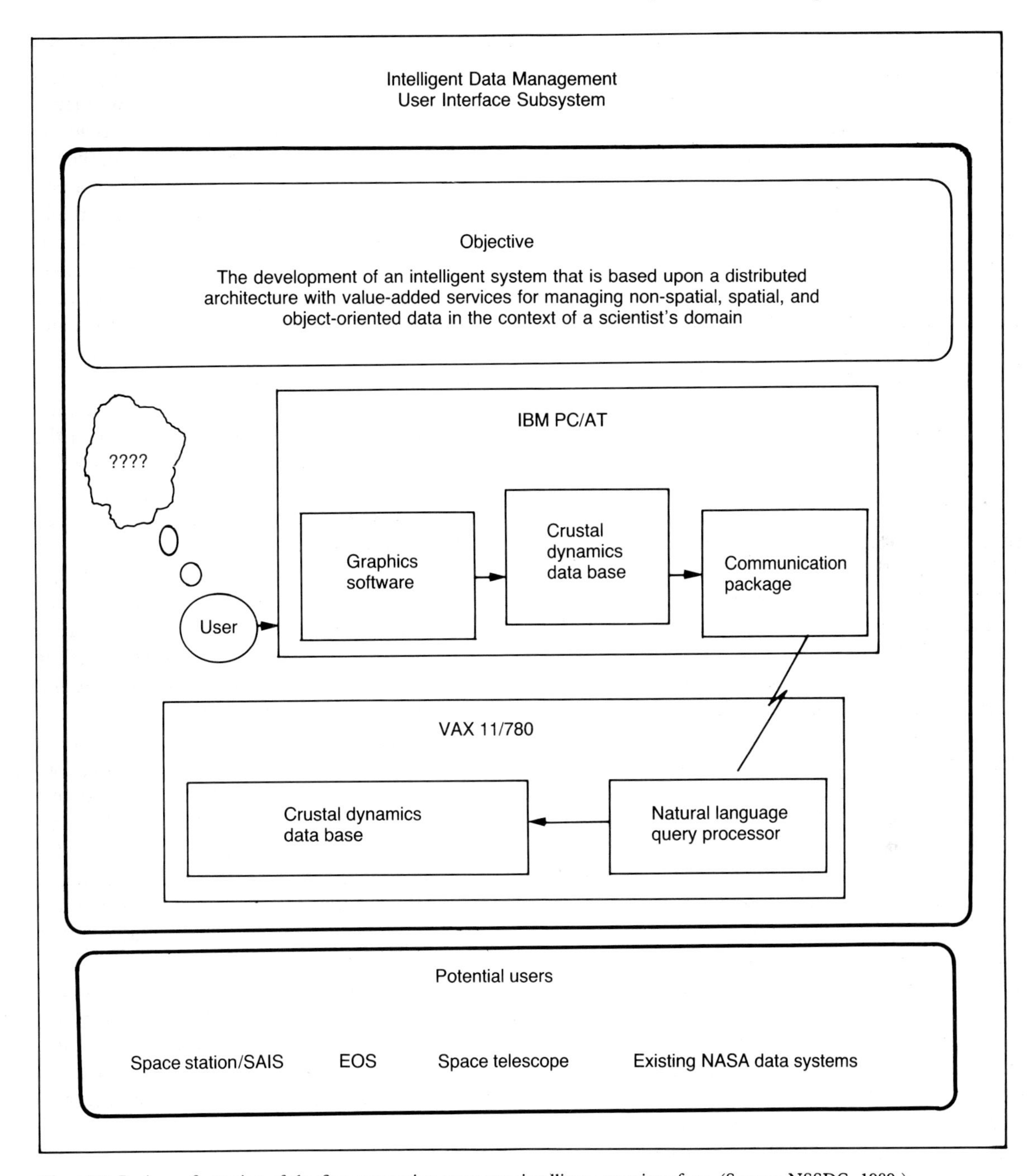

Fig. 19.2 Basic configuration of the first generation prototype intelligent user interface. (Source: NSSDC, 1989.)

retrieval is a considerable problem, for in many areas inadequate funds have been made available to ensure that all, or even a significant proportion of, the original data are kept for posterity. Unless this problem is addressed head on, it is one that will get much worse with the advent of the new higher data rate sensors expected in the mid-term future. *Data reduction* has been considered a high priority in some areas as an alternative to archiving an entire satellite data set. However, if data reduction is necessary it is important that any transformations leading to derived parameters prepared for archiving are reversible: otherwise, no subsequent analysis is possible of data at the earlier, more complete levels of detail.

In the related area of data distribution, recent developments have afforded increasing hope for the future. In the past most satellite data have been stored on computer compatible tapes. These are known to suffer from a number of limitations, including:

1. limited lifetimes (theoretically as short as about 1 year);
2. weight and bulkiness;
3. unsuitability for on-line storage;
4. lack of speed in accessing;
5. differences and inconsistencies in the formats of the included data.

In several of these respects it is fortunate that the storage and use of data on optical discs promises significant improvements on the above, although at present there are greater risks associated with them, for no standard optical disc size or format has yet been agreed.

Perhaps the most exciting new development that is taking place in this field concerns the development of 'intelligent data management systems' (Fig. 19.2), through which users will be able to interact with even the most complex data base systems, with minimal understanding of the architecture of the system, the stored data, or the query language. Thus, even as data sets become larger, more numerous and more complicated, data selection and manipulation by the research worker or commercial user should become less difficult than it has been hitherto.

19.5 Relations between technologist and user

Following on from section 19.4, a major problem area at the present time is that of the interface between remote sensing technologists on the one hand and the users of remote sensing data on the other. Defining the term 'user' is somewhat difficult in practice because there are many users with different objectives. Broadly speaking, however, we may say that there are two main groups of users, the academic and the operational. In the academic category one can include Earth scientists in universities and other research institutes. On the other hand, operational users are typified by National Parks, agricultural departments, meteorological services, water authorities, geological surveys and mineral companies. Each of these groups of users finds difficulty to a greater or lesser extent in:

1. comprehending the nature and significance of remote sensing data obtained by remote sensors;
2. defining its own requirements in terms of the engineering and physical properties of sensors used in remote sensing.

Meanwhile, it is also apparent that the space technologist does not always fully comprehend the dynamics or properties of the environmental surface being sensed. Furthermore, as increasing use is made of non-visible waveband imagery rather than conventional photography, the problems of communication between the remote sensing technologist and the general user become more and more severe. To the uninstructed person an image obtained by infrared or microwave systems looks like a photograph. Its record is, however – as we have seen earlier in this book – in most cases, an entirely different information array from that obtained by photography. For example in microwave studies the measured response (image) obtained from a land surface is affected by complex factors, which can be grouped under two headings:

1. Characteristics of the microwave sensing system: e.g. polarization direction, observation angle and frequency of bands used.
2. Characteristics of the surface sensed: e.g. electrical and thermal properties, surface roughness size and temperature and its distribution.

In such circumstances it is clearly desirable that there should be a flow of information from the physicist/technologist to the Earth science user. However, it is equally true that many physicist/technologist personnel have very limited conceptions of the physical mathematical models of the Earth phenomena being sensed. This has sometimes resulted in exaggerated claims being made for the usefulness of remote sensing systems. For instance, some early writers on air photography and soil studies led potential users to think that soil mapping could be achieved easily by air-photo interpretation. Such statements were based on a lack of knowledge of soil classification and how classification techniques affected the objectives of air-photo interpretation. Lack of environmental training has also resulted in some researchers collecting second rate data concerning the land surface conditions (e.g. in respect of soil moisture – Chapter 13) for comparison with highly accurate remote sensing data.

It is highly desirable that the teams working in remote sensing should be of a multidisciplinary character from the outset. If the eventual users are involved at the beginning they can help to solve many of the technical problems, such as the types of output required. Such participation also would allow the user to make an objective evaluation of how to make best use of the information. The need for such involvement by the users is underlined by this comment made in respect of crop inventory studies during an early symposium on results from Landsat 1: '*From the standpoint of applications almost every investigation lacked complete definitions of technique and procedures which would allow a quasi-operational project to be undertaken*'.

Whilst it was true that some of the early remote sensing systems grew without due regard to the needs of the community that might be able to use them, this has become less true of more recent systems, especially those clearly related to specific end-user needs (e.g. in meteorology), and is becoming markedly less so with systems now under discussion and development for the late twentieth and early twenty-first centuries. However, this is an area which calls for continuing care and attention.

Meanwhile some potential user agencies still do not seem to recognize, understand, and plan for appropriate uses of remote sensing data. It should be recognized that fruitful use of remote sensing data depends not only on successful scientific advances but also on the development of suitable budgetary, organizational and administrative provisions within the user organizations. This often shows itself when studies of cost-effectiveness are made. At a late stage in an investigation, rather than at its beginning, one sometimes hears the question 'would it not be cheaper and easier to send out a man in an automobile to collect information?' Quite often the answer to the question is 'no', but sometimes the economics of the case are less than clear and the old accusation of 'remote sensing technology chasing users' seems apt. It seems fair to add, however, to those who make such comments, that innovation can scarcely ever be proven in advance to be cost-effective. Thus, the first move towards progressive technology is almost certainly more expensive than the status quo.

Clearly such issues are matters of judgement and the related decisions are often difficult to make. In these circumstances it is important that facilities should be available to educate the decision makers and inform the general public about the possibilities and limitations of remote sensing techniques. At the same time educational needs are growing, and becoming even more unique to each specific area of application (Table 19.1); meanwhile facilities for remote sensing are still very limited, mostly general, and unevenly spread at the present time. In developed countries, university students are more likely to have been instructed in some aspects of remote sensing studies than school leavers. Geography departments are likely to include some formal studies of remote sensing for environmental monitoring purposes, but teaching of remote sensing is likely to be fragmentary in other disciplines. The rising tide of MSc courses is helping to produce the specialists the future development of remote sensing will need, but too much of their time is spent in 'levelling up' to the knowledge of those who have already received useful remote sensing training at undergraduate level, and the provision of 'hands-on' experience using computer and image processing systems is generally more limited than the market for MSc-trained personnel requires. Particularly in less developed countries there is often a lack of remote sensing facilities able to take the student all the way from principles through the theory of applications to what is now needed perhaps most of all, namely hands-on experience with digital data processing and analysis and there is a strong danger that the gap between theory and practice in such areas will grow.

19.6 Future developments

It is fitting that this book should conclude with a brief glimpse into the future – and that it should highlight key matters which will demand the most careful attention if environmental remote sensing is to go on playing an increasing part in scientific research and operational mapping and monitoring. The following trends are likely to be evident in the foreseeable future:

Table 19.1 Identified needs for education and training in satellite remote sensing for hydrology in the UK (Source: Barrett *et al.*, 1988)

Clientele	Postgraduate	Mid-career	Senior management	Civil engineering consultants and engineering hydrologists	Technical staff from operational organizations
Types of courses	One-year MSc	Post experience courses and workshops (5 days to 6 weeks)	Short introductory courses (1–5 days)	Disseminated training courses and workshops, within water authorities, civil engineering firms, etc. (5 days–1 month)	One-year training and retraining courses
Size of market (per year)	5–25 persons	10–100 persons	10–50 persons	5–10 courses or workshops	5–75 persons
Staff time (teaching and preparation for each course)	1 man-year for preparation and teaching	10 days–10 weeks	5–25 days	3 man-days per 1 day of course or workshop	150 man-days
Facilities required	Image processing systems; hydrological and meteorological monitoring instruments, e.g. Agro-meteorological stations; computer graphical output devices; meteorological and hydrological ground monitoring equipment	User friendly image processors; hydrological data processors and computer graphical output devices	General range; case studies illustrating the range of potential uses; image processing systems; integration with existing data sets, etc.	Image processing and hydrological data processing facilities and demonstration materials; hands-on workshop facilities and processors; hardware models; usual hardcopy materials; interactive video/digital processing facilities	As wide a range as possible of current top technology, as for postgraduates
Training data sets (e.g. NOAA, Landsat, etc.)	Landsat MSS; Seasat SAR; Nimbus, Meteosat, etc. NOAA–AVHRR, to coincide with catchment rainfall data, etc. Ground truth: all available	Basic and practical project-orientated satellite, aircraft and ground data	Proven types of data, regardless of source	NOAA/Landsat digital data and NRSC leaflets. Choice related to particular needs	As wide a range as possible, from all currently available sources
Costs (approximate)	c. £5K per person Total costs c. £20K to £100K per year + hardware	£0.5–£5K per person	£0.2–£1K per person	£0.2–£1.5K per person	£5–£7.5K per person

1. National and international commitments to remote sensing worldwide will increase, and ground receiving stations for local satellite data reception will become more numerous. It also seems certain that distributed or local data processing facilities will continue to multiply rapidly.
2. Earth observation satellite series will become more numerous, and some Earth observation platforms much larger and more complex. The first point arises from the increasing number of satellite operators, and the latter from the rise of the 'mega-satellite' concept to be developed through the Space Station series of the Earth observation system (EOS) programme for implementation by the end of this decade and into the beginning of the next. Together, these will greatly expand the range of aspects of the environment receiving systematic attention from space.
3. Satellite analysis projects and programmes will increasingly combine data from the different sensors and/satellite systems, along with conventional data, to provide maximum benefits for all. It seems very likely that these analyses will be increasingly dependent upon GIS-type organizations of data and data processing routines.

In the second edition of this volume it was suggested that 'many current aspirations of remote sensing may not be fulfilled unless present problems and difficulties can be satisfactorily overcome. It should concern the whole remote sensing community – from systems engineers through hardware designers and manufacturers, through software and data-handling specialists to analysts, interpreters, and ultimate users of the results of remote sensing programmes – that, outside meteorology, progress towards fully-operational satellite systems has been at best very erratic and fraught with doubts and uncertainties'.

Unfortunately, ten years later, the same sentiments are still largely true. Whilst the second edition was being prepared at the end of the 1970s, and during the beginning of the 1980s, we suffered a particularly inauspicious period for environmental satellites, with relatively few dependable systems in operation. Goodwill for remote sensing, built up slowly through the 1960s and 1970s in the face of considerable scepticism, suffered through those satellite famine years and took several years to recover. This recovery has been based partly upon advances in satellite algorithms, partly upon the growth of the user community, which has been educated to some degree in the practices and potentials of remote sensing, but most of all through the promise of the satellite systems being planned for the future. It will be very important for satellite remote sensing as a whole for this promise to be realized through new achievements of substance and practical value to the world community as a whole.

Increasing emphasis is being placed today on the need for environmental studies to develop an 'Earth system science', concerning the many cyclic or quasi-cyclic systems which may now be recognized and traced, capped by the 'Earth–atmosphere system' itself. The challenge for remote sensing has grown rapidly in recent years because of the new appreciation we have of the Earth–atmosphere system as a very delicate, and much-threatened, entity. In this respect the recognition of growing environmental problems, such as the dangers of global warming, the growth of polar stratospheric ozone holes, and the deleterious effects of air, land and water pollution, has greatly helped to increase scientific, public and political interest in, and concern for, the global environment. It is now widely recognized that satellites to help map, monitor and model our environment over sustained periods of time (Table 19.2) are absolutely essential. Without them, future planning of our use of our global environment will be little or no more thoughtful and intelligent than it has been in the past.

For improvements to be made in such areas, remote sensing will have a vital role to play, as Figs 19.3 and 19.4 amply testify, and for this the co-operative effort of many different types of people will be required. These include engineers, mathematicians, physicists, environmental scientists, managers, planners, and many more. Indeed, it seems reasonable to conclude this book with the assertion that it can be *only through increased team efforts* that such goals will be able to be achieved. With so many new satellite remote sensing systems approaching, there must be greatly increased opportunities for remote sensing scientists to play a worthwhile part in all this process. But to ensure

Table 19.2 Sustained, long-term measurements of global variables important for the study of global change on time-scales of decades to centuries, emphasizing the paucity of existing (pre-satellite) data analysis products presently available for such studies (Courtesy, NASA)

Variable	Importance[a]	Analysis product quality[b]	Variable	Importance[a]	Analysis product quality[b]
External forcing			**Land surface properties**		
Solar irradiance	Essential	**A**	Surface characteristics (for albedo, roughness, infrared and microwave emittance)	Substantial	**C−**
Ultraviolet flux	High	**B**	Index of land use changes (broad classification of vegetation types)	High	**F**
Index of volcanic emissions	Substantial	**D**	Index of vegetation cover	Essential	**D**
Radiatively and chemically important trace species			Index of surface wetness	Substantial	**F**
CO_2	Essential	**A**	Soil moisture	Essential	**F**
N_2O	High	**A**	Biome extent, productivity, and nutrient cycling	Essential	**F**
CH_4	High	**B**	**Ocean variables**		
Chlorofluoromethanes	High	**A**	Sea-surface temperature	Essential	**C**
Tropospheric O_3	High	**C−**	Sea-ice extent	High	**B**
CO	High	**D−**	Sea-ice type	Substantial	**D**
Stratospheric O_2	Essential	**C**	Sea-ice motion	Substantial	**C−**
Stratospheric H_2O	High	**C**	Ocean wind stress	Essential	**C**
Stratospheric NO_2	Substantial	**C**	Sea level	High	**D**
Stratospheric HNO_3	Substantial	**C**	Incident solar flux	Substantial	**D**
Stratospheric HCl	Substantial	**C−**	Subsurface circulation	Essential	**C−**
Stratospheric aerosols	High	**B**	Ocean chlorophyll	Essential	**C−**
Atmospheric response variables			Biogeochemical fluxes	High	**C**
Surface air temperature	Essential	**B**	Ocean CO_2	High	**C**
Tropospheric temperature	Essential	**B**	**Land surface properties**		
Stratosphenic temperature	High	**C**	Surface radiating temperature	Substantial	**F**
Pressure (surface)	Essential	**A−**	Incident solar flux	Substantial	**C−**
Tropical winds	High	**C−**	Snow cover	Substantial	**C**
Extratropical winds	Substantial	**B**	Snow water equivalent	Substantial	**F**
Tropospheric water vapour	High	**D**	River run-off (volume)	Substantial	**B**
Precipitation	Essential	**C−**	River run-off (sediment loading)	Substantial	**D−**
Components of Earth radiation budget	High	**B**		Substantial	**F**
Cloud amount, type, height	High	**D**	River run-off (chemical consituents)		
Tropospheric aerosols	Substantial	**D**			

[a] For documenting and understanding global change.
[b] Presently available multiyear global analyses: **A**, good quantitative, well calibrated; **B**, well discriminated, absolute accuracy doubtful; **C**, useful, poor discrimination; **D**, qualitative index, interpretation doubtful; **F**, no information; −, not global coverage.

that this is so, new methods of working will be required.

Perhaps one harbinger of the future is the 'WetNet' project of NASA, currently invovling 38 laboratories distributed across the USA and other countries of the world (Fig. 19.5). Envisaged originally as a programme to analyse data from the DSMP SSM/I sensor, with special reference to

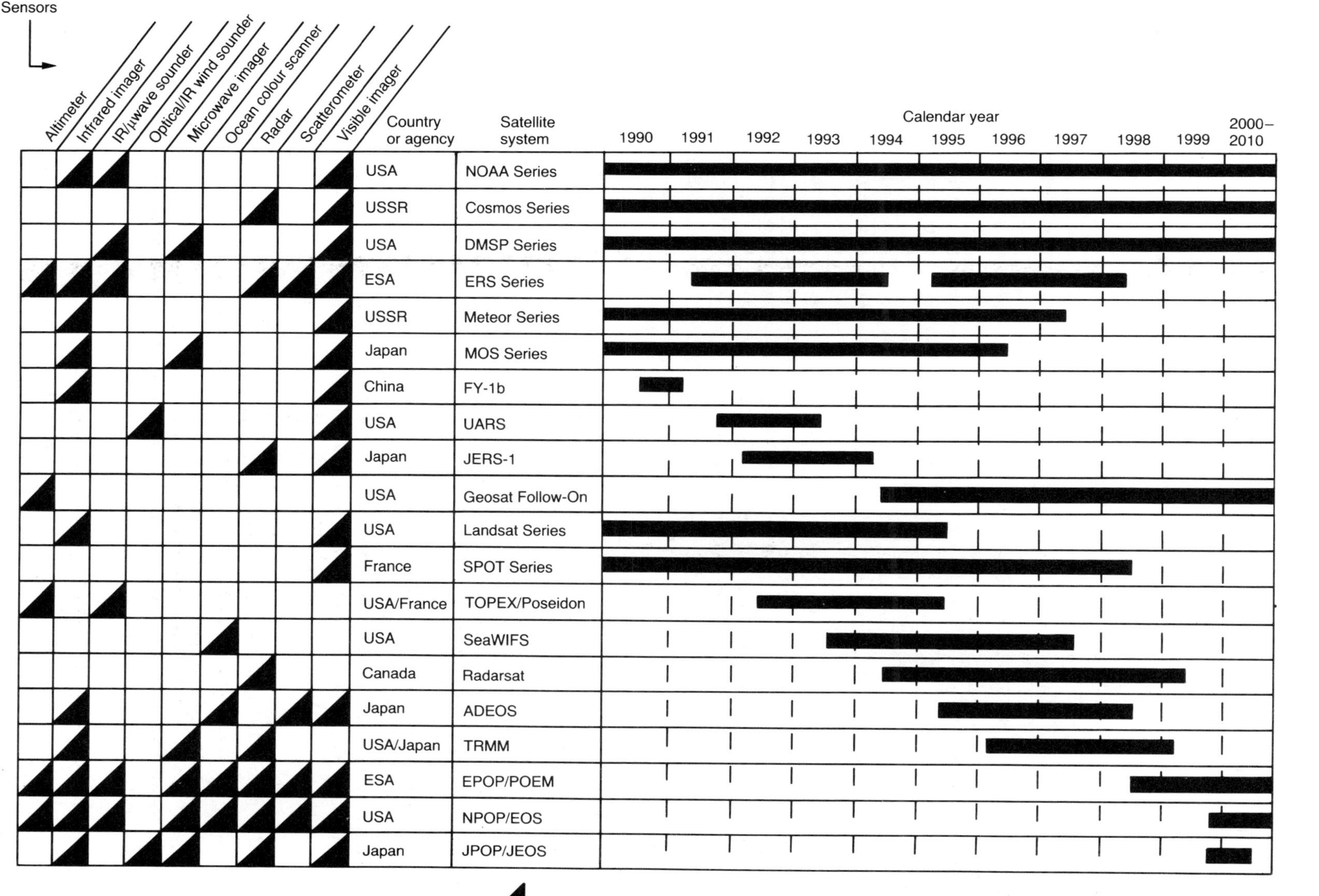

Fig. 19.3 Current and planned low-altitude satellite systems to support Earth system science, 1990 to 2010. (Courtesy, NOAA.)

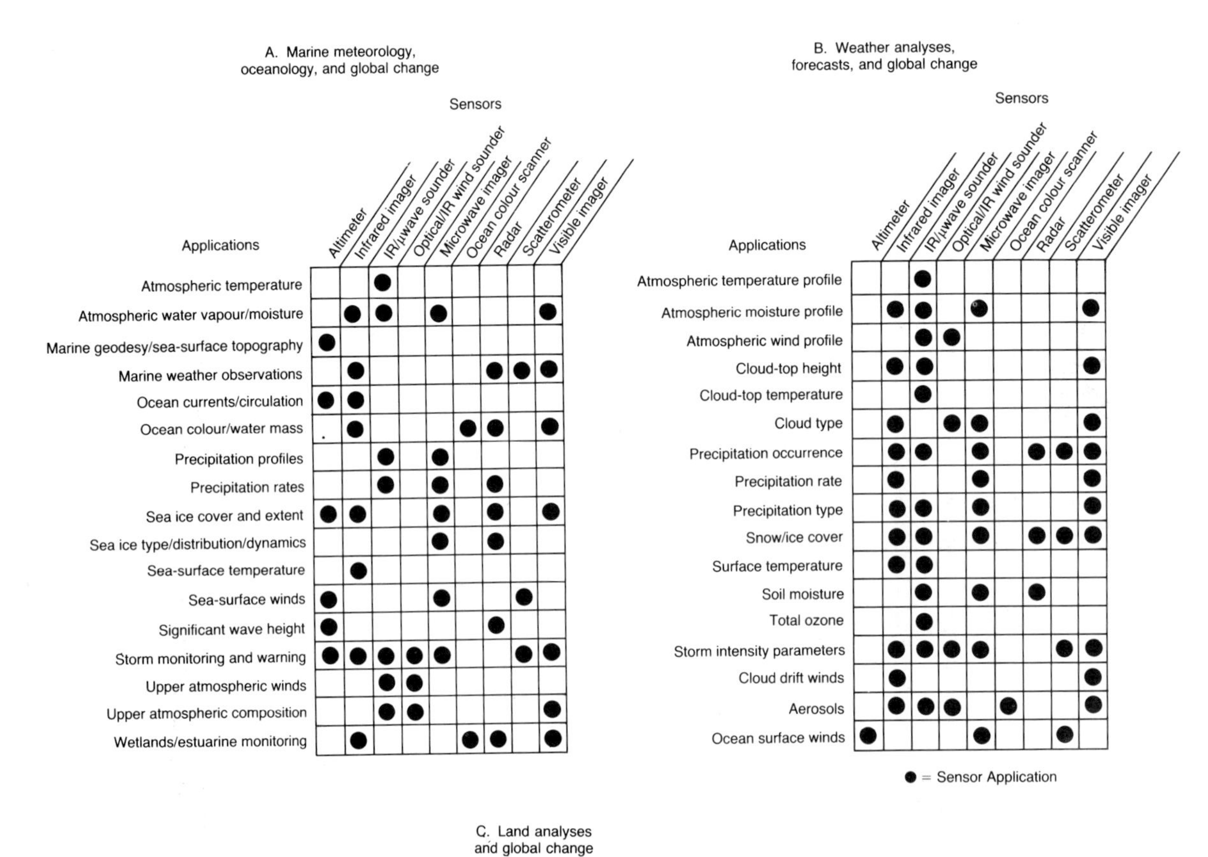

A. Marine meteorology, oceanology, and global change

Applications	Altimeter	Infrared imager	IR/μwave sounder	Optical/IR wind sounder	Microwave imager	Ocean colour scanner	Radar	Scatterometer	Visible imager
Atmospheric temperature			●						
Atmospheric water vapour/moisture		●	●		●				●
Marine geodesy/sea-surface topography	●								
Marine weather observations		●					●	●	●
Ocean currents/circulation	●	●							
Ocean colour/water mass		●				●	●		●
Precipitation profiles			●		●				
Precipitation rates			●		●		●		
Sea ice cover and extent	●	●			●		●		●
Sea ice type/distribution/dynamics					●		●		
Sea-surface temperature		●							
Sea-surface winds	●				●			●	
Significant wave height	●						●		
Storm monitoring and warning	●	●	●	●	●			●	●
Upper atmospheric winds			●	●					
Upper atmospheric composition			●	●					●
Wetlands/estuarine monitoring		●				●	●		●

B. Weather analyses, forecasts, and global change

Applications	Altimeter	Infrared imager	IR/μwave sounder	Optical/IR wind sounder	Microwave imager	Ocean colour scanner	Radar	Scatterometer	Visible imager
Atmospheric temperature profile			●						
Atmospheric moisture profile		●	●		●				●
Atmospheric wind profile			●	●					
Cloud-top height		●	●						●
Cloud-top temperature			●						
Cloud type		●		●	●				●
Precipitation occurrence		●	●		●		●	●	●
Precipitation rate		●			●				●
Precipitation type		●	●		●				●
Snow/ice cover		●	●		●		●	●	●
Surface temperature		●	●						
Soil moisture			●		●		●		
Total ozone			●						
Storm intensity parameters		●	●	●	●			●	●
Cloud drift winds		●							●
Aerosols		●	●	●		●			●
Ocean surface winds	●				●			●	

● = Sensor Application

C. Land analyses and global change

Applications	Altimeter	Infrared imager	IR/μwave sounder	Optical/IR wind sounder	Microwave imager	Ocean colour scanner	Radar	Scatterometer	Visible imager
Evapotranspiration		●			●				●
Fires/hot spots		●							●
Heat islands		●							
Ice sheet topography	●						●		
Insolation									●
Lake ice characteristics		●			●		●		●
Sea ice characteristics		●			●		●	●	●
Shore-fast ice characteristics		●			●		●		●
Snow cover		●			●		●		●
Snowpack parameters					●		●		
Soil moisture		●	●				●		
Surface albedo			●						●
Surface roughness	●						●	●	
Surface temperature		●	●						
Urbanization		●					●		●
Vegetation index					●				●
Vegetation stress		●						●	
Volcanic activity		●	●						●

Fig. 19.4 Satellite-derived data expected for environmental monitoring applications, 1990 to 2010. (Courtesy, NOAA.)

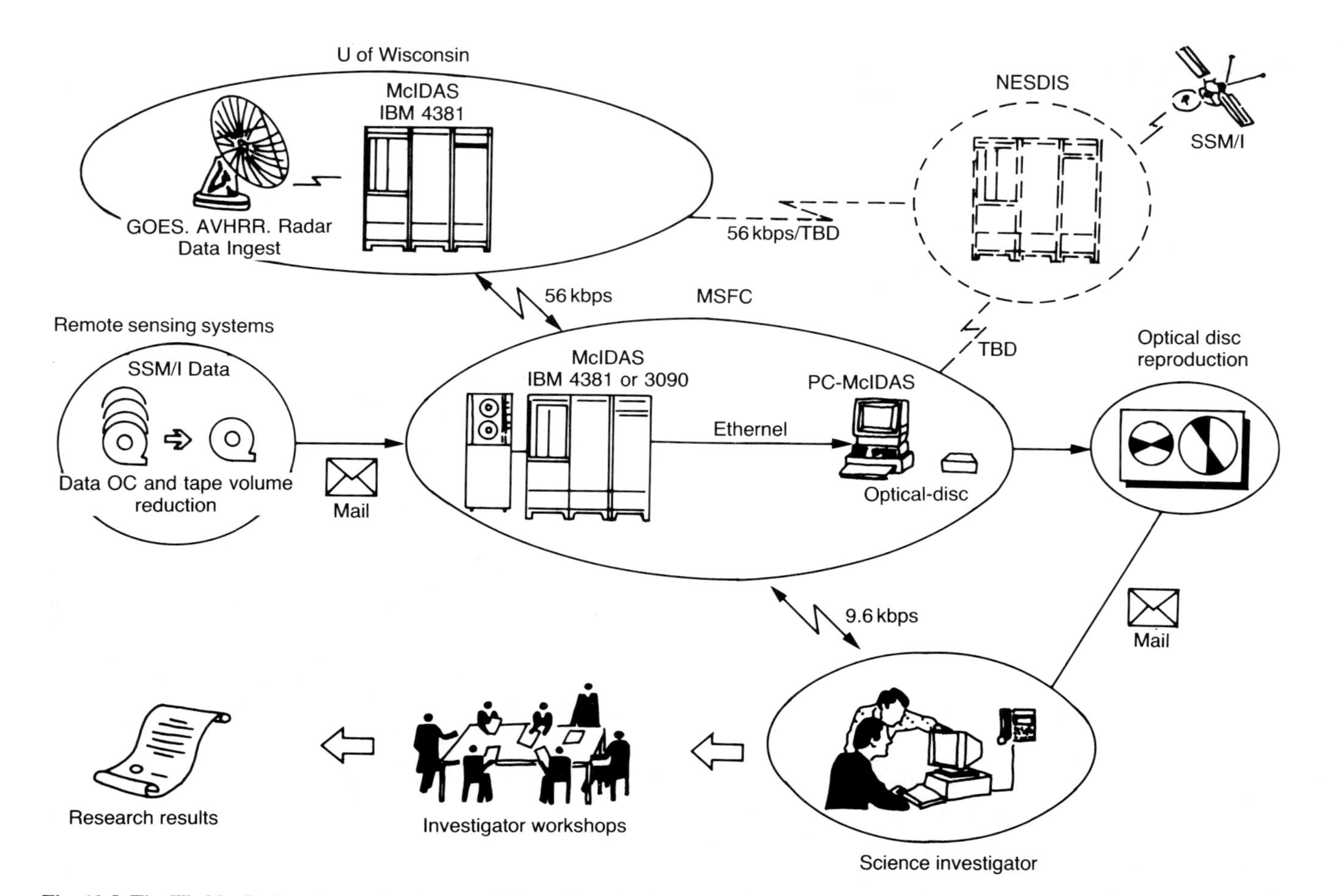

Fig. 19.5 The WetNet Project: interactive data acquisition, dissemination, interpretation and algorithm development system. Headquartered at the Marshall Space Flight Centre (MSFC), Huntsville, Alabama, this is intended as a prototype of the EOS data systems thought likely to be necessary from the late 1990s. (Courtesy, NASA.)

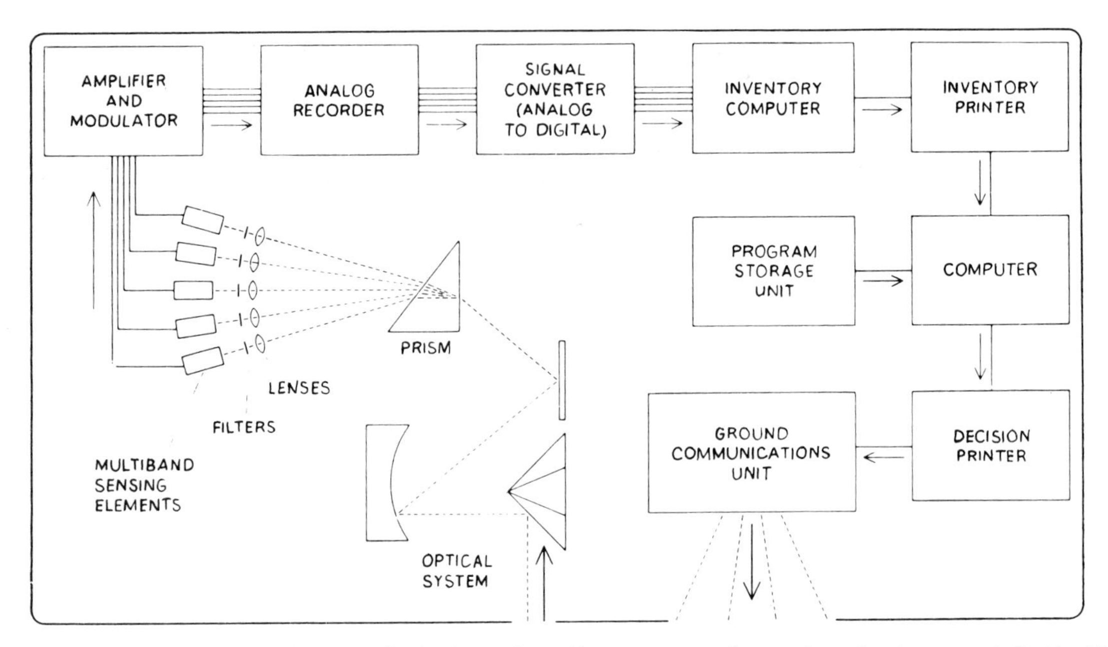

Fig. 19.6 Computerized satellite system for the future. It would sense resources in several wavebands, automatically identify them, weigh them against previously programmed data on the cost effectiveness of various management possibilities, and send to the ground a decision on what should be done. It also could be used to monitor developing situations, such as a forest fire, suggesting how ground crews might fight it, and to perform automatically such tasks as turning irrigation valves on and off as required. (Source: Holz, 1973.)

precipitation, this project is now multidisiplinary. Its intention is to provide a set of laboratories with common data sets and analytical tools, and to encourage their scientists to work much more interactively with each other than has been the case before. Through the WetNet network scientists are being encouraged to exchange their thoughts, findings, algorithms, and future plans so that the WetNet science group can together make far more rapid progress in the understanding of particular types of data, and their practical use, than they could ever expect to do if working in a more traditional scientific way – largely isolated, except when meeting for conferences and workshops, and more or less competitive. It will be through such new and imaginative arrangements that the remote sensing systems of the future may be most likely to provide the best and most cost-effective valuations of our precious terrestrial environment, and help us to manage it in the way that it, we, and our children, all deserve. Thus, the final two paragraphs, and the last illustration from the second edition of this volume seem, remarkably, still the most appropriate ways with which to end the third edition, even though these statements are based on a conceptualization from the early 1970s.

'Although it is probably inevitable that environmental remote sensing will increasingly adopt the mantle of a "big science" operation, it is essential that it listens to, and takes account of, the needs of the "small" user. In recent years, the environmentalist has been involved more in satellite design than in earlier years. For example, it was claimed that Seasat was the first satellite designed primarily by the user for the user. However, it is all too easy for the leaders in the field (the "big users") to make assumptions of the needs of others which do not match reality amongst most members of the world community of nations. The celebrated microchip should help bring the benefits of space technology within the purchasing power of many potential users less interested in multipurpose analysis systems than hard-wired ("dedicated") processors designed to give specific answers to narrowly-framed questions. Beyond that, the ultimate goal for environmental remote sensing on the global scale may be a fully-automatic, general purpose, computerized satellite system as shown in Fig.

19.6, but this will not easily be achieved. If it is, one may wonder whether man is fully-equipped himself to use the knowledge (and leisure) such an automated environmental control system might offer him.

This book has sought to provide a base from which the reader can proceed towards a greater understanding of the part that remote sensing techniques can play in our lives. If it has opened some windows on the opportunities and pitfalls that lie ahead in the use of remote sensing for Earth resource development, the authors will be well pleased.'

Selected Bibliography

Chapter 1

Abiodun, A.A. (1978), The economic implications of remote sensing from space for the developing countries, in *Earth Observation from Space and Management of Planetary Resources*, ESA SP-134, European Space Agency, Paris, 575–584.

American Society of Photogrammetry (1983), *Manual of Remote Sensing*, 2nd Edn (Vols I and II), AMS, Falls Church, VA.

Barrett, E.C. and Curtis, L.F. (eds) (1977), *Environmental Remote Sensing: Practices and Problems*, Edward Arnold, London.

ESA (1985), *Looking Down, Looking Forward*, ESA SP-1073, European Space Agency, Paris.

Frank, A.U. (1988), Requirements for a database management system for a GIS, *Photogrammetric Engineering and Remote Sensing*, **54**(11), 1537–1564.

Lagarde, J.B. (1980), Space and meteorology – an intricate cost/benefit ratio, in *Proceedings of the International Colloquium on Economic Effects of Space and Other Advanced Technologies*, Strasbourg, 28–30 April, ESA SP-151, European Space Agency, Paris, 175–183.

MacQuillan, A.K. and Clough, D.J. (1978), Benefits of spaceborne remote sensing for ocean surveillance, in *Earth Observation from Space and Management of Planetary Resources*, ESA SP-134, European Space Agency, Paris, 585–596.

NASA (1987), *From Pattern to Process: the Strategy of the Earth Observing System*, EOS Science Steering Committee Report (Vols I and II), National Aeronautics and Space Administration, Washington, DC.

NASA (1988), *Earth System Science: a Closer View*, Earth System Sciences Committee, NASA Advisory Council, National Aeronautics and Space Administration, Washington, DC.

NERC (1989), Geographic Information Systems in the Environmental Sciences, *EARSeL News*, **38**.

Peel, R.F., Curtis, L.F. and Barrett, E.C. (eds) (1977), *Remote Sensing of the Terrestrial Environment*, Butterworths, London.

US Congress (1985), *International Cooperation and Competition in Civilian Space Activities*, Office of Technology Assessment. OTA-ISC-239, US Congress, Washington, DC.

Chapter 2

Allan, T.D. (ed.) (1983), *Satellite Microwave Remote Sensing*, Ellis Horwood, Chichester.

Curran, P.J. (1985), *Principles of Remote Sensing*, Longman, London.

Feinberg, G. (1968), Light, *Scientific American*, **219**, 50–58.

Holz, R.K. (ed.) (1973), *The Surveillant Science: Remote Sensing of the Environment*, Part 1, *The Electromagnetic Spectrum – Energy for Information Transfer*, Houghton Mifflin, Boston, MA, 1–27.

Lockwood, J.G. (1974), *World Climatology: an Environmental Approach*, Edward Arnold, London.

McAllister, L.G. and Pollard, J.R. (1969), Acoustic sounding of the lower atmosphere, in *Proceedings of the Sixth International Symposium on Remote Sensing of the Environment*, Ann Arbor, Michigan, 436–450.

Rees, W.G. (1990), *Physical Principles of Remote Sensing*, Cambridge University Press, Cambridge.

Slater, P.N. (1980), *Remote Sensing: Optics and Optical Systems*, Addison-Wesley, Reading, MA.

Chapter 3

Aracon (1971), *The Best of Nimbus*, Contract No. NAS-5-10343, Aracon, Concord, MA.

Barnett, J.J. and Walshaw, C.D. (1974), Temperature measurement from a satellite, in *Environmental Remote Sensing; Applications and Achievements*, Barrett, E.C. and Curtis, L.F. (eds), Edward Arnold, London, 185–214.

ESA (1987), *Remote Sensing for Advanced Land Applications*, ESA Land Applications Working Group, ESA SP-1075, European Space Agency, Paris.

Fleagle, R.G. and Businger, J.A. (1963), *An Introduction to Atmospheric Physics*, Academic Press, New York.

Laing, W. (1971), Earth resources satellites, in *A Guide to Earth Satellites*, Fishlock, D. (ed.), Macdonald, London and Elsevier, New York, 69–91.

NASA (1987), *HIRIS: High Resolution Imaging Spectrometer, Science Opportunities for the 1990s, Earth Observing System*, Vol. IIc, EOS Instrument Panel Report, National Aeronautics and Space Administration, Washington, DC.

NOAA (1985), *The Space Station Polar Platform: NOAA Systems Considerations and Requirements*, NOAA Technical Report NESDIS 22, US Department of Commerce, Washington, DC.

Peel, R.F., Curtis, L.F. and Barrett, E.C. (eds) (1977), *Remote Sensing of the Terrestrial Environment*, Butterworths, London.

Polcyn, F.C., Spansail, N.A. and Malida, W.A. (1969), How multispectral sensing can help the ecologist, in *Remote Sensing in Ecology*, Johnson, P.L. (ed.), University of Georgia Press, Athens, GA, 194–218.

Schanda, E. (ed.) (1976), *Remote Sensing for Environmental Sciences*, Springer-Verlag, Berlin.

Schmugge, T. *et al.* (1973), *Microwave Signatures of Snow and Fresh Water Ice*, Publication No. X-652-73-335, NASA, Greenbelt, Maryland.

Sellers, W.D. (1965), *Physical Climatology*, University of Chicago Press, Chicago.

Yentsch, C.M. and Yentsch, C.S. (1984), Emergence of optical instrumentation for measuring biological parameters, *Oceanographic Marine Biology Annual Review*, **22**, 55–67.

Chapter 4

Chen, H.S. (1985), *Space Remote Sensing Systems, an Introduction*, Academic Press, Orlando, FL.

Curtis, L.F. (1973), The application of photography to soil mapping from the air, in *Photographic Techniques in Scientific Research*, Vol. 1, Cruise, J. and Newman, A.A. (eds), 57–110.

EMI Electronics Ltd. (1973), *Handbook of Remote Sensing Techniques*, Department of Trade and Industry, London.

ESA (1983), *Remote Sensing, New Satellite Systems and Potential Applications*, ESA SP-205, European Space Agency, Paris.

Gjessing, D.T. (1978), *Remote Surveillance by Electromagnetic Waves*, Ann Arbor Science Publishers Inc., Ann Arbor, MI.

Grant, K. (1974), Side looking radar systems and their potential application to Earth resources surveys, ELDO/ESRO *Scientific and Technical Review*, **6**, 117–136.

Hyatt, E. (1988), *Keyguide to Information Sources in Remote Sensing*, Mansell, London.

Laird, A.G. (1977), Passive infrared sensing of the environment, in *Remote Sensing of the Terrestrial Environment*, Peel, R.F., Curtis, L.F. and Barrett, E.C. (eds), Butterworths, London.

Lillesand, T.M. and Kiefer, R.W. (1987), *Remote Sensing and Image Interpretation*, 2nd Edn, Wiley, New York.

NASA (1984), *Earth Observing System*, Science and Mission Requirements Working Group Reports (Vols I and II), Technical Memorandum 86129, National Aeronautics and Space Administration, Greenbelt, MD.

NASA (1987), *High Resolution Multifrequency Microwave Radiometer (HMMR), Earth Observing System*, Vol. IIe, EOS Instrument Panel Report, National Aeronautics and Space Administration, Washington, DC.

Natural Environment Research Council (1974), *Remote Sensing Evaluation Flights, 1971*, Curtis, L.F. and Mayer, A.E.S. (eds), Publication Series C, no. 12.

Naval Research Laboratory (1989), *DMSP SSM/I Calibration Validation, Final Report* (Vol. 1), NRL, Washington, DC.

NOAA (1987), *Passive Microwave Observing from Environmental Satellites, a Status Report*, NOAA Technical Report NESDIS 35, US Department of Commerce, Washington, DC.

Plevin, J. and Honvault, C. (1980), The ESA remote sensing programme, *International Journal of Remote Sensing*, **1**(1), 53–67.

Schanda, E. (ed.) (1976), *Remote Sensing for Environmental Sciences*, Springer-Verlag, Berlin.

Smith, J.T. (ed.) (1968), *Manual of Color Aerial Photography*, American Society of Photogrammetry, Falls Church, VA.

Trevett, J.W. (1986), *Imaging Radar for Resources Surveys*, Chapman and Hall, London.

Chapter 5

ESA (1979), *Spacelab Users Manual*, ESA Scientific and Technical Publications Office, European Space Agency, Paris.

ESA (1986), *Europe from Space*, ESA SP-258, European Space Agency, Paris.

Harris, R.A. (1988), *Satellite Remote Sensing*, Routledge and Kegan Paul, London.

Hart, D. (1987), *The Encyclopedia of Soviet Spacecraft*, Bison Books, London.

NASA (1976), *Landsat Data Users Handbook*, Document No. 76SDS-4258, Goddard Space Flight Center, Greenbelt, MD.

NOAA (1989), *NESDIS Office of Research and Publications Research Programs*, US Department of Commerce, Washington, DC.

NRSC (1987), *UK National Remote Sensing Centre Data Users Guide*, NRSC, Farnborough.

Sabins, F.F. (1978), *Remote Sensing: Principles and Interpretation*, W.H. Freeman, San Francisco.

Velten, E. (1976), *Study on geosynchronous multidisciplinary Earth observation satellite*, Final Report, ESA Contract No. SC/127/76/HQ, Dornier/BAC/Sodetag.

Chapter 6

Becker, F., Bolle, H.J. and Rowntree, P.R. (1988), *The International Satellite Land-surface Climatology Project*, ISLSCP-Report No. 10, ISLSCP Secretariat, Free University of Berlin.

Beckett, P.H.T. (1974), The statistical assessment of resource surveys by remote sensors, in *Environmental Remote Sensing; Applications and Achievements*, Barrett E.C. and Curtis, L.F. (eds), Edward Arnold, London, 11–27.

Curran, P.J. (1985), *Principles of Remote Sensing*, Longman, London.

Curtis, L.F. (1971), *Soils of Exmoor Forest*, Soil Survey Gt. Britain, Rothamsted Experimental Station, Harpenden.

Curtis, L.F. (1973), The application of photography to soil mapping from the air, in *Photographic Techniques in Scientific Research*, Vol. 1, Cruise, J. and Newman, A.A. (eds), 57–110.

Curtis, L.F. (1974), Remote sensing for environmental planning surveys, in *Environmental Remote Sensing, Applications and Achievements*, Barrett, E.C. and Curtis, L.F. (eds), Edward Arnold, London, 88–109.

Curtis, L.F. and Hooper, A.J. (1974), Ground truth measurement in relation to aircraft and satellite studies of agricultural land use and land classification in Britain, *Proceedings Frascati Symposium on European Earth Resources Satellite Experiments*, European Space Research Organisation, 405–415.

Milton, E.J. (1987), Principles of field spectroscopy, *International Journal of Remote Sensing*, **8**(12), 1807–1827.

Sorensen, B.M. (1979), *The North Sea Ocean Colour Scanner Experiment*, 1977, Joint Research Centre, Ispra, Italy.

Townshend, J.R.G. (ed.) (1981), *Terrain Analysis and Remote Sensing*, George Allen and Unwin, London.

Chapter 7

American Society of Photogrammetry (1983), *Manual of Remote Sensing*, 2nd Edn, Vols I and II), AMS, Falls Church, Virginia.

Bryant, M. (1974), *Digital Image Processing*, Optronics International Inc., Publication No. 146, Chelmsford, MA.

Curtis, L.F. (1973), The application of photography to soil mapping from the air, in *Photographic Techniques in Scientific Research*, Vol. 1, Cruise, J. and Newman, A.A. (eds), 57–110

ESA (1987), *Remote Sensing for Advanced Land Applications*, ESA SP-1075, European Space Agency, Paris.

Muller, J-P. (ed.) (1988), *Digital Image Processing in Remote Sensing*, Taylor and Francis, London.

Ross, D.S. (1976), Image modification for aiding interpretation and additive color display, in *CENTO Workshop on Applications of Remote Sensing Data and Methods, Proceedings, Istanbul, Turkey*, US Geological Survey, 170–212.

Smith, J.T. (ed.) (1968), *Manual of Color Aerial Photography*, American Society of Photogrammetry, Falls Church, Virginia.

Warnecke, G.L., McMillin, M. and Allison, L.J. (1969), *Ocean current and sea surface temperature observations from meteorological satellites*, NASA Technical Note, D-5142, NASA, Washington, DC.

Chapter 8

Alford, M., Tuley, P., Hailstone, E. and Hailstone, J. (1974), The measurement and mapping of land resource data by point sampling on aerial photographs, in *Environmental Remote Sensing: Applications and Achievements*, Barrett, E.C. and Curtis, L.F. (eds), Edward Arnold, London.

Avery, T.E. and Graydon, L.B. (1985), *Interpretation of Aerial Photographs*, 4th Edn, Burgess, Minneapolis.

Lo, C.P. (1976), *Geographical Applications of Aerial Photography*, Crane Russak, New York; David and Charles, Newton Abbot, Devon.

Reeves, R.G. (ed.) (1975), *Manual of Remote Sensing*, American Society of Photogrammetry, Falls Church, Virginia.

Slama, C.C. (ed.) (1980), *Manual of Photogrammetry*, American Society of Photogrammetry, Falls Church, Virginia.

Swain, P.H. and Davis, S.M. (eds), (1978), *Remote Sensing: The Quantitative Approach*, McGraw Hill, New York.

Walker, A.S. (1991), Digital photogrammetry at Leica: DSPI and DVP, *Proceedings of the Conference on Photogrammetric and Analytical Digital Systems*, Sociedad Espanola de Cartografia Fotogrametria y Teledeteccion, Barcelona, 32–45.

Chapter 9

Browning, K. (1979), The FRONTIERS Plan, *Meteorological Magazine*, **108**, 161–174.

Conlan, E.F. (1973), *Operational products from ITOS*

scanning radiometer data, NOAA Technical Memorandum, NESS 52, Washington, DC.

Curran, P.J. (1985), *Principles of Remote Sensing*, Longman, London.

Ekstrom, M.P. (1986), *An Introduction to Digital Image Processing*, Prentice-Hall International, Copenhagen.

Estes, J.E. and Senger, L.W. (eds) (1974), *Remote Sensing: Techniques for Environmental Analysis*, Hamilton Publishing Co., Santa Barbara, CA.

Gonzalez, R.C. and Wintz, P. (1977), *Digital Image Processing*, Addison-Wesley, Reading, MA.

Lillesand, T.K. and Kiefer, R.W. (1987), *Remote Sensing and Image Interpretation*, 2nd Edn, Wiley, New York.

Lintz, J. and Simonett, D.S. (eds) (1976), *Remote Sensing of Environment*, Addison-Wesley, Reading, MA.

Miller, D.B. (1971), Automated production of global cloud climatology based on satellite data, *Air Weather Services* (MAC), Technical Reports 242, USAF, 291–306.

Muller, J-P. (ed.) (1988), *Digital Image Processing in Remote Sensing*, Taylor and Francis, London.

Niblack, W. (1986), *An Introduction to Digital Image Processing*, Prentice-Hall International, Copenhagen.

Sabatini, R.R., Rabchevsky, G.A. and Sissala, J.E. (1971), *Nimbus Earth Resources Observations*, Technical Report No. 2, Contract No. NAS 5-21617, Aracon, Concord, MA.

Schowengerdt, R.A. (1983), *Techniques for Image Processing and Classification in Remote Sensing*, Academic Press, London.

Swain, P.H. and Davis, S.M. (eds) (1978), *Remote Sensing: the Quantitative Approach*, McGraw Hill, New York.

Chapter 10

Allison, L.J., Wexler, R., Laughlin, C.R. and Bandeen, W.R. (1978), *Remote Sensing of the Atmosphere from Environmental Satellites*, Authorized Reprint from Special Technical Publication 653, American Society for Testing and Materials, Philadelphia.

Anderson, R.K. and Veltischev, N.F. (1973), *The use of satellite pictures in weather analysis and forecasting*, World Meteorological Organization, Technical Note No. 124, WMO, Geneva.

Barrett, E.C. (1970), Rethinking climatology: an introduction to the uses of weather satellite photographic data in climatological studies, *Progress in Geography*, **2**, 153–205.

Barrett, E.C. (1974), *Climatology from Satellites*, Methuen, London, 61–146.

Barrett, E.C. and Hamilton, M.G. (1980), *The use of Meteosat data for remote sensing applications*, Final Report, Contract E.343/1979, Department of Industry, London.

Barrett, E.C. and Watson, I.D. (1977), Problems in analysing and interpreting data from meteorological satellites, in *Remote Sensing: Practices and Problems*, Barrett, E.C. and Curtis, L.F. (eds), Edward Arnold, London, 276–303.

Brimacombe, C.A. (1981), *Atlas of Meteosat Imagery*, ESA SP-1030, European Space Agency, Paris.

Collier, C.G. (1989), *Applications of Weather Radar Systems*, Ellis Horwood, Chichester.

Fusco, L., Lunnon, R., Mason, B. and Tomassini, C. (1980), *Operational production of sea-surface temperature from Meteosat image data*, ESA Bulletin, **21**, 38–43.

Harris, R. and Barrett, E.C. (1975), An improved satellite nephanalysis, *Meteorological Magazine*, **104**, 9.

Houghton, J.T., Taylor, F.W. and Rodgers, C.D. (1984), *Remote Soundings of Atmospheres*, Cambridge University Press, Cambridge.

Liljas, E. (1987), Multispectral classification of cloud, fog and haze, in *Remote Sensing Applications in Meteorology and Climatology* (q.v.) (ed. R.A. Vaughan), Reidel, Dordrecht.

Oliver, J.E. and Fairbridge, R.W. (1987), *The Encyclopaedia of Climatology*, *Encyclopaedia of Earth Sciences Series*, Vol. XI, Van Nostrand Reinhold, New York.

Schwalb, A. (1978), *The Tiros-N/NOAA A-G Satellite series*, NOAA Technical Memorandum, NESS 95, US Department of Commerce, Washington, DC.

Scorer, R. (1986), *Cloud Investigation by Satellite*, Ellis Horwood, Chichester.

Slater, P.N. (1980), *Remote Sensing: Optics and Optical Systems*, Addison-Wesley, Reading, MA, 575.

Chapter 11

Allison, L.J. (1972), *Air–sea Interaction in the Tropical Pacific Ocean*, NASA Technical Note, TN D-6684.

Barnett, J.J. *et al.* (1973), The first year of the selective chopper radiometer on Nimbus-4, *Quarterly Journal of the Royal Meteorological Society*, **98**, 32.

Barrett, E.C. (1974), *Climatology from Satellites*, Methuen, London, 147–313.

Barrett, E.C. and Martin, D.W. (1981), *The Use of Satellites in Rainfall Monitoring*, Academic Press, London.

Barrett, E.C., Lloyd, D., Kidd, C., Beaumont, M.J. and Kilham, D. (1990), *Rainfall over the North Sea*, Final Report to the Department of the Environment, PECD Ref. No. 7/8/136, RSU, University of Bristol.

Berger, A., Schneider, S. and Duplessy, J.C. (1988), *Climate and Geo-Sciences: a Challenge for Science and*

Society in the 21st Century, Kluwer, Dordrecht.

Harwood, R.S. (1977), Some recent investigations of the upper atmosphere by remote sounding satellites, in *Remote Sensing of the Terrestrial Environment*, Peel, R.F., Curtis, L.F. and Barrett, E.C. (eds), Butterworths, London, 111–124.

Henderson-Sellers, A. (ed.) (1984), *Satellite Sensing of a Cloudy Atmosphere*, Taylor and Francis, London.

London, J. (1957), *A study of the Atmosphere Heat Balance*. Final Report, Contract Af19(122)–165, Dept. of Meteorology and Oceanography, New York University, 99 p.

Miller, D.B. (1971), *Global Atlas of Relative Cloud Cover, 1967–70, based on data from meteorological satellites*, US Air Force/US Department of Commerce, Washington, DC.

NASA (1988), *Earth System Science: a Closer View*, Earth System Sciences Committee, NASA Advisory Council, National Aeronautics and Space Administration, Washington, DC.

NERC (1989), *Air of Change*, Natural Environment Research Council, Swindon.

Ohring, G. and Gruber, A. (1983), Satellite radiance observations and climate theory, *Advances in Geophysics*, **25**, 237–304.

Planet, W.G. (ed.) (1989), *Operational Ozone Monitoring with the Global Ozone Monitoring Radiometer (GOMR)*, NOAA Technical Memorandum NESDIS 28, US Department of Commerce, Washington, DC.

Rao, P.K., Holmes, S.J., Anderson, R.K., Winston, J.S. and Lehr, P.E. (eds) (1990), *Weather Satellites: Data and Environmental Applications*, American Meteorological Society, Boston, MA.

Streten, N.A. and Troup, A.J. (1973), A synoptic climatology of satellite-observed cloud vortices over the southern hemisphere, *Quarterly Journal of the Royal Meteorological Society*, **99**, 56–72.

Tanczer, T., Göyz, G. and Major, G. (1981), First FGGE Results from Satellites, *Advances in Space Research*, **1**(4), Pergamon Press, Oxford.

Vonder Haar, T.H. and Suomi, V.E. (1971), Measurements of the Earth's radiation budget from satellites during a five year period, Part I: extended time and space means, *Journal of Atmospheric Science*, **28**, 305–314.

Chapter 12

Allison, L.J. and Schmugge, T.J. (1979), A hydrological analysis of East Australian floods using Nimbus-5 Electrically Scanning Radiometer data, *Bulletin of the American Meteorological Society*, **60**, 1414–1420.

Bagle, R.E. and Serebreny, S.K. (1962), Radar precipitation echo and satellite cloud observations of a maritime cyclone, *Journal of Applied Meteorology*, **1**, 279–286.

Barrett, E.C. and Martin, D.W. (1981), *The Use of Satellites in Rainfall Monitoring*, Academic Press, London.

Barrett, E.C., Power, C.H. and Micallef, A. (1990), *Satellite Remote Sensing for Hydrology and Water Management: the Mediterranean Coasts and Islands*, Gordon and Breach, New York.

Collenge, V. and Kirby, C. (1987), *Weather Radar and Flood Forecasting*, Wiley, Chichester.

Collier, C.G. (1989), *Applications of Weather Radar Systems*, Ellis Horwood, Chichester.

Deutsch, M., Wiesnet, D.R. and Rango, A. (1981), *Satellite Hydrology*, American Water Resources Association, Minneapolis.

Engman, E.T. and Gurney, R.J. (1991), *Remote Sensing in Hydrology*, Chapman and Hall, London.

Fraysse, G. (ed.) (1980), *Remote Sensing in Agriculture and Hydrology*, Balkema, Rotterdam.

Grinstead, J. (1974), The measurement of areal rainfall by the use of radar, in *Environmental Remote Sensing: Applications and Achievements*, Barrett, E.C. and Curtis, L.F. (eds), Edward Arnold, London, 267–284.

Legeckis, R. (1986), Long waves in the equatorial Pacific and Atlantic Oceans during 1983, *Ocean–Air Interactions*, **1**(1), 1–10.

NEXRAD (1984), *Next Generation Weather Radar*, Product Description Document R400-PD202, NEXRAD Joint Systems Program Office, US Department of Commerce, Washington, DC.

Plessey Radar (1971), *The Dee Weather Radar Project*, Publication No. 6776, The Plessey Company Limited, Weybridge.

Rango, A. (1980), Remote sensing applications in hydrometeorology, in *The Contribution of Space Observations to Water Resources Management*, Salomonson, V.V. and Bhavsar, P.D. (eds), Pergamon Press, Oxford, 59–66.

Sabins, F.J. (1978), *Remote Sensing: Principles and Interpretation*, W.H. Freeman, San Francisco.

Salomonson, V.V. and Bhavsar, P.D. (eds) (1980), *The Contribution of Space Observations to Water Resources Management*, Pergamon Press, Oxford.

Schultz, G.A. and Barrett, E.C. (1989), *Advances in Remote Sensing for Hydrology and Water Resources Management*, IHP-III Project 5.1 Report, Technical Documents in Hydrology, UNESCO, Paris.

Szekielda, K.H., Allison, L.F. and Salomonson, V. (1970), *Seasonal Sea Surface Temperature Variations in the Persian Gulf*, Goddard Space Flight Center, Report No. X-651-70-416, NASA, Greenbelt, MD.

Tsenchiya, K., Arai, K. and Igarashi, T. (1987), Marine

Observation Satellite, *Remote Sensing Reviews*, **3**(2), 59–104.

Viksne, A., Liston, T.C. and Sapp, C.D. (1969), SLR reconnaissance of Panama, *Geophysics*, **34**, 54–69.

Williamson, E.J. and Wilkinson, G.G. (1980), Sea-surface temperature measurements from Meteosat, in *Report of Second Meteosat Scientific User Meeting*, March 1980, ESA, Paris, 51–511.

Chapter 13

Barringer Research Ltd. (1975), *Final Report on Remote Sensing of Soil Moisture*, Centre for Remote Sensing, Ottawa, Canada.

Beckett, P.H.T. (1974), The statistical assessment of resource surveys by remote sensors, in *Environmental Remote Sensing: Applications and Achievements*, Barrett, E.C. and Curtis, L.F. (eds), Edward Arnold, London, 11–29.

Belyakova, G.K. *et al.* (1971), Study of microwave radiation from the satellite Cosmos 243 over agricultural landscape, *Coklady Akademic Nauk, USSR*, **201**, 837–842.

Bradley, G.A. and Ulaby, F.T. (1980), *Aircraft Radar Response to Soil Moisture*, Technical Report SM-KO-04005, NAG 5-30, Remote Sensing Laboratory, University of Kansas Center for Research.

Condit, H.R. (1970), The spectral reflectance of American soils, *Photogrammetric Engineering*, **36**, 955–966.

Curran, P.J. (1978), A photographic method for the recording of polarized visible light for soil surface moisture indication, *Remote Sensing of Environment*, **7**, 305–322.

Curran, P.J. (1981), Remote sensing: the use of polarized visible light to estimate surface soil moisture, *Applied Geography*, **1**, 41–53.

Curran, P.J. (1985), *Principles of Remote Sensing*, Longman, London.

Curtis, L.F. (1974), Remote Sensing for Environmental Planning Surveys, in *Environmental Remote Sensing: Applications and Achievements*, Barrett, E.C. and Curtis, L.F. (eds), Edward Arnold, London, 87–109.

Curtis, L.F. (1976), The Mapping of Soil Associations by Remote Sensing Techniques, in *CENTO Workshop on Applications of Remote Sensing Data and Methods, Proceedings, Istanbul, Turkey*, US Geological Survey, 40–47.

Curtis, L.F. (1977), Remote sensing of soil moisture: user requirements and present prospects, in *Remote Sensing of the Terrestrial Environment*, Peel, R.F., Curtis, L.F. and Barrett, E.C. (eds), Butterworths, London, 143–158.

Gerbermann, A.H., Gausman, H.W. and Wiegand, C.L. (1971), Color and color IR films for soil identification, *Photogrammetric Engineering*, **37**, 359–364.

Gurney, R.J. (1979), The estimation of soil moisture content and actual evapotranspiration using thermal infrared remote sensing, in *Remote Sensing and National Mapping*, Allan, J.A. and Harris, R. (eds), Remote Sensing Society, 101–109.

Kroll, C. (1973), *Remote Monitoring of Soil Moisture using Airborne Microwave Radiometers*, Texas University Remote Sensing Center.

Meyer, M.P. and Maklin, H.A. (1969), Photo-interpretation techniques for Ektachrome IR transparencies, *Photogrammetric Engineering*, **35**, 1111–1114.

Milton, E.J. and Webb, J.P. (1984) The use of narrow bandwidth spectroradiometry for soils investigations in *Satellite Remote Sensing: Review and preview*, Remote Sensing Society, 401–10.

Mintzer, O.W. (1968), Soils, in *Manual of Color Aerial Photography*, Smith, J. (ed.), American Society of Photogrammetry.

Myers, V.I. and Heilman, M.D. (1969), Thermal infrared for soil temperature studies, *Photogrammetric Engineering*, **10**, 1024–1032.

Parks, L. and Bodenheimer, R.E. (1973), Delineation of major soil associations using ERTS-1 imagery, *Symposium on Significant Results from ERTS-1*, 121–125.

Parry, D.E. (1978), Some examples of the use of satellite imagery (Landsat) for natural resource mapping in Western Sudan, in *Remote Sensing Applications in Developing Countries*, Collins, W.G. and van Genderen, J.L. (eds), Remote Sensing Society, 3–12.

Romanova, M.A. (1968), Spectral luminance of sand deposits as a tool in land evaluation, in *Land Evaluation*, Stewart, G.A. (ed.), Macmillan, Australia, 342–438.

Satten White, M.B. and Ponder Henley, J. (1987) Spectral characteristics of selected soils and vegetations in Northern Nevada and their discrimination using band ratio techniques. *Inst. J. of Remote Sensing*, **23**, 155–75.

Simakova, M.S. (1964), *Soil Mapping by Colour Aerial Photography*, Israel Programme for Scientific Translation, Jerusalem.

Steg, L. and Frost, R.T. (1971), *Visible polarization signature for remote sensing of soil surface moisture*, COSPAR Plenary Meeting, Leningrad, USSR (1970), 15.

Stockhoff, E.H. and Frost, R.T. (1972), Polarization of light reflected by moist soils, *Proceedings 7th*

Symposium on Remote Sensing, ERIM, Ann Arbor, MI, 345–364.

Townshend, J.R.G. (ed.) (1981), *Terrain Analysis and Remote Sensing*, George Allen and Unwin, London.

Trevett, J.W. (1986), *Imaging Radar for Resources Surveys*, Chapman and Hall, London.

Ulaby, F.T. (1980), Active microwave sensing of soil moisture: synopsis and prognosis, *Proceedings Workshop of Microwave Remote Sensing on Bare Soil*, European Association of Remote Sensing Laboratories, Toulouse, 2–28.

Welch, R. (1966), A comparison of aerial films in the study of the Breidamerkur glacier area, Iceland, *Photogrammetric Record*, 5, 289–306.

Werner, H.D., Schmer, F.A., Horton, M.L. and Waltz, F.A. (1973), *Application of Remote Sensing Techniques to Monitoring Soil Moisture*, Environmental Research Institute of Michigan, 1245–1258.

Weston, F.C. and Myers, V.I. (1973), Identification of soils associations in western South Dakota on ERTS-1 imagery, *Symposium on Significant Results from ERTS-1*, NASA, 965–972.

Wood, J.A., Laserre, M. and Fedsejeus, G. (1990), Analysis of mid-infrared spectral characteristic of rock outcrops and an evaluation of the Kahle Model in predicting outcrop thermal inertias, *Remote Sensing of the Environment*, **30**, 169–195.

Chapter 14

Blodget, H.W. and Anderson, A.T. (1973), A comparison of Gemini and ERTS imagery obtained over Southern Morocco, *Symposium on Significant Results from ERTS-1*, NASA, 265–272.

Bodechtel, J. and Haydn, R. (1977), Analog and digital processing of multispectral data for geological application, in *Remote Sensing of the Terrestrial Environment*, Peel, R.F., Curtis, L.F. and Barrett, E.C. (eds), Butterworths, London, 159–168.

Bodechtel, J., Nithack, J. and Haydn, R. (1974), Geologic evaluation of Central Italy from ERTS-1 and Skylab data, *Proceedings Frascati Symposium on European Earth Resources Satellite Experiments*, ESRO, Paris, 205–215.

Carter, W.D. and Meyer, R.F. (1969), Geological analysis of a multispectrally processed Apollo space photograph, *New Horizons in Colour Area Photography*, American Society of Photogrammetry, 59–64.

Carter, W.D., Rowan, L.C. and Huntingdon, J.F. (eds) (1980), *Remote Sensing and Mineral Exploration*, Pergamon Press, Oxford.

Cassinis, R., Lechi, G.M. and Tonellis, A.M. (1974), Contributions of space platforms to a ground and airborne remote sensing programme over active Italian volcanoes, *Proceedings Symposium on European Earth Resources Satellite Experiments*, ESRO SP 100, Paris, 185–197.

Cole, M.M. (1973), Geobotanical and biogeochemical investigations in the sclerophyllous woodland and scrub associations of the Eastern Goldfields area of Western Australia, *Journal of Applied Ecology*, **10**, 269–284.

Cole, M.M., Owen-Jones, E.S. and Custance, N.D.E. (1974), Remote sensing in mineral exploration, in *Environmental Remote Sensing: Applications and Achievements*, Barrett, E.C. and Curtis, L.F. (eds), Edward Arnold, London, 49–66.

Drury, S.A. (1987), *Image Interpretation in Geology*, Allen and Unwin, London.

El-Baz, F. (1984), *Deserts and Arid Lands*, Martinus Nijhoff, The Hague.

Gregory, A.F. (1973), Preliminary assessment of geological applications of ERTS-1 imagery from selected areas of the Canadian Arctic, *Symposium on Significant Results from ERTS-1*, NASA, 329–344.

Laird, A.G. (1977), Passive infrared sensing of the environment, in *Remote Sensing of the Terrestrial Environment*, Peel, R.F., Curtis, L.F. and Barrett, E.C. (eds), Butterworths, London, 26–37.

Lathram, E.H., Taillevr, I.L. and Patton, W.W. (1973), Preliminary geologic application of ERTS-1 imagery in Alaska, *Symposium on Significant Results from ERTS-1*, NASA, 257–264.

Martin-Kaye, P. (1974), Application of side looking radar in Earth-resource surveys, in *Environmental Remote Sensing: Applications and Achievements*, Barrett, E.C. and Curtis, L.F. (eds), Edward Arnold, London, 29–48.

NASA (1987), *HIRIS: High Resolution Imaging Spectrometer, Science Opportunities for the 1990s, Earth Observing System*, Vol. IIc, EOS Instrument Panel Report, National Aeronautics and Space Administration, Washington, DC.

Sabins, F.J. (1978), *Remote Sensing: Principles and Interpretation*, W.H. Freeman, San Francisco.

Short, N.M. (1973), Mineral resources, geological structure and landform surveys, *Symposium on Significant Results from ERTS-1*, Vol. 111, NASA, 30–46.

Siegal, B.S. and Gillespie, A.R. (1980), *Remote Sensing in Geology*, Wiley, New York.

Siegal, B.S. and Gillespie, A.R. (eds) (1980), *Remote Sensing for Geologists*, Wiley, New York.

Vane, G. and Goetz, A.F.H. (1988), Terrestrial imaging spectrometry, *Remote Sensing of Environment*, **24**, 1–29.

Vincent, R.K. (1973), Ratio maps of iron ore deposits,

Atlantic City District, Wyoming, *Symposium on Significant Results from ERTS-1*, NASA, 379–386.

Vinogradov, B.V., Grigoryev, A.A., Lipatov, V.B. and Chernenko, A.P. (1972), Thermal structure of the sand desert from the data of IR aerophotography, *Proceedings of the 8th International Symposium Remote Sensing of Environment*, Ann Arbour, MI, 720–737.

Chapter 15

Colwell, R.N. (1969), *Analysis of Remote Sensing Data for Evaluating Forest and Range Resources*, School of Forestry and Conservation, University of California, 207.

Curtis, L.F. (1978), Remote Sensing Systems for Monitoring Crops and Vegetation, *Progress in Physical Geography*, **2**(1), 55–79.

Curtis, L.F. and Walker, A.J. (1980), Exmoor: a problem of landscape planning and management, *Landscape Design*, **130**, 7–13.

Dethier, D.E., Ashely, M.D., Blair, B. and Hopp, R.J. (1973), Phenology satellite experiment, *Symposium on Significant Results from ERTS-1*, NASA, 157–165.

Fraysee, G. (ed.) (1980), *Remote Sensing Applications in Agriculture and Forestry*, Balkema, Rotterdam.

Fuller, R.M. and Parsell, R.J. (1989), Visual and computer classifications of remotely sensed images: a case study of grasslands in Cambridgeshire, *International Journal of Remote Sensing*, **10**(1), 193–210.

Griffiths, G.H. and Wooding, M.G. (1988), Pattern analysis and the ecological interpretation of satellite imagery, *Proceedings IGARRS 88 Symposium*, Edinburgh, ESA SP-284, European Space Agency, Paris, 917–921.

Hallikainen, M., Hyyppa, J., Tares, T. and Somersalo, E. (1989), Classification of forest types by microwave remote sensing, *Proceedings of 9th EARSeL Symposium*, Espoo, Finland, 293–298.

Heller, R.C. (1968), Large scale color photography samples forest insect damage, in *Manual of Color Aerial Photography*, Smith, J. (ed.), American Society of Photogrammetry, Falls Church, VA, 394.

Hilborn, W.H. (1978), Application of remote sensing in forestry, in *Introduction to Remote Sensing of the Environment*, Richardson, B.F. (ed.), Kendall/Hunt, Dubuque, Iowa.

Howard, J.A. (1970), *Aerial Photo-Ecology*, Faber, London.

Hunting Technical Services (1986), *Monitoring Landscape Change*, Final Report to Department of Environment and Countryside Commission of England and Wales, Hunting Technical Services Ltd., Boreham Wood.

Jensen, C.E. (1948), *Dot-type Scale for Measuring Tree Crown Diameters on Aerial Photographs*, U.S. Forest Service, Central States Forest Experiment Station, Note No. 48.

Johannsen, C.J. and Saunders, J.L. (eds), (1982), *Remote Sensing for Resource Management*, Soil Conservation Society of America, Iowa City.

Leedy, D.L. (1968), The inventorying of wildlife, in *Manual of Colour Aerial Photography*, Smith, J. (ed.), American Society of Photogrammetry, Falls Church, VA, 422.

Leeman, V., Earing, D., Vincent, R.K. and Ladd, S. (1971), *The NASA Spectral Information System: a data compilation*, NASA, Houston, TX.

Lindgren, D.T. (1985), *Land Use and Remote Sensing*, Martinus Nijhoff, Dordrecht.

Lyons, T.R. and Avery, T.E. (1977), *Remote Sensing: a Handbook for Archaeologists and Cultural Resource Managers*, National Park Service, US Department of Interior.

NRSC (1990), *Ecological Consequences of Land Use Change*, SP(90) WP46, National Remote Sensing Centre, Farnborough.

NRSC (1990), *Use of Satellite Data for the Preparation of Land Cover Maps and Statistics*, SP(90) WP33, National Remote Sensing Centre, Farnborough.

Parry, D.E. and Trevett, J.W. (1979), Mapping Nigeria's vegetation from radar, *Geographical Journal*, **145**(2), 265–281.

Pedgley, D.E. (1974), Use of satellites and radar in locust control, in *Environmental Remote Sensing: Applications and Achievements*, Barrett, E.C. and Curtis, L.F. (eds), Edward Arnold, London, 143–152.

Roberts, H. and Colwell, R.N. (1968), *The Application of Remote Sensing to the Inventory of Livestock and Identification of Crops*, School of Forestry and Conservation, University of California, 20.

Roffey, J. (1969), *Radar Studies on the Desert Locust*, Anti-Locust Research Centre Occasional Report, 17/69.

Rogers, E.J. (1949), Estimating tree heights from shadows on vertical aerial photographs, *Journal of Forestry*, **47**, 182–190.

Schaefer, G.W. (1972), Radar detection of individual locusts and swarms, *Proceedings of the International Study Conference on the Current and Future Problems of Aridology*, London, 1970, 379–380.

Shantz, H.L. (1954), The place of grasslands in the Earth's cover of vegetation, *Ecology*, **35**, 143–155.

Spurr, S.H. (1960), *Photogrammetry and Photo-interpretation*, Ronald Press, New York.

Thorley, G.A. (1968), Some uses of color aerial photography in forestry, in *Manual of Colour Aerial Photography*, Smith, J. (ed.), 393.

Watson, R.M. (1977), Air photography in East African Game management, in *The Uses of Air Photography*, St. Joseph, J.K.K. (ed.), John Baker, London.

Worley, D.P. and Meyer, H.A. (1955), Measurement of

crown diameter and crown cover and their accuracy on 1:12 000 scale photographs, *Photogrammetric Engineering*, **21**, 372–386.

You-Ching, F. (1980), Aerial photo and Landsat image use in forest inventory in China, *Photogrammetric Engineering and Remote Sensing*, **46**, 1421–1430.

Chapter 16

Anderson, J.R., Hardy, E.E., Roach, J.T. and Witmer, R.E. (1976), *A Land Use and Land Cover Classification System for Use with Remote Sensor Data*, US Geological Survey Professional Paper 964.

Bell, T.S. (1974), Remote sensing for the identification of crops and diseases, in *Environmental Remote Sensing: Applications and Achievements*, Barrett, E.C. and Curtis, L.F. (eds), Edward Arnold, London, 154–166.

Berg, A. (ed.) (1981), *Application of Remote Sensing to Agricultural Production Forecasting*, Balkema, Rotterdam.

Bush, T.F. and Ulaby, F.T. (1975), *Remotely Sensing Wheat Maturation with Radar*, Technical Report 177-55, University of Kansas Center for Research.

Curtis, L.F. and Hooper, A.J. (1974), Ground truth measurements in relation to aircraft and satellite studies of agricultural land use classification in Britain, *Proceedings Frascati Symposium on European Earth Resources Satellite Experiments*, European Space Research Organization, 405–415.

Curtis, L.F. (1978), Remote sensing systems for monitoring crops and vegetation, *Progress in Physical Geography*, **2**(1), 55–79.

Fraysse, G. (1977), Perspectives offered by remote sensing in agricultural resources management, in *Remote Sensing of the Terrestrial Environment*, Peel, R.F., Curtis, L.F. and Barrett, E.C. (eds), Butterworths, London.

Goel, N. and Strebel, D. (1983), Inversion of vegetation canopy reflectance models for estimating agronomic variables, *Remote Sensing of Environment*, **13**, 487–507.

Idso, S.B., Hatfield, J.L., Jackson, R.D. and Reginato, R.J. (1979), Grain yield prediction: extending the stress-degree-day approach to accommodate climatic variability, *Remote Sensing of Environment*, **8**, 267–272.

LARS (1968), *Remote multispectral sensing in agriculture*, Laboratory for Agricultural Remote Sensing, Purdue University, Research Bulletin, 844.

Macdonald, R.B. and Hall, F.G. (1978), The LACIE experience: a summary, *Proceedings of the 8th International Symposium on Remote Sensing for Observation and Inventory of Earth Resources and Endangered Environment*, 26 pp.

Macdonald, R.B., Bauer, M.E., Allen, R.D., Clifton, J.W., Erickson, J.A. and Landgrebe, D.A. (1973), Results of the 1971 Corn Blight Watch Experiment, *Proceedings of the 8th International Symposium on Remote Sensing of Environment*, Ann Arbor, Michigan, **1**, 157–189.

Mauser, W. (1988), Extraction of agricultural plane parameters from multitemporal TM and X-SAR data, *Proceedings of 8th EARSeL Symposium*, European Association of Remote Sensing Laboratories, Strasbourg.

Meyer, M.P. and Calpouzos, L. (1968), Detection of crop diseases, *Photogrammetric Engineering*, **35**, 554–556.

Millard, J.P., Hatfield, J.L. and Goettelman, R.C. (1979), Equivalence of airborne and ground acquired wheat canopy temperatures, *Remote Sensing of Environment*, **8**, 273–275.

Nagy, G., Shelton, G. and Tolaba, J. (1971), Procedural questions in signature analysis, *Proceedings of the 7th International Symposium on Remote Sensing of Environment*, Ann Arbor, MI.

Steven, M.D. and Clark, J.A. (1990), *Applications of Remote Sensing in Agriculture*, Butterworths, London.

Thompson, D.R. and Weymanen, O.A. (1980), Using Landsat digital data to detect moisture stress in Corn–Soybean growing regions, *Photogrammetric Engineering and Remote Sensing*, **46**(8) 484–497.

Townshend, J.R.G. (1984), Agricultural land-cover classification using Landsat temporal-spectral profiles, *International Journal of Remote Sensing*, **5**, 681–698.

Wilson, D.R. (ed.) (1975), in *Aerial Reconnaissance for Archaeology*, Council for British Archaeology, London.

Chapter 17

American Society of Photogrammetry (1983), *Manual of Remote Sensing*, 2nd Edn, Vols I and II, AMS, Falls Church, Virginia.

Bush, P.W. and Collins, W.G. (1974), The application of aerial photography to surveys of derelict land in the United Kingdom, in *Environmental Remote Sensing: Applications and Achievements*, Barrett, E.C. and Curtis, L.F. (eds), Edward Arnold, London, 167–183.

Davis, S., Tuyahov, A. and Holz, R.K. (1973), Use of remote sensing to determine urban poverty neighbourhoods, *Landscape*, 72–81.

Holz, R.K., Huff, D.L. and Mayfield, R.C. (1969), Urban spatial structure based on remote sensing imagery, *Proceedings of the 6th International Symposium on Remote Sensing of Environment*, University of Wisconsin, Ann Arbor, MI, 243–276.

Horton, F. (1974), Remote sensing techniques and urban

data acquisition, in *Remote Sensing: Techniques for Environmental Analysis*, Estes, J.E. and Senger, L.W. (eds.), Hamilton Publishing Co., Santa Barbara, 243–276.

Kernie, T.J.M. and Matthews, M.C. (1985), *Remote Sensing in Civil Engineering*, Surrey University Press, London.

Kiefer, R.W. (ed.) (1980), *Civil Engineering Applications of Remote Sensing*, American Society of Civil Engineers, New York.

Richardson, B.F. (ed.) (1978), *Introduction to Remote Sensing of the Environment*, Kendall/Hunt, Dubuque, IA.

Toll, D. (1980), Urban area update procedures using Landsat data, *Proceedings American Society of Photogrammetry*, 225–247.

Chapter 18

Barrett, E.C. (1981), Satellite-improved rainfall monitoring by cloud indexing methods: operational experience in support of desert locust survey and control, in *Proceedings of the AWRA Symposium on Satellite Hydrology*, Sioux Falls, SD, American Water Resources Association, Minneapolis, MN.

Barrett, E.C. and Hamilton, M.G. (1980), *The Use of Meteosat Data for Remote Sensing Applications*, Final Report, Contract E. 343/1979, Department of Industry, London.

Barrett, E.C. and Martin, D.W. (1981), *The Use of Satellite Rainfall Monitoring*, Academic Press, London.

Barrett, E.C., Brown, K.A. and Micallef, A. (1991), *Remote Sensing for Hazard Monitoring and Disaster Assessment: Applications to the Mediterranean Sea and its Coastlands*, Gordon and Breach, New York.

Berberian, M. (1978), Tabas-e-Golsham (Iran) catastrophic earthquake of 16 Sept. 1978: a preliminary report, *Disasters*, **4**, 207–219.

Burton, I., Kates, R.W. and White, G.F. (1978), *The Environment as Hazard*, Oxford University Press, Oxford.

Giddings, L. (1976), *Extension of Weather Data by Use of Meteorological Satellites*, Technical Memorandum, LEC-8377, Lockheed Corporation, Houston, TX.

Hielkema, J.U. (1980), *Remote Sensing Techniques and Methodologies for Monitoring Ecological Conditions for Desert Locust Population Development*, Final Technical Report, FAO-USAID, CGP/INT/349/USA, FAO, Rome.

Howard, J.A., Barrett, E.C. and Hielkema, J.U. (1978), The application of satellite remote sensing to monitoring of agricultural disasters, *Disasters*, **4**, 231–240.

Ingraham, D. and Amorocho, J. (1977), Preliminary rainfall estimates in Venezuela and Colombia from GOES satellite image, in *Preprints 2nd Conference on Hydrometeorology*, Toronto, 25–27 August, American Meteorological Society, 316–323.

UN Disaster Relief Organization (1978), *Disaster Prevention and Mitigation: a Compendium of Current Knowledge* (in several parts), United Nations, New York.

Wasserman, S.E. (1977), The availability and use of satellite pictures in recognizing hazardous weather, in *Earth Observation Systems for Resource Management and Environmental Control*, D.J. Clough and L.W. Morley (eds), Plenum Press, New York, 419–436.

World Meteorological Organization (1977), *The Use of Satellite Imagery in Tropical Cyclone Analysis*, Technical Note No. 153, WMO, Geneva.

World Meteorological Organization (1979), *Operational Techniques for Forecasting Tropical Cyclone Intensity and Movement*, WMO No. 528, Geneva.

Chapter 19

Barrett, E.C., Herschy, R.W. and Stewart, J.B. (1988), Satellite remote sensing requirements for hydrology and water management from the mid-1990s in relation to the Columbus Programme of ESA, *Journal of the Hydrological Sciences*, **33**(1), 1–17.

Clough, D.J. and Morley, L.W. (1977), *Earth Observation Systems for Resource Management and Environmental Control*, Plenum Press, New York.

Houghton, J.T., Jenkins, G.J. and Ephraums, J.J. (1990), *Climate Change, the IPCC Scientific Assessment*, Intergovernmental Panel on Climate Change, WMO/UNEP, Cambridge University Press, Cambridge.

NASA (1987), *From Pattern to Process: the Strategy of the Earth Observing System*, EOS Science Steering Committee Report, Vols I and II, National Aeronautics and Space Administration, Washington, DC.

NSSDC (1989), *The National Space Data Center*, NSSDC 88-26, NASA, Greenbelt, MD.

Sherman, J.W. (1991), The near-term ensemble of satellite remote sensors for Earth System Monitoring, in *Proceedings of the 5th AVHRR Users Meeting*, Tromso, Norway.

Sziekielda, K.H. (1986), *Satellite Remote Sensing for Resources Development*, Graham and Trotman, London.

US Congress (1985), *International Cooperation and Competition in Civilian Space Activities*, Office of Technology Assessment, OTA-ISC-239, US Congress, Washington, DC.

Index

Note: Figures and tables are shown in **bold** and *italics* respectively

Absorption
  frequency in gases **66**, 66
  of light **45**
  radiation 24, *26*, 29–34
  spectrum, atmospheric **31**, 31
Absorption spectrometers 65–6
Accuracy tests 112–13
Acid rain 11
Acoustic sounding 16, **17**
Active microwave cloud monitoring 218
Active remote sensing 17–18, 26
Active sensors
  airborne 347
  outside visible wavelengths 66–72, 72
Administrative boundary, USA-Canadian border **351**
Advanced Microwave Sounding Units (AMSU–A and B) 200
Advanced Very High Resolution Radiometer (AVHRR) 37, 58, 94, *182*, 182
  NOAA imagery 185–6
Aerial photography 12
  archaeological 344
  for civil engineering 364
  conventional 259–63
  in forestry 313–14
  ground checking 101
  for industrial analysis 356
  in land use 327–9
  in mapping and photogrammetry 12
  for settlement classifications 359
  and side-looking radar 72
  in transport studies 355
  in water distribution 352
Aerial photography
  see also Air photo interpretation
Aero Electronics Ltd **77**, 79
Aerosols 226
Africa 380, **382**, 383
Agfa-Gevaert, Contour Film 129
Agricultural holdings *338*
Agricultural surveys 10
Agriculture **106**, 325, 335
  Advanced Information System (EC) 326, **327**, *328*
Agriculture and Resources Inventory Surveys through Aerospace Remote Sensing (AGRISARS) 337, 340–2
Air circulations 221–2
Air flow, analyses from GOES–IO **221**, *221*
Air Force Target Signatures Measurement Program (USA) 330, **331**
Air photo interpretation (API) 133–8, 281–7, **282**, 285, 314, *315*
Air photography, *see* Aerial photography
Air pollution 357
Airborne Imaging Spectrometer (AIS) 66, 297
Airborne platforms 73, 75–9, **78**
Airborne radar, non-coherent 178
Airborne Visible-Infrared Imaging Spectrometer (AVIRIS) **300**, 300–1
Aircraft
  imagery **150**, 150–1
  powered 77–9
  use in disasters 368, **369**, 371–2
Aircraft-based systems, for atmospheric sensors 175, 178
Aircraft-borne support systems 247
Alaska **292**, 294
Albedo (reflection coefficient) *25*, 25–6, 205, 209–10
Algorithms 161, 330, 395
American Society for Photogrammetry: *Manual of Photographic Interpretation* 328
Analog data 119–24
Analog training-set techniques 161
Analysis
  data 124, 147, 162
    digital 165–8, 200
Angular reflectance distributions, above savannah **45**
Animals, loss of 313
Antarctica 211
Antenna temperature 27
Anticyclones 193

Aperture synthesis 69
Apollo mission 83
  Saturn-6 spacecraft **292**, 293
Applied macroclimatology 204
Applied meteorology 188–201
Arabian Sea, tropical cyclones in **222–3**
Archaeological survey 344, 347
Archiving 390–2
Arctic Circle 210
Argos monitoring systems 115
Ariane rocket 90
Arizona 273–4
Armyworm (*Spodoptera exempta*) 383–5
Array scanning **57**, 58
ARTEMIS system (FAO) **385**, 385
Atmosphere
  absorption spectrum **31**, 31
  climatology of middle and upper 224–6, **225**, **226**
  computer models of 204
  remote sensing of 175–82
Atmospheric teleconnections 222–4
Austin (Texas), housing areas comparison *360*, 361
Australia 241, 242, **288, 289**
Automatic Collection and Telemetry (ACT) 114–15
Automatic Picture Transmission (APT) 182–3
Autosat system 196
AVHRR, *see* Advanced Very High Resolution Radiometer

Back-scattering, *see* Reflection
Background brightness 162
Balloon platforms 75–7, **78**
Banded scan system 61
Bausch and Lomb Transfer Scope **143**
Bay of Bengal **37**
Bear Lake (Utah-Idaho border) *43*, 43, **44**
Beckett, P.H.T. 263
Belgium 318
Benefits, economic and practical 7–12
Bidirectional Reflectance (BDR) 305, *308*, 308–9
Bidirectional Reflectance Distribution Function (BDRF) 108
Bidirectional Reflectance Factor (BRF) 108, 305
Big Thompson Canyon flood 373
Biomass 308–9
Biosphere 3
Bispectral cloud monitoring 218
Bispectral sensing 38
Black body concept 22
Blooms 44–6
Boston (US) 361
Brightness temperature 27
Bristol airport **350**
Broad band sensing 38
Buckinghamshire **276**
Built environment 349–66

Calibration 101
California 314–15
Calluna vulgaris (Heather) 308–9
Cameras
  photographic 47–52
    space-borne 52–4
Canada 233–4, 273, **283**, 320
Carbon dioxide 30–1, **32**
Cartography, automated 260
Casagrande Soil Classification 259
Catania University (Sicily) 368–70
Census Cities Project 361
Central Dense Overcast (CDO) 377
Change, global 223, *396*
Change detection 158, 363
Check sites 113, *114*
Chernobyl nuclear power station catastrophe 386
Cherry Picker platform 75
Chester Dee weather radar project 229–31, **230**, **231**
Chlorophyll *252*, 252, **306**, 338
Cities, contemporary problems 353–5, **354**
Civil engineering 257, 364–6
Classification 330, 333
Climate
  change 11
  global monitoring 203–4
Climatology 203
  of middle and upper atmosphere 224–6, **225**, **226**
  see also Clouds
Cloudiness, mean **164**, 165
Clouds 37, 214–15, 226
  bispectral monitoring 218
  characteristics of 191–2, *193*
  classification 195
  climatology methods 215, 217, 218
  cover distribution 162–5, 204, **216**
  imagery 162, **163**, 188–90
  indexing of **217**, 217, 218
  meteorology of 177
  models 218
  monitoring 218
  patterns 193
  photography 177
  top pressures 215
  tropical cyclone **376**
Clustering algorithms 161, 330
Coastal Zone Colour Scanner (CZCS) 10, 61, 188, *252*, 252

Coasts
  mapping 53, 254
  waters 10, *11*, 252
  zones 242, 254
Collateral data 101, 105–7
Colorado, south-west 297
Colour enhancement 156
Colour photography 263, 360–1
Commission of the European Communities (CEC) 312
Commitments, national and international 395
Commonwealth of Independent States (CIS) 54, 85, 110, 289, 387, 389
Communication lines 352
Companies, multinational 389
Computer analysis 147, 260
Computer Compatible Tape (CCT) 166–7
Computers 10, 390
Conical scanning **57**, 58
Conifers 314, *315*
Conservation 4, 310–13, 323–4
  in National Parks 310–13
Construction lines 352
Construction material inventories 365
Contingency Tables 113, *114*
Contrast stretching 157, 158
Convective activity, mesoscale 196
Convective storms, severe 372–5
Convergence of evidence 136
CORINE project (CO-orRdination, INformation, Environment) 312–13
Corn Blight Watch Experiment (1971) **333**, 333–4, **334**
Corrections 150–2
Cosmos 1109 satellite 80
COST–73 system 230–2
Crop marks **345**
Crop production 325
Crop surveys 329–40
Crops
  canopy studies 338
  classifications 339, 339–40
  discrimination 329
  disease in 329
  hazards to 379–85
  infrared sensing of 341–4
  inventory studies 393
  modelling of 338–9, 341–4
  multispectral sensing of 329–40
  radar sensing of 340–1
  stage and leaf cover 106–7, **108**
  study of 105–7
  yield of 325, 341–4
Cumuliform convection 196, **197–9**
Curves, radiation emission 22, **23**
Cyclogenesis 200
Cyclones 192, **222–3**, **376**

Daedalus DS–1230 59–61
Darien region, (eastern Panama-north-west Columbia) **240**
Data
  analog 119–24
  analysis 124, 147, 162
    digital 165–8, 200
  calibration 153–5
  classifications 124–6
  collateral 101, 105–7
  collection, spectral 109–10
  compression 155
  digital to optical conversion 123–4
  distribution 97–9
  encoding 170
  'eternal triangle' of **121**
  flow **125**
  geostationary satellite 211–13
  ground 102–5, **106**, *107*
  handling large sets 390
  in situ 101, 112–13
  integration 113, **170**
  interpretation 124–6
  management 170–1
  manipulation 147–50, 171
  multispectral 195–6, 335–7, 341–4
  optical to digital conversion 120–3
  preprocessing 126–32, 330–2
  processing 13–14, 129–32, 257
  quality and quantity **149**, 150–8
  rainfall 228
  reduction 391–2
  sampling 112–13
  stages in handling and use 124–6
Data
  see also Satellites
Data Collection Platforms (DCP) 95
Data Collection Systems (DCS) 113–17, 183
Database Management Systems (DBMS) 12–13
Defense Meteorological Satellite Program (DMSP) 200, 218, 354
Deforestation 11, 257
Demography 358–64
Density slicing 155–6
Desert Locust Commission **380**, 380
Desert locust (*Schistocerca gregaria*) 379–83
Desertification 11, 304
Detectors 55–8
Developing countries, technology transfer to 13
Development, sustainable 4

Digital data 119–24, 129–32, 390
analysis 165–8, 200
Digital Elevation Models (DEM) 143–6, **145**, 272
Digital Video Plotter (DVP) 141, **144**
Digital workstations 143–6
Disasters 367–70
Disease 320
forest 320
Diseases, crop 328–9
Distribution problems 390–2
Doppler (coherent) radar 177, 178
Dot-type crown wedge 313–14
Drainage
network features 239–41
pattern classifications 274
Drainage Pattern Analysis (DPA) 275
Drone **77**, 79
Drum microdensitometers 121
Drum scanners **122**
Dugald River Lode, Australia 289, **291**
Dunkirk, (France) **253**

Earth
monitoring from orbit 387
radiation from 36–8, 38–9
resources
evolution 82–5
satellites 171, 357
scene measurements 155
Earth Observation System (EOS) 7, *8*, 33, 279, 301, 387
satellite series 11–12, 395
Earth Radiation Budget (ERB) 183, 188
Earth system science 395, **397**
Earth-atmosphere
energy budgets 204–14
radiation budgets **210–11**, 204, **205**, 210–11
Earth-atmosphere system 395
Earthnet programme **98**, 98, *99*
Earthquakes 11
Echo intensities 177
Edge enhancement 156–7
Education
needs of 393, *394*
tertiary sector 7
El Niño events 248–51
Electromagnetic energy, *see* Electromagnetism
Electromagnetism 6, 16–17, 18–20, 36, 38
Electron Beam Recorder (EBR) **123**, 123
Electronically Scanned Microwave Radiometer (ESMR) 61
Emission, of radiation 22–3, **23**, 38–9, 44
Energy
acoustical 16, **17**
electromagnetic, *see* Electromagnetism
infrared 26
visible 15
Energy budgets, earth-atmosphere 204–14
Enhancement
image
digital 155–8, **157**
photographic 128–9, *129*
radar 297
Environment
concern for 3–5, 395
monitoring **398**
pollution 357
patterns mapping 42
Environmental sciences 13–14
EOS, *see* Earth Observation System
Eosat headquarters 98
EROS Data Center (EDC) 132
Erosion, soils 228, 257
ERS satellite family 245
ERS-1 (European Resources Satellite) 10, 72, 96, **97**, 97, 255
payload instruments **71**
Estuaries 102
EUMETSAT 9–10
EURASEP programme 111–12, **112**
European Commission, Advanced Agriculture
Information System 326, **327**, *328*
European Space Agency (ESA) 98, 387; ERS satellites; Meteosat
Evaluation 101
Evaporation 236–8
Evapotranspiration 236–7
Exmoor **104**, 105, 310–11
Extinction coefficient 24
Extractive industries 357
Eye-brain system 15–16, 19, 26

Fairey multispectral camera system 51
False colour film 127
Faulting 283, **285**, **286**, **294**
Features
analysis and interpretation 158–61
recognition 124, 147, 161–5
rural 351–2
Field spectroscopy 101, 107–10
Field surveys 101–2
Field-acquired reflectance spectra 297–300, **299**
Fields
ancient systems 344–6, **346**
boundaries 352
capacity 259
patterns 352
size 337–8
FIFE 89 study 110, **111**

Film
  photographic
    colour 127–8
    panchromatic 126–7
    resolution and sensitivity *48*, 48, *50*, 51–2
Finland **317**, 317, 318, **319**
Fire fuel, wildland 320, **321**
Fire hazards 320
Flash-flood threat, US 373–4
Floating mark equipment 140–1
Flood hydrograph prediction 242–4
Flood-plain mapping 241–2
Floods *241*, 241–2, 372
Flying spot scanners **122**, 123
Folding 283, **284**, **288**, **289**
Foliation in rocks 282–3
Food and Agriculture Organization (FAO) 385
  ARTEMIS system **385**, 385
Food crops availability 385
Force fields 16
Forecasting, weather **148**, 191–5, 384
Foreshortening 70, **71**
Forestry
  infrared colour photography in 314–15, 320
  resource management example 313–20
Forests 303, 320
  classification of *314*, 314
  mapping 313–15
  radar shadows in **316**, 316–17
Forests
  see also Deforestation
Framing camera 47
France 143, **145**, 318, 389
Frequency
  absorption in gases **66**, 66
  electromagnetic spectrum 19
Fresnel zone pattern 69–70, **70**
Frontal systems 192–3, 200, 223, **224**
Fuel 303, 320, **321**
Fungi 329, **330**

Gases **66**, 66, *67*
Gemini spacecraft 83
Genetic diversity 313
Geobotanical techniques 285
Geographic Information System (GIS) 12, *13*, 168–72, **170**, 176, 390
Geography, growth of remote sensing in 12–13
Geology **68**, 281, 285–7, **288**
  infrared imaging in 293–7
  use of air photo-interpretation 281–7
  use of non-airborne sensors 287–93
Geology
  see also Hydrogeology
Geometrical rectification 361
Geostationary satellites 179, 196
  data 211–13
  imagery 373
  meteorological 94–5, 384
  steerable high-resolution sensor systems 370, *371*
Geostationary satellites
  see also GOES; Meteosat satellites
Geosynchronous satellite views **184**
Global change 223
Global Environment Monitoring System (GEMS) **4**, 4
Global hydrological cycle **228**
Global warming 53, 203, 224
GMS (Himawari) 179
GOES (Geostationary Operational Environmental Satellites) 94, 179, *190*, 213, 373
GOES–IO, air flow analyses **221**, 221
Grating spectrometer 65
Gravity 16
Grazing lands 304
Great Barrier Reef 242
Green leaf area indices (GLAI) 113
Green Wave passage 310
Greenland ice-cap 210
Ground data 102–5, **106**, *107*
Ground morphology 105
Ground observation platforms 74–5
Ground radiometry 113
Ground stations 216, 387–8
Ground-based systems, for atmospheric remote sensors 175–8
Groundwater 245
Growing degree day (GDD) 343–4
Gulf War (1990–1) 356

Haryana district (Delhi) **269**, 271
Hawker Siddeley Dynamics **58**, 59
Hazards 320, 367, 370–2, *372*, 385
Heat Capacity Mapping Mission (HCMM) 61, 95, 275
  Ground Resolution Element (GRE) 341
Heat Capacity Mapping Radiometer (HCMR) 275
Heat loss 357
Heather 308–9
Helicopters 79
High Resolution Imaging Spectrometer (HIRIS) 46, *300*, 301
High Resolution Infrared Sounder (HIRS–2) *34*
High Resolution Picture Transmission (HRPT) 182–3
High Resolution Visible (HRV) 54–5, 90–2, **92**, 318
Highway inspection 355
Holland 259–60
Housing quality 360–1
Hunting Technical Services Ltd 268–70
Hurricanes 193–5, 221, 372, **373**, 375–9, 384

Hydrogeology 245
Hydrologic basin models **243**, **244**, 244–5
Hydrological cycle, global **228**
Hydrometeorology 227–38

Ice monitoring 232–6
Image
  acquisition techniques **300**
  enhancement 155–8, 297
    digital 155–8, **157**
    photographic 128–9, *129*
  processing 130–2, *168–9*
  quality 48, **50**
  recording systems **122**
  utilization in civil engineering 364
Imagers 58–61
Imagery
  aircraft **150**, 150–1
  geostationary satellite 373
  ground ranged 70, **71**
  radar 67–9, 287, 315–17, 352
  satellite **150**, 150–1, 373
  slant ranged 70, **71**
  stereoscopic 92, 135–6, **136**
Imagery
  see also Satellites
Imaging
  infrared 293–7
  multiple-frequency microwave 61–2, *63*
Imaging radar 67–9
Imaging spectrometer 297–301, **298**, **300**
Imaging systems, and Landsat **84**
In situ data 101, 112–13
In situ sensing 5–6
Income taxation of forests 318, **319**
Incompatibility systems 390
Index of Refraction 24–5
India 85, 179, 241, 245, 257, **269**, 271
Industrial complexes 356–8
Industrial land, derelict 357–8, *358*
Infrared colour photography 127, 263, 314–15, 320, 329
Infrared energy 26
Infrared imaging 293–7
Infrared Interferometer Spectrometer (IRIS) 33, **35**
Infrared linescan imagery **290**, 357
Infrared Linescan Type 212 **58**, 59
Infrared linescanner (IRLS) **57**, 58–9, **60**
Infrared radiation 214
Infrared radiometers 55, 205
Infrared sensors 55–61, 341–4
Infrared spectrometers 65
Infrastructure Planetary Image Processing System
    (IPIPS) **166**, 166
Insat satellites 179
Insect infestations 372, 379–84
Insolation (Incoming solar radiation) 24
Instantaneous Field of View (IFOV) 83
Institute of Geological Sciences 285–7, **288**
Intelligent data management system **391**, 392
Inter-Tropical Cloud Band (ITCB) **162**, 196, 223
Interactions, vertical 224, 226
Interactive Flash-flood Analysis (IFFA) 374
International Biological Programme (IBP) 309
International Council of Scientific Unions (ICSU) 4
International Geosphere-Biosphere Programme (IGBP)
    4–5
International Imaging Systems **51**
International law 389
International Satellite Land Surface
Climatology Project (ISLCP), FIFE 89 study 110, **111**
Interpretation, data 124–6
IRS–1A satellite 85
Italy **294**, 368–70

Japan 179, 387
  see also, Marine Observation Satellites
Jet Propulsion Laboratory (JPL) 301, 347
Johnstown (Pennsylvania) flood 373–4
Joint Research Centre (Ispra) 339
  EURASEP programme 111–12, **112**
  Pilot Project 326–7
Joints 283–4
Jordan **262**, **281**, **284**, **286**
Journals 7

Kansas 374
  FIFE 89 study 110, **111**
  University 277, 340
Karakum desert 266
KERN DSP1 workstation 143, **145**
Keyguide to Information Sources in Remote Sensing 99
Keys, photographic **262**, 263, 328
Kinetic temperature 27
Kirchhoff's law 22, **23**
Kodak *48*, 129
Kosi River, (India) 241
Kursk (Soviet Union) 110

Laboratory for Agricultural Remote Sensing (LARS)
    330, **332**
LACIE (Large Area Crop Inventory Experiment)
    309–10, **335**, 335–7, **336**, **337**
Land cover 43
  classification 325, *326*
Land pollution 357
Land use
  classifications *114*, 325, *326*
  data 43

mapping **137**, 349–50
past 344–7
Landforms 69–75
non-photographic systems 69–75, 268–79
photographic systems 259–68
Landsat 1 **39**, 54, 268–72, **271**
2, and 3 52–3, 54, 293–7
Landsat 3
image **156**
Multispectral Scanner (MSS) 82, 152–3
Thematic Mapper (TM) 153
Landsat 5, Thematic Mapper (TM) 339
Landsat
data 6, 98, 268, 273, 309
collection system 113–14, **115**
marketing company 98
imagery **84 131**, 151–2, 157, 159, **160**
for settlement classifications 359
for urban land use change maps 361
use in snow mapping 232–3
information source potential 361
Multispectral Scanner (MSS) 10, 83–7, 245, 270–1, **171**, 239
geology use 293–7
processing requirements *131*, 132
Return Beam Vidicon 54, *130*
system 83–5
Thematic Mapper (TM) 83, **84**, 84, *86*, **87**
data 126, 129–32, 239, 318
for urban growth monitoring 361–2
Landscape
change 312, 361
features classification *312*
mapping 260
Landscapes, protected 310
Large Format camera (LFC) 53, **54**, 89
Laser Atmospheric Wind Sounder (LAWS) 72
Laser radar 72
Laser scanners 123
Latent photographic images 119–24
Leaf Area Index (LAI) 308
Leaf structure 305–7, **306**
Leica SD2000 141–3, **144**
Lidar (laser radar) 72
Lightsats 389
Limb Infrared Monitoring of the Stratosphere (LIMS) 188
Line arrays 300
Lineaments **283**, 283, 293
Linear discriminant functions 161
Linescan scanning **57**
Lithological interpretation, stages in 284–5
Litigation and claims 365–6
Liverstock breeds 313
Local area studies, fully integrated 110–12
Locusts 279–83
Long Ashton (near Bristol) 102, **103**

Magnetism 16
Malvern Hills **68**
Map making, *see* Mapping
Mapping 146
cloudiness 204
clouds 162–5, **216**, **235**
coastal 53, 254
conventional of soils 258–9
cyclogenesis 200
digital **235**
flood-plain 241–2
forest 313–15
from spectral data 330
fuel-grade peat 273
geological 297
land use **137**, 349–50
landscapes 260
moorland 311
radar 239
rainfall 215, 218, 383
satellite 42
snow 232, **235**
soil 258–9, 260, **261**, 263, 272, 364–5, 392–3
topographic 12
urban **355**, **354**
vegetation **137**
Mapping cameras 47
Marine environment 370; Coasts
Marine Observation Satellites (MOS) 10, 85, 96, 245, 254
Marine transportation *11*
Mariners 1–9 55
Maritime precipitation **229**
Mark Yeo (Somerset) **60**
Marseille region (France) 143, **145**
Masts, portable 75, **76**
Maya civilization 347
Measuring devices 5
Medicine 313
Memory management problems 390
Mercury satellite 82–3
Mesoscale matrix 215
Meteorological satellites 85, 93–4, *182–3*
Meteorology 7–10, 177, 200
applied 188–201
Meteorology
see also Hydrometeorology
Meteosat 4 satellite, payload 187
Meteosat 94, 179, 213
for armyworm monitoring 383–4

AVHRR data 320, **321**
cloud movement monitoring **220**, 221
Data Collection System (DCP) 114–15, **116**
geostationary satellite system 7–10
histogram analysis **195**, 195
infrared window waveband data 247, **249–50**
orbital and sensor characteristics *95*
visible and infrared monitoring 241
water vapour absorption imaging 214
Meteosat-type satellites 8–10
Metric Camera **53**, 53, 90
Microdensitometers 120–3, **291**
Microlight aircraft 77–8
Micrometer wedge 313
Microwave radar 61–5, *63*, 239, 276–7, 291–3, 297
Microwave radiometers 61–5
Microwave Scanning Radiometer (MSR) 254, 255
Microwave sensing 38, 254
Microwave Sounding Unit (MSU) *34*, 61
Microwave studies 392
Mie scatter 24
Mineral resources 293–301
Models
crop canopy radiance 338
crop yields 341–4
electromagnetic radiation 18–19
evapotranspiration estimation 236–7
flood hydrograph prediction 242–4
hydrologic basin **243**, **244**, 244–5
TELL–US 236, **237**, 275–6
vegetation index 338–9
vegetation reflectance 309
Moderate Resolution Imaging Spectrometer-Nadir (MODIS–N) *300*, 301
Modular Opto-electronic Multispectral Scanner (MOMS) 90, **92**, 300
Modulation Transfer Function (MTF) 48
Moisture
soil 243–4, 259, 276–7, **278**, 341
soils, subsurface 278–9
Monitoring Landscape Change (MLC) Project 312
Moorland, inventory of 310–11
MOS-1 satellites, *see* Marine Observation Satellites
MSc courses 393
Multispectral cloud monitoring 218
Multispectral data 195–6, 335–7, 341–4
Multispectral Electronic Self-Scanning Radiometer (MESSR) 254
Multispectral images 40–1
Multispectral photography 263–6
Multispectral Scanner (MSS) 10, 82, 297, **298**
Landsat, *see* Landsat
Multispectral sensing 38–9, 329–40
Narrow band sensing 38
NASA National Aeronautics and Space Administration, 6–7
NASA: *Earth System Science: a Closer View* 203
NASA: *Report on Active and Planned Spacecraft and Experiments* 179
NASA, 'Earth System Science' 7
National Parks 310–13
National Points of Contact (NPOC) 13
National Research Council 335
Natural Environment Research Council: *Air of Change* 203
Natural Environment Research Council, Equipment Pool for Field spectroscopy 110
Natural remote sensing 15–16, 26
Nematode attacks 328–9, **329**
Nephanalysis 191–2, **194**, 195, 214
Nevada (USA) **264**, 266, 272
New England (USA) **39**, 294–5, **295**
New Guinea **287**
New Mexico 274, **292**, 293
NEXRAD system 230–2, *232*
Nigeria 316
Nimbus 2
Persian Gulf surface temperatures **246**, 247
South Atlantic observations **120**
Nimbus 5
passive microwave data 241
Surface Composition Mapping Radiometer (SCMR) 37–8
views from the THIR **37**
Nimbus 5 Microwave Spectrometer (NEMS) 61
Nimbus 7, Coastal Zone Colour Scanner (CZCS) 96–7, *252*, 252
Nimbus 7 payload 187–8
Nimbus
imagery **292**, 294
radiometers on 218
research satellites 33, **35**, 187–8
satellites 33, 37
NOAA National Oceanic and Atmospheric Administration, 33, *34*
NOAA 2, tape recorder noise 152, **153**
NOAA 6 80
NOAA
AVHRR sensor 46
AVHRR–HRPT, multispectral data 195
HRPT multispectral data 234
images, computer-mapped polar stereographic **191–2**
near-polar orbiters 179
NESDIS 98, 373
GOSSTCOMP (Global Operational Sea Surface Temperature Computation) model 247, **248**
polar-orbiting 58, 94, 95

products 188, *189*
satellites 33, *34*, 81–2, 155
operational 182–7
Tiros-N type 179, 183–7
weather 37, 147, *182*, 233–6
tropical cyclones in Arabian Sea **222–3**
Noise reduction 152–3
Non-selective scatter 24
Normalized difference vegetation index (NDVI) 94, 338
North Sea 218, **219**
Northampton 361–2, **362**
Nowcasts 384
Numerical processing 165

Obion County (West Tennessee), Landsat–1 imagery of 271
Oblique photography 138
Observation platforms, ground 74–5
Observations, conventional 102
Ocean Colour Scanner (OCS) 111–12, **112**
Ocean monitoring 10, *11*, 96
Oceanography 102, 245–56
Oil spills 252, 370
Ontario Geological Survey 273
Operational Vertical Sounder (OVS) 94
Optical limnology 252
Optical oceanography 252
Orbits
geosynchronous **82**, 82
Landsat 83, **86**
near-polar, sun-synchronous 179
retrograde 81
satellite 80–2
sun–synchronous 81, 179
Outer Space Committee (OSC) 389
Overstocking 304
Oxygen 30
Ozone 30
holes 11, 183, 203, 224, 386

P–1000 Photoscan, scanning times *121*, 121
Pakistan, Saindak copper deposit **296**, *296*, 296
Panoramic camera 47
PARABOLA radiometer 108
Parallax measurements **140**, 140
Parking assessments 355
Passive micrometer radiometer (PMR) systems 62–4
Passive microwave
cloud monitoring 218–21
data 200–1
Passive remote sensing 17, 38–9
Passive sensors, outside visible wavelengths 55–61, 61–5
Pasture, permanent 303–4
Pathfinder data sets 98
Pattern analysis systems 332
Pelican system (Catania University) 370
Penetration depth 64
Perpendicular Vegetation Index (PVI) 338–9
Persian Gulf **246**, 247
Pest infestations 328–9, **329**
Phenology 305, 307–10
Photochemical oxidant formation 11
Photogrammetry 12, 138–46, 352, 368
Photographs
geometric characteristics 138–41, **139**
interpretation 12, 134–6
multispectral **51**, 51
polarized data 266–8
preparation of 126–8
vertical and oblique characteristics 36, **138**
Photography 26, 54
aerial, *see* Aerial photography
clouds 177
colour 263, 360–1
conventional 38
development of 6
vertical line overlap **49**
Photon detectors 55–7
Photon energy 19
Phrenology Satellite Experiment 309
Phylloxera 313
Phytoplankton 44–6
Picture brightness values 214
Pipelines 352
Plan Position Indicator (PPI), display units 177, **178**
Planck's law 22–3, **23**
Plankton populations 45–6
Plant ecology 305
Plants
indicator 285
loss of 313
Platforms
atmospheric remote sensor 175
ground observation 74–5
Landsat 83
seaborne 111
sensor 73–82
Point rainfall estimation **374**
Polar-orbiters
NOAA payload 182–7
radiation budgets 213
Polarization 64
Polarization percentages **267**, 267–8
Pollution 242, 252, 357
Population studies 3, **4**, 358–61
Portugal 313
Poverty, urban 361
Power stations, hydro-electric 232

Precipitation
  forecasting **148**
  frozen 232–6
  maritime **229**
  monitoring 228–32
  study of 178
Precipitation
  see also Rainfall; Snow
Prelaunch calibration 155
Preprocessing 124, 126–32, 129–32, 330–2
Prism spectrometer 65
Private sector, remote sensing transfer to 389
Probabilistic statements 147
Processing, data 124, 129–32, **167**, 390
Processing centres, data 13–14
Profiling, vertical (sounding) 31–3
Profiling systems 200
Prograde orbits 81
Purdue University (USA) 259
Pushbroom scanning 90, 119, **121**, 124

Quartz 39–40

Radar Radio Direction and Ranging, 17–18
  airborne, non-coherent 178
  for atmospheric remote sensing 177
  Doppler (coherent) 177, 178
  imagery 67–9, 297, 315–17, 352
  K–band, for inaccessible terrain analysis 239
  laser 72
  microwave 61–2, *63*, 228, 239, 276–7, 291–3, 297
  non-coherent, principal applications 177–8
  rainfall systems 228–32, **231**
  side-looking 72
  use in rural areas 353
  weather 178, *180–1*, **233**
Radar altimeter **71**, 72
Radar equation 228–9
Radar sensing, crop 340–1
Radar shadows, in forest areas **316**, 316–17
Radiance 20, *21*
  of crop canopy, modelling 338
  upwelling **45**, 46
Radiating (brightness) temperature 27
Radiation
  absorption of 24, 26, 29–34
  at its target 25–6, 27
  at source 20–2, 22–3
  electromagnetic 16–17, *18*, 18–19
  emission of 22–3, **23**, 38–9, 44
  from the earth, atmospheric window wavebands 36–8, 38–9
  from the sun 29–36
  infrared 214
  longwave 29
  nature of 18–20
  net **209**, 210
  observed 27–8
  in propagation 23–5
  quantities and units 20, *21*
  reflected 38–9
  shortwave 29
Radiation balance 204, **205**, **210–11**, 210–11
Radiation budget, earth-atmosphere 204–14
Radiation geometry 108, **109**
Radiation scatter 23–4
Radiometers
  in Bear Lake multispectral sensing *43*, **44**
  for cloud monitoring 218
  hand-held 75
  infrared 55, 205
  microwave 61–5, **63**
  multiband 31–3
  radar 61–5
  satellite 58
Radiometry, ground 113
Rainfall 215–21
  area identification 177
  in CDO and BF clouds **378**
  data 228
  hurricane 378–9, 379
  maps 218, 383
  mean annual over North Sea **219**
  monitoring 217–18, **381–2**, 381–3
    radar for 228–32, **231**
  run-off from 243
  types 177
Rainfall
  see also Precipitation
Rainforest **287**, 318
RAMP (Radar Mapping of Panama) 239
Range height Indicator (RHI) **178**, 178
Rango, A. 238
Rayleigh scatter 24
Real aperture radar (RAR) 66, 67
Receiving stations **88**, 387
Reconnaissance cameras 47
Reflectance
  characteristics 265, 338
  curves 305–6, **306**, **308**
  differences in surface and crop types *41*
  distributions 45
  from vegetated surfaces 305–7, **307**
  spectra **42**, 297–300, **299**
Reflected radiation 38–9
Reflection (back-scattering) 25–6, 44
Reflection coefficient, *see* Albedo
Refraction 24–5

Regional materials inventories 365
Remote camera systems 355
Remote sensing, definition 6
Remotely piloted vehicle (RPV) **77**, 79
Reprocessing 124
Research and Development satellites 179, 187–8
Research results 7
Resource management 313–20
Reststrahlen (residual ray) 40
Retrograde orbits 81
Return Beam Vidicon (RBV) camera 54, 83, **84**
Rhodesia (Zimbabwe) 327–8
River management 228
Rivers 102
Road planning 355
Rockets, high-altitude sounding 79–80
Rocks
  field examination of samples **104**
  foliation in 282–3
  space observations for study 293–301
  spectral signature of 39–41, **40**
Rocks
  see also Faulting; Folding
Run-off 242, 243
Rural change 351
Rural community settlements, small 352
Rural structures 351–3

St Lawrence River, ice reconnaissance in 233–4
Salinity assessments 251–4
Sample sites 113
Sand desert features 289
Satellite Infrared Spectrometer (SIRS) 65
Satellite meteorology 7–10, 12
Satellite oceanography 254–6
Satellites
  algorithm advances 395
  analysis, sensor-satellite combined data for 395
  approach to hazard monitoring 372
  armyworm forecasting improvement scheme 384
  data 143, 147
    constraints on widespread use 388–9
    for forestry 317–20
    for hazard monitoring 385
  for disaster assessment 370
  DMSP–block 5D–2 200
  Earth observation 12
  Earth resources 171, 357
  high-level 82–3
  imagery **150**, 150–1, 192–3, 373
    hurricane monitoring use 375
    rainfall evaluation 215–18
  Insat 179
  life and orbit altitude *80*
  low-level 80–2
  NASA 82–3
  platform-sensor combinations 387
  platforms 73, 80–5
  polar orbiting 58
  radiometers 58
  remote sensing 7–10, 203–4
  research and development 179, 187–8, 195–201
  Russian 85
  SeaWIFS 10, 96
  sensor systems 157, 273–4
  solar observatory 205
  systems
    civilian **93**, **397**
    computerized **400**, 400–1
    weather 36–7, 175, 176, 178–82
  TRMM 218
  urban area analysis 362
Satellites
  see also Geostationary satellites; Marine Observation Satellites
Scan system, banded 61
Scanners
  data, visible and near-infrared 68–75
  forms of 43, **57**, 58–61, **122**, 123
  multispectral, *see* Multispectral Scanner (MSS)
Scanning Microwave Spectrometer (SCAMS) 61
Scanning Multichannel Microwave Radiometer (SMMR) 61, 188, 254
Scatter, radiation 23–4, 36
Scatterometers **71**, 72, 254
Scofield-Oliver technique 373, 374–5
Screwworm (*Cochliomyia hominovorax*) 379
Sea defences 53
Sea ice 232
Sea state 111–12, 254–5
Sea-surface temperature (SST) 245–7, **246**, 254
Sea-Viewing, Wide Field-of-View Sensor (SeaWIFS) 10, 96
Seafloor features 254
Seasat 1 200, 245, 254
Seasat
  mission 70–2
  SAR image of Tay Estuary and Dunkirk **253**
Security issues 388
Selective Chopper Radiometer (SCR) **33**, 224–6
Senses 15
Sensing, microwave 38, 254
Sensor, Earth Observation Budget 188
Sensors
  infrared 55–61, 341–4
  passive, photographic 47–55
Shifting cultivation 303
Shortgrass prairie 309

Shuttle Imaging radar (SIR-A and SIS-B) 86–9, **88**, 347
Side-Looking Airborne Radar (SLAR) 66–9, 70
Side-looking radar systems **68**, 239, 291–3
Signatures
  crop 330
  multispectral *42*
  radar **316**
  scattering 72
  spectral 22, **23**, 38–9, 39–46, **263**, 265–6
Site investigations *365*, 365
Skylab project 72, **79**, 79–80, 84–5
Snell's law 25
SNIPE drone **77**, 79
Snow 232–6, 234, 243
Snowdonia 287, **290**
Software package 361, 390
Soil analysis, by spot data 272
Soils 68–75
  associations 271–2
  boundaries, goodness testing 260–3
  by non-photographic systems 268–79
  categories 258
  classifications 258–9
  conditions 105, **267**, 267–8
  erosion 228, 257
  mapping 258–9, 260, **261**, 263, 272, 364–5, 392–3
  markings 344
  moisture 243–4, 259, **274**, 275, 276–7, **278**, 341
    subsurface 278–9
  nature of 257
  parameters 259
  photographic systems 259–68
  pre-maps 272–3
  profile 257
  samples **104**
  spectral characteristics of **263**
  surveys 272, **273**
  temperatures **104**, 275, 289
  thermal inertia properties of 236
  water *258*, 259
Solar Back-scattered Ultraviolet-Total Ozone Mapper System (SBUV–TOMS) 188
Solar radiation 29, **30**
Sonar (Sound Navigation and Ranging) 18
Sounders 32–3, 61, 226
Sounding (vertical profiling) 31–3, 188, *189*
Soviet Union 85, 110, 289, 387, 389
Soyuzcarta 54
Space agencies, national 6
Space platform 73
Space Shuttle 85–90, 300, 347
  Challenger mission 53
Spacecraft observation systems 355–6
Spacelab **89**, 89, *90*, **91**, 114
Special Sensor Microwave Imager (SSM/I) 61
Species analysis 314
Spectral Analysis Manager (SPAM) 301
Spectral indicatrix 108
Spectral ratioing, reflected radiances 295
Spectral reflectance
  curves for urban phenomena **349**, 349–50
  field measurements in 107–10
  sands 266
Spectral signatures **306**, 333
  crop 329–30, **331**
Spectrometers 33, 43, 55
  absorption 65–6
  imaging 297–301, **298**, **300**
Spectroradiometers 43, 44–6
Spectroscopy
  field 101, 107–10
  terrestrial imaging 297–301
Spectrum
  electromagnetic 19–20, **20**, **30**
  solar radiation 29, **30**
SPOT 1 system, photogrammetric tests of 143–6
SPOT
  Daedalus AADS 1268 digital Multispectral system 274
  data 10, 272, 339–40
  DCS 119
  HRV sensors 54–5, 318
  imagery 232–3, 239
  multilinear array sensor system 119, **121**, 124
  steerable sensor 370
  system 90–3
Spring barley **343**, 343
Spruce budworm 320
Sputnik 1, 387
SSM–I200 *65*
Stand volume tables 314
Statistical sampling theory 155
Stefan's law 22
Stereoplotters **141**, 141
Stereoscopic imagery 92, 135–6, **136**
Stock-pile inventories 356–7
Storms 223, 372–5
Strahler, A.N. 4–5
Stratosphere Aerosol Measurements 11 (SAM 11) 188
Stratosphere and Mesosphere Sounder (SMS) 188
Stream network analysis 239–41
Stress degree day (SDD) 341–3
Strip camera 47
Structural geology 281
Subsidies
  European agriculture 10
  public 389
Sudan, western, land system study *268*, 268–70
Sun 81

radiation from 29–36
Surface, temperature changes 28
Surface Composition Mapping Radiometer (SCMR) 37–8
Surface drainage 239, **240**
Surface features, measured radiation of 64
Surface hydrology 238–45
Surface monitoring 10–12
Surface soils, moisture content **274**, 275
Surface water area assessment 242
Surveys
geological 105
hydrological 105
Synoptic meteorology 196
Synoptic weather systems, climatology of 222–4
Synthetic aperture radar (SAR) 66, 67, 69–72, 70–2, **71** **253**, 254
Systems, photographic and non-photographic **48**

Tabas-e-Golshan (Iran) earthquake **369**
Tasselled Cap concept 338–9
Tay Estuary (Scotland) **253**
Technologist-user relations 392–3
Technology transfer, to developing countries 13
Television cameras, satellite borne 55
TELL–US project 236, **237**, 275–6
Temperature 27
measurements 341
water 245–51
Temperature Humidity Infrared Radiometer (THIR) **37**, 37, 188
Temperatures, soil **104**, 259, 275, 289
Tennessee River Valley (USA) **359**, 359
Terrain
analysis 239, 260, 272, 364
hierarchical classification **260**, 260
classification 105
evaluation 352
interpretation 268
Terrain studies, Third World 268, 271
Terrestrial radiation **30**
Thematic Mapper (TM) 10
crop analysis 340
data 272, 311–12, 390
Landsat 83, **84**, 84, *86*, **87**
Thematic maps 171–2
Thermal capacity 27, *28*
Thermal conductivity 27, *28*
Thermal diffusivity 27, *28*
Thermal imager system **61**, 61
Thermal inertia 27–8, *28*, 289–91, **291**
Thermal infrared imagery 275–6, 320
Thermal infrared scanning 287–91
Thermal sensors 115–17
Thetford (Norfolk) **52**
Third World countries 268, 271
Three-dimensional space classification **196**
TIROS Television and Infrared Observation Satellite 94, 178–9, 387
TIROS Ice Reconnaissance Project (TIREC) 234
TIROS Operational Vertical Sounder (TOVS) 94, 183, **187**, 214
Topographic mapping 12
Towers 75, **76**
Traffic studies 355
Training sample selection 332
Training sets 159–61, 330
Training sites 113, *114*
Transfer Scope 141, **143**
Transmission
atmospheric 34–6
radiation 27
Transmittance, radiant energy 24, **25**
Transparent dot templets 314
Transportation 355–6, 365
Trees 304, 313–15, *see* Forests
Tropical climatology 223
Tropical cyclones 372, **377**
Tropical meteorology 196
Tropical patterns **376**
Tropical Rainfall Monitoring Mission (TRMM) satellite 218
Tropical rainforest **287**, 318
Tropical regions 221, 239
Tropical storms 193–5, 372
Tropics, radar mapping projects on 239
Turbidity 252
Typhoons **176**, 372

Ulaby, F.T. 277
United Kingdom (UK)
costs of remote sensing 389
Natural Environment Research Council, *Air of Change* 203
United Nations
Committee for Space Research (COSPAR) 7
Outer Space Committee (OSC) 389
United Nations
see also Food and Agriculture Organization
United Nations Disaster Relief Organization (UNDRO) 367–8
United Nations Educational Scientific and Cultural Organization (UNESCO) 4
United Nations Environment Programme (UNEP) 4
United States of America (USA) **39**, 233–4, 259, 271–2, 359–60, 387
Department of Agriculture 325–6
Geological Survey 272

National Weather Satellite Service 234
north-eastern *371*
public subsidies in 389
storm problems of 372–3
Weather Bureau 373
see also California; Colorado; Nevada; New England; Wyoming
Upper Rhine, crop studies 340–1
Urban areas 353–6
changes 361, **363**, 363–4
subregions 354–5
Urban information systems 353
USAF, Environmental Technical Applications Center 215
User, definition 392
User agencies 393
User-technologist relations 392–3

Vegetation
Earth's natural 303, *304*
mapping **137**
multispectral signature variation *42*
reflectance modelling 309
surveys 113
Vegetation index
global maps 94
models 338–9
Vertical temperature profile retrievals 188, *189*
Vidicon cameras 54–5, **55**
Visible imaging 293–7
Visible spectrum 19
Visible and Thermal Infrared Radiometer (VTIR) 254
Visible-Infrared Radiometer (VIRR), Seasat 254
Visual powers 15–16
Volcanoes 11, 287, 372
Vortices, mid-latitude 223, **224**
Vorticity 192, 200
Vosges forests (France) 318

Washington DC (USA) 359–60
Water 102, 227, 252, 352, 365
circulation patterns 245–51
pollution 357
quality 242, 251–4
temperature 245–51
types classification **251**, 252
Water vapour 31, 214
Water-balance, scheme **243**, **244**
Wavebands, radiation 31
Wavelength, electromagnetic spectrum 19
Weather 7
forecasting **148**, 191–5, 384
radar 178, *180–1*, **233**
Weather satellites 155, 182–8, 195–6, 375
data 162, **163**, 233–6, 384
NOAA 37, 147, *182*, 233–6
systems 178–82
Weather studies, remote sensing advantages 175
WEFAX (Weather Facsimile) 114
Wetlands monitoring 242
WetNet Project 218, 396–400, **399**
Wheat production **335**, 335–7, **336**, **337**, **340**, 340, **343**, 343
White body 26
Wien's (displacement) law 22
Wild A.5 Autograph plotters **142**
Wild A.8 Stereo Plotter **142**
Wildland 314–15, 320, **321**
Wildlife 304–5, 320–3
Wind
flows 221–2
satellite-derived 188, **221**
speed maps 255
Windows, atmospheric 31, 33, 34–5, 36–8
Woodlands 304, *see* Forests; Trees
Woolacombe area **154**
World Meteorological Organization 82, 94, 179
World Weather Watch (WWW) 82, 94, 179
World War 2 356
Wyoming (USA) 295–6

Yacht races, trans-Atlantic 115, **117**
Yellowstone National Park *272*, 272
Yorkshire, derelict industrial land 357–8

Zermatt, digital mapping of **235**

NORTHERN MICHIGAN UNIVERSITY LIBRARY
3 1854 004 650 258